STP 1285

Composite Materials: Fatigue and Fracture (Sixth Volume)

Erian A. Armanios, editor

ASTM Publication Code Number (PCN):
04-012850-33

ASTM
100 Barr Harbor Drive
West Conshohocken, PA 19428-2959

Printed in the U.S.A.

ISBN: 0-8031-2411-2 ✓
ISSN: 1040-3086

Photocopy Rights

Peer Review Policy

Each paper published in this volume was evaluated by three peer reviewers. The authors addressed all of the reviewers' comments to the satisfaction of both the technical editor(s) and the ASTM Committee on Publications.

The quality of the papers in this publication reflects not only the obvious efforts of the authors and the technical editor(s), but also the work of these peer reviewers. The ASTM Committee on Publications acknowledges with appreciation their dedication and contribution of time and effort on behalf of ASTM.

Printed in Ann Arbor, MI
June 1997

Foreword

The Sixth Symposium on Composites: Fatigue and Fracture was held 16–18 May 1995 in Denver, Colorado. ASTM Committee D30 on High Modulus Fibers and Their Composites sponsored the symposium. Erian A. Armanios, Georgia Institute of Technology, presided as symposium chairman and is editor of this publication.

Contents

Overview

The Sixth Symposium on Composites: Fatigue and Fracture sponsored by Committee D30 on High Modulus Fibers and Their Composites, was held in Denver on 16–17 May 1995. The symposium featured 38 paper presentations covering metal matrix composites, fatigue and damage progression, strength and residual properties, damage tolerance and fracture analysis, mode mixity and delamination, property characterization and environmental effects, and standardization and design. The symposium sessions were chaired by W. S. Johnson of Georgia Institute of Technology, R. Martin of Materials Engineering Research Laboratory, J. Masters of Lockheed Engineering and Science, L. Carlsson of Florida Atlantic University, G. Murri and J. Reeder of NASA Langley Research Center, A. Rosenberger of Wright-Patterson Air Force Base, J. Fish of Lockheed Advanced Development Co., B. Davidson of Syracuse University, P. Lagace of Massachusetts Institute of Technology, and S. Hooper of Wichita State University.

This publication includes 29 papers organized in 6 sections.

Damage Tolerance and Fracture Analysis

Lagace and Priest investigated the damage tolerance of longitudinally notched pressurized composite cylinders and explored the limitations of a methodology developed for biaxially loaded quasi-isotropic configurations. This methodology uses coupon fracture data to predict cylinder failure. In all cases, the methodology was not able to predict the failure pressures of uniaxially loaded cylinders. Their results show that failure is controlled by local damage mechanisms and underscores the importance of understanding local behavior in developing failure prediction methodologies.

Hooke, Armanios, Dancila, Thakker, and Doorbar presented an investigation into the failure of a SCS-6/Ti-6-4 thin-walled cylindrical shell subjected to internal pressure. An experimental setup was designed to allow application of an internal pressure using a hydraulic system while maintaining zero axial load. The results indicated that the failure pressure was in good agreement with theoretical predictions from an anisotropic shell and an engineering thin-walled solutions and coupon strength data.

Saczalski, Lucht, and Saczalski reviewed advanced experimental design methods related to load capacity degradation of composite recreational structures. The results of a case study dealing with a mountain bike frame failure suggested that such structures are susceptible to design load degradation due to manufacturing variations and usage. The paper illustrated the benefits of the multivariable experimental methodology as a design tool in optimizing competitive, cost effective, and safe recreational composite structures.

Chamis, Murthy, and Minnetyan presented an overview of a computational simulation approach for progressive fracture in polymer matrix composite structures. The approach is independent of stress intensity factors and fracture toughness parameters and integrate composite mechanics with finite element. Results of structural fracture in composite plates, shells, and builtup structures are presented, and parameters/guidelines are identified for use as structural fracture criteria, inspection intervals, and retirement for cause.

Marcucelli and Fish designed a torsion specimen to evaluate the transverse shear strength of composite materials. Tests of specimens made of graphite/epoxy material with six different layups exhibited linear load-deflection behavior and failure was due to transverse shear stress. A quasi-three-dimensional finite element analysis was conducted to determine the interlaminar shear stress at failure and the test specimens failed at the predicted value.

Fatigue and Life Prediction

Hermann and Hilleberry performed constant amplitude fatigue tests on unidirectional and cross-ply SCS-6/TIMETAL 21S titanium matrix to study fiber bridging of matrix cracks. Broken fibers were observed among intact, bridging fibers in the wake of the matrix fatigue crack. A model was developed to study the mechanism of the cracked composite and predictions were compared with test measurements.

Tanaka and Tanaka present a study on the effect of stress ratio on the Mode II fatigue crack propagation in unidirectional graphite/epoxy laminates. Both end-notched and end-loaded split specimens were tested under various stress ratios, and a fracture mechanics equation was proposed for predicting crack propagation rate. The effect of stress ratio on the micromechanisms of crack propagation was discussed on the basis of microscopic fracture surface observations.

Pelegri, Kardomateas, and Malik studied delamination growth of cross-ply graphite/epoxy laminates under constant amplitude cyclic compressive loads. Expressions describing the fatigue crack growth and accounting for mode dependence are derived and numerically integrated to produce delamination length versus number of cycles. Experimental results were compared to data from a previous study using unidirectional specimens.

Krüger and König used elastic fracture mechanics criteria to predict delamination growth under tension-tension and tension-compression fatigue loading. A Paris law diagram was obtained experimentally using Modes I and II specimens and correlated with computed mixed-mode results from a three-dimensional finite element analysis.

Schaff and Davidson presented a residual strength-based wearout model for predicting the life of polymeric composite laminates subjected to two-stress level and randomly ordered load spectra. The model includes a cycle mix factor and accounts for the dispersion of strength distribution during fatigue loading. The theoretical fatigue life distributions are compared to experimental results, and excellent correlation was obtained based on the 63.2 percentile of the probability of failure distribution.

Strength and Residual Properties

Chatterjee, Yen, and Oplinger examined the stress fields in tabbed unidirectional and cross-ply specimens to improve the methods for determining the axial strength of unidirectional composites. Use of soft tabs, low taper angles and ductile, tough adhesives are recommended to reduce failure near end tabs of unidirectional coupons. Data reduction schemes for cross-ply specimens are critically examined, and test results from cross-ply and unidirectional tension and compression specimens are compared.

Sawicki and Nguyen examined the performance of biaxially loaded composite joints using an apparatus designed to apply transverse bearing loads independent of longitudinal bypass loads. Two configurations were considered to compare transverse bearing effects in relatively high and low stiffness laminates. High stiffness laminates exhibited greater sensitivity to transverse hole deformation, laminate damage, and fastener torque. The results of this investigation showed the importance of using conservative bearing strengths in developing bolted joint allowables.

Wu and Wilson investigated the residual strength of aramid reinforced aluminum laminate, ARALL-3, and glass aluminum reinforced epoxy, GLARE-2, center notched panels and used the linear elastic fracture mechanics R-curve approach for the residual strength prediction. The R-curves calculated from test of different layup, and size panels with various initial crack extensions showed that they were independent of initial crack lengths and panel width. The results showed that the R-curve approach was a suitable and simple predictive method for fiber/metal laminates.

Miller, Portanova, and Johnson evaluated experimentally the impact damage resistance and residual mechanical properties of $[0/\pm45/90]_s$ SCS-6/TIMETAL 21S composites using quasi-static indentation and drop-weight tests at two nominal energy levels. The composite strength and constant amplitude fatigue response were evaluated to assess the effects of the sustained damage. Results showed that matrix cracking, characteristic of low impact energies, was not sufficient to reduce tensile strength or fatigue life. Only when the fibers were broken, as is the case at higher impact energies, the tensile strengths and failure strains were reduced.

Mode Mixity and Delamination

Trakas and Korschot explored the relationship between the mode of fracture, ply orientation, and apparent interlaminar toughness using double cantilever, end notched flexure, and modified split cantilever beam tests. Extensive use of SEM fractography was made in an attempt to correlate measured energies to the fracture surface deformation. While Modes II and III shared common fractographic features, corresponding values of critical strain energy release rate did not correlate. The work pointed to the need to quantify subsurface deformation and crazing to account for the relatively large values of critical strain energy release in Modes II and III.

Martin and Hansen presented a novel method to achieve stable delamination growth for all mixed mode ratios in the mixed mode bending specimen by maintaining constant opening displacement rate. Such a constant rate is achieved by attaching a second displacement trans-ducer to the hinges of the specimen and externally controlling the test machine using this transducer. Compliance calibration expressions were developed, and comparisons from double cantilever and edge notched flexure beams data were performed.

Beuth and Narayan addressed the problem of separating crack extension modes in oscillatory composite delamination models. Using a modified virtual crack closure technique they developed a method for obtaining energy release rate quantities independent of the virtual crack extension length. Predicted mode mix values are compared to energy release rate ratios using other methods proposed in the literature for the analysis of oscillatory delamination problem.

Palmer, Armanios, and Hooke developed an analytical model to predict the effect of internal delamination in unsymmetric laminates. The model accounts for shear deformation and was applied to a class of hygrothermally stable graphite/epoxy laminates. The model predicted a larger magnitude of shear and peel stresses at the free-edge compared to the delamination tip. Test data exhibited more coupling than analytically predicted which was attributed to the interfacial stresses associated with the nonplanar deformation of the Teflon film as a result of the cure cycle.

Friis, Hahn, Cooke, and Hooper proposed a finite element model for predicting the effect of fiber bridging, fiber properties, and fiber-matrix interface strength on the crack tip stresses and propagation potential of a chopped fiber composite. A technique for modeling variable fiber-matrix interface strength with contact element of variable coefficient of friction was presented. Results showed that fiber bridging reduced the crack tip stresses and resulted in stable crack propagation or crack arrest.

Environmental Effects

Crasto and Kim conducted an investigation to determine the effects of temperature and moisture on the initiation of free-edge delamination in a graphite/epoxy laminate under uniaxial compression. The onset of delamination was determined experimentally by monitoring axial and transverse strains. Absorbed moisture and elevated test temperature reduced the residual stresses. However, delamination initiated at significantly lower stresses with increasing test temperature due to significant decrease in interlaminar transverse strength. Comparison of stress level prediction and onset of delamination interface with experiment was performed.

Rosenberger and Nicholas examined the influence of an oxidizing environment on the isothermal and thermomechanical fatigue of unidirectional SCS-6/TIMETAL 21S composites through a comparison of tests performed in air and helium. In general, an environmental influence on fatigue life was found under conditions in which matrix crack initiation and growth are dominant. Fatigue conditions in which life is dominated by fiber were not affected by test environment.

Li and O'Brien developed a shear deformation theory including hygrothermal effects for the analysis of laminates with midplane edge delamination under torsion loading. The analysis of edge crack torsion test layup indicated no hygrothermal effects on the Mode III strain energy release rate. Another class of antisymmetric layups was investigated leading to a means of determining Mode III toughness between two dissimilar layers.

Parida, Prakash, Mangalgiri, and Vijayaraju evaluated the influence of environmental and geometric parameters on the behavior of fastener joints in carbon fiber composites using single and double shear lap configurations. Laminates made from unidirectional and fabric prepregs were tested in as-received condition at room temperature and under hot/wet condition after hygrothermal aging. The degradation of bearing strength in hygrothermally aged specimens under hot/wet condition was about 25 to 30% as compared to the corresponding room temperature values.

Kallmeyer and Stephens investigated the localized creep response of a quasi-isotropic graphite/polymer matrix composite laminate subjected to bolt-bearing loads at ambient and elevated temperature in order to assess the long term durability of advanced composite joints. Monotonic tensile and static creep tests were performed on single-hole bolted joints at various temperatures and the influence of lateral constraint on the creep response was studied. The test temperature and bolt clampup torque were found to have a substantial influence on the elongation of bolt holes subjected to bearing loads.

Testing and Failure Mechanisms

Alif and Carlsson examined the stress-strain responses and damage evolution of five harness satin weave carbon/epoxy and four harness satin weave glass/epoxy composites in tension, compression, and shear. In contrast to conventional laminates where distributed damage in the form of matrix cracks is observed, the damage in both composites was confined to the region where ultimate failure occurred. Elastic properties were in good agreement with micromechanics predictions based on uniform strain, but failure stress predictions were less accurate.

Benson, Karpur, Stubbs, and Matikas correlated the results from six nondestructive evaluation techniques with the residual tensile strength of a unidirectional metal matrix composite after being isothermally fatigued. Scanning electron microscopy and metallography were used in the correlation and verification of fatigue damage. The immersion surface wave technique proved to be the most promising method for correlating damage with the residual strength for this particular composite.

Koudela, Strait, Caiazzo, and Gipple evaluated quasi-isotropic spool and unidirectional curved-beam specimens to determine the viability of using either one or both to characterize the interlaminar static and fatigue behavior of carbon/epoxy laminates. The interlaminar tensile strength of the spool specimen was larger than the curved-beam specimens, and the data scatter, attributed to fabrication process, was significantly lower for the spool specimens. The fatigue limit for both specimen was shown to be at least 40% of the average interlaminar tensile strength. Both specimens were found to be adequate for characterizing the interlaminar tensile static and fatigue behavior of carbon/epoxy laminates.

Minnetyan and Chamis investigated the use of the compact tension specimen in laminated composites using two examples. A computational simulation methodology is used and damage initiation, growth, accumulation, progressive fracture, and ultimate fracture modes were identified. The influence of laminate configuration and composite constituent properties on the compact tension specimen test characteristics were quantified.

Wu, Reddy, and Wilson conducted a design study to determine the weight savings and performance increase in stiffened wing skin panels made of aluminum lithium 2090-T83, aramid reinforced aluminum laminate, ARALL-3 and glass aluminum reinforced epoxy, GLARE-2. Six Z-stiffened compression panels representing upper and lower wing covers were fabricated from these material systems and tested to failure. The study confirmed that a 10 to 15% weight savings could be achieved. All panels tested under compression failed in a column bucking mode and the predicted critical loads, compared to those from the tests, were conservative.

The editor wishes to thank the authors, session chairpersons, and reviewers for ensuring the quality of the papers presented in the symposium and included in this STP. Special thanks is extended to the ASTM staff, particularly to Dorothy Savini for her dedication during the organization phase of the symposium and to Kathy Dernoga, Monica Siperko, and Helen Hoersch for their hard work during the publication phase. The editor wishes to acknowledge John Masters for his help during the review phase. The invaluable help of Stefan Dancila of Georgia Tech throughout this undertaking, is greatly appreciated.

Erian A. Armanios
School of Aerospace Engineering,
Georgia Institute of Technology,
Atlanta, Georgia,
symposium chairman and editor.

Damage Tolerance and Fracture Analysis

Paul A. Lagace[1] and Stacy M. Priest[2]

Damage Tolerance of Pressurized Graphite/Epoxy Cylinders Under Uniaxial and Biaxial Loading

REFERENCE: Lagace, P. A. and Priest, S. M., **"Damage Tolerance of Pressurized Graphite/ Epoxy Cylinders Under Uniaxial and Biaxial Loading,"** *Composite Materials: Fatigue and Fracture (Sixth Volume), ASTM STP 1285*, E. A. Armanios, Ed., American Society for Testing and Materials, 1997, pp. 9–26.

ABSTRACT: The damage tolerance behavior of internally pressurized, longitudinally slit, graphite/epoxy tape cylinders was investigated. Specifically, the effects of longitudinal stress, subcritical damage, and structural anisotropy were considered including their limitations on a methodology, developed for quasi-isotropic configurations, which uses coupon fracture data to predict cylinder failure. AS4/3501-6 graphite/epoxy cylinders with $[90/0/\pm45]_s$, $[\pm45/0]_s$, and $[\pm45/90]_s$ layups were tested in a test apparatus specially designed and built for this work such that pressurization resulted in only uniaxial (circumferential) loading of the cylinders. All cylinders had a diameter of 305 mm and slit lengths ranged from 12.7 to 50.8 mm. Failure pressure was recorded and fracture paths and failure modes evaluated via post-test reconstruction of the cylinders. These results were compared to results from previous tests conducted in biaxial loading. Structural anisotropic effects were further investigated by testing cylinders with the quasi-isotropic $[0/\pm45/90]_s$ layup and comparing these with the results from the other quasi-isotropic $[90/0/\pm45]_s$ layup. In all cases, the failure pressures for the uniaxially loaded cylinders fell below those for the biaxially loaded cases and the methodology was not able to predict these failure pressures. These differences were most marked in the case of the structurally anisotropic cylinders. Differences in fracture paths and overall failure mode were found to be greatest in the cases where there was the largest difference in the failure pressures. Strain gages placed near the slit tips showed that subcritical damage occurred in all cases. These results, coupled with previous work, show that failure is controlled by local damage mechanisms and the subsequent stress redistribution and damage accumulation scenario. It is thus necessary to assess the local effects, such as subcritical damage, while accounting for the influence of global parameters such as longitudinal load and structural anisotropy. Recommendations are made as to work that should be done to better understand such phenomena and thus lead to the establishment of methodologies to better characterize the failure, and thus damage tolerance, of composite materials and their structures.

KEYWORDS: graphite/epoxy composites, damage tolerance, pressurized cylinders, fracture (materials), composite materials

One area of particular importance in the use of composites is "damage tolerance," that is, the ability of a structure to continue to perform after it experiences damage. Currently, the damage-tolerant design of composite structures relies heavily on expensive and time-consuming

[1]Professor of Aeronautics and Astronautics and MacVicar Faculty Fellow, Technology Laboratory for Advanced Composites, Department of Aeronautics and Astronautics, Massachusetts Institute of Technology, Cambridge, MA 02139.
[2]Engineer, Ford Motor Company, Dearborn, MI 48124; formerly, graduate fellow, Massachusetts Institute of Technology, Cambridge, MA 02139.

9

experimentation in a "building block approach" [1] to determine needed properties and ensure the high level of safety that is required, especially if these structures are used for aerospace applications. A better understanding of the effects of damage on the performance of composite structures and better analytical tools and associated methodologies to predict this performance are required to make composite structures more economical and trustworthy, and to therefore allow composites to be utilized to their maximum potential.

Such a methodology to predict the fracture pressure of notched internally pressurized composite cylinders has been proposed [2,3]. The internally pressurized, thin-walled cylinder is a structure of particular importance to the aerospace industry because of its similarity to a transport aircraft fuselage. But such a methodology, if proven effective, can also be utilized for other pressurized structures of this general shape including rocket motor casings, fuel tanks, and oil pipelines. The methodology uses notched coupon fracture data and material properties to determine the fracture pressure of cylindrical specimens by accounting for the increased stress intensities at the notch due to the effects introduced by the internally pressurized shell configuration.

This methodology was originally developed, and has been verified, only for cylinders with a quasi-isotropic configuration [2–5]. It is important to examine the extent to which this methodology is applicable to composite cylinders of general configuration. Recent work [6] has been undertaken to assess the applicability of this methodology to composite cylinders with a structurally anisotropic skin configuration. The results of this investigation show that the methodology, in its current form, is not applicable in the general case. This suggests that more work is needed to isolate and understand the effects that are neglected by the predictive methodology and thus cause the methodology to be invalid in the general case. The establishment of such understanding may lead to adaptive techniques to utilize the general philosophy behind the methodology and thus be able to predict structural behavior based on results obtained at the coupon level.

These effects and limitations are therefore investigated in the current work. Specifically, the effects of longitudinal stress, subcritical damage, and structural anisotropy were considered in terms of their overall effect on the damage-tolerant behavior of internally pressurized, longitudinally slit,[3] graphite/epoxy cylinders as well as on the applicability and limitations of the existing predictive methodology. These effects were assessed by testing tape cylinders with $[90/0/\pm45]_s$,[4] $[\pm45/0]_s$, and $[\pm45/90]_s$ layups in a test apparatus specially designed and built for this work such that pressurization resulted in only uniaxial (hoopwise) loading of the cylinders. These results were compared to results from previous work conducted in biaxial loading [6] as well as coupon data in that work. Structural anisotropic effects were further investigated by testing cylinders with the quasi-isotropic $[0/\pm45/90]_s$ layup that is a stacking sequence variation of the previously tested $[90/0/\pm45]_s$ layup with higher values of the bending-twisting factors (D_{16} and D_{26}) but comparable ratios of these parameters to the main bending stiffness parameter, D_{11}. In addition, the failure modes and fracture paths of these cylinders are compared to those of the coupons and biaxially loaded cylinders, with the same layups, to further examine and understand the mechanisms of failure.

Review of Failure Prediction Methodology

The basic methodology proposed by Graves and Lagace [2] has several key parts: the correlation for notched coupon fracture, assumptions concerning the applicability of coupon data to the pressurized cylinder, and analysis to account for the geometrical effects of the shell.

---

[3]Also referred to as "axial" in a number of figures.
[4]For the cylinder, the 0° direction is along the circumference (that is, hoopwise).

The flat-plate fracture correlation developed by Mar and Lin [7,8] is utilized to characterize the notched tensile fracture of uniaxially loaded coupons. The far-field fracture stress of a notched composite plate under uniaxial tension is correlated via

$$\sigma_{f\text{plate}} = H_c(2a)^{-m} \tag{1}$$

where $2a$ is the length of the notch transverse to the loading direction, H_c is defined as the laminate fracture parameter [9], and m is the value of the stress singularity at a notch tip at the fiber-matrix interface. The composite fracture parameter, H_c, is determined experimentally and is a function of layup. The exponent, m, that has been determined analytically to be 0.28 for graphite/epoxy, such as AS4/3501-6 utilized here, is a function of the shear moduli and Poisson's ratios of the fiber and matrix [10]. Experimental work has indicated that the correlation is applicable to both tape and fabric graphite/epoxy laminates containing holes or slits oriented arbitrarily relative to the loading axes [4,9,11] and that while H_c is dependent on the laminate, it is not dependent on the notch geometry.

Two important assumptions are made in order to utilize the plate (that is, coupon) data for the pressurized configuration where the stress state differs due to the different structural configuration. The first assumption is that the longitudinal stress does not contribute to failure. This is partially based on evidence that shows that the tensile fracture stress of uniaxially-loaded coupons is not reduced by a notch oriented along the direction of load [12]. Thus, the hoop stress is considered as the primary cause of failure in the pressurized cylinder with longitudinal notches. The equivalent cylinder failure pressure for a far-field coupon failure stress may thus be determined via

$$p_{f\text{plate}} = \sigma_{f\text{plate}}(t/r) \tag{2}$$

where r is the cylinder radius and t is the laminate thickness. The second assumption is that the same failure mechanisms occur in the cylinder as in the coupons. If this assumption is met, which cannot be verified until after tests have been conducted, the laminate fracture parameter, H_c, that represents the coupon data is also representative of the pressurized cylinder failure. This allows the use of Eq 1 in Eq 2 to yield

$$p_{f\text{plate}} = H_c(t/r)(2a)^{-m} \tag{3}$$

It is important to point out that if this latter assumption is not met, then the laminate fracture parameter used to characterize the notched fracture response will not be representative of the behavior of the cylinder since different fracture mechanisms would be operative.

A second factor to consider is that the shell curvature causes an intensification of the local membrane stresses due to the inherent coupling between membrane and bending effects in shells. Folias [13–15] used shell theory to determine this effect on the stress field at slit ends in isotropic shells and proposed stress intensity factors to account for this stress intensification. These factors are related to the shell geometry, slit size, material properties, and applied loading. For a longitudinal slit in a cylindrical shell loaded via internal pressure, the extensional stress intensification factor, representing the increase in local extensional stresses at the slit tip in a shell, can be approximated as

$$K_i = (1 + 0.317\lambda_i^2)^{0.5} \tag{4}$$

where the isotropic shell parameter, λ_i^2, is a function of laminate properties and specimen geometry

$$\lambda_i^2 = a^2[12(1 - v^2)]^{0.5}/rt \qquad (5)$$

where v is the Poisson's ratio of the laminate. The fracture pressure of the cylinder can then be related to the equivalent cylinder failure pressure for a far-field coupon by reducing the plate capability by the stress intensification factor for the cylinder

$$p_{fcyl} = p_{fplate}/K_i \qquad (6)$$

Therefore, this allows determination of the cylinder failure pressure, via Eqs 3 through 6, from the geometry of the cylinder, the laminate properties, and the notched fracture parameter determined from coupon specimens.

Problem Definition and Approach

As previously noted, three specific limitations to the predictive methodology were considered: the effects of longitudinal stress, subcritical damage, and structural anisotropy. Although each of these are presented as separate issues, they can often couple to yield the final effect. Furthermore, the same specimens could be utilized in a number of cases to yield information on more than one of these issues.

The main item is the consideration of the effects of longitudinal stress. This is particularly important in coupling with subcritical damage. Work with tape laminates loaded in uniaxial tension [16–18] has shown that localized damage in the form of splitting in the 0° plies and associated delamination can mitigate stresses at the notch tip and result in reduced notch sensitivity. Such a mechanism is inherently limited in fabric laminates due to the woven nature of the fibers but was possibly present in particular cases in the previous work [6] on pressurized composite cylinders where a circumferential fracture path, suggestive of such a 0° ply split, was observed at the slit tip. Thus, experiments were conducted on quasi-isotropic, $[90/0/\pm45]_s$, and structurally anisotropic, $[\pm45/0]_s$ and $[\pm45/90]_s$, composite cylinders under only hoopwise loading to assess the role of longitudinal load. Three cylinders of each configuration were tested, one each with longitudinal slits with lengths of 12.7, 25.4, and 50.8 mm. The special uniaxial loading apparatus described in the next section was designed and built for this purpose.

The second item is the consideration of the effects of subcritical damage. As noted, this can couple with the effects of longitudinal load since the presence of longitudinal load may change or contribute to the subcritical damage that forms. In all cases, the occurrence of such damage was monitored by the placement of strain gages in strategic locations near the slit tip as described in the section on experimental procedures. Furthermore, careful examination of the failed cylinders was conducted to reveal the initial fracture paths causing failure. In addition, two $[90/0/\pm45]_s$ cylinders with 12.7-mm-long slits were tested under biaxial conditions (normal pressurization) to further investigate the circumferential fracture path observed in the previous work [6].

The final item is the consideration of the effects of structural anisotropy, that is, structural coupling. In all cases but one tested to date, the predictive methodology has been successful in predicting the failure pressure of quasi-isotropic composite cylinders constructed from graphite/epoxy tape or fabric prepreg. However, no success was achieved in the case of structurally anisotropic layups. Two of the layups, $[\pm45/0]_s$ and $[\pm45/90]_s$, are structurally anisotropic and are thus used to consider this effect. In addition, a variation on the quasi-isotropic layup, $[0/\pm45/90]_s$, is utilized. The Folias curvature correction factors are the same for both quasi-isotropic laminates since they have the same in-plane properties and these factors are therefore theoretically valid for both layups. However, the use of the latter allows stacking sequence effects to be examined, as well as the effects of having a higher degree of anisotropy with respect to the bending properties of the laminates since the bending-twisting coupling terms are small, but nonzero, in quasi-isotropic laminated composite configurations. The bending properties of the four laminates considered in this work and a comparable quasi-isotropic $[0_f/45_f]_s$ graphite/epoxy fabric laminate are listed in Table 1 (the subscript, f, indicates a fabric ply). The configurations represent an array of magnitudes of the parameters and also of the ratios of the coupling parameters to the basic bending parameter, D_{11}. In order for a direct comparison to be made with the $[90/0/\pm45]_s$ laminate, the $[0/\pm45/90]_s$ laminate is tested in the biaxial configuration. Two different longitudinal slit sizes are utilized: 12.7 and 50.8 mm.

These considerations yield the test matrix summarized in Table 2.

Uniaxial Loading Test Apparatus

The uniaxial loading test apparatus is designed to provide a path for the pressure loading in the longitudinal direction to be carried such that the cylinder takes virtually no longitudinal stress. Full details of the design and manufacture are provided in Ref *19*.

The design, illustrated in Fig. 1, consists of four basic subsystems: the supports, the rods, the rod support plates, and the endplates. Each of these are subsequently described. The overall system was designed for a maximum pressure of 6.9 MPa and was kept as modular as possible.

Two independent supports were manufactured to simply support the endplates that contain the actual test specimens. The supports are identical except that one rests on rigid casters while the other rests on legs. The casters are included for two reasons: one, to allow different length cylinders to be tested (not utilized here), and two, so that the overall fixture can expand in the longitudinal direction without restriction so as to properly carry longitudinal loads in the bars.

TABLE 1—*Calculated laminate bending properties.*

Laminate[a]	Bending Property, GPa · mm³					
	D_{11}	D_{22}	D_{12}	D_{16}	D_{26}	D_{66}
$[90/0/\pm45]_s$	5.53	9.37	0.70	0.32	0.32	1.01
$[0/\pm45/90]_s$	10.40	2.72	1.58	0.64	0.64	1.90
$[0_f/45_f]_s^b$	15.79	15.81	1.84	0.00	0.00	1.87
$[\pm45/0]_s$	2.13	1.92	1.41	0.64	0.64	1.54
$[\pm45/90]_s$	1.92	2.13	1.41	0.64	0.64	1.54

[a]0° is along circumferential direction.
[b]f indicates fabric ply.

TABLE 2—*Test matrix.*

Laminate[a]	Slit Size, mm		
	12.7	25.4	50.8
$[90/0/\pm45]_s$	1/2[b]	1/—	1/—
$[0/\pm45/90]_s$	—/1	—/—	—/1
$[\pm45/0]_s$	1/—	1/—	1/—
$[\pm45/90]_s$	1/—	1/—	1/—

[a]0° is along circumferential direction.
[b]Indicates number of uniaxial tests/number of biaxial tests.

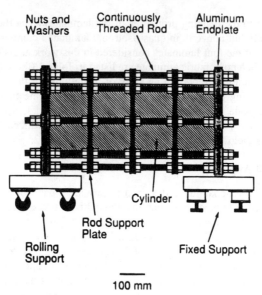

FIG. 1—*Schematic of uniaxial loading test apparatus.*

Eight continuously threaded rods were used to connect the endplates. These are made of Grade 2 steel with a root diameter of 20 mm and a length of 914 mm. These rods were sized so that an insignificant percentage (less than 1%) of the longitudinal load would be carried in the test cylinder if the cylinder were bonded to the endplates (not done for uniaxial loading in this work). The simple ratio of the longitudinal modulus times the area for the test cylinder to the eight steel bars yields this number.

The three rod support plates were added to the fixture after the initial testing since some of the initial steel rods were damaged (permanently bent) during the explosive failure of the cylinder. The plates reinforce the rods and prevent this damage during failure. These three plates are 16-mm-thick steel rings with an inner diameter 37 mm larger than the specimen outer diameter to avoid any interference during the test (prior to failure). The outer diameter of each ring is 432 mm. The rings have eight, equally spaced, drilled holes through which the steel rods are placed. The three rings are equally spaced between the two endplates.

Finally, the endplates are disks of 6061-T651 aluminum with a thickness of 32 mm and a diameter of 457 mm. The endplates, as illustrated in Fig. 2, each have a 5-mm-wide concentric groove into which the test specimens fit. A pressure bladder is utilized in the testing so that no sealing between the endplates and the test specimen is necessary. However, latex tubing is used to line the bottom of the 13-mm-deep grooves and provide a degree of cushioning between the test specimen and endplate. Each endplate has eight equally spaced holes to accept the steel bars that are attached with nuts and washers at each end. One endplate has an access panel to which the internal bladder is bonded thereby allowing insertion of the bladder into the specimen and then closure of the access panel to the endplate via a stepped arrangement. This access panel also has an access port to accept the nitrogen gas utilized in the pressurization.

Experimental Procedures

All the experimental work in this investigation was done with specimens manufactured from Hercules AS4/3501-6 graphite/epoxy tape with a nominal thickness of 0.134 mm. Full details are given in Ref *19*.

Top View

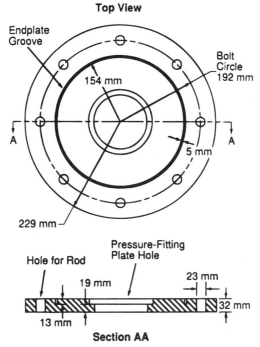

FIG. 2—*Illustration of endplate for uniaxial loading test apparatus.*

Cylinders of the configuration shown in Fig. 3 were manufactured and tested. A cylinder radius of 152 mm was chosen to match the experimental work done in previous investigations, and a cylinder length of 750 mm was selected to also match previous configurations and to assure that the stress state at the slit and associated failure were not influenced by end effects.

Cylinders were laid up on a cylindrical mandrel and cured using the standard manufacturer's cure cycle of a 1-h flow stage at 116°C and a 2-h set stage at 177°C. This was conducted in an autoclave under vacuum with a 0.59 MPa external pressure. These cylinders were then post-cured in an oven at 177°C for 8 h. Resulting average thicknesses were within 1% of the nominal thicknesses.

Slits in all the cylinders were cut with a 30 000 rpm rotary tool with a 25-mm-diameter composite blade mounted vertically on a milling machine arm. Slits were cut to the correct

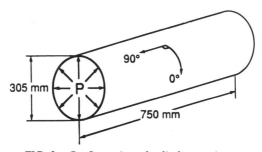

FIG. 3—*Configuration of cylinder specimen.*

length and notched at the tips by hand with a small jeweler's saw. A slit width of 0.64 mm resulted from the tools utilized, except at the tip as described earlier. Due to the explosive and destructive nature of the tests, a 50 by 50-mm grid was drawn on the entire exterior surface of each cylinder with a white paint pen. The grid squares were individually labeled to facilitate post-test reassembly of the remaining fragments.

Cylinder testing was accomplished via pressurization using bottled nitrogen. For the case of uniaxially loaded cylinders, the specially designed fixture was utilized. A consistent and repeatable setup procedure, as described in Ref *19*, was developed to ensure proper alignment of the parts of the fixture. Internal rubber bladders, made of 0.8-mm-thick gum rubber, were placed inside each cylinder to contain the pressurizing gas prior to failure. For the case of the biaxially loaded cylinders, the cylinders were sealed at both ends by bonding them in grooves cut into other aluminum endcaps. This allowed transferal of the longitudinal pressure load to the cylinder specimens. Again, the internal rubber bladders were utilized. In all cases, the cylinders were attached to pressure transducers so that internal pressure could be monitored and the internal pressure at failure determined.

All cylinders, except for one utilized in the uniaxial test apparatus verification tests (see next section), were instrumented in the same manner. One hoopwise and one longitudinal gage were placed away from the slit in order to monitor the far-field behavior. These were placed away from the slit approximately 110 mm in the longitudinal direction and 180 mm in the circumferential direction. In addition, small gages, with a 1-mm-long by 0.8-mm-wide gage element, were oriented in the circumferential (0°) direction as close as possible to each slit tip in each cylinder. The substrate on these gages were trimmed on one side almost to the gage element so that the gages could be placed as close as possible to the slit tips. Damage to the cylinder at the slit tip, that is, subcritical damage, is indicated by discontinuities and other odd behavior in this pressure/stress-strain response.

Cylinder testing was conducted in a blast chamber. A cylinder was placed horizontally within the test apparatus, or on an iron channel in the case of the two biaxially loaded cylinders. Cylinder pressurization was controlled manually with the use of a pressure regulator attached directly to a nitrogen tank at an approximate rate of 0.4 MPa/min. The pressurization rate was monitored during the test via a dial gage and a chart recording of the pressure transducer output. Data, including internal cylinder pressure, were collected at 0.5 or 1.0-s intervals.

Photographs were taken of all post-test reassembled cylinders to record failure modes. In addition, drawings of fracture patterns were made for each cylinder. These were used to identify the primary fracture path. Data reduction to stresses utilized the nominal ply thickness of 0.134 mm.

Results and Discussion

Experimental results are presented herein along with predictions and correlations. After an assessment of the uniaxial loading apparatus, the effects of longitudinal stress, subcritical damage, and structural anisotropy on the failure of pressurized composite cylinders with longitudinal slits are discussed separately. However, it should be noted that there are possibly varying degrees of interaction amongst these effects. In addition, since the current predictive methodology does not take into account these effects, application of this methodology to the structurally anisotropic or uniaxially loaded cylinders was not expected to yield accurate results, but is made to assess the quantitative difference that results.

Verification of Uniaxial Loading Test Apparatus

A series of four tests, the last to failure, were run on a [90/0/±45]$_s$ quasi-isotropic graphite/ epoxy cylinder to verify the performance of the uniaxial loading test apparatus and setup. The first three tests were run in an unnotched configuration, the last with a 50.8-mm-long longitudinal slit. Strain gage rosettes were placed at various points on the cylinder to ascertain the full state of strain. Longitudinal strain gages were placed on a steel bar on a small region where the threads and surface were machined flat. The stresses/loads in the bar and the cylinder were then calculated based on these strain measurements and a knowledge of the material and laminate moduli.

Typical far-field stress-strain results from a rosette on the cylinder during the test to failure are shown in Fig. 4. The "shear" results are reduced via standard rotation techniques from the raw rosette data. The results show that virtually no shear strain occurs in the test specimen and that all strains are linear to failure, as expected in this case. Assuming that no longitudinal stress exists in the test cylinder, the ratio of cylinder hoop stress to hoop strain should yield the laminate hoop modulus while the negative ratio of longitudinal strain to hoop strain should yield the laminate Poisson's ratio. Calculations from the experimental data yield values of 59.4 GPa and 0.34, respectively. These compare quite well with values of 55.5 GPa and 0.3 using basic ply properties and laminated plate theory. In all cases, the results from these verification tests showed that the apparatus as designed and manufactured functioned properly as virtually no longitudinal stress was present in the test cylinders.

Effects of Longitudinal Stress

The primary means of assessing the influence of longitudinal stress on the failure of pressurized composite cylinders is to compare the failure pressures and initial failure modes of the uniaxially loaded cylinders to the failure pressures and initial failure modes of the biaxially loaded cylinders with the same layups and slit lengths from the past investigation [6]. Plots of the failure pressures from the current investigations along with those of the previous investigation are provided with the predictions in Figs. 5, 6, and 7 for the $[90/0/\pm45]_s$, $[\pm45/0]_s$, and $[\pm45/90]_s$ configurations, respectively. In all cases, the uniaxial data fall below that of the biaxial data by 15 to 40%. However, these comparisons are somewhat complicated by the fact that tests run on $[90/0/\pm45]_s$ coupons in order to obtain a value of the laminate fracture parameter, H_c, for comparison to the previous work, yielded a value which is 10% lower than that in the previous work [6,19]. Correlations using both values of

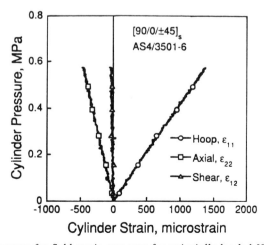

FIG. 4—*Pressure versus far-field strain response for uniaxially loaded [90/0/±45]ₛ cylinder.*

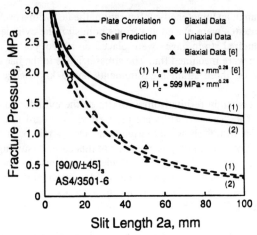

FIG. 5—*Experimental and predicted failure pressures for biaxially and uniaxially loaded [90/0/±45]ₛ cylinders.*

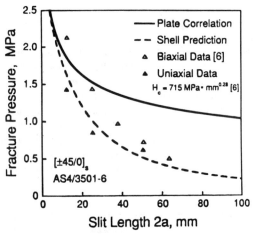

FIG. 6—*Experimental and predicted failure pressures for biaxially and uniaxially loaded [±45/0]ₛ cylinders.*

this parameter are shown in Fig. 5. This difference is likely attributable to general material and processing variability. Furthermore, this 10% difference in coupon data does not totally account for the larger difference noted between the failure pressures for the uniaxially-loaded and biaxially-loaded cases. Nevertheless, this difference should be kept in mind in evaluating the influence of the longitudinal stress in the failure of these cylinders.

The two structurally anisotropic laminate configurations show the most marked differences between the uniaxial and biaxial data. This is particularly evident for the [±45/0]ₛ configuration where the differences range between 15 and 41%. Additionally, none of the data sets follow the predicted trend, nor do the biaxial and uniaxial data sets for the same laminate configuration follow the same trend. In these cases, it is likely that the longitudinal stress plays a direct role in the failure process, and this is further ascertained by considering the failure modes.

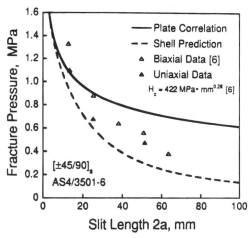

FIG. 7—*Experimental and predicted failure pressures for biaxially and uniaxially loaded* [±45/90]ₛ *cylinders.*

For both laminate configurations, the failure modes observed in the cylinders, for either uniaxial or biaxial loading, are notably different than those observed in the coupons [6]. This implies that the value of the composite fracture parameter determined from the coupons is unlikely to capture the effects observed in the cylinders. However, the failure modes for the [±45/90]ₛ cylinders are somewhat similar for the uniaxial and biaxial loading conditions. In the biaxially loaded cylinders [6], the fracture extended in the longitudinal direction from the slit ends before bifurcating to ±45° directions at a distance between 20 and 200 mm from the slit. The plies are generally cleanly fractured along these lines. In the uniaxially loaded case, there was a shorter length of longitudinal fracture prior to fracture turning to the −45° direction (perpendicular to the top ply fibers). The plies were again cleanly fractured, as can be seen in Fig. 8, for the case of a uniaxially loaded [±45/90]ₛ cylinder with a 12.7-mm slit. Although the initial longitudinal path exists and the turning is to the −45° direction, the initial longitudinal path length is shorter for the uniaxially loaded case and the fracture path turns, but does not bifurcate. This may be a sufficient difference to account for the differences observed in the failure pressures for the two different types of loading.

FIG. 8—*Photograph of failed* [±45/90]ₛ *uniaxially loaded cylinder with 12.7-mm slit.*

For the $[\pm 45/0]_s$ cylinders, there is considerable difference in the failure modes between the uniaxially and biaxially loaded cylinder, particularly at the smaller two slit lengths of 12.7 and 25.4 mm. This correlates with the largest differences observed in the failure pressures between the two loading types of 33% for the cylinders with the 12.7-mm slit and 41% for cylinders with the 25.4-mm slit. The fracture path in the $[\pm 45/0]_s$ coupons [6] was typically oriented at a $-45°$ angle, perpendicular to the direction of the surface fibers, from the slit tip to the edge of the laminate. Secondary delamination of the external $+45°$ plies along the fracture path resulted in a "butterfly" fracture pattern. In the biaxially loaded $[\pm 45/0]_s$ cylinders [6], four fracture paths originated at the slit ends and extended primarily in the $0°$ and $\pm 45°$ directions as shown in Fig. 9. In contrast, the uniaxially loaded cylinders with the two smaller slit sizes (12.7 and 25.4 mm) show a more predominant fracture path in the $+45°$ direction, parallel to the top surface fibers as can be seen in Fig. 9. As in the case of their biaxially loaded counterparts, the fracture paths are more ragged as the plies do not fail cleanly as in the cases of the other laminate configurations. However, the uniaxially loaded $[\pm 45/0]_s$ cylinder with the 50.8-mm axial slit shows a failure mode, in Fig. 10, more similar to the biaxially loaded cases with a clear hoopwise $(0°)$ fracture path coupled with angular fracture paths.

FIG. 9—*Photographs of failed $[\pm 45/0]_s$ cylinders loaded* (upper) *biaxially (from Ref 6), and* (lower) *uniaxially.*

FIG. 10—*Photograph of failed [±45/0]ₛ uniaxially loaded cylinder with 50.8-mm slit.*

In contrast, it is somewhat surprising that the quasi-isotropic $[90/0/\pm45]_s$ configuration shows a difference between the uniaxial and biaxial cases. This configuration is most consistent with the assumptions of the predictive methodology in that it is quasi-isotropic. Furthermore, since longitudinal stress is assumed to be ineffective in the methodology and the methodology correlates the biaxial data well, it would be expected to also correlate the uniaxial data well. Closer examination of the data indicates that the differences between the two loading cases are 16, 20, and 27% for the cylinders with 12.7, 25.4, and 50.8-mm slit lengths, respectively. It seems unlikely that such differences can be totally attributed to data scatter including the 10% difference observed in the coupon data, particularly since the differences are consistently larger.

This leads to a consideration of the failure modes. In the previous work [6], the fracture in the coupons traveled along the ±45° directions approximately halfway to the coupon edge from the slit ends before turning to 90° and running parallel to the slit with the 0° fibers breaking cleanly along this fracture path. The initial fracture paths in the biaxially loaded [90/0/±45]ₛ cylinders [6] are similar, as shown in Fig. 11, except in the case of the cylinder with the 12.7-mm slit as previously noted. The current uniaxially-loaded cylinders showed similar failure modes as can be seen for the case of the cylinder with a 50.8-mm slit in Fig. 11. Comparison of the failure modes in the coupons and the biaxially and uniaxially loaded cylinders thus shows the failure modes to be consistent in all three cases.

The one remaining factor from the methodology, therefore, is the shell stress intensification factor. This factor was derived by Folias for the case of biaxial loading due to pressurization. However, removal of the longitudinal load should not change the stress intensification of the circumferential stress at the notch tip unless there is some coupling of the longitudinal and circumferential stresses. Such coupling does not exist for the isotropic case. But, it is possible that structural anisotropy/orthotropy can create such coupling. Furthermore, the occurrence of subcritical damage may also affect this coupling. Although the macroscopic failure modes of the biaxially loaded and uniaxially loaded cylinders are the same, the initial damage may differ leading to a difference in failure pressure. These items are further discussed in the next section on subcritical damage.

The comparisons of fracture paths and macroscopic failure modes coupled with the failure pressures indicate that in the cases where the failure modes between the uniaxially and biaxially loaded cases are more similar, the failure pressures for the two cases are also closer. However, the exact role of the longitudinal stress in this cannot be clearly ascertained without further investigation that must include detailed analysis of the local stress state in the two cases and, thus, a better assessment of the stress intensification due to the structural configuration.

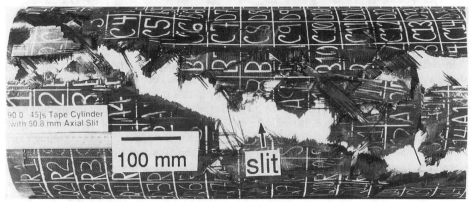

FIG. 11—*Photographs of failed [90/0/±45]ₛ cylinders loaded* (upper) *biaxially (from Ref 6), and* (lower) *uniaxially.*

Furthermore, the role of the longitudinal stress is likely linked to the formation of subcritical damage. This aspect is subsequently discussed.

Effects of Subcritical Damage

As discussed previously, the occurrence of subcritical damage was detected by discontinuities observed in the strain response of the gages at the slit tip. Although this method cannot guarantee the detection of all damage at the notch tip, the presence of discontinuities in the strain response, as shown in Fig. 12, does clearly indicate that subcritical damage has occurred. However, quantitative statements about the subcritical damage cannot be made.

In all cases, for all laminate types and for both the uniaxially and biaxially loaded cases in this work, subcritical damage occurred. It is therefore clear that subcritical damage is a part of the overall process. The formation and effects of such damage are likely tied to the overall stress state, implying that the effect of longitudinal stress is coupled to this issue. Furthermore, "competing" damage modes may change the history of the overall damage formation and, thus, final failure. This is illustrated by one example.

The biaxially loaded [90/0/±45]ₛ cylinder with the 12.7-mm slit from the past investigation [6] showed a significant circumferential fracture path at one slit tip as shown in Fig. 13. This is in contrast to that observed in the biaxially and uniaxially loaded cylinders of the same configuration and slit size in the current work where no circumferential fracture path is observed

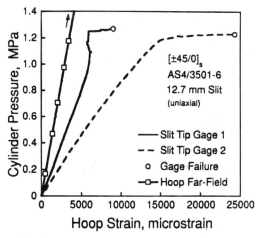

FIG. 12—*Cylinder pressure versus slit tip and far-field circumferential strains.*

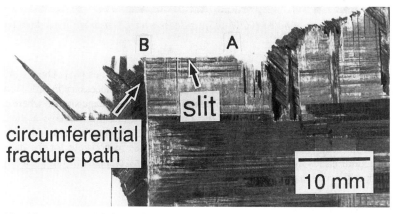

FIG. 13—*Photograph of failed [90/0/±45]ₛ cylinder with 12.7-mm slit loaded biaxially (from Ref 6).*

(Fig. 14). The former case also has a considerably higher failure pressure, by 23%, than the latter cases. It has been observed previously that the formation of a 0° split is able to mitigate local stress concentrations and increase fracture stress [16–18]. However, it is possible that this subcritical damage mode of a 0° (that is, circumferential) split is competing with other subcritical damage modes. If the stress state and critical fracture parameters are such that small variations in the material and the processing result in a change as to the subcritical damage mode that first occurs, the subsequent damage scenario can be quite different leading to a difference in final failure.

Past work on coupons [16–18,20] has shown that the effects of subcritical damage on load redistribution at the notch tips and the subsequent effects on the coupon failure behavior are indeed highly dependent on loading condition, material, and laminate. The effects of this subcritical damage can range from detrimental to beneficial in terms of increasing or decreasing the final failure stress, depending on the degree to which the localized damage relieves or

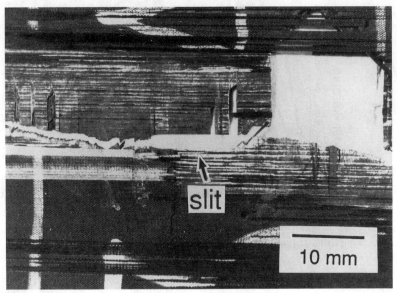

FIG. 14—*Photograph of failed [90/0/±45]ₛ cylinder with 12.7-mm slit loaded uniaxially.*

exacerbates the stress concentration in the primary load-bearing plies [*16*]. Thus, a more basic understanding of the damage mechanisms at the tip of the slit is necessary in order to properly predict the final failure. This requires experimental progressive damage studies where cylinders and coupons are loaded to points below the failure point and the damage at the slit tips investigated nondestructively or destructively, or both. This must be coupled with careful evaluation of the local stress states to ascertain the stresses responsible for the damage that occurs. Structural issues that cause an intensification or any changes in the local stress state must be properly accounted for in such modeling.

Effects of Structural Anisotropy

The issues discussed thus far clearly indicate that the specifics of the local stress state are key in the origin of damage and its accumulation scenario to the point where final failure occurs. Clearly, the structural parameters of the laminate will affect the local stress state. For example, the existence of bending-twisting coupling terms will cause local twisting and associated stresses. Since it has been identified that local considerations are key in this response, it would not be fruitful to attempt to correlate the observed behavior with global parameters such as those that characterize structural anisotropy as found in Table 1. These factors must be taken into account when analyzing such configurations, but it is their effect on the local stress state and the resulting damage that must be ascertained.

Summary

The damage tolerance of longitudinally slit pressurized composite cylinders was investigated and the limitations of a methodology that uses coupon fracture data to predict cylinder failure were explored. This predictive methodology was developed for quasi-isotropic configurations and was previously shown to be valid for such configurations but invalid for structurally anisotropic configurations. Thus, the specific effects of longitudinal stress, subcritical damage, and structural anisotropy were considered. A special test fixture was designed and manufactured to test pressurized cylinders under uniaxial (circumferential) loading only.

In all cases, the methodology was not able to predict the failure pressures of the uniaxially-loaded cylinders. This was most marked in the case of the structurally anisotropic cylinders. Furthermore, in all cases, the failure pressures for uniaxial loading fell below those for the same configuration under biaxial loading. Such differences were again most marked for the structurally anisotropic cases. Examination of the fracture paths and overall failure modes showed that the largest differences in the failure pressures occurred for those cases with the least match in failure mode. Strain gages placed near the slit tips showed that subcritical damage occurred in all cases. Previous work on coupons has indicated the importance that such damage can play in the overall failure response.

These results coupled with previous work thus show that failure is controlled by local damage mechanisms and the subsequent stress redistribution and damage accumulation scenario. It is therefore important to understand the local behavior in order to assess the role of various parameters and to predict the failure of composite structures. Careful analysis must be performed to determine the local stress state. This modeling must include the considerations of the global structural configurations and its effect on the local stresses as well as the local details including progressive damage that occurs. Such stress analyses must then be utilized with reliable failure criteria for various damage mechanisms that occur in composites and their structures. In addition, experimental progressive damage studies should be conducted so that the subcritical damage states that exist can be documented and compared with predictions from the stress and failure analyses.

This work also shows that predictive methodologies that use global/averaging parameters to try to capture this local damage behavior from coupon data cannot properly account for the details of the local behavior and thus will not be able to predict failure except in cases where the damage modes and fracture paths leading to final failure do not change between the coupons and the structure. It is time that the composites community take a renewed interest in the basic mechanisms of damage and failure in order to develop methodologies to better predict the failure response of composite structures. Correlative methodologies with "scaling" factors and analyses, such as those represented by the "building block approach" are good engineering tools for the present when the deficit in understanding demands tools that are overly conservative in order to provide a safe structure. However, such tools cannot hope to provide the design capability to utilize composite materials to their full effectiveness. Composites and their structures need to be treated at the level where their key mechanisms are operative. Only then can a true understanding and predictive capability be established for failure, and thus damage tolerance, of composite materials and their structures.

Acknowledgments

This work was sponsored by the NASA Langley Research Center under NASA Grant NAG-1-991.

References

[1] Whitehead, R. S. and Deo, R. B., "A Building Block Approach to Design Verification Testing of Primary Composite Structures," *Proceedings*, Twenty-fourth AIAA/ASME/ASCE/AHS SDM Conference, Lake Tahoe, NV, 1983, pp. 473–477.

[2] Graves, M. J. and Lagace, P. A., "Damage Tolerance of Composite Cylinders," *Composite Structures*, Vol. 4, No. 1, 1985, pp. 75–91.

[3] Lagace, P. A. and Saeger, K. J., "Damage Tolerance Characteristics of Pressurized Graphite/Epoxy Cylinders," *Proceedings*, Sixth International Symposium on Offshore Mechanics and Arctic Engineering, American Society of Mechanical Engineers, Houston, TX, March 1987, pp. 31–37.

[4]   Chang, S. G. and Mar, J. W., "The Catastrophic Failure of Pressurized Graphite/Epoxy Cylinders Initiated by Slits at Various Angles," *Journal of Aircraft*, Vol. 22, No. 6, June 1985, pp. 462–466.

[5]   Saeger, K. J. and Lagace, P. A., "Fracture of Pressurized Composite Cylinders with a High Strain-to-Failure Matrix System," *Composite Materials: Fatigue and Fracture, Second Volume, ASTM STP 1012*, P. Lagace, Ed., American Society for Testing and Materials, Philadelphia, 1989, pp. 326–337.

[6]   Ranniger, C. U., Lagace, P. A., and Graves, M. J., "Damage Tolerance and Arrest Characteristics of Pressurized Graphite/Epoxy Tape Cylinders," *Composite Materials: Fatigue and Fracture (Fifth Volume), ASTM STP 1230*, R. H. Martin, Ed., American Society for Testing and Materials, Philadelphia, 1995, pp. 407–426.

[7]   Mar, J. W. and Lin, K. Y., "Fracture of Boron/Aluminum Composites with Discontinuities," *Journal of Composite Materials*, Oct. 1977, pp. 405–421.

[8]   Mar, J. W. and Lin, K. Y., "Fracture Mechanics Correlation for Tensile Failure of Filamentary Composites with Holes," *Journal of Aircraft*, Vol. 14, July 1977, pp. 703–704.

[9]   Lagace, P. A., "Notch Sensitivity and Stacking Sequence of Laminated Composites," *Composite Materials: Testing and Design (Seventh Conference), ASTM STP 893*, J. M. Whitney, Ed., American Society for Testing and Materials, Philadelphia, 1986, pp. 161–176.

[10] Fenner, D. N., "Stress Singularities in Composite Materials with an Arbitrarily Oriented Crack Meeting an Interface," *International Journal of Fracture*, Vol. 12, No. 5, Oct. 1986, pp. 705–721.

[11] Lagace, P. A., "Notch Sensitivity of Graphite/Epoxy Fabric Laminates," *Composites Science and Technology*, Vol. 26, 1986, pp. 95–117.

[12] Brewer, J. C., "Tensile Fracture of Graphite/Epoxy with Angled Slits," TELAC Report 82-16, Massachusetts Institute of Technology, Cambridge, MA, Dec. 1982.

[13] Folias, E. S., "On the Effect of Initial Curvature on Cracked Flat Sheets," *International Journal of Fracture Mechanics*, Vol. 5, No. 4, 1969, pp. 327–346.

[14] Folias, E. S., "An Longitudinal Crack in a Pressurized Cylindrical Shell," *International Journal of Fracture Mechanics*, Vol. 1, No. 2, 1965, pp. 104–113.

[15] Folias, E. S., "On the Prediction of Catastrophic Failures in Pressurized Vessels," *Prospects of Fracture Mechanics*, G. C. Sih, H. C. van Elst, and D. Broek, Eds., Nordhoff International, Leiden, the Netherlands, 1974, pp. 405–418.

[16] Lagace, P. A., Bhat, N. V., and Gundogdu, A., "Response of Notched Graphite/Epoxy and Graphite/PEEK Systems," *Composite Materials: Fatigue and Fracture, Fourth Volume, ASTM STP 1156*, W. W. Stinchcomb and N. E. Ashbaugh, Eds., American Society for Testing and Materials, Philadelphia, 1993, pp. 55–71.

[17] Harris, C. E. and Morris, D. H., "A Fractographic Investigation of the Influence of Stacking Sequence on the Strength of Notched Laminated Composites," *Fractography of Modern Engineering Materials: Composites and Metals, ASTM STP 948*, J. E. Masters and J. J. Au, Eds., American Society for Testing and Materials, Philadelphia, 1987, pp. 154–173.

[18] Lagace, P. A. and Nolet, S. C., "The Effect of Ply Thickness on Longitudinal Splitting and Delamination in Graphite/Epoxy under Compressive Cyclic Load," *Composite Materials: Fatigue and Fracture, ASTM STP 907*, H. T. Hahn, Ed., American Society for Testing and Materials, Philadelphia, 1986, pp. 335–360.

[19] Priest, S. M., "Damage Tolerance of Pressurized Graphite/Epoxy Tape Cylinders Under Uniaxial and Biaxial Loading," TELAC Report 93-19, Massachusetts Institute of Technology, Cambridge, MA, Dec. 1993.

[20] Simonds, R. A. and Stinchcomb, W. W., "Response of Notched AS4/PEEK Laminates to Tension/Compression Loading," *Advances in Thermoplastic Matrix Composite Materials, ASTM STP 1044*, G. Newaz, Ed., American Society for Testing and Materials, Philadelphia, 1989, pp. 133–145.

David A. Hooke,[1] *Erian A. Armanios,*[1] *D. Stefan Dancila,*[1]
Ash Thakker,[2] *and Phil Doorbar*[3]

Failure of a Titanium Metal-Matrix Composite Cylindrical Shell Under Internal Pressure

REFERENCE: Hooke, D. A., Armanios, E. A., Dancila, D. S., Thakker, A., and Doorbar, P., **"Failure of a Titanium Metal-Matrix Composite Cylindrical Shell Under Internal Pressure,"** *Composite Materials: Fatigue and Fracture (Sixth Volume), ASTM STP 1285,* E. A. Armanios, Ed., American Society for Testing and Materials, 1997, pp. 27–44.

ABSTRACT: An investigation into the failure of a SCS-6/Ti-6-4 thin-walled cylindrical shell subjected to internal pressure is presented. The shell geometry consists of two end flanges, a middle circular cylindrical section with a stacking sequence of $[90/\pm30/0]_s$, and a $[90]_8$ reinforcement band within the middle section. The stress field in the middle section is predicted based on an anisotropic shell solution and the failure based on laminate strength data. An experimental setup is designed to allow the application of an internal pressure to the shell using a hydraulic system while maintaining a zero axial load. Strain gages placed at selected locations on the specimen monitor the state of strain during the test, and acoustic emission is used to monitor damage onset and progression. The results indicate that the ultimate load is in good agreement with theoretical predictions from the anisotropic shell solution and the engineering thin-walled theory. Acoustic emission provides a correlation with damage initiation and progression. Fracture surface analysis gives an insight into the initiation and the progression of failure.

KEYWORDS: metal-matrix composites, thin-walled shell, testing, internal pressure, failure, fatigue (materials), fracture (materials), composite materials

Metal-matrix composites (MMC) provide an attractive alternative to meet increased stress and temperature design requirements at reduced weight. Increased specific stiffness and strength are beneficial characteristics of silicon carbide (SiC) reinforced titanium (Ti) as compared to conventional titanium alloys. Accurate characterization of stiffness, strength, and failure behavior coupled with efficient prediction models and testing methods are needed to exploit the benefits of titanium matrix composites.

The majority of work on Ti/MMC is devoted to material characterization and response at room and elevated temperatures. However, no results are found for the testing of a Ti/MMC at the component level. Gravett and deLaneuville [1] tested three rings of SCS-6/Ti-15-3 MMC material under internal pressure. The internal pressure was applied using twelve hydraulic actuators with properly sized load shoes that rested on the internal diameter of the ring. Load was applied via the load shoes to the inside diameter. Such a method would not be feasible for the shell configuration considered in this work because of the considerable length of the

[1]Graduate research assistant, associate professor, and graduate research assistant, respectively, Georgia Institute of Technology, Atlanta, GA 30332–0150.
[2]Senior manager, Rolls-Royce Inc., Atlanta, GA 30339–3769.
[3]Group leader, Rolls-Royce plc., Derby, UK.

27

test specimen, and uniform loading would not be possible with such an arrangement. Zaretsky and deLaneuville [2] tested MMC rings at room and elevated temperatures. The method of internal pressure application was not discussed. Temperature effects on the monotonic and fatigue strengths of SCS-6/Ti-15-3 laminates were examined by Pollock and Johnson [3]. Kamiya and Fujita [4] have performed internal pressure tests on fiber-reinforced polymer (FRP) cylinders using pressurized oil to determine fracture criteria variances between cylinders made with a roving or a chopped fiber material. Teng and Rotter [5] have performed tests on isotropic cylinders under axial compression and internal pressure to help characterize the stability and buckling characteristics of cylinders with imperfections. Wang and Socie [6] have performed combination loadings on G-10 E-glass woven composite tubular specimens. Tests of pneumatically pressurized graphite/epoxy notched cylinders under uniaxial and biaxial loading were conducted by Lagace and Priest [7] in order to investigate their damage tolerance.

An internal pressure test to failure of a SCS-6/Ti-6-4 cylindrical circular shell with end flanges is presented in this work. The shell consists of a [90/±30/0]$_s$ middle circular cylindrical section with a [90]$_8$ band designed to provide an additional reinforcement at a selected location along the length, as shown in Fig. 1, and two end flanges bonded using a scarf joint.

The objective of this work is to evaluate the failure of the shell under internal pressure. An experimental setup is designed in order to allow for internal pressure loading. Damage initiation and progression is monitored using acoustic emission. An anisotropic shell analysis is used to predict the stress and strain fields in the middle section and provide an estimate of the ultimate internal pressure based on laminate strength data. The predictions from this theory are compared with the engineering thin-walled shell results. The theoretical predictions for ultimate load are compared with test data.

The analytical model provides closed-form expressions for the stress and strain fields and is presented first. This is followed by the design of the loading fixture, instrumentation, test setup, and procedure. The test results including strain data, ultimate strength, and fracture behavior are presented in a third section along with a correlation between analytical predictions and test results. Finally, a number of conclusions and recommendations are proposed.

Analytical Model

The purpose of the analytical model is to provide an estimate of the ultimate internal pressure. In the analytical model, the stress and strain fields for the middle section of the shell with a [90/±30/0]$_s$ stacking sequence are investigated. The reinforcement band section is less critical and will not be considered.

Consider the cylindrical laminated composite shell appearing in Fig. 2 with inner radius, r_1, and outer radius, r_2, subjected to an internal pressure, p. The constitutive relationship for stress and strain is

$$\begin{Bmatrix} \epsilon_{rr} \\ \epsilon_{\theta\theta} \\ \epsilon_{zz} \\ \gamma_{\theta z} \\ \gamma_{rz} \\ \gamma_{r\theta} \end{Bmatrix} = \begin{bmatrix} \bar{S}_{11} & \bar{S}_{12} & \bar{S}_{13} & \bar{S}_{14} & 0 & 0 \\ \bar{S}_{12} & \bar{S}_{22} & \bar{S}_{23} & \bar{S}_{24} & 0 & 0 \\ \bar{S}_{13} & \bar{S}_{23} & \bar{S}_{33} & \bar{S}_{34} & 0 & 0 \\ \bar{S}_{14} & \bar{S}_{24} & \bar{S}_{34} & \bar{S}_{44} & 0 & 0 \\ 0 & 0 & 0 & 0 & \bar{S}_{55} & \bar{S}_{56} \\ 0 & 0 & 0 & 0 & \bar{S}_{56} & \bar{S}_{66} \end{bmatrix} \begin{Bmatrix} \sigma_{rr} \\ \sigma_{\theta\theta} \\ \sigma_{zz} \\ \tau_{\theta z} \\ \tau_{rz} \\ \tau_{r\theta} \end{Bmatrix} \tag{1}$$

where r, θ, and z denote the radial, circumferential, and axial coordinates, respectively. The $\bar{S}_{ij}$ terms are the transformed compliance constants, obtained by rotation of the orthotopic compliances about the normal to the shell surface. For a laminated composite shell undergoing

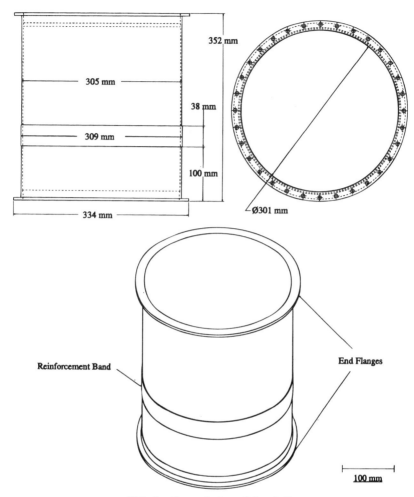

FIG. 1—*General view of the shell.*

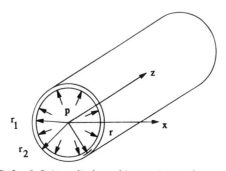

FIG. 2—*Infinite cylinder subject to internal pressure.*

membrane loading, the effective compliances, denoted by $\bar{S}_{ij}$, are estimated using a plane-strain assumption and no-slip condition at the ply interfaces. They are expressed as

$$\bar{S}_{ij} = (r_2 - r_1) A_{ij}^{-1} \tag{2}$$

where A_{ij} are the in-plane stiffness coefficients of classical lamination theory [8]. For a $[90/\pm30/0]_s$ layup, representing the middle section, the following effective compliance coefficients vanish

$$\bar{S}_{14} = \bar{S}_{24} = \bar{S}_{34} = \bar{S}_{56} = 0 \tag{3}$$

The stress field for a cylindrical shell subjected to an internal pressure, p, is axisymmetric and is given by Lekhnitskii [9]. For an uncapped shell, the nonzero stress components are expressed as

$$\begin{Bmatrix} \sigma_{rr} \\ \sigma_{\theta\theta} \end{Bmatrix} = \frac{p\alpha^{k+1}}{(1 - \alpha^{2k})} \begin{Bmatrix} \rho^{k-1} - \rho^{-k-1} \\ k(\rho^{k-1} - \rho^{-k-1}) \end{Bmatrix} \tag{4}$$

where

$$\alpha = \frac{r_1}{r_2}$$

$$\rho = \frac{r}{r_2}$$

$$k = \sqrt{\frac{\bar{S}_{11}\bar{S}_{33} - \bar{S}_{13}^2}{\bar{S}_{22}\bar{S}_{33} - \bar{S}_{23}^2}} = \sqrt{\frac{\bar{C}_{22}}{\bar{C}_{11}}} \tag{5}$$

The stiffness coefficients, obtained by inverting Eq 1, are denoted by $\bar{C}_{ij}$. Parameter k is a measure of material anisotropy and equals 1 for isotropic materials. The associated strains are

$$\begin{Bmatrix} \epsilon_{rr} \\ \epsilon_{\theta\theta} \\ \epsilon_{zz} \end{Bmatrix} = \frac{p\alpha^{k+1}}{(1 - \alpha^{2k})} \begin{Bmatrix} (\bar{S}_{11} + k\bar{S}_{12})\rho^{k-1} - (\bar{S}_{11} + k\bar{S}_{12})\rho^{-k-1} \\ (\bar{S}_{12} + k\bar{S}_{22})\rho^{k-1} - (\bar{S}_{12} + k\bar{S}_{22})\rho^{-k-1} \\ (\bar{S}_{13} + k\bar{S}_{23})\rho^{k-1} - (\bar{S}_{13} + k\bar{S}_{23})\rho^{-k-1} \end{Bmatrix} \tag{6}$$

For the SCS-6/Ti-6-4 MMC material properties provided in Table 1 and Eq 5, parameter $k = 1.151$.

TABLE 1—*Material properties for SCS-6/Ti-6-4 MMC.*

$E_{11} = 219.2$ GPa
$E_{22} = 134.7$ GPa
$G_{12} = 60.05$ GPa
$\nu_{12} = \nu_{13} = 0.29$
$G_{23} = 34.23$ GPa
$\nu_{23} = 0.36$

TABLE 2—*Predicted and measured moduli.*

	Prediction	Experimental
E_{zz}	180.6 GPa	185.1 GPa
$E_{\theta\theta}$	159.8 GPa	166.3 GPa
$G_{\theta z}$	64.80 GPa	$\cdots$
$\nu_{\theta z}$	0.27	0.27

The in-plane effective moduli predicted based on Eq 2 are provided in Table 2. The hoop stress and strain predicted by the engineering thin-walled shell theory are given by

$$\bar{\sigma}_{\theta\theta} = \frac{pr_1}{(r_2 - r_1)} \tag{7}$$

$$\bar{\epsilon}_{\theta\theta} = \frac{pr_1}{(r_2 - r_1)E_{\theta\theta}} \tag{8}$$

where $E_{\theta\theta}$ denotes the effective Young's modulus along the hoop direction.

The through-the-thickness radial and hoop stress distributions normalized by the applied internal pressure are presented in Fig. 3. Note that the constant hoop stress predicted by Eq 7 is the average hoop stress from Eq 4.

The radial, hoop, and axial strain distributions through-the-thickness normalized by the engineering value in Eq 8 are shown in Fig. 4. The radial and axial strains are a result of the respective Poisson's contractions associated with the hoop strain shown in Fig. 3.

Failure Prediction

The ultimate pressure is predicted based on laminate strength data. As shown from the stress distribution in Fig. 3, the shell is subjected to predominantly constant hoop stress. The

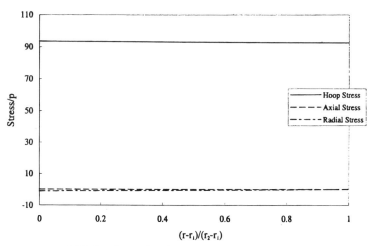

FIG. 3—*Stress distribution through-the-thickness.*

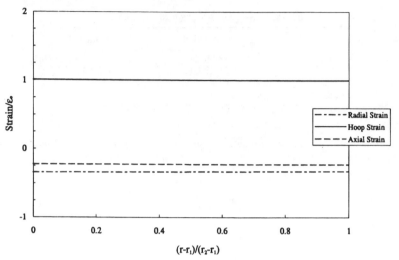

FIG. 4—*Strain distribution through-the-thickness.*

failure stress is measured from a test of a laminate with the same layup subjected to uniaxial tensile stress. A SCS-6/Ti-6-4 MMC panel with $[90/\pm30/0]_s$ was sectioned into two specimens, one along the 90° fiber direction and one transverse to the 90° fiber direction, with 19 by 152 by 1.78 mm dimensions and tested under tensile load. Each was instrumented with back-to-back "Tee" rosette strain gages to eliminate bending effects from the results. The measured axial modulus, E_{zz}, transverse modulus, $E_{\theta\theta}$, and in-plane Poisson's ratio, $\nu_{\theta z}$, are included in Table 2. The measured failure hoop stress was 1.03 GPa that results in an ultimate internal pressure of 11.02 MPa based on the anisotropic shell theory, Eq 4, and 11.08 MPa based on the engineering thin-walled shell theory.

Test Method

Fixture Design Constraints

Three main considerations are taken into account in the design of the testing apparatus. The first is that under the predicted failure pressure of 11.08 MPa the fixture should not deform in a way that would affect the test specimen. The second is that the test proceed safely. The third is that only circumferential loading be applied to the shell. The result of these constraints leads to the design presented in this work.

The basis of the design is an internal steel mandrel with steel end caps. The center mandrel has a 38-mm wall thickness, with 13-mm-thick end caps. The end caps are "light duty" because there is nearly no end loading as a result of the design loading constraint. Conversely, the mandrel is exposed to a very high pressure, therefore, a thick wall is used. The stresses in the mandrel are calculated using Eqs 4 and 5 where

$$k = 1$$

$$p = 13.8 \text{ MPa}$$

$$\sigma_{r_{max}} = 13.8 \text{ MPa}$$

$$\sigma_{\theta_{max}} = -61.5 \text{ MPa}$$

Assuming an underestimation of the failure pressure of 25%, the maximum expected internal pressure is 13.8 MPa. This estimate of the stresses show that the mandrel is stressed at levels well below the elastic limit of steel. The constraint of allowing only circumferential stress is addressed by sealing the ends of the shell with O-rings on the inner diameter. This method allows for circumferential loading, but axial loading is negligible because there is no rigid coupling between the test specimen and the mandrel. The resulting design is a safe and effective tool for the internal pressure test.

Upon failure, the shell would virtually explode if gas were used. Instead, hydraulic oil is chosen as the pressurizing medium. The oil being nearly incompressible, the pressure applied to the shell rapidly drops to zero when leakage occurs, as is the case when the shell fails.

Fixture

The test fixture in Fig. 5 consists of an internal mandrel, O-rings, and two end plates. The internal mandrel serves to occupy most of the internal volume of the test specimen and hold the end plates. The end plates hold the O-ring seals in place in a groove formed by the mandrel and the shell. The design of the fixture allows an internal pressure to be applied to the shell. Longitudinal loading is not applied because there is no rigid clamping of the shell during the test. One end of the mandrel is drilled and tapped for a pressure fitting and cross-drilled to allow flow of pressurized fluid into the volume formed by the mandrel and the test specimen.

Hydraulic pressure is applied using a hand pump typically used for hydraulic jacks. Small increments in pressure are achieved because of the short stroke of the pump. This allows for a very stable, controlled, and inexpensive method of applying internal pressure. In addition, when failure occurs, only the amount of oil used to pressurize the system is lost.

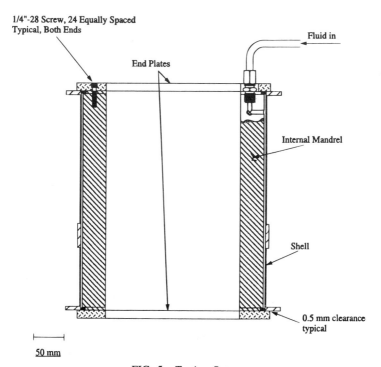

FIG. 5—*Testing fixture.*

Instrumentation Design Constraints

The constraint placed on the instrumentation is that strain and pressure data be obtained reliably and accurately. Twelve strain gage rosettes are mounted on the test specimen. Previous elastic tests on the shell incorporated all the rosettes to determine the strain fields for various loadings. For those tests, strains were measured using manual switching and balancing boxes and strain indicators. For the present test, the pressure is applied as a ramp and, therefore, strain and pressure readings must be taken in real time. This precludes the use of a manual transcription method. Realizing the uniformity of the load, four strain gages are used to monitor the test.

The gage locations used to monitor the test were determined by a small-scale internal pressure test, with incremental loading and manual strain monitoring. The specimen was pressurized to 310.3 kPa, and strains on all twelve rosettes were measured. The results showed that the circumferential strain in the reinforcement band was about one half that of the rest of the specimen. The other gages showed that the strain field was uniform around the rest of the specimen.

Instrumentation

A Keithley Series 500 analog-to-digital conversion system connected to an IBM PC was used for data acquisition. Four channels of strain and internal pressure were monitored during the test. Pressure was monitored using a pressure transducer, with a precision analog gage for confirmation. The strain gages were connected to a remote printed-circuit board containing completion resistors and circuitry for Wheatstone bridges. The output of the Wheatstone bridges was then connected to the Keithley unit where it was amplified. The amplified signal was read and converted to strain. Real-time pressure and strain data were displayed on the computer screen as the test progressed.

All four strain gages monitored circumferential strain. Two of the gages were located at a distance of 49 mm from the top flange, 180° opposed to each other. The other two gages were located at a distance of 101 mm from the other flange, 180° opposed to each other and 90° opposed from the other gages as shown in Fig. 6. Labels A, L, J, and C in the figure denote the relative locations of the strain gages on the shell. All gages had a gage length of 1.6 mm, a gage factor of 2.09, and a resistance of 120 Ω.

A Physical Acoustics Corporation acoustic emission (AE) system was used to assist in monitoring the onset of damage during the test. The AE system hardware consists of a 386-PC computer, a sensor conditioner/data acquisition unit, and piezoelectric AE sensors. The computer serves as a system control unit and operator interface by running an application software package that allows for the acquisition, storage, processing/postprocessing, and display of the monitored AE activity. For the purpose of the present test, the use of only one sensor was found sufficient.

Procedure

The specimen was placed in the fixture using the end-plates to secure the O-ring seals. Strain gages were then connected to the appropriate instrumentation channels. The chamber formed by the internal mandrel and the test specimen was then filled with hydraulic oil through the top port in the mandrel. When the chamber was filled, the pressure transducer, gage, and supply line were connected to the top port through a multi-port manifold. A small amount of pressure was applied to the specimen to dislodge any air present in the system. The system was bled of any air by cracking open the highest compression tube fitting. In this case, the

Strain Gages

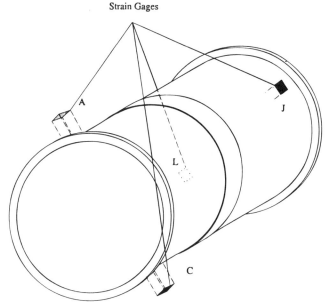

FIG. 6—*Strain gage location.*

fitting connected to the pressure gage was opened and closed. After bleeding, the system was ready for the actual test.

The test began by zeroing any initial offsets in the strain gage circuits and ensuring that the pressure gages read zero at zero applied pressure. The computer code and AE equipment was started and data acquisition began. As the pump was operated, increasing the internal pressure, the computer took data at a rate of approximately 10 Hz. Strain and pressure readings were logged and monitored by the computer. A plot of strain versus pressure was displayed on the computer screen to indicate major changes in the specimen or between strain gages. Acoustic emission data were taken and logged. A loudspeaker in the AE equipment provided an audible indication of damage progression during the test. Pressure was increased until specimen failure. Upon failure, the hydraulic oil was contained using an absorbent and the specimen was removed from the test fixture.

Results

The results of the test on the single shell are shown in Figs. 7 through 11. Figure 7 shows the pressure versus strain characteristics for the strain gages using the load sequence shown in Fig. 8. Figure 7 shows a linear behavior of the material until failure. The relative values of the slopes indicate a uniform response of the specimen to the loading. The small differences in the slopes can be attributed to local variations in wall thickness.

The failure pressure is taken as the pressure at the last reading before the transducer registered zero. The failure pressure of the specimen is 11.44 MPa with a corresponding average failure strain of 4391 $\mu\epsilon$. The measured failure pressure is within 3.67% of the anisotropic shell theory prediction and 3.15% of the engineering thin-walled shell theory.

The overall AE emission of the shell during the test can be characterized as intense. The total number of stress waves detected by the sensor (AE hits) is 2975, and a time history of

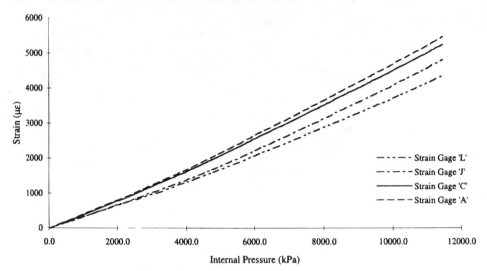

FIG. 7—*Strain trend for internal pressure test.*

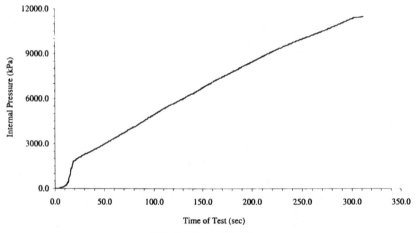

FIG. 8—*Rate of loading.*

accumulated number of hits is shown in Fig. 9. Note an initial stage with no or very reduced AE, followed by a continuously increasing amount of AE activity. When correlated with Fig. 8, which shows the pressure variation during the test to closely resemble a ramp, one can immediately recognize an AE activity rapidly increasing with load.

The histogram in Fig. 10 depicts the time history for the rate of AE hits. Note that before $t = 247$ s, corresponding to an applied pressure of 10 MPa, the accumulated number of hits is 31, representing 1.04% of the total number of hits. Also note that the AE is sparse and random. Beyond $t = 247$ s, however, there is a continuous AE being generated. Based upon this observation, the corresponding load level is identified as a threshold for the initiation of damage.

Two remarks should be made regarding Fig. 10. First, the jagged appearance of the histogram is due in part to the step-by-step pattern of the applied pressure and in part to the bin grouping

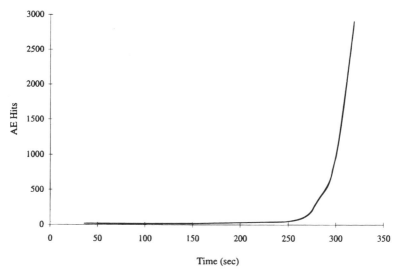

FIG. 9—*Acoustic emission hits versus time.*

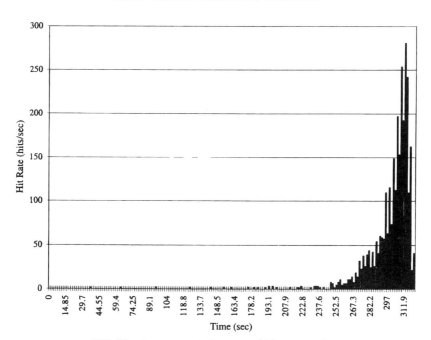

FIG. 10—*Acoustic emission rate-of-hits versus time.*

of the hits, specific to this particular type of representation. Second, the decrease in AE noticeable toward the end of the histogram can in part be related to a slower rate of loading, as observed in Fig. 8.

The amount of energy associated with the detected AE activity is represented in Fig. 11, also as a time history. One can immediately observe that the AE hits occurring at higher loading levels account for most of the energy released in the failure process.

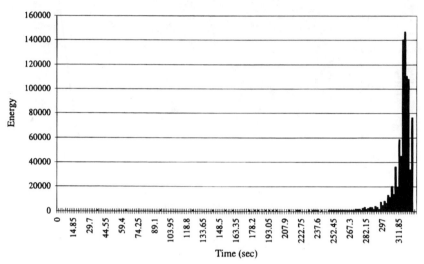

FIG. 11—*Acoustic emission energy versus time.*

Fracture Surface

The failure of the shell was almost instantaneous. The video monitoring of the test indicates that the entire failure process took place within the time span between two video frames, which is equivalent to 0.04 s. As a consequence, the progression of failure can only be inferred from an analysis of the fracture surface.

The general features of the fracture surface are shown in Figs. 12 and 13. One can observe a failure of the shell oriented generally along the generator, and, therefore, normal to the loading, for the entire length of the cylindrical section. At one of the points where this fracture surface intersects with the scarf joint rings and the end flanges, the fracture is bifurcating, with each branch following a circumferential direction. At the other intersection point, a complex fracture surface is generated and the fracture surface is reoriented in a circumferential direction. A certain amount of plastic deformation can easily be observed, leading to the typical fracture features of shell failure under internal pressure.

A closer visual inspection of the fracture surface allows the identification of specific features that enable one to reconstruct the progression of failure. Figure 14 shows a detail of the lower part of Fig. 12, and one can notice a number of distinctive characteristics of this area. First, a portion of the fracture surface has a slant orientation, measured as being approximately 30° with respect to the generator and, therefore, coinciding with one of the ply orientations of the shell. Second, one can also observe in Fig. 14 a slight difference in the reflective properties of the outer shell surface in the vicinity of the slant fracture surface. A closer look reveals local yielding of the thin layer of titanium covering the laminate.

On the inner surface of the shell, the fracture surface produced a straight line with the same orientation, as appearing at the bottom of Fig. 13. There is no noticeable sign of yielding on the inner surface. One can also observe that the fracture surface in this area is not always normal to the shell surface, revealing local interlaminar fracture. While the ends of the slant fracture gradually curve towards a generator orientation on the outer surface, the straight

FIG. 12—*Post failure shell general view.*

fracture line on the inner surface has sharp corner ends. The change of direction is achieved through a zig-zagged pattern with orientations of ±30°, as one can see in Fig. 13.

At the opposite end of the shell, the bifurcation area shows a very complex fracture surface, as shown on the top of Fig. 13 and in detail in Fig. 15. The corresponding end flange is plastically bent out-of-plane in the vicinity of this area, indicative of large bending loads applied during the failure process. The appearance of the fracture surface in the middle section of the shell is depicted in Fig. 16. One can observe fiber pullout, especially in the area of the reinforcement band, and the generally straight and normal-to-the-shell-surface nature of the fracture surface. Finally, Fig. 17 illustrates the appearance of the complex fracture surface in the opposite scarf joint area. One can also notice the straight-line appearance of the circumferentially oriented fracture.

Based upon these observations, the following interpretation of damage initiation, damage progression, and failure can be made. The rapid nature of the failure suggests a critical damage growth and the features of the slant fracture area indicate the location of damage initiation and triggering of failure. The slant orientation at an angle coinciding with the orientation of one layer suggests that matrix failure parallel to the fibers occurred, generating internal

FIG. 13—*Post failure inner surface view.*

damage. This initiation should be correlated with the load at which substantial AE started, and corresponds to 10 MPa of applied pressure.

With increasing load, the damage continued to grow both along the fibers and in the thickness direction, fracturing the adjacent layers along the same orientation. Based upon the observed characteristics of the fracture surface in this area, it appears that the damaged layer was located closer to the inner surface and reached it before the outer one. The straight line intersection with the inner surface and the lack of yielding in this region are both supportive for this failure sequence. As the applied load continued to increase, the damage progressed to a point at which its size became nearly critical, but without the damage having penetrated the wall completely, therefore preserving the sealed condition of the pressurizing medium. This stage can be recognized by the zig-zagged growth at the ends of the slant segment.

Finally, with increasing load and having reached a critical size, the damage progressed instantaneously. It simultaneously penetrated through the wall, producing the yielding of the remaining layers, progressed along the generator of the cylinder, and bifurcated in circumferen-

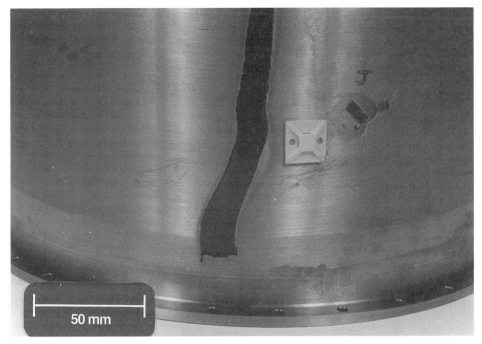

FIG. 14—*Failure initiation area.*

tial directions. The amount of plastic deformation in the walls of the shell are in agreement with this failure progression sequence.

The final stage of fracture corresponds to the point when the longitudinal fracture reached the opposite scarf joint. Due to the combined effects of internal pressure and loading produced by the outward deformation of the shell walls, the end flange is bent out-of-plane and the fracture surface is redirected in a circumferential direction, generating the complex fracture surface at the turning point.

Conclusion

An experimental setup is designed for the failure testing of a SCS-6/Ti-6-4 MMC shell under internal pressure using a hydraulic system, while maintaining zero axial load. Acoustic emission is used to monitor damage progression. Analytical predictions of failure load are obtained using closed-form solutions from an anisotropic shell theory and the engineering thin-walled shell theory, based upon laminate strength data. The analytical predictions are within 3.2% for the engineering thin-walled shell theory and 3.7% for the anisotropic shell theory. Acoustic emission monitoring proved to be an effective indicator of damage initiation and progression. An investigation of the fracture surface characteristics provides insight into the damage initiation site and growth sequence.

Acknowledgments

The authors wish to acknowledge the Georgia Space Grant Consortium and Rolls-Royce Inc. for their support of this work.

FIG. 15—*Complex failure mechanism area.*

FIG. 16—*Detail view of the failure surface near the reinforcement band.*

FIG. 17—*Detail view of the complex failure mechanism area.*

References

[1]   Gravett, P. W. and deLaneuville, R. E., "Analysis and Results of an SCS-6/Ti-15-3 MMC Reinforced Ring Structure Under Internal Radial Loading," *Proceedings*, ASME DE. Vol. 55, Conference on Reliability, Stress Analysis and Failure Prevention, 1993 ASME Design Technical Conferences, Tenth Biennial Conference on Reliability, Stress Analysis, and Failure Prevention, Albuquerque, 1993, pp. 199–212.

[2]   Zaretsky, E. V. and deLaneuville, R. E., "MMC Life System Development (Phase I): A NASA/ Pratt & Whitney Cooperative Program," *Proceedings*, ASME DE. Vol. 55, Conference on Reliability, Stress Analysis and Failure Prevention, 1993 ASME Design Technical Conferences, 10th Biennial Conference on Reliability, Stress Analysis, and Failure Prevention, Albuquerque, 1993, p. 189–198.

[3]   Pollock, W. D. and Johnson, W. S., "Characterization of Unnotched SCS-6/Ti-15-3 Metal Matrix Composites at 650°C," *Composite Materials: Testing and Design (Tenth Volume), ASTM STP 1120*, G. C. Grimes, Ed., American Society for Testing and Materials, Philadelphia, 1992, pp. 175–191.

[4]   Kamiya, O. and Fujita, H., *Fracture on Cylindrical FRP by Internal Pressure*, Transactions of the Japan Society of Mechanical Engineers, Series A, Tokyo, Vol. 52, No. 484, Dec. 1986, pp. 2604–2608.

[5]   Teng, J.-G. and Rotter, J. M., "Buckling of Pressurized Axisymmetrically Imperfect Cylinders Under Axial Loads," *Journal of Engineering Mechanics*, Vol. 118, No. 2, Feb. 1992, pp. 229–247.

[6]   Wang, J. Z. and Socie, D. F., "Biaxial Testing and Failure Mechanisms in Tubular G-10 Composite Laminates," *Composite Materials: Testing and Design, Eleventh Volume, ASTM STP 1206*, E. T. Camponeschi, Jr., Ed., American Society for Testing and Materials, Philadelphia, 1993, pp. 136–149.

[7]   Lagace, P. A. and Priest, S. M., "Damage Tolerance of Pressurized Graphite/Epoxy Cylinders Under Uniaxial and Biaxial Loading," in this volume.

[8]   Vinson, J. R. and Sierakowski, R. L., *The Behavior of Structures Composed of Composite Materials*, Martinus Nijhoff Publishers, Dordecht, The Netherlands, 1986.

[9]   Lekhnitskii, S. G., *Theory of Elasticity of an Anisotropic Body*, Holden-Day, Inc., San Francisco, 1963.

Todd Saczalski,[1] Brad Lucht,[2] and Ken Saczalski[1]

Experimental Design Methods for Assessment of Composite Damage in Recreational Structures

REFERENCE: Saczalski, T., Lucht, B., and Saczalski, K., **"Experimental Design Methods for Assessment of Composite Damage in Recreational Structures,"** *Composite Materials: Fatigue and Fracture (Sixth Volume), ASTM STP 1285*, E. A. Armanios, Ed., American Society for Testing and Materials, 1997, pp. 45–69.

ABSTRACT: This paper reviews advanced experimental design methods for use in evaluating multivariable effects related to load capacity degradation of fiber-reinforced composite recreational structures, such as bicycle frames, that may be subjected to subtle forms of small projectile impact damage. For instance, road debris impacts, caused by particles (that is, stones and pebbles) ejected from other road vehicles, can result in subtle damage to the primary composite frame structures and may lead to reduction in load capabilities and safety margins. To demonstrate the test methodology, multilayered composite tubes of varying diameter and thickness were subjected to high velocity gas gun pellet impacts in order to induce damage beneath surface layers that showed little evidence of severe impact. Bilateral eccentric column loading was then used to simulate strength degradation in normal use after projectile impact damage had been applied. The experimental design method utilizes an efficient number of test articles to examine the influence of many variables including material imperfections, manufacturing inconsistencies, and subtle impact damage. The test method combines the test data into a experimentally developed mathematical response function that identifies nonlinear interaction effects of several variables that cannot be ascertained easily by traditional single-variable test strategies and analytical approaches. A case study, dealing with an actual mountain bike frame failure, is also reviewed to demonstrate the practical need for development of fault-tolerant designs.

KEYWORDS: composite materials, recreational structures, bicycle frames, impact energy, impact testing, compression testing, buckling (materials), damage, failure (materials)

Fiber-reinforced composite materials and structures are often used for their light-weight and high-strength performance characteristics in today's world of sport and competition. Racing and mountain bicycle frames represent typical examples of such recreational composite structures. Unfortunately, these structures are susceptible to strength degrading damage, induced through normal handling and impact by road debris. This often subtle damage may appear to be no more severe than a mere surface imperfection, but when coupled with manufacturing imperfections such as incomplete bonding of plies and fiber bundles, as well as localized void sites that give rise to stress risers within the composite material, the designer is confronted with the formidable task of how to determine safe design allowables in light of the multitude of

[1]Research and consulting engineers, respectively, Environmental Research and Safety Technologists, Inc., Laguna Beach, CA 92651.

[2]Advanced technology consultant engineer, San Diego Engineering and Statistics, Midwest Office, Kansas City, MO 64131–4220.

variables that can influence the strength degradation. In many instances, due to the complex nature of the problem, these recreational composite structures are often designed with little understanding of damage-induced performance limitations. The consequence of not including adequate damage assessment investigations in the design of these products is that individuals may be injured when the structure fails under what might be considered benign or normal use conditions.

This paper reviews certain advanced experimental design methods suitable for efficiently evaluating the performance degrading effects on composites caused by surface impact damage, as well as manufacturing and material imperfections. A focus is made on factorial-based experimental strategies that provide a cost effective and statistically reliable means for assessing the influence of multivariable effects involved in the design, use, and fabrication and curing of various types of composite structures. In order to demonstrate the application of the experimental strategies to structures similar to bicycle composite frame members, multilayered composite tubes of varying diameter and thickness were subjected to small particle, high velocity, impacts so as to induce various levels of subtle damage. The composite tubes were then loaded bilaterally to failure to simulate "normal use" types of loading on a structure like a composite bicycle frame fork system. Since designs of bicycle frame members and fork structures will likely vary from one manufacturer to another, this study used tapered composite tube structures only as a means to demonstrate how the experimental methods could be applied to such structures. The experimental results of the example presented in this study demonstrate how the multivariable experimental strategies can be used to identify the influence that road debris and associated particle impact momentum may have on key structural parameters such as residual compression strength and reduction of energy-absorbing capability. An important feature of the experimental methodologies presented in this study is that the effects of many variables may be combined into a single mathematical response function that allows identification of key variables and nonlinear interaction effects of variables that cannot be easily ascertained by traditional "one-at-a-time" test strategies. In addition, the performance of the structural system for conditions not tested can be evaluated mathematically from the experimentally generated mathematical response functions. Finally, the effects of material and manufacturing imperfections (that is, porosity, resin/fiber ratio variations, etc.), while not amenable to easy quantification, can be evaluated in an implicit manner from the statistical variations determined by running random-order repeat tests in the experimental strategies and considering "pooled" standard deviations. The following case study, dealing with surface damage and manufacturing imperfections as they relate to fracture and failure of an actual composite racing bike frame, is reviewed to illustrate some of the problems discussed.

Case Study of Recreational Composite Failure

In this case study, a composite racing fork structure had failed after the race had been completed and while the rider was making a slow, gradual turn. The manufacturer of the racing bicycle suggested that the cycle fork had experienced a severe impact during the race that ultimately resulted in the collapse of the right and left fork structures. From a visual examination, the fork structure appeared to have failed while being loaded by a combination of axial compression and lateral forces. Cross-sections of the fork structure also revealed extreme variations of wall thickness within each leg of the fork. In addition, a few small chips on the surface gelcoat near the front of the forks suggested that the structure may have experienced some particle or debris impact that could account for some subsurface ply delamination damage.

Figure 1 shows the damaged composite fork after it had been dissected and cut into samples for detailed scanning electron microscope (SEM) and optical evaluation near the regions of

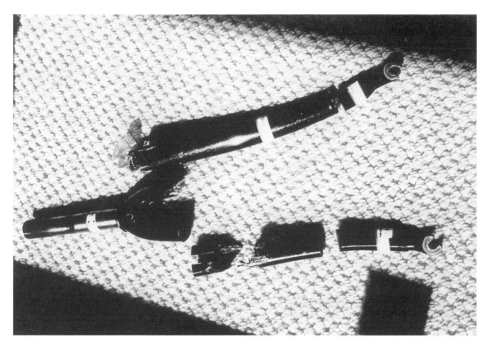

FIG. 1—*Damaged composite bicycle fork sectioned for examination.*

catastrophic failure. The results of the SEM and optical microscopic study revealed that at or near the regions of fracture there were signs of delamination, cracks, voids, and poor bonding. Figure 2 shows a SEM photograph taken near the region of fracture. Figure 3 shows unbonded cross-ply composite layers beneath the surface gel-coat (that is, black upper layer). Figure 4 shows a polished cross section of the composite at the site of localized surface impact damage that appears to have affected interior composite layers. Figures 5 and 6 show areas of incomplete ply adhesion and localized bonding of fibers at cross-ply interfaces beneath the surface layers. Figure 7 illustrates a general lack of adhesion between fibers and the resin system. Figure 8 also shows fibers with very little attachment to cross plies. Figures 9 and 10 illustrate fractured fibers with large variation of resin bonding. Figures 11 and 12 show optical microscope photographs of a polished cross-section near the failure area that included ply delaminations, voids, and resin-rich regions. A detailed study of the effects voids can have on graphite/epoxy structures can be found in the works of Uhl et al. [1] and Judd et al. [2]. A spectrum analysis showed no evidence of chemical contamination in the failure regions.

Finally, the fork was examined carefully for evidence of unusual high-impact shock load damage (that is, impact with ditches, curbs, etc.) by visual and microscopic examinations of the bearings in the crown race area (that is, the junction of the handle bars and the fork structure). Such damage usually manifests itself in the form of localized impact damage or warping of the softer bearing guides or races. No such damage was found. In light of this, it was concluded that the manufacturing imperfections found in the regions of fracture, when coupled with "normal use" operational rider/road loads and localized damage from road debris impact, contributed to degrading the margin of safety and resulted in the ultimate failure under what appeared to be benign load conditions. Additional research suggested that no adequate strength reduction considerations had been employed in the design of this type of recreational

FIG. 2—*SEM photograph taken near region of fork failure.*

FIG. 3—*Failure region showing unbonded cross plies under surface gel-coat (black layer).*

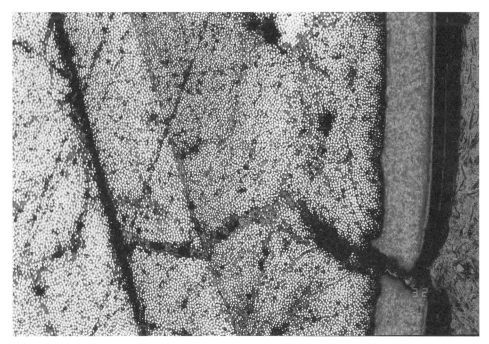

FIG. 4—*Polished cross section showing damage under surface impact location.*

composite structure so as to account for material imperfections, manufacturing variations, and the effects of subtle particle impact damage. Furthermore, the failure to account for strength reduction caused by the factors just cited was not atypical of the manufacturers of recreational composites. In fact, no specific standards or recommended practices seem to address the issue of designing to account for strength reduction due to the factors noted (that is, particle impact damage, and material and manufacturing variations). Also, as noted earlier, traditional analytical and simple experimental approaches are not easily or reliably applicable to predicting or understanding the effects of the preceding factors. In light of this, it was felt that a review of efficient multivariable experimental strategies applicable to damage assessment and associated strength reduction, along with a demonstrative example, might prove useful for improving future designs of light-weight and high-performance recreational composite structures. The following sections of this paper review the advanced experimental methods and demonstrative example.

Review of Experimental Screening Methods

Numerous variables influence how a composite structure may be weakened, leading to eventual premature failure. For instance, the size, shape, mass, and velocity of an impacting object against the surface of a complex multilayered composite can result in minimal surface damage, yet cause a serious reduction in the margin of safety and lower buckling strength and energy absorbing capacity if the structure is fatigued or shock loaded after the damage has been induced. Other factors such as naturally inherent material imperfections and variations of cure conditions further complicate the problem in that these types of variables are not easily quantifiable but, in fact, do exist and influence the ultimate system performance levels. An

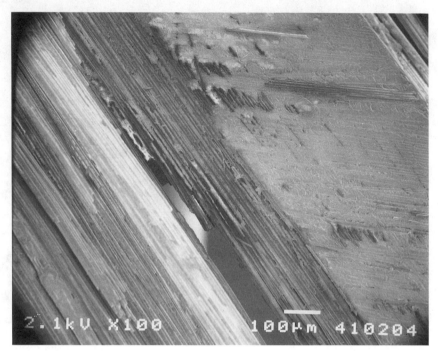

FIG. 5—*SEM photograph showing localized bonding at a cross-ply interface.*

estimate of the effects of such variables (material and manufacturing imperfections), while not easily quantifiable due to the random distribution throughout the composite structure, can be made in an implicit manner, as noted earlier, through the use of repeat tests of a given set of quantifiable variables that can be used to measure statistical variation. Unfortunately, this measure of statistical variation will also reflect experimental test imperfections and, as such, care must be taken to reduce as much as possible the experimental range of tolerances.

Identifying and sorting out which of the many variables are most significant in altering the behavior of the structure is not an easy task. Fortunately, there are several organized experimental approaches that can be utilized to "initially" screen the numerous variables and identify which are the most influential in altering structural response. One such method is the Plackett-Burman method [*3*] that is based on a factorial approach. The factorial approach will serve as the basis for the other suggested test methodologies reviewed in this paper.

In general, an experimenter faces the task of exploring the relationship between any number of variables, x_i, and a response, $\hat{y}$. The manner of how each variable relates to the output varies; it may be know precisely, in which case a mathematical function can be used to describe the system response

$$\hat{y} = f(x) + \epsilon \tag{1}$$

where ϵ represents experimental error.

In this case, $f(x)$ can be thought of as

FIG. 6—*SEM photograph taken at same region as Fig. 5 with different magnification.*

$$f(x) = x_1 z_1 + x_2 z_2 + \cdots + x_i z_i \tag{2}$$

Here the zs are known constants and, in practice, are known functions of the experimental variables, x_i. This linear model is the model used most traditionally in experimental design, and provides an estimate of the affect of a single variable (that is, load) at selected fixed conditions of other variables (that is, temperature, strain rate, etc.).

This traditional approach has resulted in the typical evaluation of just one variable, or factor, at a time, while all other factors are held constant. This approach is useful when testing isotropic materials, such as metals, which tend to react in a proportional or linear manner to applied loads.

But what if the exact nature of the relationship between variable and response is not so precisely known? In this case, an assumption might be made that the association between system response and various variables is likely to be smooth, yet perhaps not linear. Now, a graduating function, $g(x)$, might be used to describe the empirical system

$$\hat{y} = g(x) + \epsilon \tag{3}$$

It is possible to use statistical methods to choose adjustable constants in $g(x)$ that will help determine the possible nonlinear response function. A test method based on such a formulation would evaluate several test variables simultaneously and thus provide designers with a more cost effective and broader range of data to allow engineers in the field to anticipate important parameter limitations such as residual strength degradation due to variables such as those identified earlier. It is this choice of a multivariable experimental design strategy, based on

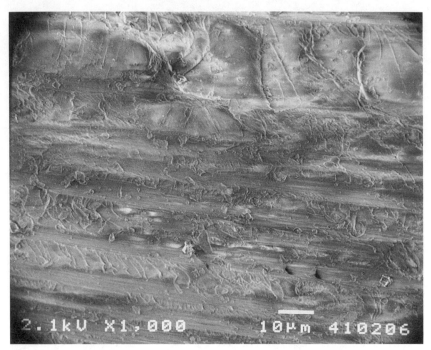

FIG. 7—*SEM photograph showing lack of adhesion between fibers and resin.*

the factorial response polynomial approach, that allows accurate system response measures that include inherent imperfections and variations, as well as easy identification of key variables and their interactions. The effect of inherent imperfections, (such as variations in the consistency of prepreg materials within a given batch) usually manifest themselves in the amount of statistical variation detected during the multivariable design test strategies. As noted earlier, these test strategies also allow for inexpensive prediction of system response for combinations of variable parameters not tested through the use of experimentally generated mathematical response functions.

Overview of Two-Level and Box-Behnken Factorial Methods

Two-Level Factorial

At the beginning of a problem evaluation, there may exist a long list of variables to be considered. By past experience, the designer may exclude some variables from consideration outright. Others may fall in a gray area, while others still may be known to have a strong effect on system response.

Often a good compromise in variable selection is the employment of a preliminary screening method, as mentioned earlier, or the use of various simple two-level fractional factorial experimental approaches to select initial ranges of parameters. Such an approach allows the identification of important input variables that may be initially overlooked in traditional test methods. In addition, the simple two-level factorial approach allows one to examine the interdependence of variables and their effect on system response. In a two-level factorial experimental design, each variable is tested at a maximum and minimum level. This initial

FIG. 8—*SEM photograph showing poor adhesion between fibers at cross-ply surface.*

two-level pattern can serve as a building block in developing more advanced experimental matrices (such an approach is demonstrated in this paper).

In general, a 2^k factorial design consists of an experiment where every possible combination of high ($+1$) and low (-1) values are considered. The minimum number of tests, n, required for statistical reliability is given as

$$n = 2(2^{N_i}) \tag{4}$$

where N_i is the number of independent variables in the test design. All tests are run in a random order with at least one repeat run for each test condition. This can be shown for a three-variable problem by considering a cube in space. Each of the vertices in the cube represents either a high or low value of a variable pairing, as show in Fig. 13.

The test results are then combined into polynomial form. Pooled standard deviations, along with the student t test are used to evaluate significant variables and interaction factors of the polynomial coefficients, as well as the effect of inherent but not easily quantifiable variables such as manufacturing and material variations. Equation 5 illustrates the general form of the multivariable response surface polynomial for a two-level factorial approach

$$\hat{y} = a_0 + a_1x_1 + a_2x_2 + \cdots + a_{12}x_1x_2 + \cdots + a_{ij}x_ix_j + \cdots + a_{123}x_1x_2x_3 + \cdots + \text{etc.} \tag{5}$$

Again, the $\hat{y}$ term represents the dependent variable response. The polynomial, or regression, coefficients are obtained from the test results. An explanation of how to calculate these coefficients can be found in Saczalski [4].

FIG. 9—*SEM photograph showing fractured fibers and poor adhesion between fibers.*

Box-Behnken

The Box-Behnken method is based on a subset of the three-level factorial approach and enables examining the nonlinear functional relationship between a certain dependent response (such as residual buckling strength) and three levels of values of several independent quantitative experimental variables (such as impact momentum). The general form of the Box-Behnken response surface polynomial is shown in Eq 6

$$\hat{y} = b_0 + \sum_{i=1}^{k} b_i x_i + \sum_{i=1}^{k} \sum_{j=1}^{k} b_{ij} x_i x_j \tag{6}$$

The $\hat{y}$ term in Eq 6 represents the dependent response variable (that is, critical failure load, vertical displacement, system energy, etc.). The x_i variables are the values of the independent parameters (that is, impact momentum, location of damage, applied load offset, etc.). Each independent variable is scaled nondimensionally such that $+1$ represents the maximum level of the test variable, -1 represents the minimum value of the test variable, and a value of 0 indicates the mean value. The terms b_0, b_j, and b_{ij} are regression coefficients. They are calculated using the results obtained from the experimental tests as specified by Box-Behnken. For the constant and linear polynomial terms, the regression coefficients are

$$b_0 = \overline{Y}_0 = \frac{1}{n_0} \sum_{u=1}^{n_0} Y_u \tag{7}$$

FIG. 10—*SEM photograph taken at same region as Fig. 9 with different magnification.*

and

$$b_i = A \sum_{u=1}^{N} X_{iu}Y_u \tag{8}$$

where $\overline{Y}_0$ is the mean value of center point test results, n_0 is the number of tests performed at the center point, N is the total number of tests, Y_u is the test result for a specific test configuration (U), and X_{iu} is the nondimensional coefficient (that is, $+1$, 0, -1) taken from the design matrix. The coefficient, A, is given in Table 1 as a function of the number of variables in the test series. The second-order nonlinear regression terms are given as

$$b_{ij} = D \sum_{u=1}^{N} X_{iu}X_{ju}Y_u \tag{9}$$

and

$$b_{ii} = B \sum_{u=1}^{N} X_{iu}^2 y_u + C \sum_{i}^{N_i} \sum_{u=1}^{N} X_{iu}^2 Y_u - \frac{\overline{Y}_0}{s} \tag{10}$$

where the coefficients, B, C, D, and s, are also given in Table 1 and N_i represents the number of independent variables in the test design. As in the discussion on the two-level factorial approach, the Box-Behnken method utilizes repeat tests of certain variables to determine

FIG. 11—*Optical microscope photo showing voids and ply delamination at failure region.*

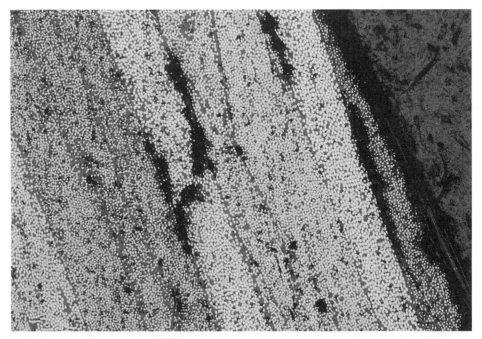

FIG. 12—*Photograph taken at same region as Fig. 9 but at different magnification.*

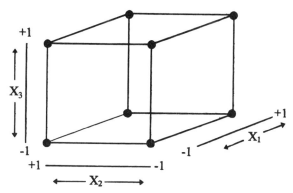

FIG. 13—*Geometric representation of two-level factorial.*

TABLE 1—*Constants for the calculation of Box-Behnken coefficients.*

N_i	A	B	C	D	s	N_0
3	1/8	1/4	−1/16	1/4	2	3
4	1/12	1/8	−1/48	1/4	2	3
5	1/16	1/12	−1/96	1/4	2	6

statistical variation that can shed some light on the role of material and manufacturing imperfections in affecting system performance. The last two columns of Table 1 contain the coefficients for calculation of standard error data. Obviously, if, after careful control of all experimental factors, large statistical variations continue to exist in the experimental output, then these variations are most likely due to the material and manufacturing imperfections, and care should be taken by the designer to account for these variations in the determinations of the necessary margins of safety.

In the discussion of the demonstrative example that follows in this section, a three-variable Box-Behnken experimental design is utilized, requiring 15 tests to be run. The specific combinations of independent parameter levels X_1, X_2, and X_3 for each of these tests are shown in Table 2 under the heading, Test Matrix Configuration. Notice that the last three specimens listed in this table (that is, BB-13 through BB-15) are repeat test articles. These test articles are used as a basis for establishing statistical variations. All tests (that is, BB-01 through BB-15) are run in a random order to help remove certain test biases. A more detailed discussion of this method can be found in Box [5]. An analysis of the Box-Behnken method as it is applied to composite materials in the aerospace industry can be found in Saczalski [6].

Illustrative Example of Experimental Design Method

In order to demonstrate the multivariable approach on a composite structure similar in size and manufacturing characteristics to a composite bicycle frame structure, a tapered multilayered graphite/epoxy golf shaft was selected. As noted earlier in this paper, it is recognized that each recreational composite structural designer will have different goals and competitive design features that will result in unique variations from one manufacturer to another. It is not the intent of this study to critique any specific manufacturer, or to show exactly what happened to the composite frame member of the case study, but rather to present an efficient, reliable,

TABLE 2—*Box-Behnken test matrix configuration and test results.*

Specimen	Test Matrix Configuration			Output	
	Impact Momentum, X_1	Damage Location, X_2	Lateral Offset, X_3	Buckling Load, P_{cr}	Energy Before Failure, N · m
BB-01	+1	+1	0	3.155	12.5
BB-02	+1	−1	0	4.089	21.7
BB-03	−1	+1	0	3.062	19.2
BB-04	−1	−1	0	3.066	22.0
BB-05	+1	0	+1	2.089	9.9
BB-06	+1	0	−1	4.667	18.3
BB-07	−1	0	+1	1.889	22.1
BB-08	−1	0	−1	6.667	13.8
BB-09	0	+1	+1	3.111	24.1
BB-10	0	+1	−1	3.489	7.8
BB-11	0	−1	+1	2.711	26.7
BB-12	0	−1	−1	5.089	27.4
BB-13	0	0	0	3.644	36.6
BB-14	0	0	0	3.378	13.2
BB-15	0	0	0	3.999	17.1

and cost effective approach to assist the designer of such products in understanding the limitations imposed on "normal use" conditions by factors such as load degradation caused by projectile or debris impact, as well as material and manufacturing variations. The golf shaft structure was chosen because it had similar size and material characteristics as the bicycle frame structure, but did not focus attention on any one bicycle design. This structure was only selected to demonstrate the test methodology and to enable others to recognize the potential of using such methods in the design of light-weight, thin-section, high-performance recreational composites.

Basically, the golf shaft test specimens were first subjected to projectile impact so as to induce localized surface damage at various locations along the shaft length to demonstrate simulation of road debris impact. Next the shafts were loaded in compression to demonstrate how one might go about identifying degradation of ultimate load carrying capability, or energy absorbing capability, under load conditions similar to those likely to be applied to composite fork structures during normal use compressive loading from impact with road hazards such as ditches or pot holes.

The composite shafts selected for this demonstrative example were originally 107 cm in length, with a 1.58-cm outside diameter at the butt end and a 0.85-cm outside diameter at the tip end. A trim cut removed 15 cm from the butt end and 38 cm from the tip, leaving a resultant shaft length of 54 cm. The final outside diameter at the butt end was 1.5 cm, with a 0.142-cm wall thickness. The final outside diameter at the tip end was 1.1 cm, with a 0.165-cm wall thickness. The butt end was then epoxied into an aluminum block 1.27 cm thick for the purpose of clamping one end of the structure. A 1.1-cm screw head was epoxied into the tip end and positioned into a countersunk plate. The circular head was used to simulate a pinned end boundary condition such as at the junction of a fork structure and a wheel axial. A sketch of the specimen configuration in the vertical compression test fixture, identifying the test variables to be evaluated, can be found in Fig. 14. Figure 15 shows the test specimen in the actual test fixture. Preliminary impact testing was performed to determine the type of impactor and method of impact that would best simulate actual impact damage. A 0.450-cm

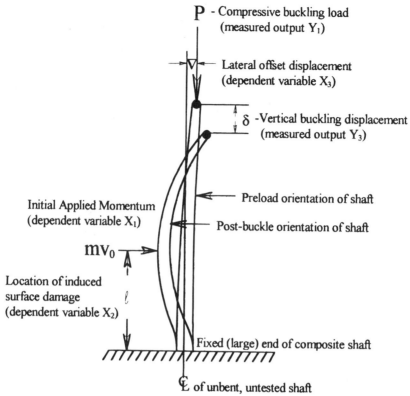

$\mathbf{P}$ - Compressive buckling load
(measured output Y_1)

Lateral offset displacement
(dependent variable X_3)

δ -Vertical buckling displacement
(measured output Y_3)

Preload orientation of shaft

Initial Applied Momentum
(dependent variable X_1)

Post-buckle orientation of shaft

mv_0

Location of induced
surface damage
(dependent variable X_2)

ℓ

Fixed (large) end of composite shaft

$\mathbf{\mathcal{C}}$ of unbent, untested shaft

FIG. 14—*Compression loading test setup for impacted composite shaft.*

diameter metal ball weighing 380 g was chosen as a representative debris particle or high-velocity impact projectile.

The rational this selection was that a small mass, impacting the shaft at a high velocity, would allow the projectile to remain in contact with the structure for a much shorter period of time, resulting in localized damage before the remainder of the composite structure had time to react inertially to the impact load. Impact momentum was chosen as one of the independent variables because various size projectiles at various impact velocities could be evaluated with such a parameter. A diagram of the high-velocity impact test setup can be found in Fig. 16. It was determined from several impacts at a wide range of velocities that minor visual damage could be imparted to the system without incurring apparent catastrophic damage. A view of the typical impact damage can be seen in Fig. 17, along with a comparison of the actual impact projectile size relative to the test specimen cross section.

The first two variables for the Box-Behnken test matrix developed in this example were selected from the general type of damage and particle impact locations shown on the surface of the case study composite. Thus, impact momentum and damage location along the length of the shaft were chosen as two independent variables of potential interest. A third variable was identified through evaluation of the normal use loads applied to the damaged case study composite. It was noted that the failed composite bicycle fork was being operated under a combined load path of axial compression and lateral bending. Along with applying an axial compressive load to the impacted shaft, it was decided to approximate the lateral bending by

FIG. 15—*Demonstrative example test specimen mounted in compression test fixture.*

offsetting the shaft tip from the vertical compression axis. Lateral offset, or eccentricity of axial compression on the shaft, became the third independent variable. The lateral offset (eccentricity) variable, X_3, is identified near the top of Fig. 14. Obviously, many other variables could have been studied (that is, effects of fatigue loading, etc.) or identified if a preliminary screening test would have been performed. However, for the purpose of simplifying the demonstration of the multivariable test methodology, only the three variables identified here were selected. In addition, the variable effects of material and manufacturing imperfections were also not evaluated in this demonstrative example. As noted earlier, these effects can be assessed in an implicit manner through the determination of the statistical variations calculated from the randomly run repeat configuration tests, and examples of such calculations are contained in the work of Box [5].

Determining the three test parameter levels for the X_2 variable (that is, location of projectile damage along the shaft length) was arbitrary and, therefore, impacts were performed at three equidistant points between the butt and tip ends of the shaft. The values for X_2 were selected as 1/4, 1/2, and 3/4 length distances along the shaft.

Before reasonable levels of the other two variables (that is, impact momentum and lateral offset) for the Box-Behnken test matrix could be determined, however, a two-level factorial test method was performed to assist in identifying what range of variable levels were most appropriate. The initial high and low values of impact momentum, X_1, and lateral offset, X_3, used for this preliminary two-level factorial study can be found in Table 3. In this preliminary study, the impact momentum was applied only at the midshaft length using the gas gun setup shown in Fig. 16. Next the impacted shafts were compressively loaded to estimate normal use loading limits. The compressive buckling results of the two-level test series, shown in Table 4, were used to assist in establishing the three levels of the impact momentum and lateral offset variables used in the Box-Behnken analysis that are shown in Table 5.

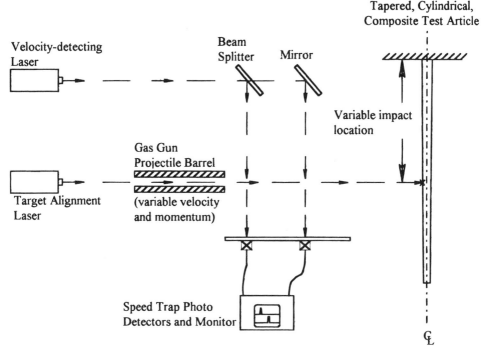

FIG. 16—*High-velocity projectile impact test setup.*

With regard to the preliminary two-level factorial study, the resulting polynomial equation at the bottom of Table 4 indicates that, at the levels tested, the impact momentum linear coefficient (X_1 = 0.491) and lateral offset linear coefficient (X_3 = 0.480) had similar weighting effects on the resultant critical buckling load polynomial. The fact that the X_1X_3 interaction coefficient (0.102) was of similar magnitude indicates that the coupling effects between the two variables are worth considering. Note that an undamaged shaft, tested to failure with no lateral offset, had a critical buckling value of 4.61 kN, while an undamaged shaft, submitted to a compression load with an offset of 2 cm, failed at a value of 3.44 kN (see Table 4 results).

As noted, the results from the two-level factorial study helped in the selection of the final ranges for the remaining Box-Behnken variables. For instance, the maximum level of impact momentum was left at the same maximum level tested in the two-level factorial approach, while the offset variable was increased by a factor of two. In part, these choices were subjectively made from observations of the surface damage induced by the gas gun apparatus, and by observing the ultimate buckling damage levels. Also, preliminary calculations of road debris particle velocity and observations of windshield damage assisted in the final choice of variable maximum levels.

After selection of the high, low, and midrange values for each of the three variables chosen for this demonstrative example, the Box Behnken multivariable test method was then performed. First, as noted before, each test specimen was impacted with the level of momentum and location indicated by the first two columns shown the Table 2 under Test Matrix Configuration. These impacts were applied in a random order by using the gas gun setup shown in Fig. 16. Next, the specimen was mounted in the compression test device with the corresponding level of offset (that is, eccentricity) and then subjected to a constant compressive stroke load of 0.1

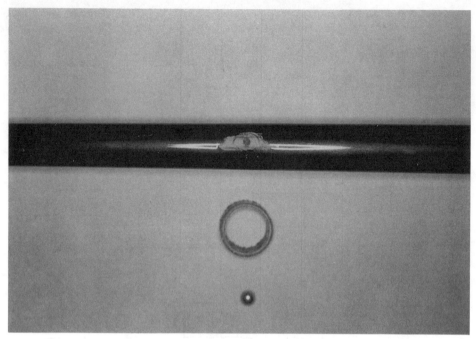

FIG. 17—*Typical impact damage on test specimen and comparison of projectile size.*

TABLE 3—*Two-level factorial example dependent variables.*

Variables	Min (−1)	Max (+1)
X_1 = impact momentum, N · s	0.000	0.052
X_3 = lateral offset, cm	0.000	2.030

TABLE 4—*Two-level factorial example test matrix and buckling results.*

Trial	Coefficients				Output, $P_{cr}(kN)^a$
	A_0	X_1	X_3	X_1X_3	
1	+	...	...	+	4.609
2	+	+	...	...	3.422
3	+	...	+	...	3.444
4	+	+	+	+	2.667
Sum	+14.142	−1.964	−1.920	+0.409	
Sum ÷ 4	+3.536	−0.491	−0.480	+0.102	

$^aP_{cr} = 3.536 - 0.491X_1 - 0.480X_3 + 0.102X_1X_3$.

TABLE 5—*Values of Box-Behnken example nondimensional parameters.*

Independent Variables	Variable Levels		
	Min (−1)	Mean (0)	Max (+1)
X_1 = impact momentum, N · s	0.017	0.035	0.052
X_2 = impact location, from shaft base, cm	12.95	25.90	38.85
X_3 = lateral offset, away from damage, cm	0.00	2.03	4.06

in./in. The ultimate buckling load and energy before failure, as determined by the area under the load-deflection curve, were measured. Figure 18 shows a closeup of a test specimen under buckling load. Figure 19 shows the extent of the damage incurred by the buckled shaft. High tensile loads on the outer surface of the shaft have caused fiber pullout, fiber failure, and delaminations to occur. The results for each test configuration, labeled BB-01 through BB-15 are shown in the last two columns of Table 2.

Application of the Box-Behnken methodology, with the Table 2 results, yields the Eq 11 that describes the buckling load resultant polynomial as a function of impact momentum and lateral offset for the case, $X_2 = 0$ (that is, impact at the shaft midlength).

$$P_{cr} = 3.67 - 0.09X_1 - 1.26X_3 + 0.55X_1X_3 - 0.05X_1^2 + 0.21X_3^2 \qquad (11)$$

The linear terms clearly demonstrate, for the range of variables evaluated, that lateral offset has a significantly greater effect on critical buckling load than does impact momentum

FIG. 18—*Typical compression buckling failure at region of projectile damage.*

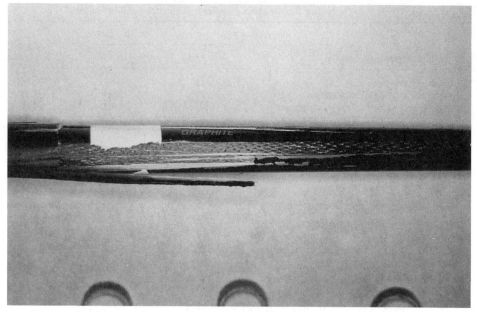

FIG. 19—*Typical post-buckling failure damage of demonstrative example test specimen.*

$$\|-1.26\| \gg \|-0.09\|$$

The magnitude of the X_1X_3 coupling term indicates that there are definite interactive effects taking place. The coefficients of the nonlinear terms again demonstrate that lateral load offset has a greater effect on buckling strength than impact momentum. In fact, the value of the nonlinear term for load offset indicates that there is a slight increase in buckling strength when offset is increased. However, this term is negated by the larger linear term

$$\|-1.26\| > \|0.21\|$$

A surface map of the Eq 11 polynomial is shown in Fig. 20. It is clearly evident that the largest buckling load will result when minimum values of particle momentum and lateral load offset are considered. The warping of the surface demonstrates the nonlinear and interactive variable effects at work. In fact, the coupling term dominates any linear particle momentum influence. Thus, the minimum value of buckling load does not take place at the maximum values of X_1 and X_3, as expected. Instead, it occurs at the minimum value of particle momentum.

Given these results, the buckling load was next evaluated as a function of damage location and lateral load offset while using the mean value of particle impact momentum. This resultant polynomial is shown in Eq 12

$$P_{cr} = 3.67 - 0.27X_2 - 1.26X_3 + 0.50X_2X_3 - 0.28X_2^2 + 0.21X_3^2 \tag{12}$$

Load offset is again the dominate linear term, but not to the extent that it was over impact momentum. The coupling term is quite similar to that found in Eq 11. The magnitude of the two nonlinear coefficients are quite similar. The plot of the resultant polynomial is shown in Fig. 21.

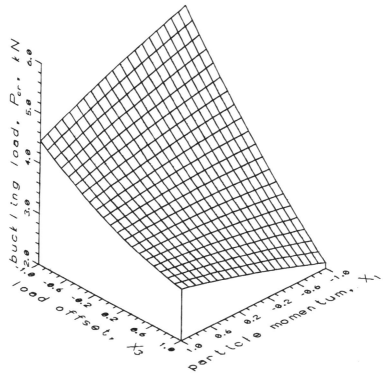

FIG. 20—*Buckling load as a function of impact momentum and load offset.*

Examination of the variation of buckling load at the minimum value of damage location (near the tip), and again at the maximum value (near the shaft), indicates results that are inverse of each other. That is, at a negligible load offset, the buckling load is highest when damage is located at the tip end of the shaft. However, when a large degree of lateral offset is evaluated, the buckling load is highest when damage is located toward the butt end of the shaft.

A second dependent variable considered was energy absorbed by the test specimen before buckling failure. This was determined by measuring the area under the load-deflection curve, as shown in Figs. 22 and 23.

Holding damage location constant at the mean value, the resultant polynomial for energy as a function of impact momentum and load offset is given in Eq 13

$$E_{cr} = 22.30 - 1.84X_1 + 1.94X_3 - 4.18X_1X_3 - 4.46X_1^2 - 1.81X_3^2 \qquad (13)$$

The coefficient on the X_1 term indicates that impact momentum has a much larger effect on energy than it did on buckling load, and in fact is nearly equivalent in effect as the load offset term, X_3. Yet the coupling coefficient actually has a greater effect than either of the linear terms. A plot of this polynomial is found in Fig. 24. Next, the energy equation polynomial is developed for the case of impact momentum held constant at the mean value, as shown in Eq 14

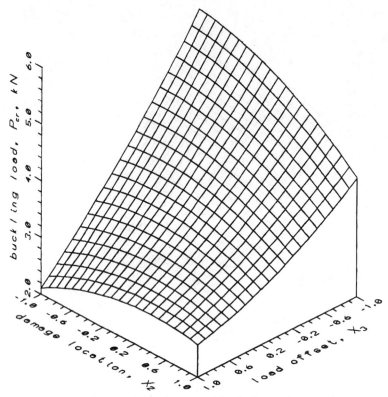

FIG. 21—*Buckling load as a function of load offset and impact location.*

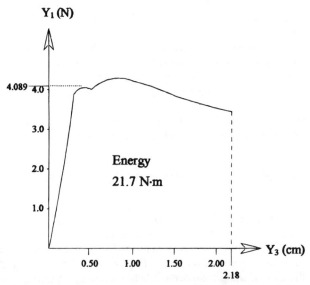

FIG. 22—*Load versus deflection curve for Specimen BB-02.*

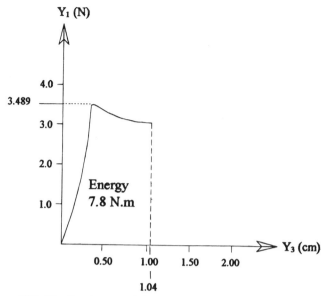

FIG. 23—*Load versus deflection curve for Specimen BB-10.*

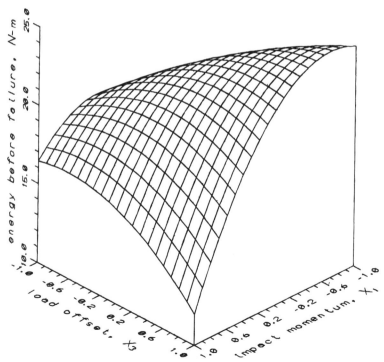

FIG. 24—*Energy before failure as a function of impact momentum and load offset.*

$$E_{cr} = 3.67 - 4.27X_2 + 1.94X_3 + 4.25X_2X_3 + 1.01X_2^2 - 1.81X_3^2 \qquad (14)$$

The magnitude of the linear coefficient of the damage location variable, X_2, indicates that the shaft can withstand substantially less energy imparted into the system before failure when damage is located toward the tip end of the shaft. The surface plot for this polynomial is shown in Fig. 25. As illustrated by Saczalski et al. [6], these plots and polynomial response functions can be used to interpolate between actual test values. For the most part, interpolated values usually come out within about 15 to 20% of actual test measures for points other than those used to generate the response polynomials. Extrapolation should be limited to no more than 10% beyond the range of maximum ($+1.0$) and minimum (-1.0) values of parameters tested (that is, ± 1.1).

The results for buckling load response and energy absorbed before failure indicate that certain design parameters may be less sensitive to particle or debris (projectile) impact than others. The selection of which design parameters are most critical and where certain regions of the structure need improvement is left to the designer. However, without the use of an efficient experimental approach, such as that presented in this paper, it would be difficult to identify margins of safety and other critical design parameters, as well as areas of needed structural improvement in a given prototype design. Finally, as stated earlier, the effects of inherent material and manufacturing variations and flaws can be evaluated by the multivariable methods through the use of the statistical variations developed from the random repeat tests run during each of these methods.

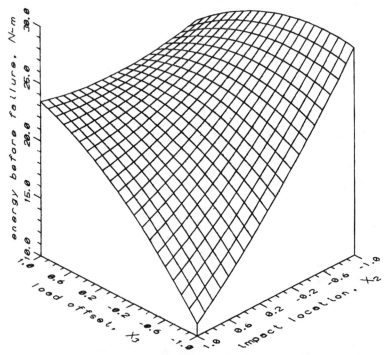

FIG. 25—*Energy before failure as a function of load offset and impact location.*

Conclusions

The results of the case study suggest that high-performance, light-weight, thin-section composite structures such as bicycle frames are susceptible to design load degradation due to factors such as normal use road hazards, and material and manufacturing variations, to name a few. Analyzing the effects of such factors and sorting out areas of needed improvement in such a manner as to develop a competitive light-weight and cost-effective composite recreational frame structure requires more than the traditional "one at a time" or "random" single variable test or analytical approaches. As pointed out, these simple approaches provide only limited information and cannot easily take into account multivariable interactions and random effect variables caused by material and manufacturing imperfections or inconsistencies. What this study and the demonstrative example have attempted to illustrate is that the multivariable experimental methodology can be a valuable tool to aid the designer in optimizing and designing safer recreational composite structures that may be subjected to load or energy degrading environments. Specifically, the Plackett-Burman method can provide a useful "first-look" tool for efficiently screening many variables with a limited number of test specimens, and provide insight into which variables are most dominant in the influence of important structural response parameters. Then, the factorial-based experimental strategies, such as the Box-Behnken method, can be used to systematically examine, with some statistical reliability, the interactions and influence of the more dominant variables and inherent imperfections. Finally, the resulting mathematical response polynomials allow for study of the influence of levels of the variables that were not explicitly tested.

Acknowledgments

The authors wish to thank the following individuals for their contributions and support: Alan Nettles, composite materials engineer, NASA Marshall Space Flight Center, and George Burghart, Steve Burghart, and Mark West, Mission Research Corporation, Costa Mesa, California.

References

[1] Uhl, K. M., Lucht, B. Q., Jeong, H., and Hsu, D. K., "Mechanical Strength Degradation of Graphite Fiber Reinforced Thermoset Composites Due to Porosity," *Review of Progress in Quantitative Nondestructive Evaluation*, Vol. 7B D. O. Thompson and D. E. Chimenti, Eds., Plenum Publishing Corp., 1988, pp. 1075–1082.

[2] Judd, N. C. W. and Wright, W. W., "Voids and Their Effects on the Mechanical Properties of Composites—An Appraisal," *SAMPE Journal*, Jan./Feb. 1978, pp. 10–14.

[3] Plackett, R. L. and Burman, J. P., "The Determination of Optimal Multifactorial Experiments," *Biometrika*, Vol. 33, 1946, pp. 305–323.

[4] Saczalski, T. K., "Determination of Multivariable G_{Ic} and G_{IIc} Fracture Toughness Response Functions for Fiber/Resin Composites," *Proceedings*, Thirty-fourth International SAMPE Symposium and Exhibition, G. A. Zakrzewski, M. Mazenko, S. T. Peters, and C. D. Dean, Eds., Society for the Advancement of Materials and Processes Engineering, Vol. 34, 1989, pp. 726–736.

[5] Box, G. W. P. and Behnken, D. W., "Some New Three Level Designs for the Study of Quantitative Variables," *Technometrics*, Vol. 2, No. 4, Nov. 1960, pp. 455–475.

[6] Saczalski, T. K., Lucht, B. Q., and Steeb, D., "Advanced Experimental Design Applied to Damage Tolerance of Composite Materials," *Proceedings*, Twenty-third International SAMPE Technical Conference, R. L. Carri, L. M. Poveromo, and J. Gauland, Eds., Society for the Advancement of Materials and Process Engineering, Vol. 23, 1991, pp. 38–50.

Christos C. Chamis,[1] Papper L. N. Murthy,[1] and Levon Minnetyan[2]

Progressive Fracture in Composite Structures

REFERENCE: Chamis, C. C., Murthy, P. L. N., and Minnetyan, L., **"Progressive Fracture in Composite Structures,"** *Composite Materials: Fatigue and Fracture (Sixth Volume), ASTM STP 1285*, E. A. Armanios, Ed., American Society for Testing and Materials, 1997, pp. 70–84.

ABSTRACT: In this paper an overview of the research activities related to an approach that has been developed independent of stress intensity factors and fracture toughness parameters is presented for the computational simulation of progressive fracture in polymer-matrix composite structures. The damage stages are quantified based on physics via composite mechanics while the degradation of the structural behavior is quantified via the finite element method. The approach accounts for all types of composite behavior, structures, load conditions, and fracture processes starting from damage initiation, to unstable propagation and to global structure collapse. Results of structural fracture in composite plates, shells and built-up structures are presented to demonstrate the effectiveness and versatility of this approach. Parameters/guidelines are identified that can be used as criteria for structural fracture, inspection intervals, and retirement for cause. Generalization to structures made of any or combinations of materials are outlined, and lessons learned in undertaking the development of computational simulation approaches, in general, are summarized.

KEYWORDS: damage initiation, growth, accumulation, stability, propagation, collapse, beams, panels, plates, shells, fracture modes, hygrothermal environment, frequencies, buckling resistance, strain-energy release rates, fatigue (materials), fracture (materials), composite materials

It is generally accepted that flaw-damaged structures fail when the damage grows or coalesce in a critical region such that (1) the structure cannot safely perform as designed and as qualified, or (2) catastrophic global fracture is imminent. This is true for structures made from traditional homogeneous materials as well as fiber composites. The difference between fiber composites and traditional materials is that composites have multiple fracture modes that initiate local damage compared to only a few for traditional materials. Any predictive approach for simulating structural fracture in fiber composites needs to formally quantify: (1) all possible fracture modes, (2) the types of damage they initiate, and (3) the coalescence and propagation of the damage to critical dimensions for imminent structural failure.

One of the ongoing research activities at NASA Lewis Research Center is directed toward the development of a methodology for the "Computational Simulation of Structural Fracture in Fiber Composites." A part of this methodology consists of step-by-step procedures to simulate individual and mixed-mode fracture in a variety of generic composite components [1–3]. Another part has been to incorporate these methodologies into an integrated computer

[1]Senior aerospace scientist and aerospace engineer, respectively, NASA Lewis Research Center, Cleveland, OH 44135.

[2]Associate professor, Department of Civil Engineering, Clarkson University, Potsdam, NY 13699-5710.

[3]This is an invited paper honoring the induction of Mr. Chamis as ASTM Fellow. It summarizes the work of the authors over the past decade.

code identified as CODSTRAN for composite durability structural analysis [4–6]. The objective of this article is to summarize the results and lessons learned to date pertaining to this activity. Specifically, the fundamental aspects of the methodology integrated into CODSTRAN are discussed herein with applications to generic composite structural components. All the results presented here have appeared before in other reports and for complete details of any specific problem, the reader is advised to consult one of the cited references. It is noted that CODSTRAN consists mainly of finite elements and composite mechanics. Both of these have been verified extensively as is well known in the structures and composite communities. Furthermore, experimental verifications for CODSTRAN were described in previous work by the authors and that work is cited in the references herein. Consequently, the present overview article focuses only on computational simulation results.

The generic types of composite structures considered are: (1) plates and shells and (2) built-up structures. These structural components are assumed to be made from graphite—fiber/epoxy-matrix composites with typical constituent properties listed in Tables 1 and 2. Structural fracture is assessed by one or all of the following indicators: (1) the displacements increase very rapidly, (2) the natural frequencies decrease very rapidly, (3) the buckling loads decrease

TABLE 1—*T300 graphite fiber properties.*

Number of fibers per end = 3000
Fiber diameter = 0.00762 mm (0.300E − 3 in.)
Fiber density = 1772 Kg/m³ (0.064 lb/in.³)
Longitudinal normal modulus = 221 GPa (32.0E + 6 psi)
Transverse normal modulus = 13.8 GPa (2.0E + 6 psi)
Poisson's ratio (ν_{12}) = 0.20
Poisson's ratio (ν_{23}) = 0.25
Shear modulus (G_{12}) = 8.97 GPa (1.3E + 6 psi)
Shear modulus (G_{23}) = 4.83 GPa (0.7E + 6 psi)
Longitudinal thermal expansion coefficient = −1.0E − 6/°C (−0.55E − 6/°F)
Transverse thermal expansion coefficient = 1.0E − 5/°C (0.56E − 5/°F)
Longitudinal heat conductivity = 43.4 J-m/h/m²/°C (580 BTU-in./h/in.²/°F)
Transverse heat conductivity = 4.34 J-m/h/m²/°C (58 BTU-in./h/in.²/°F)
Heat capacity = 712 J/Kg/°C (0.17 BTU/lb/°F)
Tensile strength = 2413 MPa (540 ksi)
Compressive strength = 2069 MPa (486 ksi)

TABLE 2—*HMHS epoxy-matrix properties.*

Matrix density = 1265 Kg/m³ (0.0457 lb/in.³)
Normal modulus = 4.27 GPa (620 ksi)
Poisson's ratio = 0.34
Coefficient of thermal expansion = 0.72E − 4/°C (0.4E − 4/°F)
Heat conductivity = 1.25 BTU-in./h/in.²/°F
Heat capacity = 0.25 BTU/lb/°F
Tensile strength = 84.8 MPa (12.3 ksi)
Compressive strength = 423 MPa (61.3 ksi)
Shear strength = 148 MPa (21.4 ksi)
Allowable tensile strain = 0.02
Allowable compressive strain = 0.05
Allowable shear strain = 0.04
Allowable torsional strain = 0.04
Void conductivity = 16.8 J-m/h/m²/°C (0.225 BTU-in./h/in.²/°F)
Glass transition temperature = 216°C (420°F)

very rapidly, or (4) the strain energy release rates associated with the multiple progressive fractures increase very rapidly. These rapid changes are herein assumed to denote imminent global structural failure. Based on these rapid changes, parameters/guidelines are identified that can be used as criteria for (1) structural failure, (2) inspection intervals, and (3) retirement for cause. Because of the reasons that the method does not utilize any of the conventional fracture mechanics and also due to a lack of such complete treatment reported by others in the past, the authors refer to this approach as a "new approach." However, the ideas behind the concept are quite well known.

In the present approach, computational simulation is defined in a specific way. Also general remarks are included with respect to (1) applications to large structures or structural systems, or both, (2) lessons learned about conducting such a long duration research activity, with regard to increasing computational efficiency, gaining confidence, and enhancing its application.

Fundamentals

This "new" approach to structural fracture is based on the following concepts.

1. Any structure or structural component can sustain a certain amount of damage prior to structural fracture (collapse).
2. During damage propagation, the structure exhibits progressive degradation of structural integrity as measured by "global" structural behavior variables such as loss in frequency, loss in buckling resistance, or excessive displacements.
3. The critical damage can be characterized as the amount of damage beyond which the structural integrity degradation is very rapid, induced by either (1) small additional damage or (2) small loading increase.
4. Structural damage is characterized by the following sequential stages: (1) initiation, (2) growth, (3) accumulation, (4) stable or slow propagation (up to critical amount), and (5) unstable or very rapid propagation (beyond the critical amount) to collapse.

These concepts are fundamental to developing formal procedures to (1) identify the five different stages of damage, (2) quantify the amount of damage at each stage, and (3) relate the degradation of global structural behavior to the amount of damage at each stage.

The formal procedures included in this new approach are as follows:

1. Damage–stage identification—(1) Damage initiates when the local stress state exceeds the corresponding material resistance as measured by the strength allowables. (2) Damage grows when the stress exceeds the corresponding material resistance on the damaged periphery for every possible failure mode. (3) Damage accumulates when multiple sites of damage coalesce. (4) Damage propagation is stable or slow when small increases, in either the damage propagation or loading condition, produce insignificant or relatively small degradation in the structural behavior (frequencies, buckling resistance, and displacements). (5) Damage propagation is unstable or very rapid when small increases in the damage propagation or in loading conditions produce significant or very large changes in the global structural behavior variables (frequencies, buckling resistance, and displacements).
2. Damage quantification—The amount of damage is formally quantified by suitable modeling of the physics in the periphery of the damaged region in order to keep the structure in equilibrium for the specified loading conditions, structural configuration, and boundary conditions. This part of the procedure is most conveniently handled by

using computational simulation in conjunction with incremental/iterative methods as will be described later.

3. Structural Behavior Degradation—This part of the procedure is quantified by using composite mechanics in conjunction with the finite element analysis. The damage stages are quantified by the use of composite mechanics while degradation of the structural behavior is quantified by the finite element method where the damaged part of the structure does not contribute to the resistance but is carried along as a parasitic material. It is very important to note that nowhere in this approach was there any mention of either stress intensity factors or fracture toughness parameters. This computational simulation approach by-passes both of them. However, use is made of the structural fracture toughness in terms of global strain energy release rate (SERR) because it is a convenient parameter to identify the "critical damage amount." The global strain energy release rate is defined as the ratio of strain energy released due to additional damage in the structure to the amount of additional damaged volume of the structure. In this respect, it is similar to the classical definition of strain energy release rate where it is taken as the ratio of strain energy released to amount of additional crack opening area. The critical global SERR in the context of the present approach is described subsequently.

The combination of composite mechanics with the finite element method to permit formal description of local conditions to global structural behavior is normally handled through an integrated computer code as shown schematically in Fig. 1. The bottom of this figure describes the constituent properties as functions of environmental and mechanical loading conditions. Based upon the ply stresses and the ply strength allowables, the criteria for damage initiation, growth and accumulation, and propagation are examined. The constituent properties are updated

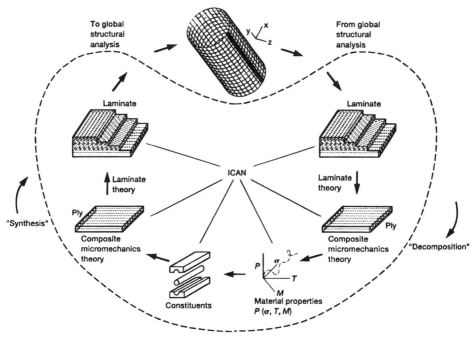

FIG. 1—*CODSTRAN simulation cycle.*

as a result at every load increment. For example, if the ply longitudinal stress exceeds the allowable strength, then the fiber/matrix longitudinal moduli are replaced by negligible values so that the ply essentially does not carry any load and the stresses get redistributed to the surrounding plies. However, if a particular ply's transverse strength exceeds its allowable, then only the matrix is assumed to have failed and therefore the matrix modulus is replaced with a negligible value. Once the current constituent properties are determined, as shown in the left part of the Fig. 1, through repeated applications of micromechanics, macromechanics, and laminate theory, the global structural stiffness matrix is assembled and fed to the finite element analysis. Thus, the left part integrates (synthesizes) local damage conditions to global structural behavior (response). The results of finite element analysis are the nodal stress resultants. These are used to decompose global response changes (laminate stresses/strains, for example, as a result of any increments in loads or stiffness updates) on the local (micro) material stress/resistance. The load is incremented only if no further damage is noticed due to changes in ply level stresses. Otherwise, only the material properties at the constituent level are updated at every iteration till a balance between the applied loading and the local response is reached. Overall structural equilibrium is maintained by iterations around the "cart-wheel" until a specified convergence is reached. This procedure is illustrated in Fig. 2. The final result in terms of load versus global displacement is shown in Fig. 3. The schematics in Figs. 1 to 3, collectively summarize the fundamentals and implementation of this computational simulation approach to composite structural failure and also to structural failure in general. The details of the process and various failure criteria are explained in Ref 7. Applications to specific structures/components are described in subsequent sections.

Plates

A laminated plate with a rectangular slit is selected to illustrate the effects of damage propagation on vibration frequencies and buckling resistance as well as the effects of hygrothermal environments. The dimensions of the plate are shown in the figure. The material is T300/

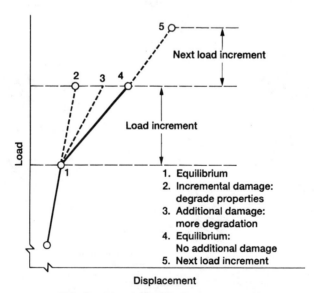

FIG. 2—*CODSTRAN load incrementation.*

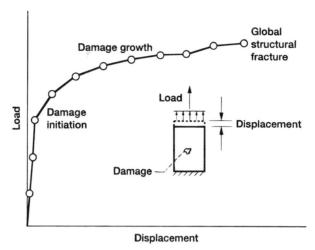

FIG. 3—*Overall CODSTRAN simulation.*

Ep (graphite fiber in epoxy matrix) and the laminate is a $[\pm 15]_{2s}$. Typical results obtained by using CODSTRAN are shown in Fig. 4 [7–9]. Collectively, these figures show how the damage propagates in the structure as a result of increasing load under various moisture/temperature combinations. Furthermore, how the global stiffness-related properties, vibration frequency, and buckling load resistance degrade due to increasing damage are also shown. The damage percent in the figures is defined as the ratio of the amount of damaged volume as determined by the nodal and elemental fractures to the total structural volume. If the nodes are only partially damaged (only few plies show damage at a node), then the damaged volume is computed proportional to the ratio of damaged to total number of plies. The details of how the percentage of damaged volume is determined are given in Ref 8. The important observations are: (1) the reference case, at room temperature and without moisture, exhibits the least amount of damage accumulation compared to the other cases; (2) moisture alone has a negligible effect on fracture load but increases the damage extent to fracture; (3) combined temperature and moisture (hygrothermal) decrease the load to fracture but permit substantial damage accumulation to fracture; (4) both the vibration frequency and the buckling resistance decrease very rapidly as the failure load (structural collapse) is approached; and (5) the hygrothermal environments degrade the structural behavior of the plate.

The important conclusion is that this approach provides a formalism to simulate complex environmental effects on structural behavior. That is, the temperature and moisture affect the matrix locally while the composite mechanics and the finite element method integrate these local effects to structural behavior (buckling resistance in this case).

Shells

CODSTRAN is used to simulate the damage initiation, growth accumulation, and propagation to global failure of a composite shell with through-the-thickness as well as partial initial defects and subjected to internal pressure with hygrothermal environment.

1. Through-the-Thickness Defect—Typical results for a through-the-thickness initial defect are shown in Fig. 5 [9]. The dimensions of the shell and the defect are given in the figure. The material chosen is a T300/epoxy system and the layup is $[90/\pm15/90/\mp$

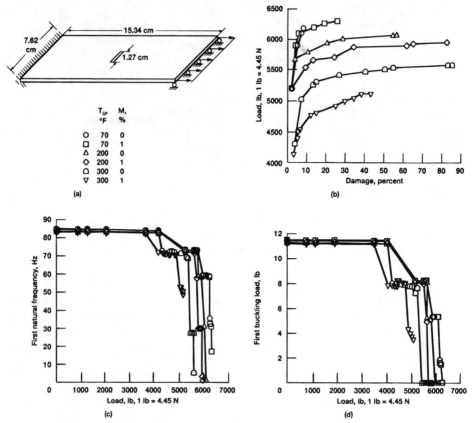

FIG. 4—*Load-induced progressive damage and effects on composite (T300/Ep[±15]₂ₛ) plate structural response including hygrothermal environment. Plate is 15.34 by 7.62 cm (6.0 by 3.0 in.) with 1.27-cm (0.50-in.) long slit: (a) geometry and environment, (b) damage, (c) vibration frequency, and (d) buckling load.*

15/90]. Such a layup is typical for a pressure vessel type structure. The results in this figure show that: (1) shells subjected to internal pressure sustain relatively low damage accumulation to fracture compared to other structural components, (2) shells are less tolerant to hygrothermal effects compared to other structural components, (3) the vibration frequencies of the shell do not degrade rapidly as the fracture pressure is approached, and (4) hygrothermal environments have a significant effect on the vibration frequencies of the shell. An important observation is that composite shells with through-the-thickness defects subjected to internal pressure, exhibit a brittle type behavior to fracture. This explains, in part, the successful application of linear elastic fracture mechanics to these types of structures.

2. Partial-Thickness Defects—The composite shell shown in Fig. 6 is investigated with initial fiber defects in two adjacent hoop plies occurring as (1) surface ply defects and (2) internal ply defects, as depicted in Fig. 7 [10]. The defects are simulated by replacing the corresponding ply stiffnesses with negligible values thereby preventing them from carrying any load. Such defects could occur due to some inadvertent damage to the structure. Computational simulation results for these two cases are summarized in Fig.

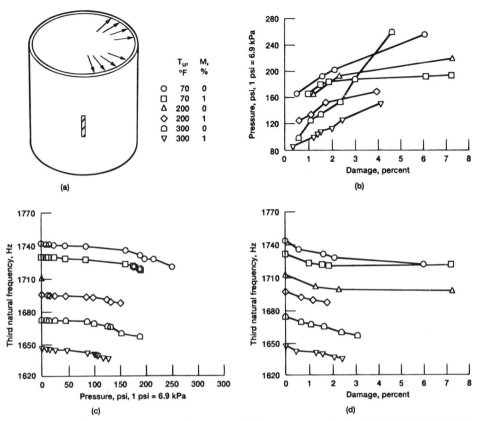

FIG. 5—*Load-induced progressive damage and effects on composite shell (T300/Ep[90/±15/90/ ∓15/90]) structural behavior including hygrothermal environment. This shell is 4.88 cm (1.91 in.) in diameter, 7.62 cm (3 in.) long with 1.27-cm (0.50-in.) long slit: (a) geometry and environment, (b) pressure, (c) vibration frequency, and (d) buckling load.*

8. Case 1 exhibits results in a gradual damage growth and propagation with local degradation. There is sufficient local distortion of the shell geometry during the damage propagation to serve as a warning of approaching structural fracture. On the other hand in Case 2, damage propagation to structural fracture occurs without warning as a sudden catastrophic fracture of the shell. Figure 9 summarizes damage initiation and structural fracture pressures for the two cases with reference to the fracture pressure of a defect-free shell. It is noteworthy that surface ply defects reduce the ultimate fracture pressure by 15% whereas interior or mid-thickness ply defects reduce the ultimate fracture pressure by 23%. The important conclusion is that the complex structural behavior of shells with damage accumulation can be computationally simulated for any type of defect as well as for defect-free shells.

Built-Up Structure

The built-up structure consists of a stiffened composite cylindrical shell panel with imposed boundary conditions to represent the behavior of a segment of the entire cylindrical shell. The

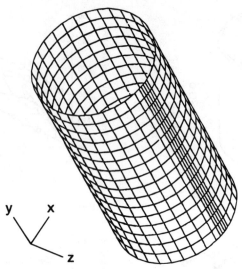

FIG. 6—*Shell structure evaluated for composite shell T300/Ep[90₂/±15/90₂/±15/90₂/∓15/90₂];* $shell\ diameter = 101.6\ cm\ (40\ in.),\ length = 203.2\ cm\ (80\ in.),\ 612\ nodes,\ 576\ quadrilateral$ *elements, initial fiber defect in two adjacent hoop plies, and defect extends 12.7 cm (5 in.) along axial direction of shell.*

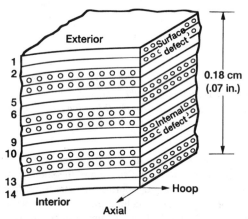

FIG. 7—*Shell laminate structure schematic indicating initial defects for composite shell T300/ Ep[90₂/±15/90₂/±15/90₂/∓15/90₂]. Pre-existing defect before loading was assumed. Cases considered were surface defect (Plies 1 and 2 or Plies 13 and 14) and mid-thickness defect (Plies 9 and 10).*

shell is subjected to (1) axial tension, (2) axial compression, (3) shear, (4) internal pressure (with the associated axial and hoop generalized stresses), and (5) combinations of these four fundamental loads. The axial loads are applied along the longitudinal direction of the stiffeners. The in-plane shear load is applied along the skin. The pressure loading is applied on the inside of the shell skin. The appropriate boundary conditions and the loads with coordinate axes are indicated in Fig. 10. The built-up structure is made of Thornel-300 graphite fibers in an epoxy-matrix (T300/epoxy) composite system. The outer shell laminate consists of fifty 0.127 mm.

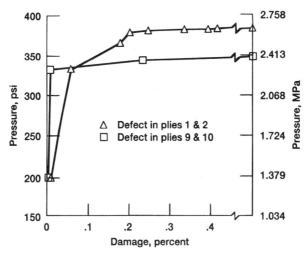

FIG. 8—*Damage propagation with pressure for composite shell T300/Ep([90₂/±15]₂/90₂/∓15/90₂).*

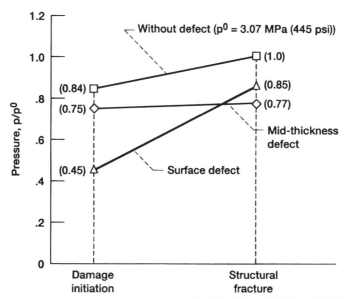

FIG. 9—*Summary of results for composite shell T300/Ep[90₂/±15/90₂/±15/90₂/∓15/90₂].*

(0.005 in.) plies resulting in a composite shell thickness of 6.35 mm. (0.25 in.). The laminate configuration for the outer shell is $[90_2/\pm15]_s$. The 90° plies are in the hoop direction and the $\pm15°$ plies are oriented with respect to the axial direction of the shell. The cylindrical shell panel has a constant radius of curvature of $R = 2.286$ m (90 in.). The subtended angle of the shell panel arc is $\theta = 30°$ or $\pi/6$, resulting in an arc length of $s = R\theta = 1.197$ m (47.12 in.). The length of the stiffened panel along the shell axis is 1.219 m (48 in.).

The stiffener elements are made from the same T300/epoxy composite as the outer shell. The stiffeners are glued to the outer shell at all surfaces of contact. The adhesive properties

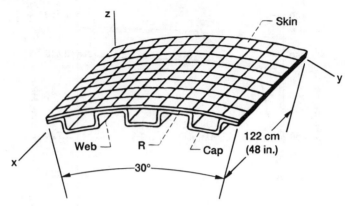

FIG. 10—*Stiffened composite shell panel conditions and loads: axial tension/compression:* x = l; *shear: skin only; internal pressure: Inside skin; damage initiation; damage accumulation; and boundary conditions: fixed at* x − 0; *simply supported at* y = 0 *and* y = 30°, *arc,* R = 1.286 m.

between the outer shell and the stiffeners are the same as those of the epoxy matrix. In general, the stiffener laminate configuration consists of 20 plies of $[\pm45]_s$ composite structure for the webs and for the elements that attach to the outer shell. Stiffener flanges have an additional 30 plies of 0° (axial) fibers. Figure 11 indicates laminate configurations in the structural elements of the stiffened shell.

The finite element model contains 168 quadrilateral thick shell elements, of which 96 are utilized to represent the outer shell, as indicated by the grid lines shown in Fig. 10. The remaining 72 elements are used to represent the stiffener webs and flanges.

Because the finite element properties and resulting generalized stresses are specified at each node, duplicate nodes are needed where there are discontinuities in the finite element properties (Ref 11). Duplicate nodes have the same degree of freedom coordinates but allow the definition of different structural properties. At points where duplicate nodal definitions are required, the node with the smallest number is designated as the master node and the other nodes are designated as the slave nodes that are assigned exactly the same degree of freedom coordinates

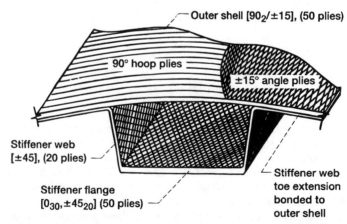

FIG. 11—*Laminate structure schematic (dimensions-in Fig. 10).*

as the master node. In the actual finite element model, however, corresponding master and slave nodes coincide at a point. The finite element model for the investigated stiffened shell panel requires 333 nodes of which 171 are master nodes and the remaining 162 are slave nodes.

Loading on the stiffened shell panel that is of interest for design purposes may include one or more of the following components: (1) axial tension or (2) axial compression, (3) shear, and (4) internal pressure. Composite structural durability is first investigated under each one of these four loading cases. In addition, four combined loading cases are also investigated as follows: (5) axial tension and shear, (6) axial compression and shear, (7) axial tension and shear under internal pressure, and (8) axial compression and shear under internal pressure. Boundary conditions, as shown in Fig. 10, are the same for all eight loading cases. In each case, computational simulation of structural durability under loading is carried out through the stages of damage initiation (local fracture in any of the modes), damage growth, and damage accumulation, up to the stage of damage propagation and structural fracture. Note that local fracture is simulated by stiffness deletion in that mode.

The results for combined loading with axial tension are summarized in Fig. 12. The loads are incremented in a proportional manner keeping the ratios of axial load to in-plane shear and internal pressure intact. As can be seen, the internal pressure has a stabilizing effect compared to that for shear with axial tension. With the addition of internal pressure, both the ultimate load capacity and damage tolerance of the structure show significant increases. The results for combined loading with axial compression are shown in Fig. 13. It is interesting to note that application of internal pressure has increased the damage tolerance almost by 25% even though the ultimate load capacity remained the same as for the case where only compressive loads are applied. The case where in-plane shear load is combined with compressive loads is by far the most adverse loading condition for this type of structure. A summary of the fracture progression sequence is given in Table 3. Three principal regions, the cap, the web,

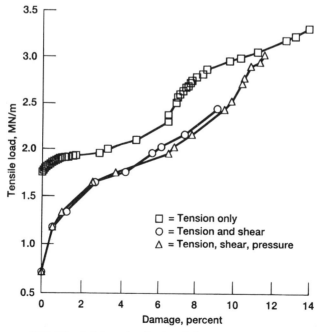

FIG. 12—*Axial tension load and damage progression.*

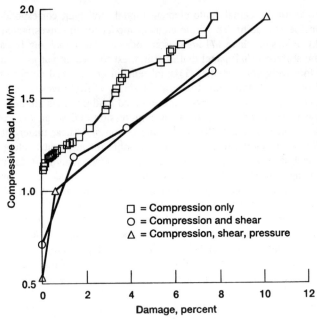

FIG. 13—*Axial compression load and damage progression.*

TABLE 3—*Summary of the fracture progression sequence.*

Loading Components	Damage Initiation	Damage Growth	Damage Progression	Fracture in
Tension, T	cap	web	skin	cap
Compression, C	cap	web		cap
Shear, S	web	skin	cap	skin
Pressure, P	cap	skin		cap
S + T	web	cap	skin	cap
S + C	web	cap		web
S + T + P	web	cap	skin	cap
S + C + P	web	cap		web

and the skin of the structure are identified in Fig. 10. These are referred to in the identification of various events relating to damage progression for individual as well as combined loading situations. The four events, that is, damage initiation, damage growth, damage progression, and global fracture, are tracked with respect to each of the loading conditions. The details present in this figure cast CODSTRAN into a "virtual desk-top virtual" laboratory.

Generalization and Lessons Learned

The discussion of the present approach focussed on its application to composite structures that are far more complex than conventional metallic structures. However, the approach is readily adaptable to structures made from any material or any combination of materials. Based on the experience and success to date, it can be readily generalized as is outlined in Fig. 14. The steps in the outline are the same for any structure. The difference is only in the description and history-tracking of the material behavior. Specifically, once the structure and the loading conditions are identified, a finite element model is constructed. Based upon the analysis results, "hot spots" (places that are severely stressed to cause initiation of damage) are identified. Flaws are introduced here and are allowed to grow with the applied loading. The structural performance, as characterized by such quantities as natural frequency or buckling load, is then monitored with the increase in load/damage. The flaw size that causes a predetermined unacceptable level in the global performance is then determined and considered as the critical amount of damage beyond which the structure must be retired for repair/inspection/or replacement.

The important lessons learned in developing this new approach are generic and should be instructive for undertaking the development of new approaches in general. These lessons are summarized here:

1. continuity in research activity;
2. participants' composite knowledge: that is,

    (a) structural mechanics principles,
    (b) finite element analysis,
    (c) composite mechanics,
    (d) fracture mechanics concepts, and
    (e) software development;

3. participants willing to question traditional approaches adopt/invent new ones;

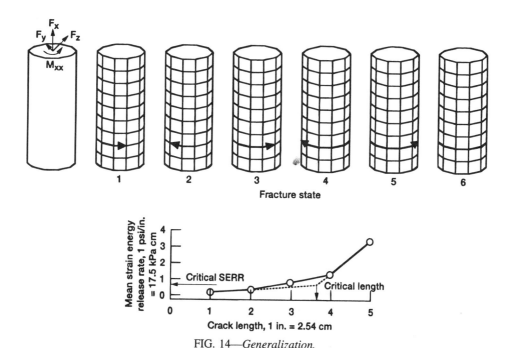

FIG. 14—*Generalization.*

4. management support; and
5. availability of computer facilities and support.

The authors firmly believe that all the items in this summary are necessary for the successful development of new approaches that differ from traditional ones. The authors are also convinced that the development of any new approach is not and should not be a short-term activity because the developers grow smarter with the continuous feedback and mature intuitively with the accumulation of experience during the development period.

Summary

A "new" approach independent of stress intensity factors and fracture toughness parameters has been developed and described for the computational simulation of the fracture of composite structures. This approach is inclusive in that it integrates composite mechanics (for composite behavior) with finite element analysis (for global structural response). The integration of these two disciplines permits: (1) quantification of the fracture progression from local damage initiation to structural fracture (collapse), (2) accommodation of any loading conditions including temperature and moisture, and (3) the effects of material degradation due to hygrothermal environments. The versatility of the approach is demonstrated by using it to computationally simulate fracture in typical structures (plates, shells, and built-up components) in a variety of fracture conditions. Parameters/guidelines are identified that can be used as criteria for structural fracture, inspection intervals, and retirement for cause. Generalization of structural systems and structures made from other components is outlined. Important lessons learned from undertaking the development of new approaches are summarized.

References

[1] Murthy, P. L. N. and Chamis, C. C., "Interlaminar Fracture Toughness: Three-Dimensional Finite-Element Modeling for End-Notch and Mixed-Mode Flexure," NASA TM 87138, NASA Lewis Research Center, Cleveland, 1985.

[2] Murthy, P. L. N. and Chamis, C. C., "Composite Interlaminar Fracture Toughness: 3-D Finite Element Modeling for Mixed Mode I, II, and III Fracture," NASA TM 88872, NASA Lewis Research Center, Cleveland, 1986.

[3] Wilt, T. A., Murthy, P. L. N., and Chamis, C. C., "Fracture Toughness, Computational Simulation of General Delamination in Fiber Composites," NASA TM 101415, NASA Lewis Research Center, Cleveland, 1988.

[4] Chamis, C. C. and Smith, G. T., "Composite Durability Structural Analysis," NASA TM 79070, NASA Lewis Research Center, Cleveland, 1978.

[5] Chamis, C. C., "Computational Simulation of Progressive Fracture in Fiber Composites," NASA TM 87341, NASA Lewis Research Center, Cleveland, 1986.

[6] Irvine, T. B. and Ginty, C. A., "Progressive Fracture of Fiber Composites," Journal of Composite Materials, Vol. 20, March 1986, pp. 166–184.

[7] Minnetyan, L., Chamis, C. C., and Murthy, P. L. N., "Structural Behavior of Composites with Progressive Fracture," NASA TM 102370, NASA Lewis Research Center, Cleveland, Jan. 1990.

[8] Minnetyan, L., Murthy, P. L. N., and Chamis, C. C., "Composite Structure Global Fracture Toughness via Computational Simulation," Computers–Structures, Vol. 37, No. 2, 1990, pp. 175–180.

[9] Minnetyan, L., Murthy, P. L. N., and Chamis, C. C., "Progressive Fracture in Composites Subjected to Hygrothermal Environment," Proceedings, Thirty-second SDM Conference (Part 1), Baltimore, MD, 8–10 April 1991, pp. 867–877.

[10] Minnetyan, L., Chamis, C. C., and Murthy, P. L. N., "Damage and Fracture in Composite Thin Shells," NASA TM 105289, NASA Lewis Research Center, Cleveland, Nov. 1991.

[11] Minnetyan, L., Rivers, J. M., Murthy, P. L. N., and Chamis, C. C., "Structural Durability of Stiffened Composite Shells," Paper AIAA 92-2244, Proceedings, Thirty-third AIAA/ASME/ASCE/AHS/ASC, Structures, Structural Dynamics and Materials Conference, Dallas, 13–15 April 1992.

Koi T. Marcucelli[1] and John C. Fish[1]

Determination of Transverse Shear Strength Through Torsion Testing

REFERENCE: Marcucelli, K. T. and Fish, J. C., **"Determination of Transverse Shear Strength Through Torsion Testing,"** *Composite Materials: Fatigue and Fracture (Sixth Volume), ASTM STP 1285*, E. A. Armanios, Ed., American Society for Testing and Materials, 1997, pp. 85–100.

ABSTRACT: The in-plane characterization of composite materials is, in general, well understood and widely utilized throughout the aerospace industry. However, the use of composites in structural elements such as fuselage frames and rotorcraft flexbeams place large out-of-plane or through-the-thickness stresses for which there is little data. Efforts to determine the interlaminar shear strength of laminated composites have been hampered due to the nonlinear behavior of test specimens and the limitations of current analysis tools.

An inexpensive rectangular torsion test specimen was designed to determine the interlaminar shear strength, s_{23}, of composite materials. Six different layups were fabricated of AS4/2220-3 carbon/epoxy unidirectional tape and tested in pure torsion. All of the specimens failed abruptly with well-defined shear cracks and exhibited linear load-deflection behavior. A quasi-three-dimensional (Q-3-D) finite element analysis was conducted on each of the specimen configurations to determine the interlaminar shear stress at failure. From this analysis, s_{23} was found to be 107 MPa for this material.

KEYWORDS: composite materials, fracture (materials), delamination, fatigue (materials), interlaminar shear strength, quasi-three-dimensional finite element analysis, shear cracks, torsion

Determination of the out-of-plane shear properties of composite materials has shown limited success. The anisotropic character of composites makes it difficult to design a test that isolates the desired failure mode [1,2]. In addition, classical (linear) analyses are of limited use when the load-deflection behavior of test specimens become very nonlinear.

Torsion testing has been used recently to determine the intralaminar and interlaminar shear properties, as well as different failure modes, in complex layups [3–5]. In addition, both closed-form solutions and finite element analysis (FEA) has also been used to study the complex stress states due to the application of a torsional load [6–12]. The primary difficulty encountered is the nonlinear behavior of the test specimens resulting in poor correlation with linear FEA solutions. This has led to the use of very thick (and expensive) laminates, on the order of 100 plies, to minimize this effect.

This study focused on the design of a test specimen that exhibited linear load-deflection behavior to specimen failure. This allows the use of linear analyses to determine the shear stress distribution at failure. In addition, the effects of ply orientation on the torsional load carrying capability was investigated.

[1]Stress engineer and research and development engineer, respectively, Lockheed Martin Skunk Works, Palmdale, CA 93599.

TABLE 1—*Test specimen layups.*

Configuration	Layup
1	$[\pm 45_5/0_6]_s$
2	$[\pm 45_5/0_5/15]_s$
3	$[\pm 45_5/0_5/30]_s$
4	$[\pm 45_5/0_5/60]_s$
5	$[\pm 45_5/0_5/75]_s$
6	$[\pm 45_5/0_5/90]_s$

Specimen Design

Thirty-six specimens comprised of six different layups were fabricated from AS4/2220-3 unidirectional pre-preg tape. The specimens were symmetric with respect to the midplane and were constructed with three distinct sublaminates. The outer plies consisted of five groups of $\pm 45°$ plies. Adjacent to these were five $0°$ plies. Finally there was a two-ply midplane sublaminate with the ply orientation varying from $0°$ to $90°$ in $15°$ increments. The outer $\pm 45°$ sublaminate maximizes the torsional rigidity of the laminate, and the $0°$ plies isolate the failure mode to the midplane sublaminate. This design allows the midplane orientation angle to vary greatly with very little change in the torsional stiffness, allowing the direct comparison of the twist angles of different specimen layups. These specimen layups are shown in Table 1. The test specimens were 229 mm long, 38 mm wide, and 4.8 mm thick. Glass/epoxy tabs (38 mm long) were secondarily bonded on the specimens resulting in a test length of 152 mm (Fig. 1).

Test Procedure

Six specimens of each laminate configuration was tested at room temperature ambient (RTA) conditions. The through-the-thickness edges of the test specimens were coated with water-soluble typewriter correction fluid to help visualize failures. The test specimens were installed in a tension-torsion bi-axial test frame equipped with hydraulic grips. Rotation, torque, and axial tension were recorded by a computerized data acquisition system continuously during

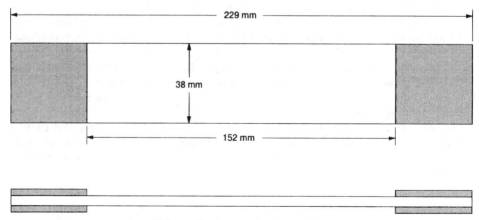

FIG. 1—*Torsion specimen geometry.*

the test. The specimens were loaded in pure torsion and the axial tension was monitored to verify that it was zero. A constant twist rate of 10°/min was used for all specimens. Failure was associated with a large drop in load, and the test was stopped after this occurred.

Experimental Results

Load-deflection plots of the torsion specimens verified that the laminate design was linear to failure (Fig. 2). All failures were characterized by the appearance of well-defined shear cracks accompanied by an audible emission. All of the test specimens demonstrated a large reduction in load carrying capability at failure as shown in Fig. 2.

Two different failure modes were observed for all of the specimen configurations. Failure for the 90°, 75°, and 60° midplane specimens consisted of a well-defined shear crack through the midplane sublaminate. The 0° and 15° specimens failed with a shear crack through the 45° ply adjacent to the 0°/45° interface. The 30° specimens demonstrated both types of failures. In all cases, the shear cracks became delaminations upon reaching the interfaces with the 0° plies. Photographs of the two failure modes are shown in Figs. 3 and 4.

The average strengths for each of the specimen configurations are represented in Fig. 5. Two observations can be observed from the data:

1. The 0°, 15°, and 30° specimens exhibited failures in the outer ±45° plies, and demonstrated similar failure loads (COV = 7.5%) and twist rates (COV = 9.9%) (COV = coefficient of variation).

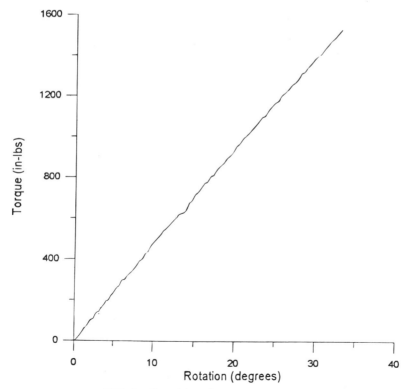

FIG. 2—*0° midplane load deflection plot.*

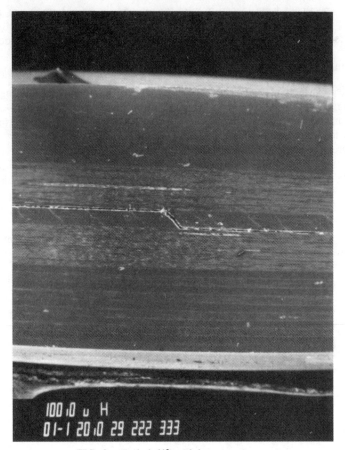

FIG. 3—*Failed 60° midplane specimen.*

2.   The torsional strength of the test specimens decreases as the midplane orientation angle of the specimens is increased from 30° to 90°.

Finite Element Analysis

A quasi-three-dimensional (Q-3-D) finite element analysis [*13*] was conducted on each of the layups to evaluate the three-dimensional stress state of the laminate cross sections. This analysis assumes a state of generalized plane deformation that requires that each cross section deforms to the same shape [*14*].

The coordinate system for the finite element analysis is shown in Fig. 6. Torsion is defined as

$$T = \int \int \{y\sigma_{xz} - z\sigma_{xy}\} \, dy \, dz \qquad (1)$$

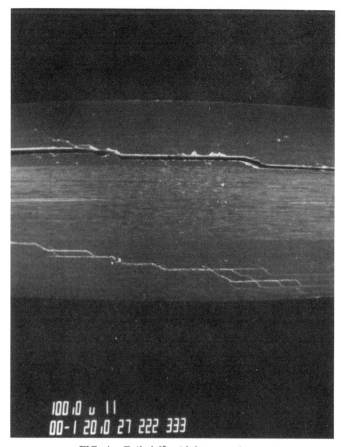

FIG. 4—*Failed 0° midplane specimen.*

where

T = applied torque, N-m;
y = horizontal coordinate, m;
z = vertical coordinate, m;
σ_{xz} = shear stress on the x-z plane, MPa; and
σ_{xy} = shear stress on the x-y plane, MPa.

The Saint Venant torsion theory (SVTT) [14] defines the displacements corresponding to the rotation of the cross section as

$$v(x, y, z) = -\phi xz + V(y, z)$$
$$w(x, y, z) = \phi xy + W(y, z) \tag{2}$$

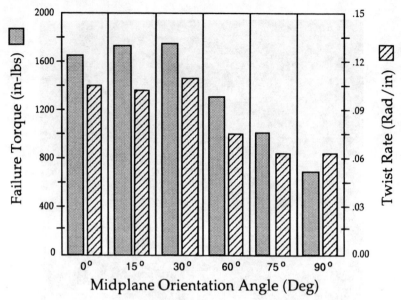

FIG. 5—*Specimen torsional performance.*

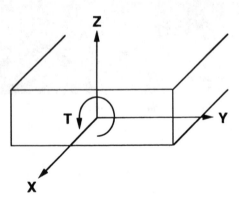

FIG. 6—*Finite element coordinate system.*

where

ϕ = torsional twist rate, rad/m;
x = distance along the length of the specimen, m;
$V(y, z)$ = transverse displacement as a function of y and z, m; and
$W(y, z)$ = vertical displacement as a function of y and z, m.

The shear strains can then be related as follows

$$\epsilon_{xz} = \frac{\partial w}{\partial x} + \frac{\partial u}{\partial z} = \phi y + \frac{\partial u}{\partial z}$$

$$\epsilon_{xy} = \frac{\partial v}{\partial x} + \frac{\partial u}{\partial y} = -\phi z + \frac{\partial u}{\partial y} \qquad (3)$$

where

ϵ_{xz} = shear strain in the x-z plane,
ϵ_{xy} = shear strain in the x-y plane,
$\partial w/\partial x$ = change in w displacement with respect to x,
$\partial u/\partial z$ = change in u displacement with respect to z,
$\partial v/\partial x$ = change in v displacement with respect to x, and
$\partial u/\partial y$ = change in u displacement with respect to y.

The ϕy and $-\phi z$ terms represent the prescribed strains, and the $\partial u/\partial z$ and $\partial u/\partial y$ terms correspond to the out-of-plane warping of the cross section. Thus, the strain state in a three-dimensional solid under twisting deformation can then be represented as

$$\begin{Bmatrix} \epsilon_{yy} \\ \epsilon_{zz} \\ \epsilon_{yz} \\ \epsilon_{xz} \\ \epsilon_{xy} \end{Bmatrix} = \begin{bmatrix} R_{22}R_{23}R_{24}R_{25}R_{26} \\ R_{32}R_{33}R_{34}R_{35}R_{36} \\ R_{42}R_{43}R_{44}R_{45}R_{46} \\ R_{52}R_{53}R_{54}R_{55}R_{56} \\ R_{62}R_{63}R_{64}R_{65}R_{66} \end{bmatrix} \begin{Bmatrix} \sigma_{yy} \\ \sigma_{zz} \\ \sigma_{yz} \\ \sigma_{xz} \\ \sigma_{xy} \end{Bmatrix} + \begin{Bmatrix} 0 \\ 0 \\ 0 \\ \phi y \\ -\phi z \end{Bmatrix} \qquad (4)$$

$$R_{ij} = S'_{ij} - \frac{S'_{i1}S'_{j1}}{S'_{11}} \qquad (i, j = 2, 3, 4, 5, 6) \qquad (5)$$

where

R_{ij} = modified compliance coefficients [14] and
S'_{ij} = stress-strain relationship coefficients, m²/N.

$\{0\ 0\ 0\ \phi y - \phi z\}$ is a vector representing the applied torsional strains on the specimen. The axial strain, ϵ_{xx}, is treated as a constant in this formulation, and allows the reduction of the six stress-strain relationships down to five [13].

The finite element formulation used in the analysis is based on the Hellinger-Reissner principle [15]. The resulting mixed formulation element is more rigorous than the traditional displacement type in that both stress and strain are solved as independent variables, thus ensuring stress equilibrium within each element [16].

The finite elements themselves are four-noded rectangular elements that are free to warp out of the plane of the modeled cross section. Each node of the element has three translational degrees of freedom, yielding twelve degrees of freedom for each element.

The results from the finite element program are displacements and stresses for each element. The stresses are transformed back into the ply (1-2-3) coordinate system for each element.

An example of the finite element mesh is shown in Fig. 7. The area modeled is one quarter of the cross section. The smallest elements are one third of a ply thickness in both width and thickness to capture the high gradients at the free edge. Larger elements are used where the stress gradients are less severe.

The material properties used in the analysis are:

$$E_1 = 150.3 = \text{extensional modulus in 1 direction, GPa}$$
$$E_2 = E_3 = 9.7 = \text{extensional modulus in 2 and 3 direction, GPa}$$
$$v_{12} = v_{13} = 0.27 = \text{Poisson's ratio for the 1–2 and 1–3 planes}$$
$$v_{23} = 0.50 = \text{Poisson's ratio for the 2–3 plane}$$
$$G_{12} = G_{13} = 5.2 = \text{shear modulus for the 1–2 and 1–3 planes, GPa}$$
$$G_{23} = 3.4 = \text{shear modulus for the 2–3 plane, GPa}$$
$$H = 0.15 = \text{average ply thickness, mm}$$

The 1, 2, and 3 subscripts represent the fiber, long transverse, and short transverse directions, respectively, for a unidirectional ply.

Each of the laminate configurations was modeled independently. Loading for each layup was a prescribed twist per unit length equal to the average value obtained from testing for each configuration.

Finite Element Analysis Results

Figures 8, 9, 10, and 11 show the stress state at the free edge of the 0°, 30°, 60°, and 90° specimens, respectively, at failure. As expected, for all of the configurations, σ_{22} and σ_{33} are near zero at the midplane, verifying that the finite element model is accurately representing the stress distribution. The small nonzero values are due to the elemental stresses being recovered at the gauss points and not the actual free edge. Away from the midplane sublaminate, the normal stresses are compressive and thus do not promote delaminations.

The finite element model predicts the values of σ_{23} and σ_{13} to be equal for each ply in the ±45° sublaminate. (Intuitively, a 45° ply under shear must have equal values of shear stress on each of these planes in order to maintain equilibrium.) All of the specimen configurations exhibit the same distribution in this region.

The shear stress, σ_{13}, is dominant in the 0° sublaminate, as would be anticipated. A slight compressive value of σ_{22} is also evident, and decays to zero at the midplane.

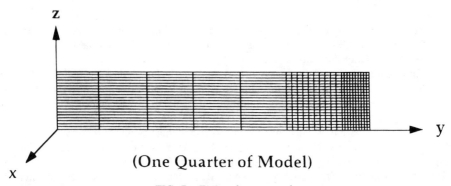

(One Quarter of Model)

FIG. 7—*Finite element mesh.*

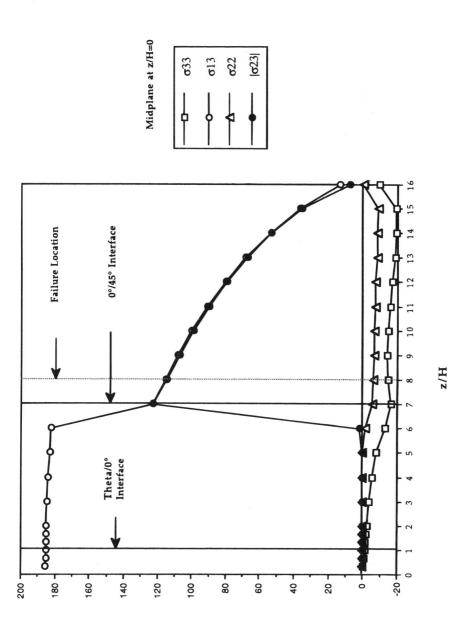

FIG. 8—0° midplane sublaminate stress distribution.

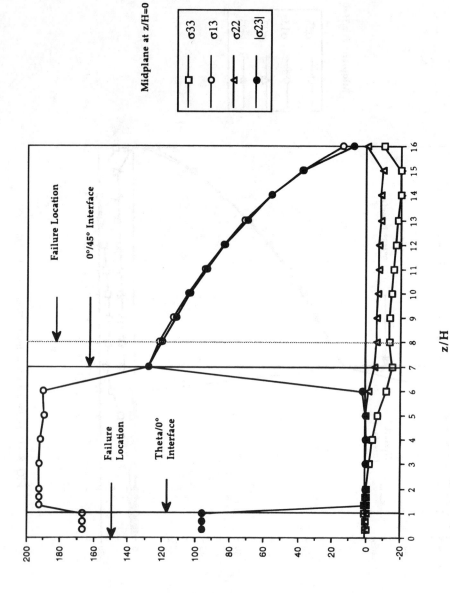

FIG. 9—30° midplane sublaminate stress distribution.

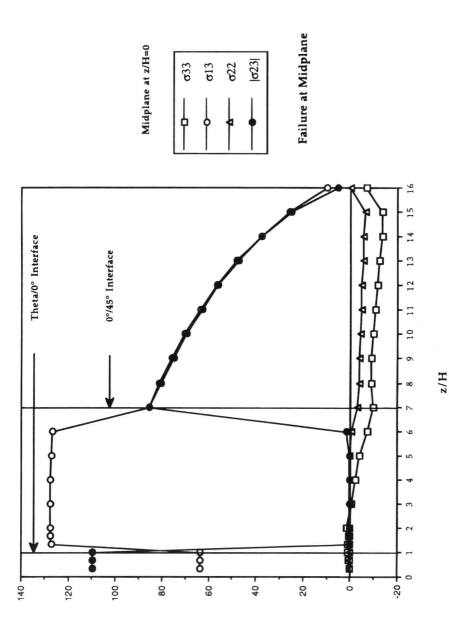

FIG. 10—*60° midplane sublaminate stress distribution.*

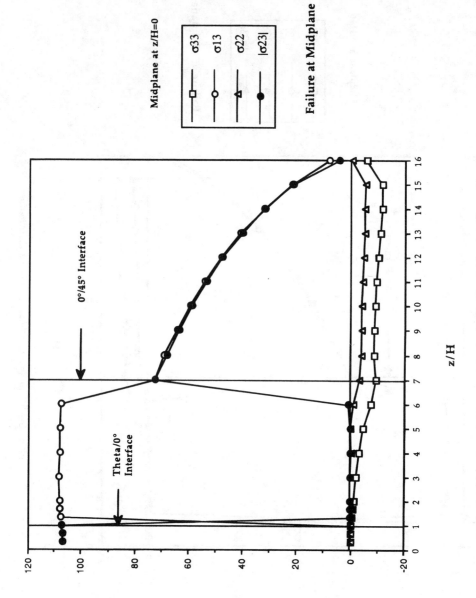

FIG. 11—90° midplane sublaminate stress distribution.

The midplane sublaminates are the only region where the stress distribution differs greatly between the layups. Increasing the midplane orientation angle from 0° to 90° will decrease the value of σ_{13} and increase σ_{23} for a given ply. The bounding conditions are pure σ_{13} for the 0° orientation and pure σ_{23} for 90°. For unidirectional tape, the interlaminar shear strength value, s_{13}, is driven by the shear strength of the reinforcing fiber, while s_{23} is a matrix-dominated property. Thus, the value of s_{13} will be greater than s_{23}. As the midplane ply orientation angle increases from 0° to 90°, the applied shear stress due to torsion is distributed more on the 2–3 plane of the laminate. Test specimen failure will occur when the applied shear stress reaches the shear strength of the laminate on the 2–3 plane.

A similar failure mode has been noted in Minguet and O'Brien [17] and was driven by transverse tensile stress. If this same failure mechanism occurs for the torsion testing, the stress at failure should be equal to the transverse tensile strength of the pre-preg; in this case 41 MPa. Rotating the stress state for the torsion specimen 45°, as shown in Fig. 12, results in an equivalent transverse tensile stress of 107 MPa. Here, the transverse tensile stress is three times the strength allowable, verifying that the failures could not be caused by transverse tension. Therefore, these failures must be caused by the state of pure shear. The testing conducted in Ref 17 was three- and four-point bending of skin/stringer reinforced panels. This testing induced through-the-thickness tensile stresses of 37.2 MPa at failure, which compares well with the 41 MPa strength. Although the failure modes for these two tests were similar, the failure mechanisms are not related.

Sen and Fish [18] conducted torsion testing on glass/epoxy laminates and experienced the same failure mode as found in this work. Transverse shear stress was determined to cause the shear cracks that formed in the "grouped" plies in various locations in the laminates.

Figure 13 depicts the peak transverse shear stress (σ_{23}) at the midplane sublaminate and the 45° ply at the 0°/45° interface at failure. The midplane sublaminate shear stress increases as the ply angle increases. The 60°, 75°, and 90° sublaminate specimens failed when σ_{23} reached an average value of 107 MPa with a COV of 9.9%. Conversely, the stresses at the 45° ply decrease with increasing midplane orientation angle, with the transition between failure modes occurring at 30°. The shear stresses at the two locations for the 30° case are very close in value, explaining the presence of two failure locations for this specimen configuration.

The test results verify that the transverse shear strength, s_{23}, is not a function of ply orientation. Therefore, only one laminate design need be tested to determine this property. The author

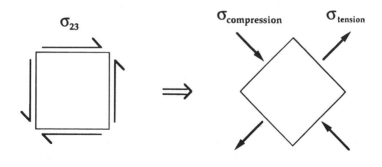

$$\sigma_{23} = \sigma_{\text{compression}} = \sigma_{\text{tension}}$$

FIG. 12—*Equivalent shear and tensile stress distributions.*

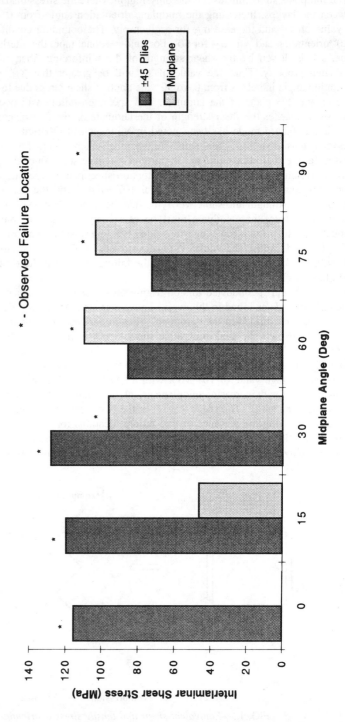

FIG. 13—*Interlaminar shear stress at failure.*

recommends the $[\pm45_s/0_s/90]_s$ layup that minimizes the amount of scrap and reduces the chance of midplane ply misalignment.

Concluding Remarks

Torsion testing can be used effectively to evaluate the transverse shear strength (s_{23}) of composite materials. An elastically tailored laminate was designed to produce linear load-deflection behavior in torsion-to-specimen failure. A linear finite element analysis was then used to determine the free-edge stress state of the test specimens at failure.

The results represented in this work do not include hygrothermal effects. Subsequent analysis has shown that for this specimen design the change in the stress distribution due to thermal conditions was small.

Failure of the test specimens occurred at two locations: at the midplane sublaminate and at the outer $\pm45°$ sublaminate. Failure at both locations was due to transverse shear stress in the 2–3 plane of the material.

The finite element analysis shows that the transverse shear strength, s_{23}, is 107 MPa for the AS4/2220-4 graphite/epoxy material system. All of the test specimens failed at the same value of s_{23}, demonstrating that the shear strength is indeed a material parameter, independent of the midplane ply orientation angle.

Using the data obtained from the 0° midplane sublaminate tests, a lower bound on s_{13} is 186.2 MPa. This again is only a lower bound on the strength since there were no shear failures in the 0° region. This value, however, suggests that s_{13} is indeed significantly greater than s_{23}.

References

[1] Knight, M., "Three-Dimensional Elastic Moduli of Graphite/Epoxy Composites," *Journal of Composite Materials*, Vol. 16, 1982, pp. 153–159.

[2] Kim, R. Y., Abrams, F., and Knight, M., "Mechanical Characterization of a Thick Composite Laminate," *Proceedings*, American Society of Composites Third Technical Conference, 25–29 Sept. 1988, Seattle, pp. 711–718.

[3] Kurtz, R. D. and Sun, C. T., "Composite Shear Moduli and Strengths from Torsion of Thick Laminates," *Composite Materials: Testing and Design, Ninth Volume, ASTM STP 1059*, S. P. Garbo, Ed., American Society for Testing and Materials, Philadelphia, 1990, pp. 508–520.

[4] Hinkley, J. A. and O'Brien, T. K., "Delamination Behavior of Quasi-Isotropic Laminates Subjected to Tension and Torsion Loads," *Proceedings*, Rotorcraft Structures Specialists' Meeting, American Helicopter Society, Williamsburg, VA, 29–31 Oct. 1991.

[5] Ferent, B. and Vautrin, A., "Torsion Response Analysis of T300/914 and T800/914 Unidirectional Specimens," *Mechanical Identification of Composites*, A. Vautrin and H. Sol, Eds., Elsevier Applied Science, London, 1991, pp. 141–148.

[6] Armanios, E. A. and Li, J., "Interlaminar Stress Predictions for Laminated Composites Under Bending, Torsion and Their Combined Effect," *Composite Engineering*, Vol. 1, No. 5, 1991, pp. 277–291.

[7] Li, J. and Armanios, E. A. "Analysis of Unidirectional and Cross-Ply Laminates Under Torsion Loading," *Fracture Mechanics: Second Symposium (Vol. II), ASTM STP 1131*, S. N. Atluri, J. C. Newman, Jr., I. S. Raju, and J. S. Epstein, Eds., American Society for Testing and Materials, Philadelphia, 1992, pp. 421–435.

[8] Sen, J. K. and Fish, J. C., "Fracture of Glass-Epoxy Laminates Under Torsion and Combined Tension-Torsion Loads," *Composite Materials: Fatigue and Fracture (Fifth Volume), ASTM STP 1230*, R. H. Martin, Ed., American Society for Testing and Materials, Philadelphia, 1995, pp. 440–466.

[9] Chan, W. and Ochoa, O., "Assessment of the Free-Edge Delamination due to Torsion," *Proceedings*, Second Technical Conference on Composite Materials, American Society for Composites, Sept. 1987, pp. 469–478.

[10] Murthy, P. L. N. and Chamis, C. C., "Free-Edge Delamination: Laminate Width and Loading Conditions Effects," *Journal of Composites Technology and Research*, Vol. 11, 1989, pp. 15–22.

[11] Hooper, S. J., Hageneier, R., and Ramaprasad, S. "Tension/Torsion Loading of Composite Laminates with Free-Edge Boundary Conditions," *Proceedings*, Rotorcraft Structures Specialists' Meeting, American Helicopter Society, Williamsburg, VA, 29–31 Oct. 1991.

[12] Fish, J. C., "Torsion and Twisting of Symmetric Composite Laminates," *Proceedings*, Thirty-third Structures, Structural Dynamics and Materials Conference, American Institute of Aeronautics and Astronautics, 1992, pp. 736–744.

[13] Fish, J. C., "Stress Analysis of Composite Rotor Systems," *Proceedings*, American Helicopter Society Forty-seventh Annual Forum, 6–8 May 1991, pp. 615–627.

[14] Leknitskii, S. G., *Theory of Elasticity of an Anisotropic Body*, MIR Publishers, Moscow, 1981 (English translation of the 1977 Russian edition).

[15] Washizu, K., *Variational Methods in Elasticity and Plasticity*, Pergamon Press Ltd., Oxford, UK, 1968.

[16] Pian, T. H. H., "Derivation of Element Stiffness Matricies by Assumed Stress Distributions," *AIAA Journal*, Vol. 2, No. 7, July 1964, pp. 1333–1336.

[17] Minguet, P. J. and O'Brien, T. K., "Analysis of Test Methods for Characterizing Skin/Stringer Debonding Failures in Reinforced Composite Panels," *Composite Materials: Testing and Design (Twelfth Volume), ASTM STP 1274*, C. R. Saff and R. B. Deo, Eds., American Society for Testing and Materials, West Conshohocken, PA, 1996, pp. 105–124.

[18] Sen, J. K. and Fish, J. C., "Failure Prediction of Composite Laminates Under Torsion," *Fracture of Composites*, E. A. Armanios, Ed., Transtec Publications, Ltd., Zuerich-Uetikon, Switzerland, 1996, pp. 285–306.

Fatigue and Life Prediction

Douglas J. Herrmann[1] and Ben M. Hillberry[2]

Effects of Fiber Bridging and Fiber Fracture on Fatigue Cracking in a Titanium-Matrix Composite

REFERENCE: Herrmann, D. J. and Hillberry, B. M., **"Effects of Fiber Bridging and Fiber Fracture on Fatigue Cracking in a Titanium-Matrix Composite,"** *Composite Materials: Fatigue and Fracture (Sixth Volume), ASTM STP 1285*, E. A. Armanios, Ed., American Society for Testing and Materials, 1997, pp. 103–125.

ABSTRACT: Constant amplitude fatigue tests were performed on $[0]_4$ and $[0/90]_s$SCS-6/ TIMETAL 21S titanium-matrix composite specimens to study fiber bridging of matrix fatigue cracks. Crack length was monitored throughout the tests, and displacements near the crack surface were measured periodically by placing an Elber gage (1.5-mm gage length point extensometer) across the crack at a number of positions. Specimens were removed prior to failure and mechanically polished to the first layer of fibers and the extent of fiber bridging observed. While some cracks were fully bridged, other cracks contained broken fibers among the intact, bridging fibers. A model has been developed to study the mechanics of a cracked unidirectional composite with any combination of intact and broken fibers in the wake of the matrix crack. Displacements near the crack surface predicted by the model agree with the Elber gage measurements for cracks that were fully bridged. Predictions were also performed for bridged cracks with discrete fiber breaks in the crack wake based on the geometry observed after polishing with good correlation between the predictions and the experimental measurements.

KEYWORDS: crack propagation, fatigue (materials), fiber failure, notches, silicon-carbide fibers, titanium-matrix composites, fracture (materials), composite materials

Continuous-fiber titanium-matrix composites (TMCs) are being considered for a number of aerospace structures where high specific strength and stiffness are required at elevated temperatures. Potential applications include advanced gas turbine engine components and hypersonic aircraft structures [1]. To design lightweight, damage-tolerant TMC structures, an understanding of fatigue crack propagation in TMCs is necessary.

Fatigue experiments performed on titanium-matrix composites such as SCS-6/Ti-15-3, SCS-6/Ti-24-11, and SCS-6/TIMETAL 21S show that cracks from notches in 0° laminates initiate in the titanium matrix and grow perpendicular to the loading direction leaving the fibers intact [2–7]. The intact fibers left in the crack wake shield the matrix crack tip, thus reducing the fatigue crack propagation rate. The reduction in the crack propagation rate can be several orders of magnitude and, in some cases, the fatigue crack may arrest [2]. As the crack propagates, stresses in the fibers that bridge the crack increase until either a steady-state condition such as described by Bao and McMeeking [8] is obtained or fibers begin to fail. The failure of the bridging fibers increases the stress intensity factor and thus the fatigue crack

---

[1]Senior project engineer, Allison Engine Company, Indianapolis, IN 46206-0420; formerly, graduate student, School of Mechanical Engineering, Purdue University, West Lafayette, IN 47907-1288.
[2]Professor, School of Mechanical Engineering, Purdue University, West Lafayette, IN 47907-1288.

propagation rate. Based on extensive experimental research in the fatigue of titanium-matrix composites, Bowen [9] has concluded that to accurately model the fatigue crack propagation behavior of titanium-matrix composites the effect of discrete fiber failures must be considered.

In this study, constant amplitude fatigue tests were performed on $[0]_4$ and $[0/90]_s$ SCS-6/ TIMETAL 21S titanium matrix composite specimens to study fiber bridging. During these tests, crack length was monitored and displacements near the crack surface were measured. To observe the extent of fiber bridging during fatigue, a number of specimens were removed prior to failure and mechanically polished to the first layer of fibers. The experimental results were compared to the discrete composite model developed by Herrmann and Hillberry [10] to model fiber bridging in titanium-matrix composites. The discrete composite model differs from other crack-bridging models in that it permits discrete fiber breaks among intact, bridging fibers.

Materials and Experimental Methods

Materials

The titanium-matrix composite studied was SCS-6/TIMETAL 21S. TIMETAL 21S[3] (previously called Ti-β21S) is a metastable beta titanium alloy with the composition Ti-15Mo-3Al-2.7Nb-0.2Si (percent by weight). The matrix is reinforced with continuous SCS-6[4] fibers. The fibers are 0.14 mm in diameter. The matrix and fiber material properties used in this study are listed in Table 1.

The $[0]_4$ and $[0/90]_s$ SCS-6/TIMETAL 21S laminates were fabricated by hot-pressing foils of TIMETAL 21S between unidirectional tapes of the SCS-6 fibers at a temperature near 1000°C. The plates were then cooled to 621°C and aged for 8 h to stabilize the matrix material, and finally cooled to room temperature. The materials used in this study were manufactured by Textron Specialty Materials Division.

The five $[0]_4$ and two $[0/90]_s$ specimens tested were center-cracked tension specimens, M(T), with a nominal width of 25.4 mm, length of 150 mm, and notch length, $2a_0$, of 5.0 mm. The notch was made by electron discharge machining (EDM). The nominal thickness was 0.88 mm for the $[0]_4$ specimens and 0.91 mm for the $[0/90]_s$ specimens. The fiber volume fractions were measured by counting fibers and resulted in mean fiber volume fractions of 0.36 and 0.35 for the unidirectional and cross-ply specimens, respectively. All specimens were polished to a mirror finish.

Experimental Methods

The fatigue tests were performed in air at room temperature using a closed-loop, servohydraulic test system with hydraulic friction grips. All tests were performed in load control with a

TABLE 1—*Mechanical properties of the fiber and matrix.*

Material	E, GPa	ν	α^a, mm/mm/°C
SCS-6	393	0.25	3.6×10^{-6}
TIMETAL 21S	112	0.35	8.6×10^{-6}

^{a}Average α (coefficient of thermal expansion) for ΔT from 621 to 21°C.

[3]TIMETAL 21S is a registered trademark of the Timet Corporation, Henderson, NV.
[4]Textron Specialty Materials Division, Lowell, MA.

stress ratio, R, of 0.1 and at a frequency of 10 Hz. The applied stress ranges for the [0]$_4$ specimens were 200, 300, and 392 MPa. The [0/90]$_s$ specimens had an applied stress range of 150 MPa.

Crack initiation and propagation were monitored with a closed-circuit video system mounted on a precision lead screw slide with a linear displacement transducer. The specimens were precracked 0.5 mm from each EDM notch before recording data for crack growth rate, *da/dN*, calculations. Specimens were removed from the test machine periodically for viewing with a microscope to ensure that cracks were symmetric through the thickness. Fatigue crack growth rates and the applied stress intensity factor ranges, ΔK, were calculated per ASTM Test Method for Measurements of Fatigue Crack Growth Rates (E 647-93).

Displacement ranges near the crack surface were measured using an Elber gage [*11*] (Fig. 1). The Elber gage is a point extensometer with a 1.5 mm gage length that was originally designed to perform crack closure measurements on aluminum fatigue specimens. The gage was positioned by mounting the gage assembly on the end of the video camera lens. A lead screw slide with motion perpendicular to the specimen surface was used to move the camera towards the specimen until the points of the Elber gage contacted the specimen, one point on each side of the crack (Fig. 1). To ensure that the gage did not slip during the loading cycle, two cycles were applied to the specimen at 0.1 Hz and the load-displacement results were monitored in real time. This procedure was repeated for a number of locations along the crack

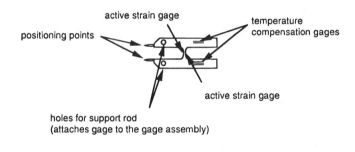

(a)

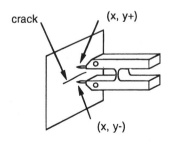

(b)

FIG. 1—*Schematic of the* (a) *Elber gage and* (b) *the placement of the Elber gage on a cracked specimen.*

resulting in a profile of Elber gage displacement range, ΔCOD^*, at a given crack length over a small number of cycles. The locations of the Elber gage points on the specimen were verified by viewing the indentations left by the Elber gage points with the video system. It is important to note that the displacement range measured by the Elber gage is not the crack opening displacement range, ΔCOD, but rather the change in displacement between two points near the crack surface on opposite sides of the crack.

Several specimens were removed from the test machine prior to specimen fracture for metallographic examination. These specimens were mechanically polished to the first ply of fibers to observe the extent of fiber bridging in the crack wake.

Fiber-Bridging Model

To model the effect of the fibers bridging a matrix crack, the tractions carried by the bridging fibers are approximated as a pressure distribution applied to the crack surface, $p(x)$ [12]. For crack surfaces subject to a remote stress, S, and a bridging pressure, $p(x)$, the composite stress intensity factor, K, is

$$K = 2 \int_0^a [S - p(x)]G(a, x) \, dx \tag{1}$$

where a is the length of the matrix crack and G is the weight function for the given geometry. Similarly for fatigue, the composite stress intensity factor range, ΔK, is

$$\Delta K = 2 \int_0^a [\Delta S - \Delta p(x)]G(a, x) \, dx \tag{2}$$

where ΔS is the applied stress range and Δp is the bridging pressure range. In this paper, K and ΔK given by Eqs 1 and 2 are assumed to be equal to K_m and ΔK_m, the stress intensity factor and the stress intensity factor range experienced by the matrix material at the crack tip [13]. The stress at the crack plane carried by a bridging fiber at location x is

$$\sigma_f(x) = \frac{p(x)}{V_f} \quad \text{and} \quad \Delta \sigma_f(x) = \frac{\Delta p(x)}{V_f} \tag{3}$$

The discrete composite model developed by Herrmann and Hillberry [10] was used to calculate the bridging pressure. The discrete composite model differs from continuum bridging models [12,14] in that the bridging pressure for any combination of intact and broken fibers in the crack wake can be calculated. Currently, the continuum models are limited to an unbridged notch followed by a fully bridged crack. The discrete composite model is briefly reviewed in Appendix I.

The discrete composite model requires a bridging law that relates the crack opening displacement to the bridging pressure. When applying the maximum stress, S_{max}, to the composite, the relationship given by Cox and Marshall [15] that includes residual stresses was used

$$\frac{COD_{max}}{2} = \lambda(p_{max} + \sigma_R)^2 \tag{4}$$

where COD_{max} is the crack opening displacement and p_{max} is the bridging pressure when S_{max} is applied to the composite. λ and σ_R are constants given by

$$\lambda = \frac{(1 - V_f)r_f E_m}{4 V_f^2 E_f E_c \tau}$$

$$\sigma_R = \frac{E_c}{E_m} \sigma_m^{res} \tag{5}$$

where E_f, E_m, and E_c are the fiber, matrix, and composite elastic moduli, respectively, with E_c calculated by the rule of mixtures. r_f is the fiber radius, τ is the fiber/matrix sliding stress, and σ_m^{res} is the matrix residual stress in the fiber direction.

For fatigue loading, the bridging relationship of McMeeking and Evans [13], which is based on the work of Marshall and Oliver [16], was used

$$\frac{\Delta COD}{2} = \frac{\lambda}{2} (\Delta p)^2 \tag{6}$$

where ΔCOD is the crack opening displacement range. The implementation of these bridging relationships, Eqs 4 through 6, into the discrete composite model is shown in Appendix I.

Experimental Results

After fatigue precracking, each specimen experienced a decrease in the crack growth rate (Fig. 2). Three of the specimens were removed from the test machine before any acceleration

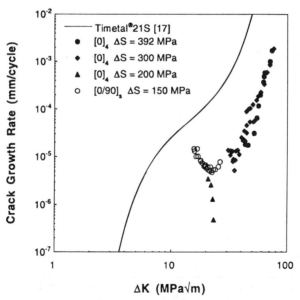

FIG. 2—*Crack growth rate, da/dN, versus stress intensity factor range, ΔK, for SCS-6/TIMETAL 21S compared to unreinforced TIMETAL 21S [17] for R = 0.1.*

in crack growth and were polished to the first row of fibers (Table 2). All of the fibers in the crack wake were intact for Specimen 3L, a $[0]_4$ subject to $\Delta S = 200$ MPa, and Specimen 10C, a $[0/90]_s$ subject to $\Delta S = 150$ MPa (Fig. 3). Specimen 6L, a $[0]_4$ subject to $\Delta S = 300$ MPa, had complete fiber bridging on one side of the EDM notch and two broken fibers among four bridging fibers on the other side of the notch (Fig. 4).

The applied stress, S, versus Elber gage displacements, COD*, curves for fatigue cracks in the region of decreasing da/dN displayed significant hysteresis during fatigue cycling for both the $[0]_4$ and $[0/90]_s$ specimens (Fig. 5). The hysteresis was observed over the entire crack length, not just near the crack tip. No evidence of crack closure was apparent in the S versus COD* hysteresis loops.

The crack growth rate increased after the initial decline for the four specimens that were not removed previously for polishing (Table 2). The transition from decreasing da/dN to increasing da/dN occurred at $a \approx 4.9$ mm and $a \approx 4.6$ mm for the $[0]_4$ specimens with ΔS = 300 and 392 MPa, respectively. The transition occurred in the $[0/90]_s$ specimen at $a \approx 5.7$ mm.

To determine if all of the bridging fibers had failed after da/dN began to increase, Specimen 4L, a $[0]_4$ specimen subject to $\Delta S = 300$ MPa, was removed from the test machine at $a = 7.14$ mm and was polished to the first row of fibers. Most of the fibers in the crack wakes had broken within a fiber diameter of the matrix fatigue crack plane. However, some fibers bridged the matrix fatigue crack near the crack tips and other fibers were broken several fiber diameters from the crack plane (Fig. 6). S versus COD* curves observed along cracks in the increasing da/dN region displayed hysteresis (Fig. 7). The change in slope of the S versus COD* curve at low S suggests that crack closure was present.

Model Predictions

The fatigue bridging relationship, Eq 6, requires knowledge of the fiber and matrix elastic moduli, fiber volume fraction, fiber radius, and the sliding stress, τ. In this paper, τ was determined using the pragmatic method first proposed by McMeeking and Evans [13]. This method determines τ by correlating the composite crack growth rate data and bridging model predictions to the matrix crack growth rate relationship. In the McMeeking and Evans approach, τ becomes a parameter that fits the model predictions to the composite crack growth rate data. The specific procedure that was used is as follows.

A value for τ was chosen for a given applied load. The bridging pressure range was calculated using the discrete composite model, and ΔK_m was calculated using Eq 2 for each composite

TABLE 2—*Summary of the fatigue crack growth tests (R = 0.1).*

Specimen ID	Lay-Up	ΔS, MPa	a_{init},[a] mm	a_{trans},[b] mm	Cycles from a_{init} to a_{trans}	a_{end},[c] mm	Cycles from a_{init} to a_{end}	End of Test
2L	$[0]_1$	392	3.1	4.6	81 500	7.3	99 000	fracture
3L	$[0]_4$	200	3.0	[d]	···	3.86	440 000	removed for polishing
4L	$[0]_4$	300	3.1	4.9	121 000	7.14	177 000	removed for polishing
5L	$[0]_4$	300	3.0	4.9	214 000	9.3	240 000	fracture
6L	$[0]_4$	300	3.3	[d]	···	3.93	84 000	removed for polishing
10C	$[0/90]_s$	150	3.1	[d]	···	5.63	350 000	removed for polishing
12C	$[0/90]_s$	150	3.1	5.7	384 000	6.8	576 000	fracture

[a] a_{init} is the initial crack length that includes the EDM notch plus a precrack of approximately 0.5 mm.
[b] a_{trans} is the crack length when crack growth begins to accelerate.
[c] a_{end} is the crack length at the end of the test.
[d] a_{trans} is not given for specimens that were removed for polishing before crack growth accelerated.

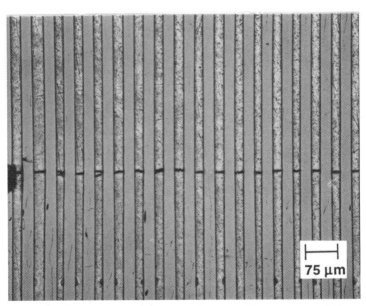

FIG. 3—*Sections polished through the outer ply of (a) a [0]₄ specimen subject to ΔS = 200 MPa (Specimen 3L) and (b) a [0/90]ₛ specimen subject to ΔS = 150 MPa (Specimen 10C). The fibers are intact along both matrix fatigue cracks.*

FIG. 4—*Section polished through the outer ply of a $[0]_4$ specimen subject to $\Delta S = 300$ MPa (Specimen 6L). Note the two broken fibers among the intact, bridging fibers.*

crack length. The ΔK_m values from the model and the corresponding measured composite crack growth rates were compared to the closure-corrected, effective matrix crack growth rate relationship, $da/dN - \Delta K_{eff}$. The τ values that provided the best correlation between the matrix and composite data were chosen. The τ values listed in Table 3 are used for all of the subsequent model predictions. For the purpose of determining τ values, cracks were assumed to be fully bridged and only composite crack growth rate data in the decreasing crack growth rate region was considered. Because the discrete composite model is directly applicable to only unidirectional composites, the predictions for the $[0/90]_s$ layup assumed that the transverse plies were matrix material.

The best-fit correlation is shown in Fig. 8. A perfect correlation would be if all of the composite data points were on the matrix $da/dN - \Delta K_{eff}$ curve. The TIMETAL 21S $da/dN - \Delta K_{eff}$ curve is for material that was processed in the same manner as the SCS-6/TIMETAL 21S, except no fibers were placed between the matrix foils [17]. The correlation was made between the composite and the matrix $da/dN - \Delta K_{eff}$ relationship because no crack closure was observed in the composite tests for the decreasing da/dN region (Fig. 5).

After determining the τ values, the bridging model was used to predict measured Elber gage displacement range profiles. So that the bridging geometry (locations of broken and unbroken fibers in the crack wake) would be known, comparisons were made to ΔCOD^* profiles measured immediately before removing specimens for polishing.

For the case of $[0]_4$ and $[0/90]_s$ specimens with fully bridged cracks (Fig. 3), the predictions are in reasonable agreement with the measurements (Fig. 9). Predictions were also performed for the crack shown in Fig. 4, Specimen 6L, with two broken and four unbroken fibers in the

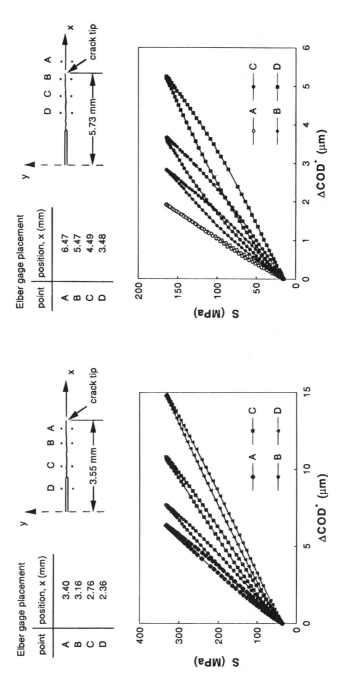

Fig. 5—*Typical applied stress versus Elber gage displacement hysteresis for bridged cracks in* (left) *a [0]₄ specimen and* (right) *a [0/90]ₛ specimen.*

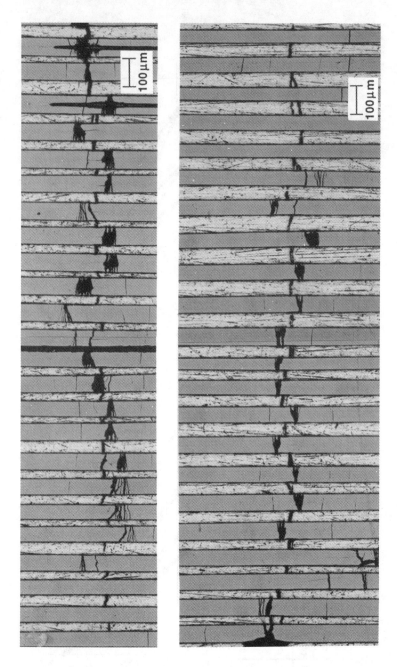

FIG. 6—*Section polished through the outer ply of a [0]₄ specimen subject to ΔS = 300 MPa with a crack length of 7.14 mm (Specimen 4L). This specimen was removed from the test machine after the crack growth accelerated. Note the broken and unbroken fibers along the crack (a) left and (b) right of the EDM notch.*

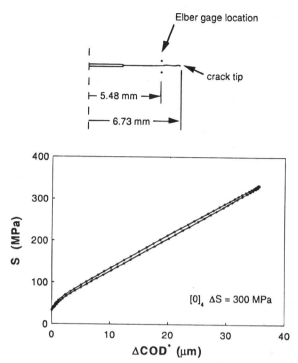

FIG. 7—*Applied stress versus Elber gage displacement hysteresis for a partially bridged crack in a [0]4 specimen.*

TABLE 3—*Fiber/matrix sliding stress, τ, values determined by fitting the bridging model and the SCS-6/TIMETAL 21S da/dN-ΔK results to the TIMETAL 21S da/dN-ΔK_{eff} relationship.*

Layup	ΔS, MPa	τ, MPa
[0]4	200	12
[0]4	300	17
[0]4	392	25
[0/90]s	150	12

crack wake. The prediction including the broken fibers is in excellent agreement with the experiment (Fig. 10).

The ΔCOD* profile for the crack in Specimen 4L (Fig. 5), the [0]4 specimen removed after *da/dN* had increased, was predicted for the cracks on both sides of the EDM notch (Fig. 11). Because the discrete composite model assumes symmetry about $x = 0$, the predictions were performed separately for the crack on each side of the EDM notch assuming symmetry about $x = 0$. This is the reason for the discontinuity in the predicted ΔCOD* at $x = 0$. The two fibers that failed approximately two fiber diameters from the crack plane (Fig. 5b) were considered to be bridging fibers in the analysis due to the significant load transfer that can occur over that distance. The model underpredicted ΔCOD* on the "left" side and near the crack tip on the "right" side.

The model was also used to predict the fiber stresses. Calculations were performed with and without the residual stresses that develop during composite processing due to the fiber/

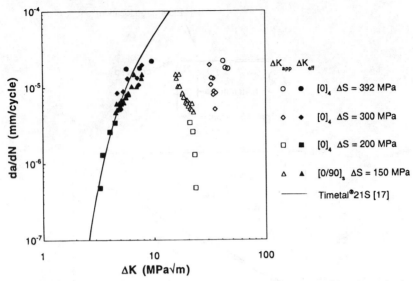

FIG. 8—*Correlation of matrix and composite crack growth rates resulting from the best-fit* τ *values used in the bridging model.*

matrix coefficient of thermal expansion mismatch. The analysis without the residual stresses assumes that the matrix residual stress near the crack tip is relieved due to the fatigue cycling. The matrix residual stress in the fiber direction, σ_m^{res}, was calculated using a one-dimensional two-bar model

$$\sigma_m^{res} = \frac{V_f E_f E_m}{E_c} (\alpha_f - \alpha_m)\delta T \qquad (7)$$

where α_f and α_m are the fiber and matrix coefficients of thermal expansion, respectively, and ΔT is the temperature change. For SCS-6/TIMETAL 21S and SCS-6/Ti-15-3, this simple approach results in residual stress predictions within 3% of those obtained using a thermoviscoplastic analysis [18]. It was assumed that all residual stresses were relieved during the age heat treatment so that $\Delta T = -600°C$. For the $[0]_4$, $\sigma_m^{res} = 223$ MPa.

Because it is reasonable to assume that failure of bridging fibers is the cause of the increase in the composite da/dN, fiber stresses were predicted for the crack length at which da/dN began to increase, a_{trans}. Complete bridging was assumed in the model. The predicted fiber stress for the most highly stressed fiber, the bridged fiber closest to the EDM notch, is significantly below the mean fiber strength, σ_f^{ult}, of 3665 MPa [19] for all of the applied stress levels (Table 4).

Discussion

The Elber gage displacements provided insight into the mechanics of crack propagation. The S-COD* curve for fully bridged cracks displayed hysteresis over the entire crack length. This behavior qualitatively supports the fatigue-bridging process formulated by Marshall and Oliver [16] and applied to the fatigue of bridged cracks by McMeeking and Evans [13]. The excellent agreement between the experimental and predicted ΔCOD* profiles for both fully

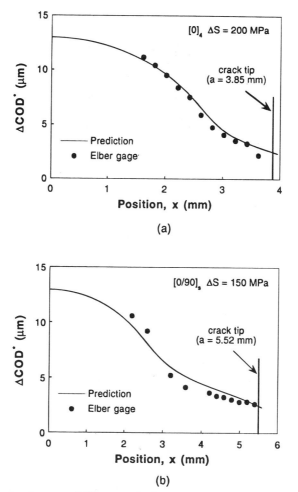

FIG. 9—*Predicted and measured Elber gage displacement ranges for (a) a [0]₄ specimen subject to* $\Delta S = 200$ *MPa (Specimen 3L) and (b) a [0/90]ₛ specimen subject to* $\Delta S = 150$ *MPa (Specimen 10C).*

bridged and partially bridged cracks demonstrates that the discrete composite model with the McMeeking and Evans bridging relationship for fatigue reasonably models the mechanics of fatigue crack propagation in SCS-6/TIMETAL 21S.

After the crack growth rate began to increase, the S-COD* relationship showed some hysteresis and evidence of crack closure (Fig. 7). The crack closure is likely due to broken fibers contacting each other. Such behavior was viewed during cycling with the video camera. The hysteresis is likely due to fiber/matrix sliding due to fiber pullout and partial crack bridging.

The method for determining τ that was employed in this paper was based on the observation that the S-COD* hysteresis loops did not indicate the presence of crack closure while crack growth decelerated. To verify that the bridging model predicts that a fully bridged matrix crack is open for the entire load cycle, the matrix stress intensity ratio, R, was calculated. R is defined by

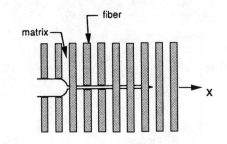

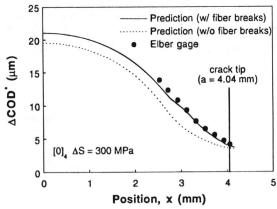

FIG. 10—*Predicted and measured Elber gage displacement ranges for a [0]₄ specimen subject to ΔS = 300 MPa with broken and bridging fibers in the crack wake (Specimen 6L).*

$$R = 1 - \frac{\Delta K_m}{K_m} \tag{8}$$

After the precrack ($a > 3$ mm), R was greater than 0.5 when the residual stresses were included in the analysis and near or above 0.5 when the residual stresses were neglected. An example of R as a function of crack length is shown in Fig. 12 for a $[0]_4$ specimen with $\Delta S = 300$ MPa and $R = 0.1$. Crack closure for the matrix material was observed for $R = 0.1$ but not at $R = 0.5$ [*17*]. Therefore, the model predicts that the crack was fully open as was observed experimentally with and without the residual stresses.

For an applied stress level, the fiber/matrix sliding stress, τ, was considered constant. The significant variation of τ with the applied stress level (Table 3) may be in part due to the number of fatigue cycles applied to each specimen. Specimens at lower applied stresses are subject to more fatigue cycles because of the lower crack growth rates. The additional fatigue cycles may wear the fiber/matrix interface and thus reduce the effective value of τ. While the pragmatic method used here to determine τ reduces it to a curve-fitting parameter, the excellent agreement between the model and measured ΔCOD* profiles suggests that an average sliding stress can reasonably model the mechanics of a bridged crack.

A limitation to using the discrete composite model in the fatigue crack growth analysis of a composite laminate is that the model considers only one ply. The ΔCOD* predictions with fiber failures (Figs. 9 and 10) assume that the same bridging geometry exists for all of the other composite plies. In practice, this will not be the case. However, since the Elber gage

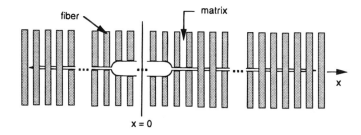

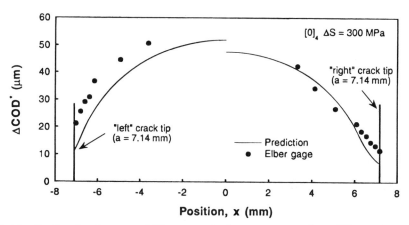

FIG. 11—*Predicted and measured Elber gage displacement ranges for a [0]₄ specimen subject to ΔS = 300 MPa with broken and bridging fibers in the crack wake (Specimen 4L). The predicted ΔCOD* is discontinuous at x = 0 because the prediction for each side of the EDM notch assumes symmetry about x = 0.*

TABLE 4—*Predicted maximum fiber stresses at a_{trans}. Complete bridging was assumed.*

Layup	ΔS, MPa	a_{trans},[a] mm	$\sigma_f^{max}/\overline{\sigma}_f^{ult}$ [b,c] With σ_m^{res}	Without σ_m^{res}
[0]₄	300	4.9	0.44	0.58
[0]₄	392	4.6	0.64	0.78
[0/90]ₛ	150	5.7	0.21	0.39

[a] a_{trans} is the crack length when crack growth begins to accelerate.
[b] σ_f^{max} is the maximum fiber stress for the most highly stressed bridging fiber.
[c] $\overline{\sigma}_f^{ult}$ is the mean fiber strength. $\overline{\sigma}_f^{ult} = 3665$ MPa [19].

measured ΔCOD* at the surface of the ply containing the broken fibers, the monolayer prediction should compare reasonably well to the measured ΔCOD* profiles. The model would be improved significantly if extended to include multiple composite plies.

The initial reduction in the crack growth rate for the matrix fatigue crack in the composite compared to the unreinforced matrix (Fig. 2) is due to fiber bridging (Fig. 3). The increase in crack growth rate is due to a reduction in the number of bridging fibers in the crack wake

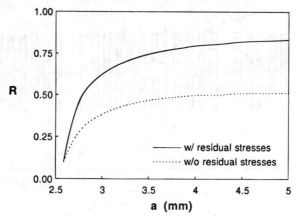

FIG. 12—*Prediction of the matrix stress intensity factor ratio, R, with and without the matrix residual stress for a [0]₄ specimen subject to* $\Delta S = 300$ *MPa and R = 0.1.*

(Fig. 6). Not all of the fibers in the crack wake must be bridging the crack to provide significant shielding of the matrix crack tip and decelerate crack growth (Fig. 4). After fatigue crack growth begins to accelerate, the composite still has more fatigue crack growth resistance than the matrix material (Fig. 2). One mechanism for this result is that the fatigue crack can be partially bridged (Fig. 6). Studies of other titanium-matrix composites [20,21] concluded that once the most highly stressed fiber failed, usually the fiber closest to the notch or hole, all of the other bridging fibers failed as well. For SCS-6/TIMETAL 21S, this is not the case. Rather, fibers may fail among intact, bridging fibers and partial bridging of several fibers near the crack tip may occur after a number of fibers close to the unbridged notch have failed.

Predictions made by Walls et al. [20] and John et al. [21] for fiber failure in SCS-6/Ti-15-3 and SCS-6/Ti-24-11 TMCs were in agreement with experimental observations. These analyses used a bridging relationship similar to Eq 4 but the residual stress term was neglected. In this study, the fiber stresses that were predicted at the crack length that the most highly stressed bridging fiber was assumed to fail were significantly lower than the reported mean strength of SCS-6 (Table 4). Similar errors in fiber strength predictions for SCS-6/TIMETAL 21S were reported by Larsen et al. [17] using a model similar to that used in [20,21]. The SCS-6 fiber has a significant variability in fiber strength [19] but this strength variability is not the cause of the difference between predictions and experimental observation since similar discrepancies did not occur for SCS-6/Ti-15-3 and SCS-6/Ti-24-11 [20,21].

Because partial fiber bridging can significantly effect ΔK_m and thus da/dN, predicting fiber failures is critical to SCS-6/TIMETAL 21S fatigue life prediction. Based on the ΔCOD^* results, the discrete composite model with the fatigue bridging relationships of McMeeking and Evans [13] accurately models the mechanics of a partially or fully bridged matrix fatigue crack. However, the locations of the bridging and broken fibers in the crack wake must be known a priori to perform the analysis for the fatigue case. To predict the total life of an SCS-6/TIMETAL 21S panel, the fiber failures must be predicted. The discrete composite model with the bridging relationships of Cox and Marshall [15] did not accurately predict these failures. Further study of fiber failure in SCS-6/TIMETAL 21S is required to develop the predictive capability required for total fatigue life prediction.

Evaluation of the fracture surfaces of the [0/90]ₛ specimen with a scanning electron microscope showed significant fatigue cracking in the transverse plies ahead of the main fatigue crack tip [22]. These cracks initiate at the transverse fibers causing additional damage not accounted for by the bridging model. This additional matrix cracking contributes to the error in the [0/90]$_s$ specimen fiber stress predictions.

Concluding Remarks

Constant amplitude fatigue tests were performed on [0]$_4$ and [0/90]$_s$ SCS-6/TIMETAL 21S titanium-matrix composite specimens to study fiber bridging of matrix fatigue cracks. Specimens were removed prior to failure and were mechanically polished to the first layer of fibers to determine the extent of fiber bridging. An Elber gage was used to measure displacements near the crack surface for comparison to predictions made using the discrete composite model. The following conclusions were drawn from this investigation:

1. Broken fibers were observed among intact, bridging fibers in the wake of the matrix fatigue crack. Partial fiber bridging was observed for specimens in the regions of decelerating and accelerating composite crack growth. To model this behavior, the failure of discrete fibers must be taken into account.
2. Measured applied stress versus Elber gage displacement (COD*) hysteresis loops support the shear-lag frictional shear stress concept for bridged cracks subject to fatigue loading developed by Marshall and Oliver [16].
3. Using the shear-lag bridging law with τ determined by a best-fit of crack growth rate data, ΔCOD* profiles predicted by the discrete composite model were in good agreement with the experimental results for fully and partially bridged cracks.
4. Fiber stresses predicted by the discrete composite model at initial fiber failure were significantly lower than the reported tensile strength of the SCS-6 fiber. Additional investigation is required before fiber-bridging models can be used to predict the failure of fibers bridging a SCS-6/TIMETAL 21S matrix fatigue crack.

Acknowledgments

Support and SCS-6/TIMETAL 21S material was provided by NASA Langley Research Center under Grant NAG-1-1316 directed by Dr. W. Steven Johnson.

APPENDIX I
Discrete Composite Model

The discrete composite model is described in detail in Ref *10*. A brief overview is presented here. Consider a unidirectional composite monolayer loaded in the fiber direction by a uniform tensile stress, S_{max}, with a crack perpendicular to the fibers (Fig. 13). Performing a force balance on a differential cell in the fiber (y) direction (Fig. 14) under the assumption of plane stress results in

$$\frac{d\sigma_{yy}^n}{dy} + \frac{1}{b}(\tau_n - \tau_{n-1}) = 0 \qquad (9)$$

where n is the cell index, b is the width of the cell, σ_{yy} is the composite normal stress in the y-direction, and τ is the shear stress on the cell. This force balance assumes that the strain in

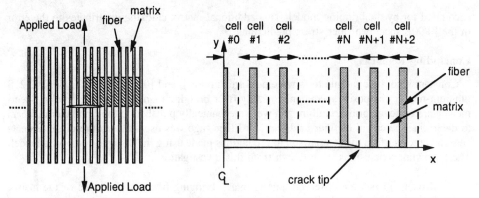

FIG. 13—*Composite monolayer and the definition of a cell in the discrete composite model.*

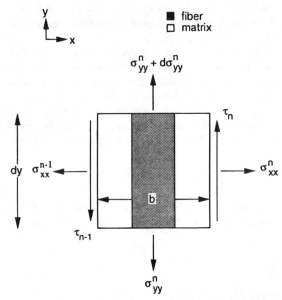

FIG. 14—*Free body diagram of a composite cell.*

the fiber direction remains constant over the composite cell for a given position, y. Therefore, the fiber and matrix stresses in the fiber direction can be related for a given cell by

$$\epsilon_{yy}^n = \frac{\sigma_{yy}^n}{E_c} = \frac{(\sigma_{yy}^n)_f}{E_f} = \frac{(\sigma_{yy}^n)_m}{E_m} \tag{10}$$

where ϵ_{yy} is the cell (composite) strain in the y-direction, and $(\sigma_{yy})_f$ and $(\sigma_{yy})_m$ are the normal fiber stresses in the y-direction of the fiber and matrix. E_c, E_f, and E_m are the elastic moduli of the composite, fiber, and matrix, respectively. The equilibrium equation, Eq 9, can be simplified using the elastic stress-strain-displacement relationships

$$\sigma_{yy}^n = E_c \epsilon_{yy}^n = E_c \frac{dv_n}{dy} \tag{11}$$

$$\tau_n = H(v_{n+1} - v_n) \qquad \text{where} \quad H = \frac{G_m}{b} \tag{12}$$

where v_n is the displacement of cell n in the y-direction, H is the shear lag material constant, and G_m is the shear modulus of the matrix. Combining and nondimensionalizing [23]

$$\frac{d^2V_n}{d\eta^2} + (V_{n+1} - 2V_n + V_{n-1}) = 0 \tag{13}$$

where

$$y = \phi\eta \quad \text{and} \quad \phi = \left[\frac{bE_c}{H}\right]^{1/2}$$

$$v_n = \psi V_n \quad \text{and} \quad \psi = S_{\max}\left[\frac{b}{HE_c}\right]^{1/2} \tag{14}$$

where η is the dimensionless coordinate in the y-direction, ϕ is the characteristic length constant, V_n is the dimensionless displacement in the y-direction, and ψ is the characteristic displacement constant.

Equation 13 can be solved using an even-valued transform [23] and the appropriate boundary conditions. Leaving the details to Ref 10, the result is

$$[D]\{A\} = \{-1\} + \{\bar{p}(A)\} \tag{15}$$

where $\bar{p}$ is the dimensionless bridging pressure, $\bar{p} = p/S_{\max}$, and

$$A_0 = \frac{V_0(0)}{2}$$

$$A_m = V_m(0) \qquad \text{for} \quad m \neq 0 \tag{16}$$

The elements of $[D]$ are given by

$$D_{mn} = \frac{1}{\pi}\left\{\frac{1}{m-n}\left[\frac{1}{2m-2n-1} + \frac{1}{2m-2n+1}\right]\right.$$

$$\left. + \frac{1}{m+n}\left[\frac{1}{2m+2n-1} + \frac{1}{2m+2n+1}\right]\right\} \qquad \text{if} \quad m \neq n \tag{17}$$

$$D_{nn} = -\frac{2}{\pi}\left[2 - \frac{1}{4n-1} + \frac{1}{4n+1}\right]$$

for an infinite plate and by

$$D_{mn} = D_{mn}^{\infty} + \sum_{j=1}^{J} [D_{mk+}^{\infty} + D_{mk-}^{\infty}] \tag{18}$$

for a finite width plate where

$$D_{mn}^{\infty} = D_{mn} \qquad \text{value for an infinite plate (Eq 17)}$$

$$k^{+} = jN_w + n$$

$$k^{-} = jN_w - n$$

N_w is the number of cells between crack centerlines (number of cells in the specimen). D_{mk+}^{∞} and D_{mk-}^{∞} values are calculated from Eq 17. J was chosen to be equal to 100 [10]. $\{A\}$ can be found from Eq 15 for any fiber-bridging relationship that defines $\{\bar{p}\}$ as a function of $\{A\}$.

Maximum Loading

The fiber-bridging relationship given by Cox and Marshall [15] that includes residual stresses, Eq 4, can be written in terms of the dimensionless variables as

$$\bar{p}_n = \frac{1}{S_{max}} \left[\frac{\psi A_n}{\lambda} \right]^{1/2} - \frac{\sigma_R}{S_{max}} \qquad n \neq 0$$

$$\bar{p}_0 = \frac{1}{S_{max}} \left[\frac{\psi A_0}{2\lambda} \right]^{1/2} - \frac{\sigma_R}{S_{max}} \qquad n = 0 \tag{19}$$

Equation 19 was used if cell n was bridged. If fiber n in the crack wake is broken

$$\bar{p}_n = 0 \tag{20}$$

Based on the bridging geometry (EDM notch size, intact and broken fibers in the crack wake), the $\bar{p}$ values given by Eqs 19 and 20 are substituted into Eq 15. Equation 15 can be solved for $\{A\}$ by iteration and the bridging pressures calculated for the intact cells by Eq 19. The stress for fiber n, σ_f^n, is given by

$$\sigma_f^n = \frac{S_{max}}{V_f} \bar{p}_n \tag{21}$$

Fatigue Loading

Equation 15 can be written for fatigue in terms of the dimensionless displacement range, ΔA, and the dimensionless bridging pressure range, $\Delta \bar{p}$, as

$$[D]\{\Delta A\} = \{-1\} + \{\Delta \bar{p}(\Delta A)\} \tag{22}$$

where

$$\Delta A_n = \frac{\Delta COD_n}{2\Delta\psi} \qquad n \neq 0$$

$$\Delta A_0 = \frac{\Delta COD}{4\Delta\psi} \qquad n = 0 \tag{23}$$

$$\Delta\bar{p}_n = \frac{\Delta p_n}{\Delta S} \tag{24}$$

$$\Delta\psi = \psi\frac{\Delta S}{S_{max}} \tag{25}$$

where ΔCOD_n is the crack opening displacement range for cell n, Δp_n is the bridging pressure range for cell n, and ΔS is the remote stress range.

The bridging relationship of McMeeking and Evans [13] can be written in terms of the dimensionless variables as

$$\Delta\bar{p}_n = \frac{1}{\sigma_\infty}\left[\frac{2\psi\Delta A_n}{\lambda}\right]^{1/2} \qquad n \neq 0$$

$$\Delta\bar{p}_0 = \frac{1}{\sigma_\infty}\left[\frac{\psi\Delta A_0}{\lambda}\right]^{1/2} \qquad n = 0 \tag{26}$$

As before, if fiber n in the crack wake is broken

$$\Delta\bar{p}_n = 0 \tag{27}$$

Based on the bridging geometry, the $\Delta\bar{p}_n$ values given by Eqs 26 and 27 are substituted into Eq 22. Equation 22 can be solved for $\{\Delta A\}$ by iteration and $\{\Delta\bar{p}\}$ calculated for the intact cells by Eq 26. $\{\Delta\bar{p}\}$ can then be used to determine the stress intensity factor range by Eq 2 where

$$\Delta p(x) = \Delta S\Delta\bar{p}_n \qquad \text{for} \quad n = \text{Int}\left[\frac{x}{b} + \frac{1}{2}\right] \tag{28}$$

APPENDIX II

Prediction of the Elber Gage Displacements

The discrete composite model directly calculates the displacements on the crack surface. However, the change in displacement between two specific points must be calculated to compare the model results to the Elber gage measured displacements. The discrete composite model is capable of determining the displacement or displacement range in the fiber direction at any point in the composite panel. From Refs 23 and 24, the dimensionless displacement range in the fiber direction for cell n a dimensionless distance η from the crack surface, $\Delta V_n(\eta)$, is

$$\Delta V_n(\eta) = \frac{2}{\pi} \int_0^\pi e^{-\delta(\theta)\eta} \sum_{m=0}^N \Delta A_m \cos(m\theta)\cos(n\theta) \ d\theta + \eta \tag{29}$$

where

$$\delta(\theta) = 2 \sin\left[\frac{\theta}{2}\right] \tag{30}$$

and the ΔA_m have been determined by solving Eq 22 with respect to Eqs 26 and 27. Equation 29 can be written in matrix form as

$$\{\Delta V(\eta)\} = [Q(\eta)]\{\Delta A\} + \{\eta\} \tag{31}$$

where

$$Q_{mn}(\eta) = \frac{2}{\pi} \int_0^\pi e^{-\delta(\theta)\eta}\cos(m\theta)\cos(n\theta) \ d\theta \tag{32}$$

The evaluation of Q_{mn} requires numerical integration. The dimensionless positions of the Elber gage points on cell n are defined as $\eta+$ and $\eta-$ and correspond to the locations above and below the crack that the Elber gage points were located, $(x, y+)$ and $(x, y-)$ (Fig. 1). The dimensionless and actual positions are related by

$$n = \text{Int}\left[\frac{x}{b} + \frac{1}{2}\right] \tag{33}$$

$$\eta+ = \frac{y+}{\phi} \quad \text{and} \quad \eta- = \frac{y-}{\phi} \tag{34}$$

where ϕ is defined in Eq 14, b is the width of a unit cell, and n is the cell index. The Elber gage displacement range, $\{\Delta COD^*\}$, is

$$\{\Delta COD^*\} = \Delta\psi[[Q(\eta+)] + [Q(\eta+)]]\{\Delta A\} + \Delta\psi\{\eta+\} + \Delta\psi\{\eta-\} \tag{35}$$

The predicted Elber gage displacement ranges determined from Eqs 34 and 35 were compared to the experimental measurements.

References

[1]   Johnson, W. S., "Damage Development in Titanium Metal Matrix Composites Subjected to Cyclic Loading," NASA Technical Memorandum 107597, National Aeronautics and Space Administration, Washington, DC, 1992.

[2]   Ibbotson, A. R., Beevers, C. J., and Bowen, P., "Stable and Unstable Crack Growth Transitions Under Cyclic Loading in a Continuous Fibre Reinforced Composite," *Scripta Metallurgica et Materialia*, Vol. 25, 1991, pp. 1781–1786.

[3]   Walls, D., Bao, G., and Zok, F., "Effects of Fiber Failure on Fatigue Cracking in a Ti/SiC Composite," *Scripta Metallurgica et Materialia*, Vol. 25, 1991, pp. 911–916.

[4]   Ghosn, L., Kantzos, P., and Telesman, J., "Modeling of Crack Bridging in a Unidirectional Metal Matrix Composite," *International Journal of Fracture*, Vol. 54, 1992, pp. 345–357.

[5]   Hillberry, B. M. and Johnson, W. S., "Prediction of Matrix Fatigue Crack Initiation in Notched SCS-6/Ti-15-3 Metal Matrix Composites," *Journal of Composites Technology and Research*, Vol. 14, 1992, pp. 221–224.

[6]   Jeng, S. M., Alassoeur, P., and Yang, J. M., "Fracture Mechanisms of Fiber-Reinforced Titanium Alloy Matrix Composites V: Fatigue Crack Propagation," *Materials Science and Engineering*, Vol. 154, 1992, pp. 11–19.

[7]   Jira, J. R. and Larsen, J. M., "Crack Bridging Behavior in Unidirectional SCS-6/Ti-24A1-11Nb Composite," *FATIGUE 93*, Vol. 2, J. P. Bailon and I. J. Dickson, Eds., Engineering Materials Advisory Services, Ltd., UK, 1993, pp. 1085–1090.

[8]   Bao, G. and McMeeking, R. M., "Fatigue Crack Growth in Fiber-Reinforced Metal-Matrix Composites," *Acta Metallurgica et Materialia*, Vol. 42, 1994, pp. 2415–2425.

[9]   Bowen, P., "Characterization of Crack Growth Resistance Under Cyclic Loading in the Presence of an Unbridged Defect in Fibre Reinforced Titanium Metal Matrix Composites," *Life Prediction Methodology for Titanium Matrix Composites, ASTM STP 1253*, W. S. Johnson, J. M. Larsen, and B. N. Cox, Eds., American Society for Testing and Materials, 1996, pp. 461–479.

[10]   Herrmann, D. J. and Hillberry, B. M., "A New Approach to the Analysis of Unidirectional Titanium Matrix Composites with Bridged and Unbridged Cracks," *Engineering Fracture Mechanics*, accepted for publication.

[11]   Elber, W., "Fatigue Crack Closure Under Cyclic Tension," *Engineering Fracture Mechanics*, Vol. 2, 1970, pp. 37–45.

[12]   Marshall, D. B., Cox, B. N., and Evans, A. G., "The Mechanics of Matrix Cracking in Brittle-Matrix Fiber Composites," *Acta Metallurgica et Materialia*, Vol. 33, 1985, pp. 2013–2021.

[13]   McMeeking, R. M. and Evans, A. G., "Matrix Fatigue Cracking in Fiber Composites," *Mechanics of Materials*, Vol. 9, 1990, pp. 217–227.

[14]   McCartney, L. N., "Mechanics of Matrix Cracking in Brittle-Matrix Fibre-Reinforced Composites," *Proceedings*, Royal Society of London, Vol. A 409, 1987, pp. 329–350.

[15]   Cox, B. N. and Marshall, D. B., "Crack Bridging in the Fatigue of Fiberous Composites," *Fatigue and Fracture of Engineering Materials and Structures*, Vol. 14, 1991, pp. 847–861.

[16]   Marshall, D. B. and Oliver, W. C., "Measurement of Interfacial Mechanical Properties in Fiber-Reinforced Ceramic Composites," *Journal*, American Ceramic Society, Vol. 70, No. 8, 1987, pp. 542–548.

[17]   Larsen, J. M., Jira, J. R., John, R., and Ashbaugh, N. E., "Crack Bridging Effects in Notch Fatigue of SCS-6/TIMETAL 21S Composite Laminates," *Life Prediction Methodology for Titanium Matrix Composites, ASTM STP 1253*, W. S. Johnson, J. M. Larsen, and B. N. Cox, Eds., American Society for Testing and Materials, 1996, pp. 114–136.

[18]   Bakuckas, J. G., Jr., and Johnson, W. S., "Implementation of Thermal Residual Stresses in the Analysis of Fiber Bridged Matrix Crack Growth in Titanium Matrix Composites," NASA Technical Memorandum 109082, National Aeronautics and Space Administration, Washington, DC, 1994.

[19]   Gambone, M. L. and Wawner, F. E., "The Effect of Elevated Temperature Exposure of Composites on the Strength Distribution of the Reinforcing Fibers," *Intermetallic Composites III*, Materials Research Society, Pittsburgh, 1994, pp. 111–118.

[20]   John, R., Jira, J. R., Larsen, J. M., and Ashbaugh, N. E., "Analysis of Bridged Fatigue Cracks in Unidirectional SCS-6/Ti-24A1-11Nb Composite," *FATIGUE 93*, Vol. 2, J. P. Bailon and I. J. Dickson, Eds., Engineering Materials Advisory Services, Ltd., UK, 1993, pp. 1091–1096.

[21]   Walls, D. P., Bao, G., and Zok, F. W., "Mode I Fatigue Cracking in a Fiber Reinforced Metal Matrix Composite," *Acta Metallurgica et Materialia*, Vol. 41, 1993, pp. 2061–2071.

[22]   Johnson, W. S., Miller, J. L., and Mirdamadi, M., "Fractographic Interpretation of Failure Mechanisms in Titanium Matrix Composites," *Proceedings*, TMS/ASM Symposium on Mechanisms and Mechanics of MMC Fatigue, 2–6 Oct. 1994, Rosemont, IL.

[23]   Goree, J. G. and Wolla, J. M., "Longitudinal Splitting in Unidirectional Composites Analysis and Experiments," NASA Contract Report 3881, National Aeronautics and Space Administration, Washington, DC, 1985.

[24]   Herrmann, D. J., "Fatigue Crack Growth in Unidirectional and Cross-Ply SCS-6/TIMETAL 21S Titanium Matrix Composite," Ph.D. thesis, Purdue University, West Lafayette, IN, Aug. 1994.

Keisuke Tanaka[1] and Hiroshi Tanaka[1]

Stress-Ratio Effect on Mode II Propagation of Interlaminar Fatigue Cracks in Graphite/Epoxy Composites

REFERENCE: Tanaka, K. and Tanaka, H., "**Stress-Ratio Effect on Mode II Propagation of Interlaminar Fatigue Cracks in Graphite/Epoxy Composites,**" *Composite Materials: Fatigue and Fracture (Sixth Volume), ASTM STP 1285*, E. A. Armanios, Ed., American Society for Testing and Materials, 1997, pp. 126–142.

ABSTRACT: The effect of the stress ratio on the propagation behavior of Mode II interlaminar fatigue cracks was studied with unidirectional graphite/epoxy laminates, Toray T800H/#3631. End-notched flexure (ENF) specimens were used for fatigue tests under the stress ratios of $R = 0.2$, 0.5, and 0.6; and end-loaded split (ELS) specimens were used for tests under $R = -1.0$, -0.5, and 0.2. For each stress ratio, the crack propagation rate was given by a power function of the stress intensity range, ΔK_{II}, in the region of rates above 10^{-9} m/cycle. Below this region, there exists the threshold for fatigue crack propagation. The threshold condition is given by a constant value of the stress intensity range, $\Delta K_{IIth} = 1.8$ MPa$\sqrt{m}$. The crack propagation rate is determined by ΔK_{II} near the threshold, while by the maximum stress intensity factor, K_{IImax}, at high rates. A fracture mechanics equation is proposed for predicting the propagation rate of Mode II fatigue cracks under various stress ratios. The effect of the stress ratio on the micromechanisms of Mode II fatigue crack propagation was discussed on the basis of the microscopic observations of fracture surfaces and near-crack-tip regions.

KEYWORDS: fatigue (materials), composite materials, graphite/epoxy, delamination, crack propagation rate, fracture mechanics, stress ratio, Mode II, end-notched flexure, end-loaded split, fracture (materials)

Graphite/epoxy composites have been used for primary members of weight critical structures. Since interlaminar fracture, or delamination, represents one of the most prevalent life-limiting failure modes, the assessment of the propagation behavior of interlaminar cracks is inevitable to establish the damage tolerance criteria for composite structures.

Interlaminar crack propagation usually takes place under mixed-mode loading. Although most previous attention has been devoted to Mode I cracks propagating under static and cyclic loadings [1–4], not much work has been conducted on Mode II propagation of interlaminar cracks. In particular, the effect of the stress ratio on Mode II propagation of interlaminar cracks under cyclic loading is not well understood.

Russell and Street [5] conducted the tests of delamination propagation under Mode II shear loading at stress ratios of $R = -1$ and 0 with four kinds of graphite fiber composites. They correlated the crack propagation rate to the maximum value of the strain energy release rate, G_{IImax}. In all four materials, the propagation rate was higher at $R = -1$ than that at $R = 0$ when compared at the same G_{IImax} value. Martin and Murri [6] reported similar results for the

---

[1]Professor and research associate, respectively, Department of Mechanical Engineering, Nagoya University, Furo-cho, Chikusa-ku, Nagoya 464-01, Japan.

Mode II propagation of delamination in AS4/PEEK composites; the propagation rate was higher at $R = 0.1$ than that at $R = 0.5$. These results suggest that the load amplitude as well as the maximum load (or G_{IImax}) plays a significant role in interlaminar crack propagation. Other works on Mode II fatigue crack propagation dealt with only one stress ratio in their tests to evaluate the effects of the matrix toughness and fiber strength on the composite resistance to Mode II delamination [7–12]. It is very important for both damage tolerance and material evaluation purposes to establish a fracture mechanics parameter that controls the Mode II propagation rate under various stress ratios.

In the present paper, the effect of the cyclic stress ratio on the propagation behavior of Mode II interlaminar fatigue cracks was studied with unidirectional graphite/epoxy laminates, Toray T800H/#3631. End-notched flexure (ENF) specimens were used for fatigue tests under the stress ratios of 0.2, 0.5, and 0.6; and end-loaded split (ELS) specimens were used for tests under the stress ratios of -1.0, -0.5, and 0.2. A fracture mechanics equation is proposed for predicting the propagation rate of interlaminar cracks at different stress ratios under Mode II cyclic loading.

Experimental Procedure

Material

The material used in this study is unidirectional, graphite fiber epoxy matrix composite made of Toray T800H/#3631. The unidirectional laminate is made of 44 plies of prepregs and has a nominal thickness of 6.2 mm. The curing temperature was 180°C and the fiber volume fraction was 60.3%. Table 1 summarizes the elastic constants and the fracture toughness of the laminate used for fatigue tests. The Mode I interlaminar fracture toughness expressed in terms of the stress intensity factor, K_{Ic}, and the energy release rate, G_{Ic}, represents the value at the onset of interlaminar crack extension from a Kapton film of 12 μm in thickness inserted in double-cantilever-beam specimens [13]. Under Mode I loading, the fracture toughness increases with crack extension because of fiber bridging. On the other hand, the fracture toughness under Mode II loading measured by using ELS specimens is fairly constant during crack growth except at the onset of a crack from an insert film [14]. The values of K_{IIc} and G_{IIc} shown in Table 1 represent the constant value during crack growth.

Specimens

The ENF and ELS specimens were used for Mode II fatigue tests. Figure 1 depicts the ENF specimen. The specimen has a width of $B = 20$ mm and a height of $2h = 6.2$ mm. A Kapton film of 12 μm in thickness is inserted between the plies at the midplane of the specimen for delamination initiation. The total span is $2L = 120$ mm. The initial delamination length is $a_0 = 30$ mm.

TABLE 1—*Mechanical properties of T800H/#3631 graphite/epoxy laminate.*

					Interlaminar Fracture Toughness			
	Elastic Constants				Mode I		Mode II	
E_1, GPa	E_2, GPa	G_{12}, GPa	ν_{12}	ν_{23}	K_{Ic}, MPa$\sqrt{m}$	G_{Ic}, J/m^2	K_{IIc}, MPa$\sqrt{m}$	G_{IIc}, J/m^2
137	8.1	4.8	0.31	0.55	1.4	140	6.2	820

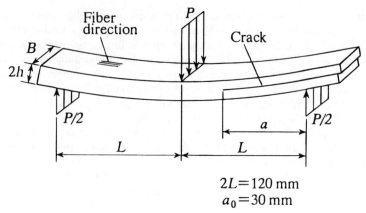

$$2L = 120 \text{ mm}$$
$$a_0 = 30 \text{ mm}$$

FIG. 1—*End-notched flexure specimen.*

The ELS specimen is illustrated in Fig. 2. The width is $B = 20$ mm and the height is $2h = 6.2$ mm. The distance between the fixed grip and the loading line is $L = 60$ mm. The initial delamination length is $a_0 = 35$ mm. To apply the cyclic load with negative stress ratios to the ELS specimen, a special loading device was developed. Figure 3 presents the experimental setup and its schema. The specimen is held between roller pins, and the pin holders swing in accordance with the deflection of the specimen. Thin copper foils are placed between the pin and the specimen to adjust the clearance.

Stress Intensity Factor

The stress intensity factor of the ENF and ELS specimens was calculated from the boundary element method (BEM) by using the Green's function for an infinite anisotropic plate containing a straight crack given by Snyder and Cruse [15]. The use of this special function makes it possible to remove the crack faces from the boundary element model of the cracked geometry. Figure 4 shows examples of the boundary element model. The isoparametric quadratic boundary

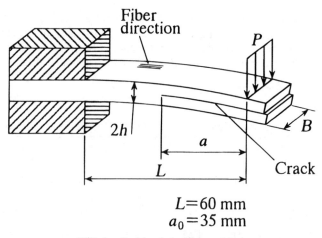

$$L = 60 \text{ mm}$$
$$a_0 = 35 \text{ mm}$$

FIG. 2—*End-loaded split specimen.*

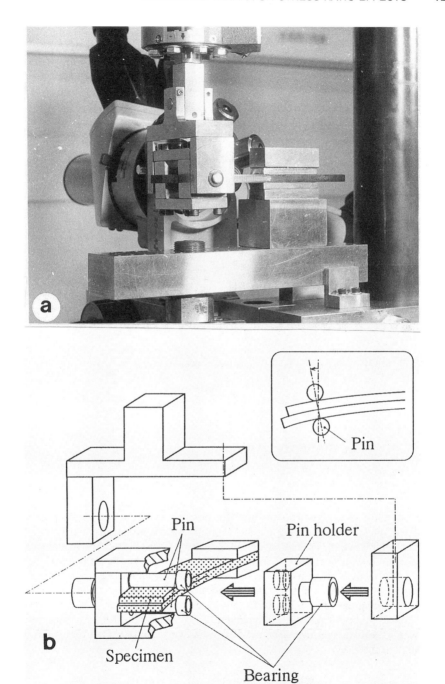

FIG. 3—*Loading device for ELS tests: (a) test setup and (b) schema of loading device.*

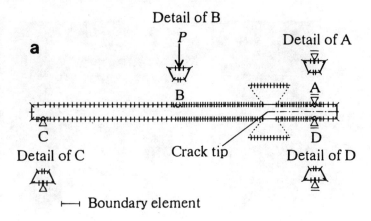

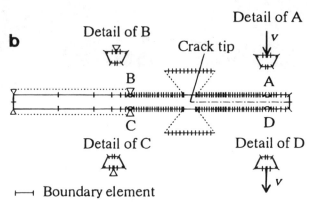

FIG. 4—*Boundary element model: (a) ENF specimen and (b) ELS specimen.*

element of 303 to 308 in number was used for discretization. As the boundary condition of the ENF specimen, the load, P, is applied at Point B, and the displacement at Point A as well as Points C and D is fixed as seen in Fig. 4a. For the ELS specimen, the same amount of the forced displacement is applied to Points A and D as in Fig. 4b. The problem of the contact between crack faces is circumvented by these boundary conditions. The trapezoidal regions containing Points A, B, C, and D are partitioned off from the cracked region to minimize the error in calculating the stress intensity factor.

Figure 5 shows the change of the compliance, C, of the load point displacement with crack length, a, for the ENF specimen. The solid line represents the following regression equation of the numerical values

$$E_1'BC = 2034 + 3.02(a/2h + 0.462)^3 \qquad (1)$$

where $E_1' = E_1/(1 - v_{12}^2 E_2/E_1)$. The experimental data obtained at crack lengths between 20 and 50 mm fall closely on the line. This agreement confirms the accuracy of the present BEM analysis.

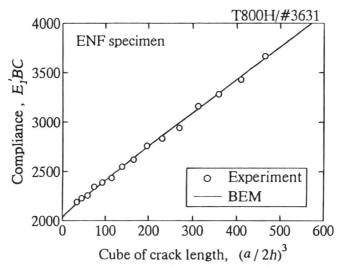

FIG. 5—*Comparison between experimental and numerical values of compliance (ENF specimen).*

The stress intensity factor, K_{II}, was computed from the path-independent integral proposed by Snyder and Cruse [15]. The K_{II} value is expressed as a function of crack length, a, as follows

$$K_{II} = \frac{P}{B\sqrt{2h}}\left(\alpha_1 \frac{a}{2h} + \alpha_0\right) \tag{2}$$

where P is the applied load, and B and $2h$ are the width and the height of the specimen, respectively. The constants in Eq 2 are $\alpha_1 = 1.24$, $\alpha_0 = 0.57$ for the ENF specimens, and $\alpha_1 = 2.49$, $\alpha_0 = 0.90$ for the ELS specimens.

The range of the stress intensity factor is defined by

$$\Delta K_{II} = K_{IImax} - K_{IImin} \tag{3}$$

where K_{IImax} and K_{IImin} are the stress intensity factors corresponding to the maximum and minimum loads, respectively. The value of K_{IImin} is negative for negative stress ratios.

The energy release rate, G_{II}, can be calculated from K_{II} as [16]

$$G_{II} = H_{II}K_{II}^2 \tag{4}$$

where

$$H_{II} = \frac{1 - \nu_{12}^2 E_2/E_1}{\sqrt{2E_1}}\left\{\sqrt{\frac{(1 - \nu_{23}^2)E_1}{(1 - \nu_{12}^2 E_2/E_1)E_2}} - \frac{2(1 + \nu_{23})\nu_{12} - E_1/G_{12}}{2(1 - \nu_{12}^2 E_2/E_1)}\right\}^{1/2} \tag{5}$$

For the present laminates, $H_{II} = 2.136 \times 10^{-2}$ GPa^{-1}. The range of the energy release rate is defined by

$$\Delta G_{\mathrm{II}} = G_{\mathrm{IImax}} - G_{\mathrm{IImin}} = \begin{cases} H_{\mathrm{II}}(K_{\mathrm{IImax}}^2 - K_{\mathrm{IImin}}^2) & \text{(for } R \geq 0) \\ H_{\mathrm{II}}K_{\mathrm{IImax}}^2 & \text{(for } R < 0) \end{cases} \tag{6}$$

that is, $\Delta G_{\mathrm{II}} = G_{\mathrm{IImax}}$ for $R \leq 0$.

Fatigue Testing

Fatigue tests were conducted in a servohydraulic fatigue testing machine in air at room temperature. The frequency of stress cycling was 10 Hz. The delamination propagation in ENF specimens was examined under either the ΔK_{II}-constant or ΔK_{II}-decreasing condition. For the ELS specimens, the ΔK_{II}-decreasing tests were conducted. The stress ratio, R, was 0.2, 0.5, and 0.6 for the ENF tests, and -1.0, -0.5, and 0.2 for the ELS tests.

For each ΔK_{II}-decreasing test, a precrack was grown from a starter film under Mode II cyclic loading, while for the ΔK_{II}-constant test, no precrack was created. The crack length was measured with a traveling microscope at a magnification of $\times 200$. White marking ink was painted on the specimen surfaces to detect the crack tip clearly.

Experimental Results and Discussion

Crack Propagation Behavior

Figure 6 shows the variation of the crack propagation rate, da/dN, with crack extension in the ΔK_{II}-constant test with $R = 0.5$. The Mode II interlaminar crack propagates at a nearly constant rate under a constant ΔK_{II}-value, while Mode I cracks decelerate even under a constant ΔK_{I}-value because of fiber bridging as shown in the figure with the dashed line. The effect of fiber bridging is negligible for Mode II crack propagation under cyclic loading as well as static loading [14,17], in contrast to Mode I cracks [3,13]. The crack propagation rate is determined by the current values of ΔK_{II} and K_{IImax}, independent of crack extension history.

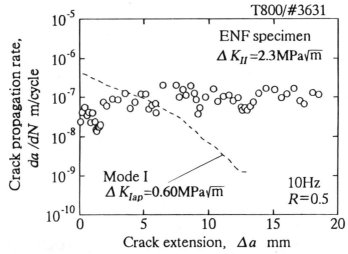

FIG. 6—*Change of crack propagation rate with crack extension under* ΔK_{II}-*constant condition.*

Figure 7 shows the da/dN value against ΔK_{II} for $R = 0.2$ obtained by the ENF and ELS tests. The relationship for the ENF specimen coincides well with that for the ELS specimen. This result confirms the independence of the da/dN-ΔK_{II} relationship of the specimen geometry.

The propagation rate of Mode II fatigue cracks obtained by ΔK_{II}-decreasing tests is plotted against ΔK_{II} in Fig. 8. For each R value, in the region of rates higher than 10^{-9} m/cycle, the relationship between da/dN and ΔK_{II} is approximated by the following power function

$$da/dN = D(\Delta K_{II})^n \qquad (7)$$

where D and n change with the R-values. The value of n increases with increasing R-value as indicated in the figure. The crack propagation rate is higher for larger R when compared at the same ΔK_{II} value.

Below $da/dN = 10^{-9}$ m/cycle, the crack propagation rate is lower than that given by Eq 7. There is a threshold value of the stress intensity range, ΔK_{IIth} for Mode II crack propagation. The ΔK_{IIth} value is equal to 1.8 MPa$\sqrt{m}$, irrespective of R-values. Since ΔK_{IIth} is constant at the threshold, the threshold value of K_{IImax}, G_{IImax}, or ΔG_{II} is not constant, but increases with increasing R-value.

The same data of da/dN are plotted against the maximum stress intensity factor, K_{IImax}, in Fig. 9. The crack propagation rate is higher for lower R values. The loading at negative K_{II}-values contributes to crack propagation as well as that at positive K_{II}-values. At higher rates, the difference in rates due to changing the R-value is reduced.

On the basis of the results presented in Figs. 8 and 9, it can be concluded that the crack propagation rate is controlled more or less by ΔK_{II} at low rates, while by K_{IImax} at high rates.

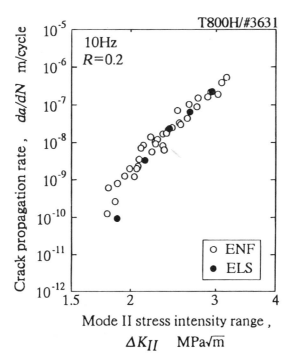

FIG. 7—*Relationships between crack propagation rate and stress intensity range obtained by ENF and ELS tests.*

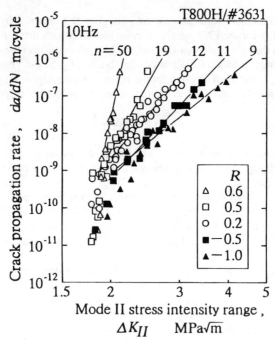

FIG. 8—*Relationship between crack propagation rate and stress intensity range.*

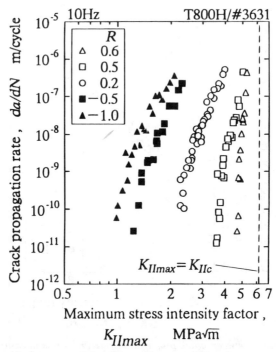

FIG. 9—*Relationship between crack propagation rate and maximum stress intensity factor.*

In Fig. 10, da/dN is plotted against the range of the energy release rate, ΔG_{II}. In comparison with the da/dN-K_{IImax} relationship shown in Fig. 9, the data for positive R-values come closer, while those for negative R-values come apart. This results from neglecting negative loading in the definition of ΔG_{II} by Eq 6.

Equation for Crack Propagation Rate

The effect of the cyclic stress ratio on Mode I propagation of interlaminar fatigue cracks has been studied by several investigators [4,6,7,18]. Hojo et al. [4,18] showed that the crack propagation rate in several kinds of graphite fiber composites was not a unique function of the range of the energy release rate, ΔG, at different R-values. They proposed the following fracture mechanics parameter that uniquely determined the crack propagation rate under Mode I cyclic loading

$$\Delta K_{eq} = \Delta K^{1-\gamma} K_{max}^{\gamma} = \Delta K(1 - R)^{-\gamma} \tag{8}$$

where γ indicates the contribution of K_{max} on Mode I crack propagation and takes a value between 0 and 1. For $\gamma = 1$, $\Delta K_{eq} = K_{max}$; da/dN is a unique function of K_{max} (or G_{max}). For $\gamma = 0$, $\Delta K_{eq} = \Delta K$; da/dN is determined by ΔK irrespective of R-values.

In the present case, the γ-value is not a material constant, but changes depending on the crack propagation rate. The ΔK_{II} value is plotted against $1 - R$ for each crack propagation

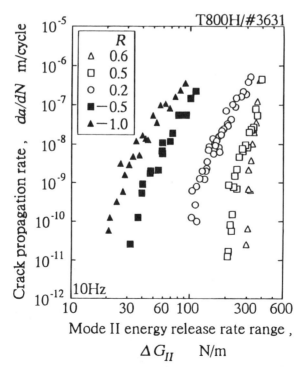

FIG. 10—*Relationship between crack propagation rate and energy release rate range.*

rate in Fig. 11, where the slope of the straight line gives the γ value for each rate. The γ-value increases with increasing crack propagation rate. Therefore, ΔK_{eq} is not applicable for the present case.

On the basis of the results presented in Figs. 8 and 9, a fracture mechanics equation for the crack propagation rate under various R-values is derived. Figure 12 depicts the relationship between da/dN and ΔK_{II}. It can be concluded that the threshold of crack propagation takes place at a constant growth rate, $da/dN = V_L$, as seen in Fig. 8. We have

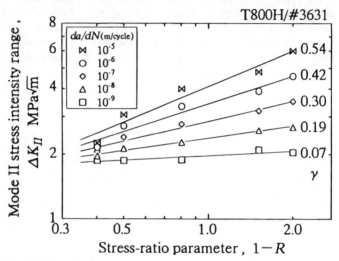

FIG. 11—*Relationship between stress intensity range and stress-ratio parameter.*

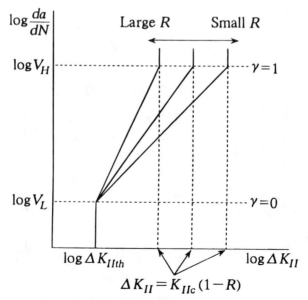

FIG. 12—*Schematic diagram of the relationship between crack propagation rate and stress intensity range.*

$$da/dN = V_L(\Delta K_{II}/\Delta K_{IIth})^n = V_L[(1 - R)K_{IImax}/\Delta K_{IIth}]^n \qquad (9)$$

On the other hand, the instability condition is given by $K_{IImax} = \Delta K_{II}/(1 - R) = K_{IIc}$. We assume that the instability takes place at a constant crack growth rate, V_H. Therefore

$$da/dN = V_H(K_{IImax}/K_{IIc})^n \qquad (10)$$

Since Eqs 9 and 10 are identical, the exponent, n, is given as a function of the stress ratio, R, as follows

$$n = \frac{\log(V_H/V_L)}{\log[(1 - R)K_{IIc}/\Delta K_{IIth}]} \qquad (11)$$

For the present case, we have the material constants, $V_L = 10^{-9}$ m/cycle, $V_H = 10^{-3}$ m/cycle, $\Delta K_{IIth} = 1.8$ MPa$\sqrt{m}$, and $K_{IIc} = 6.2$ MPa$\sqrt{m}$. The crack propagation rate predicted from the preceding equations is shown in Fig. 13 by the solid line for each R-value. The agreement between the experimental results and the predictions is very good in the wide range of the R-value between -1.0 and 0.6. Also, Eq 11 predicts that the instability takes place at once for the stress ratio above the critical value, R^*, where $R^* = 1 - \Delta K_{IIth}/K_{IIc} = 0.71$.

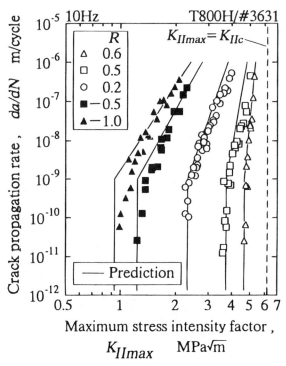

FIG. 13—*Comparison between predicted and measured values of crack propagation rate.*

Microscopic Observation

Figure 14 presents the scanning electron micrographs of interlaminar fatigue cracks propagating under $R = 0.2$ and -1.0. The arrow in each micrograph indicates the growth direction of the main crack. For positive R-values, a main crack is accompanied by small cracks in the matrix resin formed at approximately 45° to the plane of plies. These small cracks seem to be caused by the principal normal stress. The direction of the principal stress axis does not change during one stress cycling with positive R-values. On the other hand, for the case of $R = -1.0$, a main crack propagates along a zigzag path. For negative R-values, the direction of the principal stress axis changes during one stress cycling, forming X-shaped small cracks at the ±45° planes to the plies. The appearance of the main crack is very much different

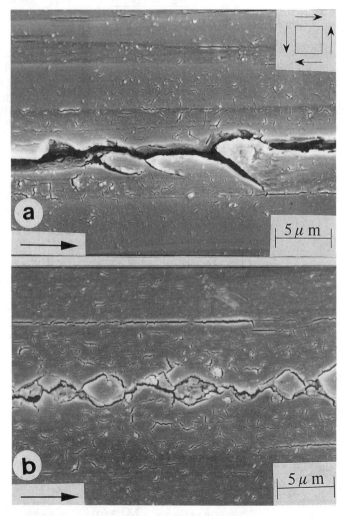

FIG. 14—*Scanning electron micrographs of Mode II fatigue crack:* (a) R = 0.2, da/dN = 1 × 10⁻⁸ m/cycle and (b) R = −1.0, da/dN = 6 × 10⁻⁹ m/cycle.

between positive and negative R-values, although the main crack propagates by connecting small cracks perpendicular to the principal stress axis in the matrix resin for both cases.

Figure 15 shows the scanning electron micrographs of the fatigue fracture surfaces for the cases of $R = 0.2$, -0.5, and -1.0, where Figs. a to c were taken near the threshold and d to f were taken at higher crack propagation rates. In Figs. b and f, the dotted line indicates the front shape of the Mode II fatigue crack observed on the fracture surface of the specimen that was first cracked by Mode II cyclic loading and subsequently broken by Mode I static loading.

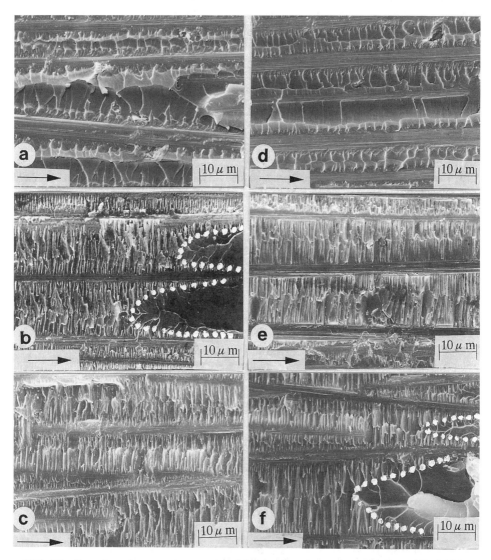

FIG. 15—*Scanning electron micrographs of fatigue fracture surface:* (a) R = 0.2, da/dN = 1×10^{-10} *m/cycle;* (b) R = -0.5, da/dN = 3×10^{-11} *m/cycle;* (c) R = -1.0, da/dN = 3×10^{-10} *m/cycle;* (d) R = 0.2, da/dN = 2×10^{-8} *m/cycle;* (e) R = -0.5, da/dN = 4×10^{-8} *m/cycle; and* (f) R = -1.0, da/dN = 4×10^{-8} *m/cycle.*

In comparison with Mode II static fracture surfaces [14], hackles are rubbed off for fatigue fracture surfaces. The appearance of the fracture surface is remarkably different between positive and negative R-values. The pitch of river markings on the matrix resin is much smaller for negative R-values than that for positive R-values.

The shape of the crack front shown in Figs. 15b and f suggests that the fracture of the resin starts near fibers and propagates toward the central region of the resin between fibers as illustrated in Fig. 16.

The appearance of the fracture surface does not change much with the crack propagation rate. This does not correspond to the change of the rate-controlling parameter from ΔK_{II} at low rates to K_{IImax} at high rates. Further study is necessary for micro- or meso-mechanisms of interlaminar fatigue crack propagation.

Conclusions

1. The propagation rate of Mode II interlaminar fatigue cracks is nearly constant when the stress intensity range, ΔK_{II}, is kept constant. The effect of fiber bridging is minimal for Mode II cracks.

2. For each stress ratio, R, the crack propagation rate, da/dN, is given by a power function of ΔK_{II} in the region of rates above 10^{-9} m/cycle. The exponent of the power function increases with increasing R-value.

3. The threshold condition of crack propagation is given by a constant value of the threshold stress intensity range, $\Delta K_{IIth} = 1.8$ MPa$\sqrt{m}$, for $R = -1.0$ to 0.6. The loading at negative K_{II}-values contributes to crack propagation as well as that at positive K_{II}-values.

4. The propagation rate of Mode II cracks is determined by ΔK_{II} near the threshold, while by the maximum stress intensity factor, K_{IImax}, at high rates.

5. The crack propagation rate da/dN for $R = -1.0$ to 0.6 is predicted by

$$da/dN = V_L(\Delta K_{II}/\Delta K_{IIth})^n = V_H(K_{IImax}/K_{IIc})^n$$

where

$$n = \frac{\log(V_H/V_L)}{\log[(1 - R)K_{IIc}/\Delta K_{IIth}]}$$

For the present material, $V_L = 10^{-9}$ m/cycle and $V_H = 10^{-3}$ m/cycle. The propagation rate given by the preceding equations agreed very well with the experimental results.

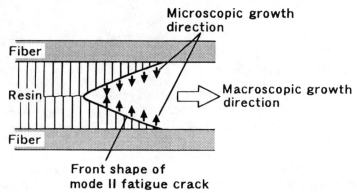

FIG. 16—*Directions of macroscopic and microscopic growth of Mode II fatigue crack.*

6. The appearance of the fatigue fracture surface was remarkably different between positive and negative R-values, showing smaller pitches of river markings on the resin fracture surfaces for negative R-values.

7. The fatigue fracture of the matrix resin starts near fibers and propagates toward the central region of the resin between fibers.

References

[1]   Wilkins, D. J., Eisenmann, J. R., Camin, R. A., Margolis, W. S., and Benson, R. A., "Characterizing Delamination Growth in Graphite-Epoxy," *Damage in Composite Materials, ASTM STP 775*, K. L. Reifsnider, Ed., American Society for Testing and Materials, Philadelphia, 1982, pp. 168–183.

[2]   Jordan, W. M., Bradley, W. L., and Moulton, R. J., "Relating Resin Mechanical Properties to Composite Delamination Fracture Toughness," *Journal of Composite Materials*, Vol. 23, 1989, pp. 923–943.

[3]   Russell, A. J. and Street, K. N., "A Constant ΔG Test for Measuring Mode I Interlaminar Fatigue Crack Growth Rates," *Composite Materials: Testing and Design, Eighth Conference, ASTM STP 972*, J. D. Whitcomb, Ed., American Society for Testing and Materials, Philadelphia, 1988, pp. 259–277.

[4]   Hojo, M., Tanaka, K., Gustafson, C.-G., and Hayashi, R., "Effect of Stress Ratio on Near-threshold Propagation of Delamination Fatigue Cracks in Unidirectional CFRP," *Composite Science and Technology*, Vol. 29, 1987, pp. 273–292.

[5]   Russell, A. J. and Street, K. N., "The Effect of Matrix Toughness on Delamination: Static and Fatigue Fracture under Mode II Shear Loading of Graphite Fiber Composites," *Toughened Composites, ASTM STP 937*, N. J. Johnston, Ed., American Society for Testing and Materials, Philadelphia, 1987, pp. 275–294.

[6]   Martin, R. H. and Murri, G. B., "Characterization of Mode I and Mode II Delamination Growth and Thresholds in AS4/PEEK Composites," *Composite Materials: Testing and Design, Ninth Volume, ASTM STP 1059*, S. P. Garbo, Ed., American Society for Testing and Materials, Philadelphia, 1990, pp. 251–270.

[7]   Bathias, C. and Laksimi, A., "Delamination Threshold and Loading Effect in Fiber Glass Epoxy Composite," *Delamination and Debonding of Materials, ASTM STP 876*, W. S. Johnson, Ed., American Society for Testing and Materials, Philadelphia, 1985, pp. 217–237.

[8]   Trethewey, B. R., Jr., Gillespie, J. W., Jr., and Carlsson, L. A., "Mode II Cyclic Delamination Growth," *Journal of Composite Materials*, Vol. 22, 1988, pp. 459–483.

[9]   O'Brien, T. K., Murri, G. B., and Salpekar, S. A., "Interlaminar Shear Fracture Toughness and Fatigue Threshold for Composite Materials," *Composite Materials: Fatigue and Fracture, Second Volume, ASTM STP 1012*, P. A. Lagace, Ed., American Society for Testing and Materials, Philadelphia, 1989, pp. 222–250.

[10] Prel, Y. J., Davies, P., Benzeggagh, M. L., and de Charentenay, F.-X., "Mode I and Mode II Delamination of Thermosetting and Thermoplastic Composites," *Composite Materials: Fatigue and Fracture, Second Volume, ASTM STP 1012*, P. A. Lagace, Ed., American Society for Testing and Materials, Philadelphia, 1989, pp. 251–261.

[11] Mall, S., Yun, K.-T., and Kochhar, N. K., "Characterization of Matrix Toughness on Cyclic Delamination Growth in Graphite Fiber Composites," *Composite Materials: Fatigue and Fracture, Second Volume, ASTM STP 1012*, P. A. Lagace, Ed., American Society for Testing and Materials, Philadelphia, 1989, pp. 296–310.

[12] Murri, G. B. and Martin, R. H., "Effect of Initial Delamination on Mode I and Mode II Interlaminar Fracture Toughness and Fatigue Fracture Threshold," *Composite Materials: Fatigue and Fracture, Fourth Volume, ASTM STP 1156*, W. W. Stinchcomb and N. E. Ashbaugh, Eds., American Society for Testing and Materials, Philadelphia, 1993, pp. 239–256.

[13] Tanaka, K., Tanaka, H., and Yamagishi, K., "Evaluation of Fiber-Bridging Effect on Mode I Interlaminar Fracture Toughness of Unidirectional CFRP Based on Measurements of Crack Opening Profiles," *Transactions*, Japan Society for Mechanical Engineers, Vol. 60, 1994, pp. 1176–1182.

[14] Tanaka, H., Tanaka, K., and Ikai, Y., "Propagation of Mode II Interlaminar Cracks in Carbon/Epoxy Laminates under Static and Cyclic Loadings," *Transactions*, Japan Society for Mechanical Engineers, Vol. 61, 1995, pp. 530–536.

[15] Snyder, M. D. and Cruse, T. A., "Boundary-Integral Equation Analysis of Cracked Anisotropic Plates," *International Journal of Fracture*, Vol. 11, 1975, pp. 315–328.

[16] Sih, G. C., Paris, P. C., and Irwin, G. R., "On Cracks in Rectilinearly Anisotropic Bodies," *International Journal of Fracture Mechanics*, Vol. 1, 1965, pp. 189–203.

[*17*]   Russell, A. J. and Street, K. N., "Moisture and Temperature Effects on the Mixed-Mode Delamination Fracture of Unidirectional Graphite/Epoxy," *Delamination and Debonding of Materials, ASTM STP 876*, W. S. Johnson, Ed., American Society for Testing and Materials, Philadelphia, 1985, pp. 349–370.
[*18*]   Hojo, M., Ochiai, S., Gustafson, C.-G., and Tanaka, K., "Effect of Matrix Resin on Delamination Fatigue Crack Growth in CFRP Laminates," *Engineering Fracture Mechanics*, Vol. 49, 1994, pp. 35–47.

Assimina A. Pelegri,[1] *George A. Kardomateas,*[1]
and Basharat U. Malik[1]

The Fatigue Growth of Internal Delaminations Under Compressive Loading of Cross-Ply Composite Plates

REFERENCE: Pelegri, A. A., Kardomateas, G. A., and Malik, B. U., **"The Fatigue Growth of Internal Delaminations Under Compressive Loading of Cross-Ply Composite Plates,"** *Composite Materials: Fatigue and Fracture (Sixth Volume), ASTM STP 1285*, E. A. Armanios, Ed., American Society for Testing and Materials, 1997, pp. 143–161.

ABSTRACT: This study focuses on the mode dependence of delamination growth under cyclic compressive loads in cross-ply composite plates. The model proposed makes use of an initial postbuckling solution derived from a perturbation procedure. A mode-dependent crack growth criterion is introduced. Expressions describing the fatigue crack growth are derived in terms of the distribution of the mode adjusted energy release rate. The resulting crack growth laws are numerically integrated to produce delamination growth versus number of cycles diagrams. The model does not impose any restrictive assumptions on the relative thickness of the delaminated and the base plates, although transverse shear stress effects are not considered. Experimental results are presented for cross-ply graphite/epoxy specimens, and the results are compared with experimental results for unidirectional specimens. The test data are obtained for different delamination locations and for different values of applied compressive strain.

KEYWORDS: composite materials, fatigue (materials), delamination, energy release rate, mode mixity, stress intensity factor, crack growth law, fracture (materials)

The applicability of laminated composite materials may become limited due to the frequent presence of delaminations, that is, partial debonding of the plies of the laminate. This partial debonding at the interface is most commonly a result of manufacturing imperfections, low velocity impacts on the surface of the composite component, or even a consequence of vibrations of the structure's propulsion system. The effect of the delamination growth can be manifested in the form of stiffness and strength degradation as well as a change in the energy absorption capacity of composite beam structures [1,2].

The study of the delamination buckling under compression in composite plates has attracted a substantial number of researchers who have studied extensively one-dimensional and two-dimensional configurations [3–7]. However, most of this work is concerned with the estimation of the bifurcation point and little attention is paid to the study of the postbuckling behavior of the composite structure. Furthermore, most investigations are focused on theoretical aspects in the absence of experimental studies especially on fatigue loading; some experimental studies have been focused on monotonic (static) loading [8].

[1]Graduate research assistant, associate professor, and graduate research assistant, respectively, School of Aerospace Engineering, Georgia Institute of Technology, Atlanta, GA 30332-0150.

The cyclic loading of delaminated plates has been studied separately under the light of pure Mode I and pure Mode II fatigue loading [9,10] and a few studies refer to mode-dependent cyclic delamination growth [11]. During the compressive fatigue loading, the delaminated layer undergoes sequential buckled/unbuckled geometrical configurations, as the composite structure is loaded and unloaded, respectively. The repeated cyclic loading of the structure results in the decrease of the material's interlayer resistance. As a consequence, there is damage accumulation in front of the crack-tip that might lead to growth and even potential loss of structural integrity.

The characteristics of delamination growth behavior beyond the bifurcation point can be determined once a postbuckling solution is available. The post critical behavior of delaminations of arbitrary size has been investigated in Refs 12 and 13. The result of these studies was an analytical formulation for the initial postbuckling behavior, using the theory of elastica to represent the deflections of the buckled layer. This work resulted in a system of nonlinear equations. This model was further developed into a closed-form solution using a perturbation procedure based on the asymptotic expansion of the load and deformation quantities in terms of the distortion parameter of the delaminated layer. This analysis led to closed-form expressions for the force and moment quantities near the crack-tip field in terms of the applied compressive displacement.

The growth characteristics of a postbuckled delaminated composite structure is of major concern. To this extent, we are interested in whether the delamination grows in a stable or unstable manner and, if the growth is stable, the growth rate. These growth characteristics can be investigated once a postbuckling solution is available. The stability of laminated structures is sensitive to the in-plane dimensions (for example, delamination length), the flexural and in-plane stiffnesses, and the loading conditions [14–18]. Kardomateas and Pelegri [16] investigated the effect of the applied strain and the delamination length on the stability of the delamination growth. Combinations of delamination length and applied strain inducing unstable growth were examined. Comparing the results with the "thin film" model [3], one concludes that the delamination growth is more likely to be stable than predicted by this model.

A mode-dependent crack growth law is developed by using the bi-material interface crack solutions for the mode mixity and the energy release rate. For a buckled configuration in compressed films, Evans and Hutchinson [5] derived a formula for the energy release rate using an asymptotically valid solution. Yin calculated the energy release rate for a circular delamination using a path independent integral approach [19]. In the present study, the energy release rate at the delamination tip is derived in closed-form through a perturbation analysis in the initial postbuckling phase.

Another aspect of delamination growth is the mode dependence of the growth process. Delaminations in composites are interlaminar cracks that are constrained to move in a specific plane (along the layer interface). Therefore, delaminations grow under mixed-mode conditions, that is, a combination of Mode I (opening mode) and Mode II (shearing mode). Traditionally, the growth behavior of delaminations was studied by using the Irwin-Griffith concept of critical fracture energy, utilizing a mode independent energy release rate criterion. Recent experimental studies on several composite materials have shown the inadequacy of mode-independent models [20,21]. In this study, a fatigue growth mode-dependent law introduced by Kardomateas et al. in Ref 11 constitutes the basis for the research presented in this paper.

Specifically in this study, Kardomateas et al. [11] examined the fatigue growth of delaminations under constant amplitude compressive loads for unidirectional graphite/epoxy plate specimens. This investigation is extended herewith by studying cross-ply construction of graphite/epoxy specimens, again under constant amplitude cyclic compression. The structure is analyzed by being separated into three different parts, that is, the delaminated, substrate, and base parts. The analytical model was proposed for isotropic material by Kardomateas [12] and has been modified to accommodate cross-ply composite models. Although the model does not include classical laminate theory, for example, Refs 22 and 23, in order to account for layup variations, the rule of mixtures is used in order to account for the material properties in each different part. This approach is being adopted in order to reduce calculation complexity in the initial postbuckling solution. Since the focal point of the study is to examine the trends that exist in various parameters during delamination growth and to correlate analytical and experimental results, the aforementioned analysis is justified. The model utilizes the elastica theory [24] by considering all three plates in each configuration as parts of three different compressive elasticae. During the perturbation procedure, the loading and geometric quantities are asymptotically expanded in terms of the distortion parameter of the delaminated plate. The analysis results in closed-form solutions for the applied load and the near tip resultant moments and forces as a function of the applied compressive displacement. Next, the interface crack solutions, as given in Refs 25 and 26, are employed for the mode mixity and the energy release rate in terms of the resultant moments and forces. The mode-dependent cyclic growth law was employed and correlated with experimental results. Finally, the dependence of the energy release rate and the mode mixity on the delamination growth is examined. The results extracted for the unidirectional graphite/epoxy plate are compared with cross-ply configurations of the same material.

Theoretical Analysis

Postbuckling Solution for a Delaminated Beam/Plate

A delaminated composite structure can be considered an aggregate of three parts, as in Fig. 1. The first part of the thickness, h, consists of the plies laying above the delamination and is referred to as the "delaminated part." The second part of the thickness, $H = T - h$, consists of the plies laying below the delamination and is referred to as the "substrate part." The remaining intact laminate of the thickness, T, is the "base plate." Without loss of generality, it may be assumed that $h < T/2$. The length of the structure is $2L$, and the initial length of

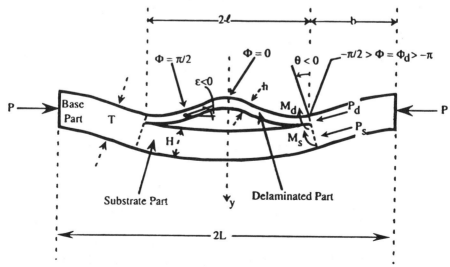

FIG. 1—*Model configuration in the post-buckled state under compressive loading. The delamination separates the configuration into the delaminated, substrate and base plates.*

the delamination is 2ℓ. The delamination is positioned symmetrically along the length of the specimen. The length of the base plate is b, where $b = L - \ell$. The subscript, $i = d, s, b$, refers to the delaminated part, the substrate part, and the base plate, respectively. The material is considered to be linearly elastic and orthotropic.

For the general case of a delaminated composite system, the initial postbuckling solution was studied by Kardomateas [12] and led to a closed-form solution. The results produced by this model applied to a unidirectional composite plate were presented in Ref 11. In the present study, the material layup is $[(0/90)_{14}, 0]$ and $[(0/90)_{28}, 0]$. In addition, due to the complexity of calculations in the initial postbuckling solution, the delaminated plate is considered to be a laminate with averaged material properties determined by the rule of mixtures [27]. Similar assumptions are made for the substrate and the base plates. The averaged composite Young's modulus, E, is the corresponding primary composite modulus of elasticity, as defined in Agarwal [27], and is valid when the fibers are intact. The bending stiffness of each corresponding plate is denoted by D_i, $D_i = E_i t_i^3/[12(1 - \nu_{12}\nu_{21})]$, where t_1 is the thickness of the ith plate, ν_{12} is the Poisson's ratio in the longitudinal direction, and ν_{21} is the Poisson's ratio in the transverse direction of the material. A brief description for the initial postbuckling solution is presented, and the closed-form solution derived will be used to determine the characteristics of the cyclic growth.

The theory of elastica [24], provides the exact laws that govern the behavior of single compressive elements elastically restrained at the ends by means of concentrated forces and moments. The generalized coordinates of deformation are the distortion parameter, α_1 (where $i = d, s, b$ for the delaminated, substrate, and base plates, respectively), that represents the tangent rotation of each individual plate at an inflection point from the straight position, and the amplitude variable, $\Phi(x)$. The deformations from the initial postbuckling configuration are assumed to be relatively small. The expressions describing the model are expanded in Taylor series in terms of the distortion parameter of the delaminated plate, α_d. For the initial postbuckling solution, a perturbation analysis is utilized with a perturbation parameter, ϵ, such that $\epsilon = \alpha_d$.

Consider the configuration shown in Fig. 1. The model assumes that the buckled delaminated plate is a part of an inflectional elastica with an amplitude, Φ_d, and a distortion parameter, α_d. At the critical state, the end amplitude is Φ_d^0. Although the substrate and the base plate undergo moderate bending with no inflection point, we may use the elastica theory to describe their nonlinear deformation. In this case, the inflection points are outside the actual elastic curve. For a slightly buckled configuration, the amplitude, Φ_d, can be expanded in terms of the distortion parameter of the delaminated layer, ϵ, as follows

$$\Phi_d = \Phi_d^0 + \phi_d^{(1)}\epsilon + \phi_d^{(2)}\epsilon^2 + O(\epsilon^3) \tag{1}$$

The end rotation at the common section, θ, at the critical state is $\theta^0 = 0$. The common section of the configuration is where the delamination starts or ends, that is, where the three parts of the structure meet. The end rotation can be expanded in Taylor series in terms of ϵ. The relevant expression in Ref 24 gives

$$\theta = (\sin \Phi d)\epsilon - \frac{1}{24} (\sin \Phi d \cos^2 \Phi d)\epsilon^3 + \cdots = (\sin \Phi_d^0)\epsilon + (\cos \Phi_d^0)\phi_d^{(1)}\epsilon^2$$

$$+ \left[(\cos \Phi_d^0)\phi_d^{(2)} - (\sin \Phi_d^0)\frac{\phi_d^{(1)2}}{2} - \frac{1}{24} \sin \Phi_d^0 \cos^2\Phi_d^0 \right]\epsilon^3 \tag{2}$$

$$= \theta^{(1)}\epsilon + \theta^{(2)}\epsilon^2 + \theta^{(3)}\epsilon^3 + O(\epsilon^4)$$

At the common section, conditions of geometrical continuity should be fulfilled. Therefore, the end rotation at the common section is the same for the delaminated, the substrate and the base plate.

Similar to the end rotation, θ, the axial force, P_d, the moment, M_d and the flexural contraction, f_d, of the delaminated plate can be expanded in Taylor series in terms of the ϵ, after their substitution in the relevant Britvek's [24] formulas. At the critical state $P_d \equiv P_d^0$ and $M_d \equiv M_d^0 = 0$. For example, the asymptotic expansion of the axial force, P_d, is

$$\frac{\ell^2 P_d}{D_d} = \frac{\ell^2}{D_d} (P_d^0 + P_d^{(1)}\epsilon + P_d^{(2)}\epsilon^2) + O(\epsilon^3) \tag{3}$$

For the substrate plate, the amplitude, Φ_s, and the distortion parameter, α_s, are expanded in terms of the distortion parameter of the delaminated plate, ϵ

$$\Phi_s = \Phi_s^0 + \phi_s^{(1)}\epsilon + \phi_s^{(2)}\epsilon^2 + O(\epsilon^3) \tag{4}$$

$$\alpha_s = \alpha_s^{(1)}\epsilon + \alpha_s^{(2)}\epsilon^2 + \alpha_s^{(3)}\epsilon^3 + O(\epsilon^4) \tag{5}$$

At the common section of the base plate, $\Phi = \Phi_b$. The amplitude and the distortion parameter of the base plate are expanded in terms of the distortion parameter of the delaminated plate, ϵ, in a similar manner as the substrate plate in Eqs 4 and 5.

Furthermore, the axial forces, P_s and P_b, the end moments, M_s and M_b, the flexural contractions, f_s and f_b, of the substrate and the base plate, respectively, along with the end rotation at the common section, θ, can be expanded similarly in Taylor series in terms of ϵ.

The next step in this analysis is the definition of the nonlinear critical path. To do so, conditions of force and moment equilibrium at the common section, along with compatibility of the shortening of the delaminated and substrate plates, are required. These conditions will be imposed in the force and deformation asymptotic expressions separately for the first-, second-, and third-order terms. The system of equations to be solved in order to derive one equation with only the distortion parameter, ϵ, as unknown, includes: (a) one nonlinear equation for the zero order terms, which defines the critical point (characteristic equation), (b) two linear algebraic equations for $\phi_d^{(1)}$ and $\phi_s^{(1)}$ that determine the first order forces, and (c) two linear algebraic equations for $\phi_d^{(2)}$ and $\phi_s^{(2)}$ that determine the second-order forces [12].

Next, the initial postbuckling solution briefly described earlier is used in conjunction with the interface crack solution given by Hutchinson and Suo [26]. This solution for a general bimaterial crack is based on the Dundurs [28] material mismatch parameters, $\tilde{\alpha}$ (a measure of the mismatch in the plane tensile modulus across the interface) and $\tilde{\beta}$ (a measure of the mismatch in the in-plane bulk modulus), and the bimaterial constant, $\tilde{\epsilon}$. For the structure under consideration, the crack grows in the resin content between two plies in the direction of the zero degree plies. Therefore, this case is treated as two isotropic materials joined along the longitudinal axis (loading direction). During growth, the h/T ratio for each specimen remains constant because the crack grows in the interface between the delaminated and the substrate plates. Delamination branching is not considered in the model. At all times, the crack front is straight and perpendicular to the structure's loading direction. The specimen is loaded symmetrically, and the delamination is placed symmetrically along the article's length. For the type of material, specimen geometry, and type of loading considered in this study, the parameters $\tilde{\alpha}$, $\tilde{\beta}$, and $\tilde{\epsilon}$ are equal to zero. In such case, for the plane-strain interface crack shown in Fig. 1, the energy release rate is given by [26]

$$G = \frac{1 - \sqrt{\nu_{12}\nu_{21}}}{2E_1} \left[\frac{P^{*2}}{Ah} + \frac{M^{*2}}{Ih^3} + 2\frac{P^*M^*}{\sqrt{AIh^2}} \sin \gamma \right] \qquad (6)$$

where P^*, M^* are linear combinations of the loads from the previous postbuckling solution

$$P^* = P_d - C_1 P - C_2 \frac{M_b}{h} \qquad (7)$$

$$M^* = M_d - C_3 M_b \qquad (8)$$

The angle, γ, is constrained such that $\gamma < \pi/2$. Moreover, A and I are positive dimensionless numbers that depend on the dimensions of the configuration.

When $\tilde{\beta} = 0$ the Mode I component of the stress intensity factor, K_I, is the amplitude of the singularity of the normal stresses in front of the crack-tip and the associated normal separation of the crack flanks; the Mode II component of the stress intensity factor, K_{II}, governs the shear stress on the interface and the relative shearing displacement of the flanks. The following expressions describe the Mode I and Mode II stress intensity factors

$$K_I = \frac{1}{\sqrt{2}} \left[\frac{P^*}{\sqrt{Ah}} \cos \omega + \frac{M^*}{\sqrt{Ih^3}} \sin(\omega + \gamma) \right] \qquad (9a)$$

and

$$K_{II} = \frac{1}{\sqrt{2}} \left[\frac{P^*}{\sqrt{Ah}} \sin \omega + \frac{M^*}{\sqrt{Ih^3}} \cos(\omega + \gamma) \right] \qquad (9b)$$

The stress intensity factors, K_I and K_{II}, are expressed with respect to the linear combinations of the loads given from the initial postbuckling solution, and therefore they are known functions of ω. The accurate determination of ω requires the numerical solution of an integral equation and has been reported by Suo and Hutchinson [25].

The mode mixity expresses the relative amounts of Mode I (opening) and Mode II (shearing) components and is defined by

$$\psi = \tan^{-1} \frac{K_{II}}{K_I} = \tan^{-1} \left[\frac{b \sin \omega - \cos(\omega + \gamma)}{b \cos \omega + \sin(\omega + \gamma)} \right] \qquad (10)$$

where b measures the loading combination as

$$b = \sqrt{\frac{I}{A}} \frac{P^* h}{M^*} \qquad (10a)$$

In order to obtain the energy release rate, G, and the stress intensity factors, K_I and K_{II}, the results derived from the initial postbuckling solution are employed. The asymptotic expressions for the forces and the moments derived from this solution are

$$P* = \epsilon P*^{(1)} + \epsilon^2 P*^{(2)} + \cdots \tag{11}$$

$$M* = \epsilon M*^{(1)} + \epsilon^2 M*^{(2)} + \cdots \tag{12}$$

Notice that the zero-order terms cancel out. More explicitly, the kth order terms, that is, $k = 1, 2$ denote the first- and the second-order terms for the load and the moment as follows

$$P*^{(k)} = \frac{H}{T} P_d^{(k)} - \frac{h}{T} P_s^{(k)} - \frac{6hH}{T^3} M_b^{(k)} \tag{13}$$

$$M*^{(k)} = M_d^{(k)} - \frac{h^3}{T^3} M_b^{(k)} \tag{14}$$

The force and moment quantities on the right hand side of these equations are known from the initial postbuckling solution.

In order to find a closed-form solution, the energy release rate and the stress intensity factors are expressed in an asymptotic form with respect to the distortion parameter, ϵ.

$$G = \epsilon^2 G^{(2)} + \epsilon^3 G^{(3)} + \cdots \tag{15}$$

$$K_{\mathrm{I,II}} = \epsilon K_{\mathrm{I,II}}^{(1)} + \epsilon^2 K_{\mathrm{I,II}}^{(2)} + \cdots \tag{16}$$

The quantity needed to complete the initial postbuckling solution is the applied strain, ϵ_0, that represents the "loading" quantity.

$$\epsilon_0 = \epsilon_0^{(0)} + \epsilon_0^{(1)}\epsilon + \epsilon_0^{(2)}\epsilon^2 \tag{17}$$

with

$$\epsilon_0^{(0)} = \frac{P^0}{ET} \quad \text{and} \quad \epsilon_0^{(1)} = \frac{P^{(1)}b}{ET} + \frac{P_d^{(1)}l}{Eh} + \frac{H}{2}\theta^{(1)}$$

$$\epsilon_0^{(2)} = \frac{f_d^{(2)}}{2} + \frac{P_d^{(2)}l}{Eh} + f_b^{(2)} + \frac{P_d^{(2)}b}{ET} + \frac{H}{2}\theta^{(2)}$$

From the preceding expression for ϵ_0 and the perturbation expressions, the computation of ϵ (distortion parameter) is possible for a given applied displacement. Thereafter, all the load and deformation quantities can be expressed with respect to ϵ and evaluated accordingly.

Cyclic Growth Law

The energy release rate, G, has been used by numerous researchers in order to predict the crack growth. In this respect, a Griffith type of criterion is used for the prediction of crack growth. According to this criterion, the energy release rate, G, is compared with the fracture toughness of the specimen, Γ_0. The growth of the delamination is assured when the following condition exists

$$G(\epsilon_0, \ell) > \Gamma_0 \tag{18}$$

The energy release rate, G, depends only on the length of the delamination, ℓ, and the applied strain, ϵ_0, for the present specimen configuration. The fracture toughness, Γ_0, can be expressed in terms of the Mode I and Mode II stress intensity factors, K_I and K_{II}, respectively [29,30].

A brief description of the parameters influencing the crack growth in the area in front of the crack-tip is given in the sequence. Kardomateas [8], in his initial postbuckling solution model, showed the dependence of the mode mixity on the applied strain and the position of the delamination inside the specimen. In addition, it was proved that the mode mixity changes as the delamination grows. To support these results, well-correlating experiments on graphite/epoxy unidirectional specimens were conducted [11].

The stress field in the area surrounding the crack-tip for a specimen subjected to fatigue is characterized by three parameters that describe its intensity and variation due to loading and geometry configuration. The three parameters are the mode mixity, ψ, the load ratio, α, with $\alpha = G_{min}/G_{max}$, and the variation of the energy release rate from G_{min} to G_{max}. Therefore, the amount of crack extension per cycle of loading can be completely described by these parameters, and in a functional form this will be

$$\frac{da}{dN} = f(G_{max}, \alpha, \psi) \tag{19}$$

The mode dependence of the delamination growth process is not yet fully understood and is a subject of research [20,31,32]. O'Brien and Kevin [32] examined the stacking sequence effect on the local delamination onset in fatigue loading. In their study, a strain energy release rate solution for a local delamination growing from an angle ply matrix crack was used in order to identify these local delaminations and their fracture mode dependence. The results showed the need for detailed analysis on the delamination fracture modes, and a mixed-mode delamination fatigue onset criteria was demonstrated.

For the time being, let us assume that the fracture toughness, Γ_0, depends on the mode mixity; it is generally increasing when $|\psi|$ is increasing, that is, increasing Mode II component. Then the mode-mixity-adjusted fracture toughness is described by Hutchinson and Suo [26] as

$$\Gamma_0(\psi) = G_I^c[1 + (\lambda - 1)\sin^2 \psi]^{-1} \tag{20}$$

where $\lambda = G_I^c/G_{II}^c$.

The values of G_I^c and G_{II}^c are the pure Mode I and Mode II toughness, respectively. The parameter, λ, adjusts the influence of the Mode II contribution in the criterion and should be determined experimentally by obtaining mode interaction curves as in Ref 20. Notice that for pure Mode I, $\psi = 0$ and $\Gamma_0 = G_I^c$, and for pure Mode II, $\psi = 90^0$ and $\Gamma_0 = G_{II}^c$. A typical value for λ for graphite/epoxy is $\lambda = 0.30$. When $\lambda = 1$ then $G_I^c = \Gamma_0$ for all mode combinations.

In order to include the effect of mode-dependent toughness in the crack growth in the initial postbuckling solution discussed before, we define the mode-adjusted crack driving force, $\tilde{G}$, as follows

$$\tilde{G} = \frac{G}{\Gamma_0(\psi)} = \tilde{G}(\epsilon_0, \psi) \tag{21}$$

For crack advance

$$G/\Gamma_0(\psi) \geq 1 \Leftrightarrow \tilde{G}(\epsilon_0, \psi) \geq 1 \tag{22}$$

To derive the mode mixity, ψ, in a closed-form solution, the postbuckling solution suggested by Kardomateas [12] was used. This theoretical model assumes cyclic loading from an unloaded position to a maximum compressive strain, ϵ_{max}. Since the stress intensity factors for pure Mode I and II are known, the mode mixity is given by the following equivalent with Eq 10

$$\psi = \tan^{-1} \frac{\epsilon K_{II}^{(1)} + \epsilon^2 K_{II}^{(2)}}{\epsilon K_I^{(1)} + \epsilon^2 K_I^{(2)}} \qquad (23)$$

The stability of the crack growth can be assessed from the study of a energy release rate, G, versus delamination length, ℓ, diagram for a specific applied strain, ϵ_{max}. For the specified strain $\epsilon_0 = E_0$, the corresponding ϵ is determined from the minimum (negative) root of the equation

$$E_0 = \epsilon_0^{(0)}(\ell) + \epsilon_0^{(1)}(\ell)\epsilon + \epsilon_0^{(2)}(\ell)\epsilon^2 \qquad (24)$$

Then the corresponding energy release rate, G, is

$$G(\ell) = G^{(2)}(\ell)\epsilon^2 + G^{(3)}(\ell)\epsilon^3 \qquad (25)$$

The delamination growth is stable when the $G - \ell$ curve has a negative slope, otherwise the delamination growth is unstable [16].

Based on the preceding discussion, the assumption that slow growth takes place in the interface between the 0 and 90 plies and by applying minimum load close to zero, therefore, load ratio $\alpha = 0$, the following crack growth law is derived

$$\frac{da}{dN} = C(\psi) \frac{(\Delta \tilde{G})^{m(\psi)}}{1 - \tilde{G}_{max}} \qquad (26)$$

where $\Delta \tilde{G}$ is the variation mode adjusted crack driving force (variation of the energy release rate normalized with the fracture toughness), $\Delta \tilde{G} = \tilde{G}_{max} - \tilde{G}_{min}$. The $1 - \tilde{G}_{max}$ in the denominator was introduced to account for very short life (less than 10^3 cycles). A similar power growth law is seen in the Paris fatigue law. In contrast, the Paris formulation uses the stress intensity factor, ΔK, for a single pure mode instead of $\Delta \tilde{G}$.

The mode dependency of the $C(\psi)$ and $m(\psi)$ parameters has been demonstrated experimentally by Russell and Street [9]. Although a slightly different growth law was used for graphite/epoxy these parameters were found to be $C_I = 0.0325$ and $m_I = 5.8$ for pure Mode I, and $C_{II} = 0.285$ and $m_{II} = 9.4$ for pure Mode II. The $C(\psi)$ and $m(\psi)$ parameters can be expressed in terms of the mode mixity following the same format as Hutchinson and Suo's [26] formula for the fracture toughness, Γ_0. Therefore, $m(\psi)$ and $C(\psi)$ are described by the following equations, respectively

$$m(\psi) = m_I[1 + (\mu - 1)\sin^2\psi] \qquad (27)$$

and

$$C(\psi) = C_I[1 + (\kappa - 1)\sin^2\psi] \qquad (28)$$

where μ is defined as the ratio of the exponents, m, at Modes II and I, and κ is defined as the ratio of the constant, C, at Modes II and I. In general, μ is less than 1 and κ is greater than 1. The present experimental study uses values of μ and κ very close to the values that

Russell and Street [9] determined by conducting pure Mode I and II tests for double cantilever beam (DCB) and end-delaminated flexure specimens. Equations 27 and 28 are semi-empirical and characterize the mode dependency of the delamination cyclic growth law.

Experimental Approach

In deriving the cyclic growth law presented in the preceding sections, it was assumed that for a given level of mode mixity, ψ, the parameters that control the rate of the delamination growth are: (a) the maximum value of the energy release rate near the tip of the delamination, and (b) the variation (spread) energy release rate in the loading/unloading cycle. We validate this assumption by using the growth law to predict the growth rate for the specimen configurations with: (1) different positions of the delamination through the thickness, and (2) different values of the applied maximum external compression, ϵ_{max}. It will be seen that, despite the wide range of combinations, the correlation between experiment and prediction is reasonable. Therefore, from simple laboratory tests, we can use this formulation to predict the number of cycles required for a delamination to grow by a specific amount, and a reliable analysis for the near-tip state of mode mixity and energy release rate can be obtained.

The material used in this study is C30/922 graphite/epoxy. The material was provided in the form of unidirectional prepreg tapes. The elastic constants of the C30/922 graphite/epoxy are: Young's modulus, $E_L = 137.90$ GPa and $E_T = 8.98$ GPa in longitudinal and transverse direction respectively; shear modulus, $G_{LT} = 7.10$ GPa; and in-plane Poisson's ratio, $\nu_{LT} = 0.30$. The thickness of the prepeg tape is 8.89×10^{-5} m (0.0035 in.). The thermoset resin content by weight of C30/922 is $33 \pm 3\%$.

The use of prepeg tapes is advantageous because it allows easy manufacturing of laminates with various ply orientations. In this study, cross-ply specimens are manufactured by hand layup of each ply following the desired stacking sequence. The delamination is introduced in each specimen with a Teflon film insert. A primary concern in the film selection is the film's thickness. Thick insert films can distort the experimental results by introducing resin pockets in front of the crack-tips. Murri and Martin [33] have studied the effect of insert thickness on fracture toughness testing. Their study concludes that inserts of 13×10^{-6} m (0.00051 in.) do not produce resin pockets while inserts of 75×10^{-6} m (0.00295 in.) and thicker do produce resin pockets. The insert is DuPont Teflon FEP film of 2.7×10^{-6} m (0.0005 in.) thickness. Once the delamination is implanted symmetrically along the length of each specimen and the plies are laid up, then the specimens are cured in an autoclave following the manufacturer's standardized curing cycle for C30/922 graphite/epoxy.

The test articles were cut to the desired dimensions: $w = 12.7 \times 10^{-3}$ m (0.5 in.) wide and $\ell = 152.4 \times 10^{-3}$ m (6.0 in.) total length. The free length of the specimen (length measured between the grips) was $2L = 96.5 \times 10^{-3}$ m (3.8 in.). The thickness of the delaminated plate was $h = 0.4 \times 10^{-3}$ m (0.0525 in.). The thickness of the base plates was $T = 1.3 \times 10^{-3}$ m (0.05 in.) for the 4/15 specimen configuration and $T = 2.6 \times 10^{-3}$ m (0.10 in.) for the 4/29 specimen configuration. The thickness measurements were taken after the curing of each specimen, and at eight different locations along its length in order to assure its constant thickness and reveal any geometric distortions.

Monotonic compressive displacement tests were conducted to open the delaminations and release the Teflon inserts. Furthermore, the applied cyclic strain that each specimen was subjected to during fatigue tests was determined based on the displacement level where no further static growth of delamination is observed under static compressive experiments.

The fatigue tests were conducted on cross-ply graphite/epoxy specimens with the aforementioned dimensions in an Instron servohydraulic testing machine. The boundary conditions imposed on each specimen when inserted to the Instron testing machine are clamped-clamped

[34]. Monitoring of the delamination growth was conducted with a Questar remote video measurement system that allows tri-axial motion of a microscope and simultaneous digital measurement of distances. The experiments were conducted under 3-Hz frequency and under constant ϵ_{max} (displacement control).

Two specimen configurations were tested. The first one consisted of a total of 15 plies with the delamination positioned between the fourth and the fifth ply. This configuration was tested for two different values of the applied strains, ϵ_{max}. The second configuration consisted of 29 plies with the same delamination position as before, so the same delamination thickness existed in all experiments. Each specimen had the same initial delamination length and was subjected to different applied maximum compressive strains. The cross-ply graphite/epoxy configurations were:

(a) 15 plies, delamination of half length, $\ell_0 = 25.4 \times 10^{-3}$ m (1.0 in.); $h/T = 4/15 = 0.2667$; maximum compressive strain, $\epsilon_{max} = 1.579 \times 10^{-3}$, noted as 4/15b.

(b) 15 plies, delamination of half length, $\ell_0 = 25.4 \times 10^{-3}$ m (1.0 in.); $h/T = 4/15 = 0.2667$; maximum compressive strain, $\epsilon_{max} = 2.632 \times 10^{-3}$, noted as 4/15c.

(c) 29 plies, delamination of half length, $\ell_0 = 25.4 \times 10^{-3}$ m (1.0 in.); $h/T = 4/29 = 0.1379$; maximum compressive strain, $\epsilon_{max} = 1.316 \times 10^{-3}$, noted as 4/29a.

The fatigue growth parameters of the cross-ply graphite/epoxy specimens that were tested were: exponent ratio $\mu = m_{II}/m_I = 0.5$ and constant ratio $\kappa = C_{II}/C_I = 10$. The critical energy release rate for Mode I, that is, Mode I interlaminar fracture toughness, G_I^C, of the cross-ply graphite/epoxy specimens, was determined by static DCB experiments that were conducted according to the standard testing method ASTM Test Method for Mode I for Interlaminar Fracture Toughness of Unidirectional Fiber-Reinforced Polymer Matrix Composites (D 5528-94A). Although the DCB tests have been used extensively and are recommended for unidirectional glass fiber reinforced structures, they can be used for structures with different material, architecture, and layup than the one proposed in the standard method, provided some caution and corrections, if needed, are used. For nonunidirectional configurations, like the cross-ply graphite/epoxy structures discussed in this paper, the values of the observed initiation values of G_I^C established by DCB tests can be affected from one, or both, of the following cases: (1) branching of the delamination away from the midplane through matrix cracks in off-axis plies that results in coupling between extension and shear due to the formation of asymmetric sublaminates as the delamination grows (in such cases pure Mode I fracture is not achieved); and (2) anticlastic bending effects that result in nonuniform delamination growth along the specimen's width. A number of tests was conducted in order to ensure the validity of the results obtained from the DCB tests. The delamination opening and growth was closely monitored and video taped through a Questar tele-microscope. A mirror devise attached to the tele-microscope enabled the observation of the delamination through the width of the specimen. After the completion of each test, each specimen was separated catastrophically along its midplane, and the delamination front was examined to ensure uniform delamination growth along the specimen's width. The method of data reduction used to calculate the G_I^C was the modified beam theory (MTB) that is the one recommended in ASTM D 5528-94A. The value determined was $G_I^C = 200$ N/m. The Mode II interlaminar fracture toughness, G_I^C, was determined by assuming the value of the ratio of Mode I and Mode II interlaminar fracture toughness to be $\lambda = 0.30$, which is a typical value for graphite/epoxy.

Discussion of Results

Experimental Results

The experimental results from the fatigue tests are presented in Fig. 2 as distinct points. The experiments were carried under displacement control, and the cyclic loading was always compressive. In Fig. 2 it is apparent that the two different configurations present different delamination growth behavior. Also, the effect of the applied strain in the 4/30 configuration changes the growth behavior of the delamination. By comparing the 4/15b and 4/15c configurations, it seems, that the higher the applied strain, for example, 4/15c, the faster the delamination growth.

Discussion of the experimental and analytical correlation for the cross-ply graphite/epoxy specimens follows in the next section.

Analytical Results

The theory described earlier will now be applied to the specimen configuration used in experiments. From this operation, the graphical representations of the mode mixity, the energy release rate, as well as the delamination growth versus the number of cycles are derived.

In the present study, we are going to compare the theoretical results obtained from the analysis of the cross-ply specimens with results obtained from the same model but for unidirectional specimens from Ref *11*.

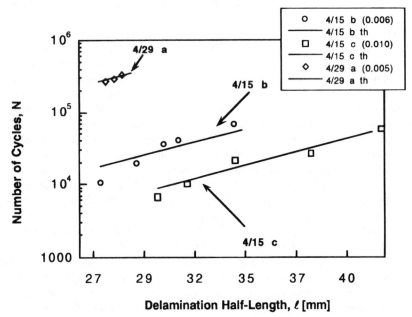

FIG. 2—*Fatigue delamination growth, N-ℓ (semilogarithmic plot), for cross-ply specimens. The number in the parenthesis denotes the maximum displacement applied to each specimen (1 in. = 0.0254 m). In each case, the experimental data are denoted with discrete empty marks, whereas the lines represent the predictions of theory. Proposed crack growth law parameters:* $m_I = 4.6$, $C_I = 0.696$ *m/cycle.*

Results from the Cross-Ply Specimens—The mode mixity, ψ and the energy release rate spreads, $\tilde{G}$ that the 4/29a, 4/15b, and 4/15c configurations exhibit are shown in Fig. 3 and Fig. 4, respectively. In Fig. 3, the vertical axis is the mode mixity. As seen at Fig. 3, the mode mixity versus the delamination half-length changes for each specimen configuration. As the delamination grows, the mode mixity reduces giving rise to its Mode II component; pure Mode II is $-90°$. The mode mixity exhibited by the 4/29 specimen is lower than that exhibited by the two 4/15 specimens. The 4/15 specimen with the larger ϵ_{max} has the smallest mode mixity of the two 4/15 specimens, going to pure Mode II ($\psi = 90°$) when delamination half-length equals 3.28×10^{-2} m. From Fig. 3, it is obvious that the mode mixity depends not only on the h/T ratio but also on the maximum applied strain, ϵ_{max}.

All the specimens are subjected to less than 15% of the critical energy release rate, $\Gamma_0(\psi)$. From Fig. 4, it can be concluded that the energy release rate is higher for higher h/T ratio, that is, the deeper the delamination is in the specimen. All the configurations exhibit reduction of the energy release rate as the delamination length increases during the fatigue tests.

The parameters, m_I and C_I, for the mode-dependent cyclic growth law were obtained from three data points (ℓ, N) of the 4/15c specimen. The data points used were: 29.972 mm, 5000 cycles; 31.344 mm, 10 000 cycles; and 34.163 mm, 30 000 cycles, see Table 1. The values obtained were $m_I = 4.6$ and $C_I = 0.696$ m/cycle.

Based on these values, the theoretical number of cycles has been evaluated for the three specimens. Figure 2 shows the experimental data (distinct points) and the predicted growth behavior (solid lines). This semi-logarithmic diagram shows the good correlation of experimental and predicted results for the 4/15c specimen. Since the 4/15c specimen's data were used

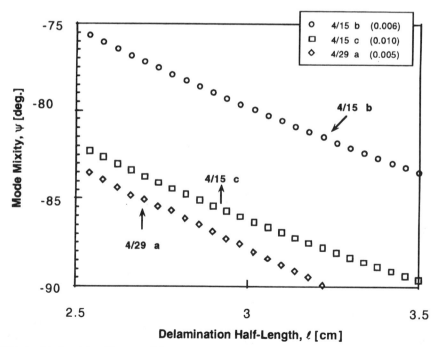

FIG. 3—*Mode mixity, ψ^0 versus delamination half-length, ℓ, for the cross-ply specimen configurations tested. The number in the parenthesis denotes the maximum displacement applied to each specimen (1 in. = 0.0254 m). The pure Mode II is for $\psi = \pm90°$.*

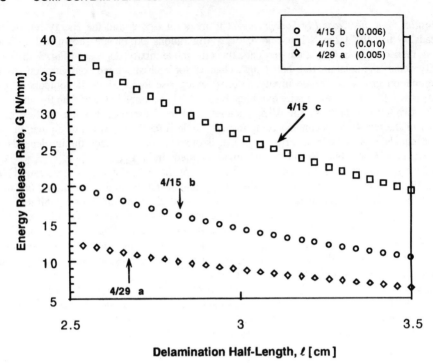

FIG. 4—*Energy release rate, G, versus delamination half-length, ℓ, for the cross-ply specimen configurations tested. The number in the parenthesis denotes the maximum displacement applied to each specimen (1 in. = 0.0254 m).*

TABLE 1—*Comparison with experiments, cross-ply, graphite/epoxy.*

h/T Specimen Type	Delamination Half-Length, mm	N Theoretical Cycles, predicted	N Experimental Cycles, from tests
4/15b	27.076	15 000	10 426
ℓ_0 = 25.4 mm	31.204	24 979	23 897
T = 1.3 mm	32.423	31 233	33 480
max applied strain = 1.579E-3	33.450	35 374	48 057
4/15c	29.972	5 000	5 000
ℓ_0 = 25.4 mm	31.344	10 003	10 000
T = 1.3 mm	34.163	29 343	30 000
max applied strain = 2.632E-3	41.808	31 246	38 000
4/29a	27.330	281 700	253 503
ℓ_0 = 25.4 mm	27.800	319 940	292 703
T = 2.6 mm	28.207	344 390	329 903
max applied strain = 1.316E-3	28.461	378 640	354 931

[a]G_I^c = 200 N/m; λ = G_I^c/G_{II}^c = 0.30; m_I = 4.6; C_I = 0.696 m/cycle; μ = m_{II}/m_I = 0.5; κ = C_{II}/C_I = 10.

to compute m_I and C_I, this behavior was expected. Also, there is reasonable correlation between the experimental and predicted results for specimens 4/15b and 4/29a. The trend to require a higher number of cycles for the delamination growth of specimens with lower h/T ratio is apparent. Another observation made from this diagram is that for specimens with the same thickness, that is, same number of plies, the one with the lower initial strain needs a larger number of cycles to exhibit delamination growth.

Comparison Cross-Ply and Unidirectional Configurations—Unidirectional C30/922 graphite/epoxy configurations from Ref *11* are used in order to compare the mode mixity and the energy release rates with the cross-ply configurations of the present study. The unidirectional configurations are 4/15b, 4/15c, and 4/29a. Figure 5 presents the mode mixity versus the delamination half-length for both layups. As seen for all configurations, the mode mixity is higher when the delamination length is smaller. When the delamination opens (increases in length), the mode mixity reduces in favor of the Mode II component. Eventually, 4/15c and 4/29a cross-ply and 4/29a unidirectional reach pure Mode II condition. The trend existing in the cross-ply laminates is the same with the unidirectional ones. The higher the h/T ratio in the specimens the higher the Mode I component in the mode mixity. Also, high values of mode mixity are favored in configurations with smaller comparatively applied strain.

The energy release rate comparison is shown in Fig. 6. From this diagram, it is apparent that the values of the energy release rate for the unidirectional specimens are higher than the

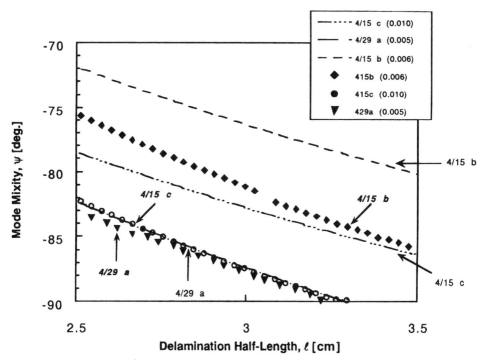

FIG. 5—*Mode mixity, ψ^0, versus delamination half-length, ℓ, for unidirectional specimen configurations (noted by dashed lines) and cross-ply ones (noted by full marks and open circles in the case of 4/15c). The number in the parenthesis denotes the maximum displacement applied to each specimen (1 in. = 0.0254 m). The pure Mode II is for $\psi = \pm 90^0$.*

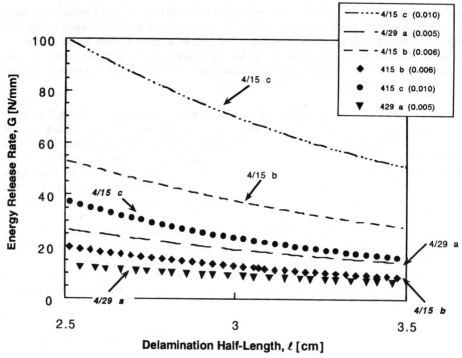

FIG. 6—*Energy release rate, G, versus delamination half-length, ℓ, for unidirectional specimen configurations (noted by dashed lines) and cross-ply ones (noted by full marks). The number in the parenthesis denotes the maximum displacement applied to each specimen (1 in. = 0.0254 m).*

corresponding cross-ply values. For both layups the energy release rate is higher for higher h/T ratios with the smaller applied strain. A possible explanation for this trend is given through the initial postbuckling solution model. As seen in this model, the energy release rate is directly analogous to the forces and moments applied at the common cross section at each time. The absolute values of these loads are smaller in the case of the cross-ply specimens.

The plot of the number of cycles versus the delamination length for the unidirectional layup is shown in Fig. 7. The m_I and C_I parameters for the mode-dependent cyclic growth law for the unidirectional specimens were obtained from three data points (ℓ, N) of the 4/30b specimen. The data points used were: 24.130 mm, 86 280 cycles; 24.968 mm, 151 012 cycles; and 25.781 mm, 231 985 cycles. The values obtained were $m_I = 10.385$ and $C_I = 0.0435$ m/cycle.

The experiments made for this case used the 4/30 and 4/15 configurations with different initial strains than the ones used in the cross-ply layup. Therefore, the comparison regarding the growth rate is treated through the proposed crack growth law. From the proposed law, it is concluded that the delamination grows faster in the unidirectional specimens than in the cross-ply ones, since the m_I exponent is larger in the unidirectional case, that is, $m_I = 10.385$ for the unidirectional and $m_I = 4.6$ for the cross-ply specimens. In addition, from Figs. 2 and 7, it is apparent that similar trends relating the number of cycles required to grow the delamination and the h/T ratio hold for both specimen types. The number of cycles needed for delamination growth is higher for smaller ratios in both cases. This conclusion comes from the study of the energy release rate diagram. The low h/T ratio specimens have a lower energy release rate than the specimens with high h/T ratios. Therefore, they require a larger number of cycles

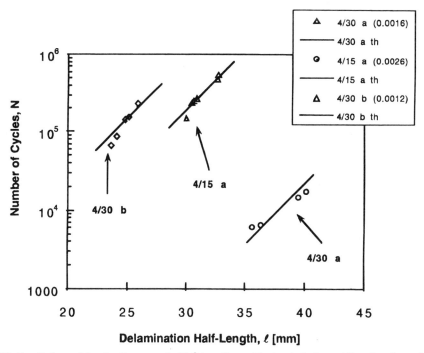

FIG. 7—*Fatigue delamination growth, N-ℓ (semilogarithmic plot), for unidirectional specimens [11]. The number in the parenthesis denotes the initial displacement applied to each specimen (1 in. = 0.0254 m). In each case, the experimental data are denoted with discrete empty marks, whereas the lines represent the predictions of theory. Proposed crack growth law parameters: m_I = 10.385, C_I = 0.0435 m/cycle.*

to grow the same amount of delamination. Another parameter related to the number of cycles required to grow a delamination is the initial applied strain. Figures 2 and 7 show that the number of cycles required for the same specimen configuration is smaller if the initial strain applied is larger. The unidirectional configurations seem to be more sensitive to this parameter than the cross-ply ones, since for a small difference in the initial strain value the number of cycles required for the 4/30a specimen decreases dramatically.

Conclusions

In this study, C30/922 cross-ply graphite/epoxy configurations are experimentally and analytically examined under compressive fatigue loading at constant displacement amplitude. Delaminations are introduced in the specimens and the effects of applied strain, and delamination length on the energy release rate, mode mixity, and delamination fatigue growth behavior is investigated. The delaminated specimens undergo buckling. The results from the cross-ply specimens of the present study are compared with results of unidirectional specimens from a previous study.

From the tests and the analysis conducted, the following results are summarized:

1. Experimental and analytical results indicate that the number of cycles required for delamination growth is larger for cross-ply specimens than for unidirectional specimens at the same applied maximum strain.

2. Analytical results indicate that the energy release rate reduces with delamination growth for both material configurations.

3. The growth of delaminations takes place under mode-mixity conditions characterized by a relatively low Mode II component in the mode mixity for unidirectional, and a relatively high value of Mode II component for cross-ply configurations. For both specimen configurations, the mode mixity decreases as the delamination grows.

4. From the analysis conducted, it was found that the h/T ratio is an important parameter for the energy release rate and the mode mixity in both materials. The trend appears to be that specimens with higher h/T ratios exhibit higher energy release rate values and higher values of Mode I component in the mode mixity.

5. Finally, the correlation between analytical and experimental results is reasonable. The values for the m_1 and C_1 parameters of the mode-dependent cyclic growth law are in good correlation with values established in previous studies [9,11].

Acknowledgments

The financial support of the Office of Naval Research, Ship Structures and Systems Division, S & T Division Grant N00014-91-J-1892, and the interest and encouragement of the grant monitor, Dr. Y. Rajapakse, are both gratefully acknowledged. The authors are also grateful to Prof. R. L. Carlson for useful discussions and comments.

References

[1] Kardomateas, G. A. and Schmueser, D. W., "Buckling and Postbuckling of Delaminated Composites under Compressive Loads Including Transverse Shear Effects," *Journal of American Institute of Aeronautics and Astronautics*, Vol. 26, No. 3, 1988, pp. 337–343.

[2] Hong, S. and Liu, D., "On the Relationship Between Impact Energy and Delamination Area," *Experimental Mechanics*, Vol. 29, No. 2, 1989, pp. 115–120.

[3] Chai, H., Babcock, C. D., and Knauss, W. G., "One Dimensional Modeling of Failure in Laminated Plates by Delamination Buckling," *International Journal of Solids and Structures*, Vol. 17, 1981, pp. 1069–1083.

[4] Chai, H. and Babcock, C. D., "Two Dimensional Modeling of Compressive Failure in Delaminated Laminates," *Journal of Composite Materials*, Vol. 19, 1985, pp. 67–97.

[5] Evans, A. G. and Hutchinson, J. W., "On the Mechanics of Delamination and Spalling in Compressed Films," *International Journal of Solids and Structures*, Vol. 20, No. 5, 1984, pp. 455–466.

[6] Shivakumar, K. N. and Whitcomb, J. D., "Buckling of a Sublaminate in a Quasi-Isotropic Composite Laminate," *Journal of Composite Materials*, Vol. 19, 1985, pp. 2–18.

[7] Simitses, G. J., Sallam, S., and Yin, W. L., "Effect on Delamination on Axially Loaded Homogenous Laminated Plates," *Journal of American Institute of Aeronautics and Astronautics*, Vol. 23, No. 9, 1985, pp. 1437–1444.

[8] Kardomateas, G. A., "Postbuckling Characteristics in Delaminated Kevlar/Epoxy Laminates: An Experimental Study," *Journal of Composites Technology and Research*, Vol. 12, No. 2, 1990, pp. 85–90.

[9] Russell, A. J. and Street, K. N., "Predicting Interlaminar Fatigue Crack Growth Rates in Compressively Loaded Laminates," *Composite Materials: Fatigue and Fracture*, Vol. 2, *ASTM STP 1012*, P. Lagace, Ed., American Society for Testing and Materials, Philadelphia, 1989, pp. 162–178.

[10] Whitcomb, J. D., "Strain-Energy Release Rate Analysis of Cyclic Delamination Growth in Compressively Loaded Laminates," *Effects of Defects in Composite Materials, ASTM STP 836*, K. L. Reifsnider, Ed., American Society for Testing and Materials, Philadelphia, 1984, pp. 175–193.

[11] Kardomateas, G. A., Pelegri, A. A., and Malik, B., "Growth of Internal Delaminations Under Cyclic Compression in Composite Plates," *Failure Mechanics in Advanced Polymeric Composites*, AMD Vol. 196, Y. T. D. S. Rajapakse and G. A. Kardomateas, Eds., American Society of Mechanical Engineers, 1994; *Journal of the Mechanics and Physics of Solids*, Vol. 43, No. 6, 1995, pp. 847–868.

[12] Kardomateas, G. A., "The Initial Postbuckling and Growth Behavior of Internal Delaminations in Composite Plates," *Journal of Applied Mechanics*, Vol. 60, 1993, pp. 903–910.

[13]   Kardomateas, G. A., "Large Deformation Effects in the Postbuckling Behavior of Composites with Thin Delaminations," *Journal of American Institute of Aeronautics and Astronautics*, Vol. 27, 1989a, pp. 624–631.

[14]   Wang, S. S., Zahlan, N. M., and Suemasu, H., "Compressive Stability of Delaminated Random Short Fiber Composites, Part I—Modeling and Methods of Analysis," *Journal of Composite Materials*, Vol. 19, 1985, pp. 296–316.

[15]   Sheinman, I. and Adan, M., "The Effect of Shear Deformation on Postbuckling Behavior in Laminated Beams," *Journal of Applied Mechanics*, Vol. 54, 1985, pp. 558–562.

[16]   Kardomateas, G. A. and Pelegri, A. A., "The Stability of Delamination Growth in Compressively Loaded Composite Plates," *International Journal of Fracture*, Vol. 65, No. 3, Feb. 1994, pp. 261–276.

[17]   Whitcomb, J. D., "Parametric Analytical Study of Instability-Related Delamination Growth," *Composite Science and Technology*, Vol. 25, 1986, pp. 19–48.

[18]   Rothschilds, R. J., Gillespie, J. W., Jr., and Carlsson, L. A., "Instability-Related Delamination Growth in Thermoset and Thermoplastic Composites," *Composite Materials: Testing and Design (Eighth Conference), ASTM STP 972*, J. D. Whitcomb, Ed., American Society for Testing and Materials, Philadelphia, 1988, pp. 161–179.

[19]   Yin, W. L., "Axisymmetric Buckling and Growth of a Circular Delamination in a Compresses Laminate," *International Journal of Solids and Structures*, Vol. 21, No. 5, 1985, pp. 503–514.

[20]   Chai, H., "Experimental Evaluation of Mixed Mode Fracture in Adhesive Bonds," *Experimental Mechanics*, Dec. 1992, pp. 296–303.

[21]   O'Brien, T. K., "Towards a Damage Tolerance Philosophy for Composite Materials and Structures," *Composite Materials: Testing and Design, Ninth Volume, ASTM STP 1059*, S. P. Garbo, Ed., American Society for Testing and Materials, Philadelphia, 1990, pp. 7–33.

[22]   Sheinman, I. and Kardomateas, G. A., "Energy Release Rate and Stress Intensity Factors for Delaminated Composites," *International Journal of Solids and Structures*, in print.

[23]   Sheinman, I., Kardomateas, G. A., and Pelegri, A. A., "Delamination Growth During Pre- and Post-Buckling Phases of Delaminated Composite Laminates," *International Journal of Solids and Structures*, submitted for publication.

[24]   Britvek, S. J., The Stability of Elastic Systems, Pergamon, New York, 1973.

[25]   Suo, Z. and Hutchinson, J. W., "Interface Crack between Two Elastic Layers," *International Journal of Fracture*, Vol. 43, 1990, pp. 1–18.

[26]   Hutchinson, J. W. and Suo, Z., "Mixed Mode Cracking in Layered Materials," *Advances in Applied Mechanics*, Academic Press, New York, Vol. 29, 1992, pp. 63–191.

[27]   Agarwal, B. D. and Broutman, L. J., Analysis and Performance of Fiber Composites, 2nd ed., Wiley Interscience, New York, 1990.

[28]   Dundurs, J., Mathematical Theory of Dislocations, American Society of Mechanical Engineering, New York, 1969, pp. 70–115.

[29]   Suo, Z., "Delamination Specimen for Orthotropic Materials," *Journal of Applied Mechanics: Transactions, American Society of Mechanical Engineers*, Vol. 57, No. 3, Sept. 1990, pp. 627–634.

[30]   Hellan, K., Introduction to Fracture Mechanics, Mc Graw Hill, New York, 1984.

[31]   Hong, C. S., and Jeong, K. Y., "Stress Intensity Factor in Anisotropic Sandwich Plate with a Part Through Crack Under Mixed Mode Deformation," *Engineering Fracture Mechanics*, Vol. 21, No. 2, 1985, pp. 285–292.

[32]   O'Brien, T. K. and Kevin, T., "Mixed-Mode Strain-Energy-Release Rate Effects on Edge Delamination Composites," *Effects of Defects in Composite Materials, ASTM STP 836*, K. L. Reifsnider, Ed., American Society for Testing and Materials, Philadelphia, 1984, pp. 125–142.

[33]   Murri, G. B. and Martin, R. H., "Effect of Initial Delamination on Mode I and Mode II Interlaminar Fracture Toughness and Fatigue Threshold," *Composite Materials: Fatigue and Fracture, Fourth Volume, ASTM STP 1156*, W. W. Stinchcomb and N. E. Ashbaugh, Eds., American Society for Testing and Materials, Philadelphia, 1993, pp. 239–256.

[34]   Kardomateas, G. A. and Pelegri, A. A., "Growth Behavior of Internal Delaminations in Composite Beam/Plates Under Compression: Effect of the End Conditions," *International Journal of Fracture*, Vol. 75, 1996, pp. 49–67.

Ronald Krüger[1] and Manfred König[1]

Prediction of Delamination Growth Under Cyclic Loading

REFERENCE: Krüger, R. and König, M., **"Prediction of Delamination Growth Under Cyclic Loading,"** *Composite Materials: Fatigue and Fracture (Sixth Volume), ASTM STP 1285,* E. A. Armanios, Ed., American Society for Testing and Materials, 1997, pp. 162–178.

ABSTRACT: The growth of delaminations in carbon fiber-reinforced epoxy (CFRE) specimens during $R = 0.1$ and $R = -1$ fatigue loading has been studied. Artificial circular and square delaminations as well as ply cuts have been introduced at various interfaces during manufacturing to simulate a pre-damaged structure and to cause delamination growth. Criteria based on fracture mechanics will be used to describe the delamination failure. Predicting delamination growth with this approach requires the distribution of the local energy release rate along the delamination front. For obtaining this energy release rate distribution, the virtual crack closure method was found to be most favorable for three-dimensional finite element analysis as the separation of the total energy release rate into the contributing modes is inherent to the method and only one complete finite element analysis is necessary. Plots of measured delamination progression per load cycle (*da/dN*-values) versus computed energy release rates have been included in a Paris law diagram as obtained experimentally using double cantilever beam (DCB) specimens to characterize Mode I and end-notched flexure (ENF) and transverse crack tension (TCT) specimens to characterize Mode II failure, respectively. Computed mixed-mode results lie well within the scatter band of the experimentally determined Paris law for Mode I and Mode II failure.

KEYWORDS: delamination growth, delamination buckling, three-dimensional finite element analysis, energy release rate, virtual crack closure method, mixed-mode failure, Paris law, fatigue (materials), fracture (materials), composite materials

Motivated by the increasing use of composites in primary structural components, research has been focused on the disbond of two adjacent fiber-reinforced layers of a laminate that is a prevalent state of damage commonly known as "delamination." A recent survey on problems concerning composite parts of civil aircraft [1] shows that delamination, mainly caused by impact, presents 60% of all damage observed. Up to now, failure caused by delamination is prevented by using empirically determined design criteria—based on maximum strains—during layout and construction of components made of fiber-reinforced materials [2]. For an optimal utilization of the potential offered by those materials as well as for the determination of inspection intervals, however, it is essential to be able to predict delamination growth under cyclic loading.

In this investigation, criteria based on fracture mechanics are used to describe the delamination failure. Propagation therefore is to be expected when a function of the mixed-mode energy release rates (G_I, G_{II}, G_{III}) along the delamination front locally exceeds a certain value, G_c. This value can be regarded as a property of the interface and depends on the material and on the ply orientations of the layers adjacent to the plane of delamination. The goal of the

[1]Research associate and senior scientist, respectively, Institute for Statics and Dynamics of Aerospace Structures, University of Stuttgart, 70550 Stuttgart, Germany.

investigation presented is to obtain information on the dependence of delamination growth on the mixed-mode energy release rates for static and cyclic loading in real structures using a combined experimental-numerical procedure as illustrated in Fig. 1 and to verify the approach considered.

For the experimental determination of critical energy release rates, several simple test specimens have been developed, such as the double cantilever beam (DCB) specimen for determination of G_{Ic}, the end-notched flexure (ENF) specimen, and the transverse crack tension (TCT) specimen for G_{IIc}. Following a "global/local testing and analysis approach" as described in Ref 3, failure criteria need to be verified and improved at each stage, that is, from the level of material characterization via coupon and subelement level up to the substructure and full

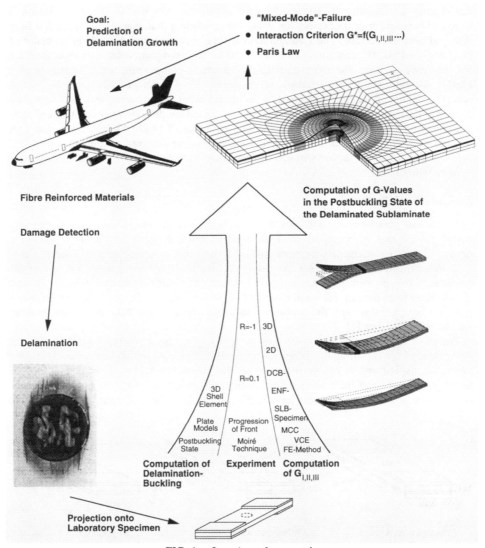

FIG. 1—*Overview of approach.*

structure level. Therefore, more complex specimens as shown in Fig. 2 were used on a subelement level. Looking at the fracture toughness data for untoughened carbon/epoxy materials, we notice that the Mode II values are about three to eight times higher than the corresponding Mode I values. It is therefore necessary in investigating failure mechanisms to consider the individual mode contributions of the energy release rate along a delamination contour. A simple analytical evaluation of the test data (as, for example, beam theory for DCB- and ENF-specimens) is no longer possible for the specimens considered. Computational methods based on three-dimensional finite element modeling are a meaningful and efficient tool, by which the energy release rate at delamination growth along the entire delamination front can be evaluated.

The specimens shown in Fig. 2, all made of unidirectionally reinforced prepreg material (Ciba Geigy T300/914C), were designed to simulate specific states of damage. The Configuration A type specimen with $[\pm5/+45/\pm5/-45/0/\pm85/0/-45/\mp5/+45/\mp5]$ layup and an artificial circular delamination of 10-mm diameter at Interface 2/3 was designed to simulate a structural component subject to tension-compression loading that, due to an impact, has been damaged near the surface. The Configuration B type specimen with a stacking sequence of $[0_2/+45/0_2/-45/0/90]_S$ containing an artificial 10 by 10-mm-square delamination at Interface 3/4 was used to serve the same purpose for structural components under tension-tension loading. An additionally introduced 10-mm cut through Ply 4 directly beneath the delamination serves to simulate fiber fracture caused by an impact event. Experiments with both specimen types will provide input data to set up a predictive model for delamination onset and growth. Specimens of Configuration C with a stacking sequence of $[\pm2/\pm45/\pm45/\pm88]_S$ and cuts through the width of both surface plies (Plies 1 and 16) were designed to simulate a structural part with ply drops and further will serve for verification tests, once the predictive model has been completed.

Computational Tools

An obvious possibility for obtaining the contributions of the individual modes to the total energy release rate is to employ three-dimensional finite element (FE) analysis. A newly developed "layered element with eight nodes" formulated according to a continuum-based three-dimensional shell theory [4] has been used for the three-dimensional models of the specimens. This element has been employed in order to reduce computing time, to prevent the elements from locking for small element thickness to element length ratios and to assure complete compatibility with the contactor and target elements that have been used to avoid structural overlapping [5].

The most significant step for the current approach is the accurate computation of the distribution of the mixed-mode energy release rates along delamination fronts that developed during the experiment. It has been found that the virtual crack closure method—based on nodal point forces and the corresponding relative displacements—is most favorable for the computation of energy release rates because the separation of the total energy release rate into

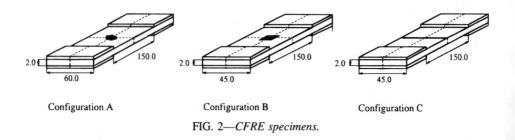

Configuration A   Configuration B   Configuration C

FIG. 2—*CFRE specimens.*

the contributions by the different crack opening modes is possible in a straight forward manner [6–9]. When using this method, only one FE-computation is necessary for a given delamination front that is beneficial especially when solving large geometrically nonlinear problems. Preliminary investigations employing DCB and ENF specimens assured the reliability of this technique as used for computation of energy release rate distributions along straight and curved delamination fronts [10]. In addition, convergence studies for ENF and single leg bending (SLB) specimens [11–13] did not show the nonconvergence of the virtual crack closure method associated with the oscillatory singularity as reported in the literature, for example, in Refs 14 and 15. This may be due to very small bimaterial constants of the considered interfaces [16] or due to finite element meshes that are in the range of relatively constant results for G_I, G_{II}, and G_{III} [15]. The issue will be investigated further.

Delamination Growth Under Tension-Compression Fatigue Loading

Experimental Program

In preparation of the experimental part of the program, Configuration A specimens have been cut from prefabricated plates with a stacking sequence of [±5/+45/±5/−45/0/±85/0/ −45/∓5/+45/∓5]. Artificial circular delaminations have been introduced at Interface 2/3 during manufacturing by embedding a double foil of a 20-μm-thick release film to simulate a pre-damaged structure. An antibuckling guide with rectangular windows on both sides of the specimen is clamped to the test specimen, suppressing its global buckling. Loading frequency is 10 Hz at $R = -1$ with stress maximums varying between 220 to 240 N/mm^2. These stress maximums have been shown to yield stable delamination growth. The out-of-plane (that is, buckling) deformation of the delaminated region has been measured via Moiré technique and has been displayed as grey levels of constant displacements. An example is given in Fig. 3.

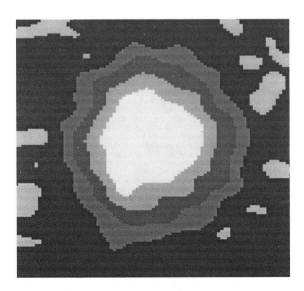

10.0 mm

FIG. 3—*Example of grey level distribution of out-of-plane (buckling) displacements.*

Using numerical post-processing procedures, the size and shape of the delaminated sublaminate is determined from this information, yielding the delamination contours [17]. An example of delamination contours, obtained by this procedure from four different specimens, is presented in Fig. 4. The obtained fronts for 100 000, 200 000, 300 000, 400 000, 500 000, and 1 million load cycles have been averaged and used as input to the corresponding finite element model.

Computation of Delamination Buckling Behavior

In a preliminary study, quadrilateral layered plate elements (with four nodes) that allow first-order transverse shear deformation have been used to keep computing times low. The investigations during this phase of the project focused on the boundary conditions to be applied along the modeled section of the specimen, the influence of the delamination size modeled, and the global performance of the nonlinear FE-code used [18,19]. Only the region inside the window of the antibuckling guide was modeled by two subplates with the nodes located at the plane of delamination over the entire region modeled. It was found that the boundary conditions applied to the edges of the region modeled have no significant influence on the computed deformations, provided we focus our interest on the region of stable delamination growth (for example, load levels of 220 to 240 N/mm²). This has been checked by applying first a clamped and then a simply supported condition. The same holds for the value of the linear buckling load for sublaminate buckling [20,21].

Due to the existence of ±α plies, models of the entire window section are necessary for accurately computing mixed-mode energy release rates along the delamination fronts, even in the case of geometrically symmetric specimens. Using the newly developed layered "three-

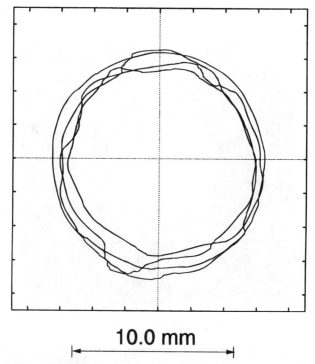

10.0 mm

FIG. 4—*Example of delamination contours, obtained from four specimens.*

dimensional shell element" with eight nodes [4], the top two [±5] layers of the sublaminate have been modeled by one element over the thickness of each layer. Also, the [+45] layer of the base laminate that is adjacent to the plane of delamination has been modeled by one element over the thickness. The remaining 13 layers of the base laminate have been joined in two elements of [±5/−45/0/±85] and [0/−45/∓5/+45/∓5] layup, respectively. A simply supported condition has been applied as a boundary condition along the edges of the modeled section. This boundary condition represents the edges of the antibuckling guide. As already mentioned earlier, the boundary conditions chosen do not have a significant influence on the buckling behavior of the sublaminate.

The deformed geometry of a specimen with the 10-mm initial delamination is shown in Fig. 5 for an applied compression load of $\sigma = 280.0$ N/mm^2. This compressive load is close to the bifurcation point for global buckling as computed by linear buckling analysis. The buckling of the sublaminate is shown in detail in Fig. 6 for an applied compression load of $\sigma = 220.0$ N/mm^2 that corresponds to the load level where stable delamination growth is observed for this geometry.

Energy Release Rates

Computed mixed-mode energy release rates for Specimen A with an embedded circular delamination of 10-mm diameter are shown in Fig. 7 for an applied compression load at which stable delamination growth has been observed during the fatigue experiments. The computed energy release rates in the tension loading phase are negligible, confirming previous assumptions that delamination growth occurs in the compression phase of the test. Looking at the distribution of G_I, $G_{shear} = G_{II} + G_{III}$, and G_{tot}, we expect the delamination to start growing into the load direction, whereas the potential to grow perpendicular to the load direction seems to be very low, as computed results in this area are significantly below the threshold values G_{Ith} and G_{IIth} (see the section on the Paris law). This is in agreement with experimental results, as shown in Fig. 8. It is, however, different from the observations in Refs 22 and 23 where an isotropic material behavior and a homogeneous quasi-isotropic laminate were assumed, respectively. This indicates that delamination growth depends on the stacking sequence of the delaminated

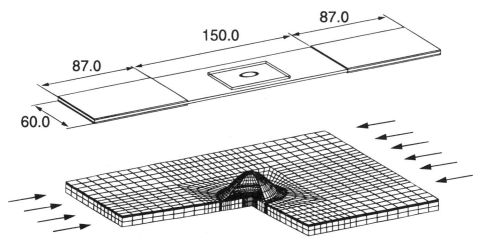

FIG. 5—*Computed post-buckling deformation (magnified) for a 10-mm diameter delamination (compression loading 280 N/mm^2).*

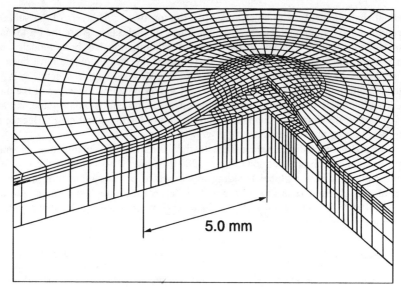

FIG. 6—*Detail of post-buckling deformation (magnified) for a 10-mm diameter delamination (compression loading 220 N/mm².)*

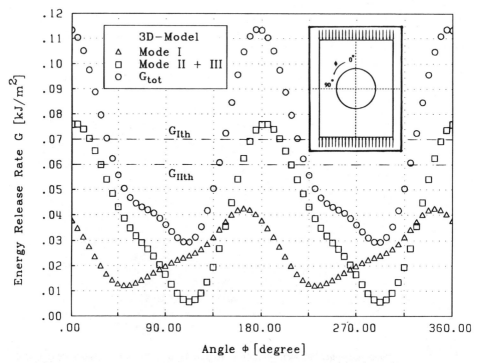

FIG. 7—*Computed energy release rates along delamination front for Specimen A.*

FIG. 8—*X-ray photo of growing delamination in Specimen A (500 000 load cycles).*

sublaminate and the base laminate. Due to the fact that G_I, G_{shear}, and G_{tot} have their maximums at about the same location, an identification of the dependency of the growth rate on the mixed-mode ratio, G_I/G_{shear}, is not possible in this example.

It should be mentioned that an only slight modification of the initial delamination diameter, (that is, 10.6 mm instead of nominally 10.0 mm) results in a 53% increase in computed total energy release rate. Analyses without and with the release foil (40 μm) show an increase in computed energy release rates up to 290% for the model with foil. Therefore, the release foil needs to be included in the model [24].

Delamination Growth Under Tension-Tension Fatigue Loading

Experimental Program

Next, the growth of delaminations under $R = 0.1$ fatigue loading has been studied. The Configuration B specimens (as shown in Fig. 2) have been cut from prefabricated plates made of the same material (T300/914C) with an altered stacking sequence of $[0_2/+45/0_2/-$

45/0/90]$_S$. Artificial 10 by 10-mm-square delaminations have been introduced at Interface 3/4 during manufacturing by embedding a double foil of release film to simulate a pre-damaged structure. For the $R = 0.1$ fatigue loading, a frequency of 3 Hz was selected in order to avoid any heat generation that might alter the material behavior. Stress maximums were chosen to be 30% of the ultimate tensile strength (UTS). For this setup, delamination growth could not be observed even after several hundred thousand load cycles.

In order to induce even more severe damage and to force the delamination to grow, new specimens have been manufactured where the [0]-ply directly beneath the deliberate delamination was damaged by a 10-mm cut through the fibers during layup and before curing. Using C-scan for detection, delamination growth was then observed at ply Interface 4/5 while the deliberate delamination at Interface 3/4 remained unaffected and did not start growing before the new delamination beneath had reached the same size and shape. As an example of a C-scan, the situation after 4000 load cycles is shown in Fig. 9. Eight different delamination fronts have been averaged, yielding the smoothed contours as given in Fig. 10. Selected contours have been used as input to the finite element analyses. At a stress level of 30% UTS, no further delamination growth could be observed once the growing delamination had reached the same size and shape of the deliberate delamination at the interface above. Increasing the stress level to 40 and 50% UTS caused both delaminations to grow simultaneously.

**Deliberate delamination
in interface 3/4**

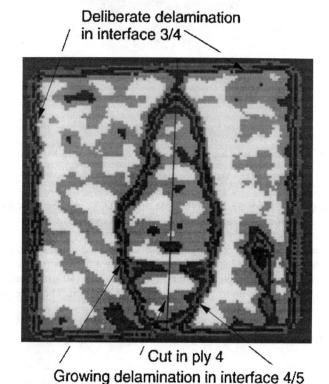

**Cut in ply 4
Growing delamination in interface 4/5
after 4,000 load cycles**

FIG. 9—*Delamination growth in Specimen B, determined by C-scan after 4000 load cycles.*

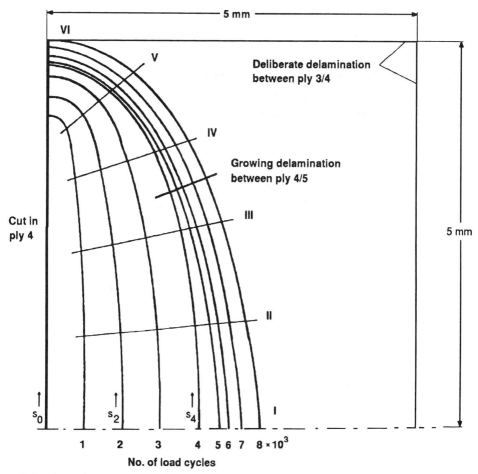

FIG. 10—*Delamination growth in Specimen B after smoothing the delamination contours (one quarter of delaminated region shown).*

Computation of the Deformation Behavior

From the delamination fronts illustrated in Fig. 10, the s_1, s_2, s_3, s_4, and s_7 fronts that developed after 1000 up to 7000 load cycles, respectively, were selected for three-dimensional analyses. Due to the structural asymmetry of the geometrically symmetric specimen caused by the $\pm\alpha$ plies, models extending over the entire width of the specimen are necessary, as shown in Fig. 11. The two top [0]-layers of the sublaminate have been grouped into one three-dimensional shell element, followed by one element that models the [45]-ply adjacent to the artificial square delamination. Two elements over the thickness were used for the cut [0]-layer, one for the [0]-ply to follow, which is adjacent to the new growing delamination. The remaining eleven layers of the base laminate were joined into two elements of $[-45/0/90_2/0/-45]$ and $[0_2/+45/0_2]$ layup, respectively. This configuration was used for the entire three-dimensional investigation of all fronts. A penetration of the layers inside the delaminated regions was prevented by using a contact processor that utilizes a contactor target concept applying the

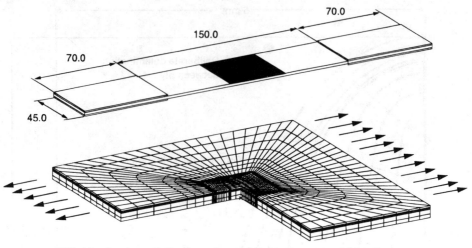

FIG. 11—*Specimen B Configuration and finite element model for Front* s_4.

penalty method [5]. Computed results show that the contact between the delaminated surfaces occurs only locally (Fig. 12), resulting in a Mode I opening.

Energy Release Rates

As an example for the typical distribution of energy release rates along the deliberate delamination (Interface 3/4), the energy release rates for the situation after 3000 load cycles are shown in Fig. 13. Along the entire front, energy release rates are far below any threshold values, G_{Ith}, G_{IIth}, for delamination growth (see the following section) so that the contour is expected to remain constant. Significant peaks are being observed in the immediate vicinity

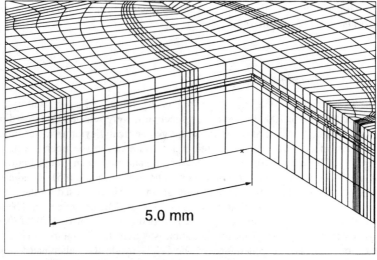

5.0 mm

FIG. 12—*Detail of deformed geometry for Front* s_3.

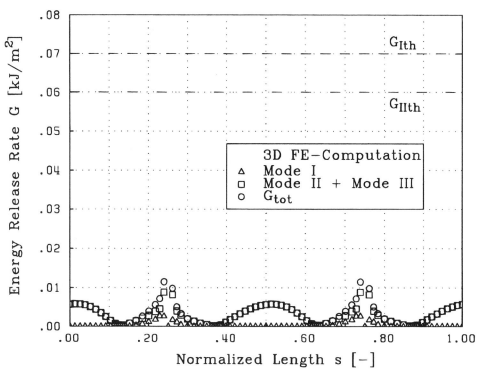

FIG. 13—*Computed energy release rates along the square delamination (Interface 3/4) of Specimen B after 3000 load cycles.*

of the ply cut ($s = 0.25$ and $s = 0.75$) due to the stress concentration caused by the cut. For the growing front between Plies 4 and 5 (s_3 obtained after 3000 load cycles), results are plotted in Fig. 14. Energy release rates are distributed evenly along a considerable part of the growing front, with significant deviations again only in the immediate vicinity of the ply cut. A significant Mode I contribution caused by a crack opening is noticeable. The shear mode contribution, G_{shear}, is considerably higher than the experimentally obtained threshold value G_{IIth} for delamination growth (see the following section) so that delamination growth is to be expected along the entire front. This is in agreement with experimental observations.

Summarizing, the total energy release rates are fairly constant along the entire front of the propagating delamination, dropping progressively with growing delamination as shown in Fig. 15. Mode I contribution decreases until it finally becomes negligible. This is illustrated in Fig. 16 for the growing s_1, s_2, s_3, s_4, and s_7 fronts by plotting the mode mixity

$$r = \frac{G_I}{G_{II} + G_{III}}$$

along the front.

Comparing the results to those obtained using a quarter model [25], we notice a remarkable difference, concluding that the asymmetry caused by the ± 45 plies has to be taken into account, yielding the necessity of a full-width model. Additionally, it becomes obvious that simply

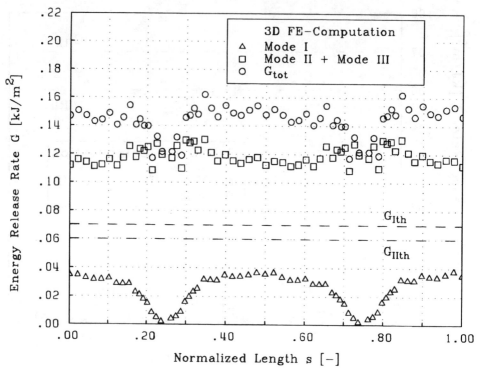

FIG. 14—*Computed energy release rates along the propagating Front s_3 (Interface 4/5) of Specimen B after 3000 load cycles.*

suppressing any Mode I contribution to avoid layer interpenetration is insufficient and contact analysis becomes essential.

Comparison with Paris Law

Paris law fits for fatigue loading, as obtained experimentally using DCB specimens to characterize Mode I and ENF and TCT specimens to characterize Mode II failure, respectively, are found in Ref 26 and have been plotted in Fig. 17. There is quite a difference between Paris law lines for Mode I and Mode II, indicating that the delamination growth rate in general depends on the mixed-mode ratio, r. This is consistent with the differences observed for fracture toughness. The threshold values, G_{Ith}, G_{IIth}, below which delamination growth does not occur have also been taken from Ref 26. For Configuration B specimens, the delamination growth, Δa, was measured between two consecutive fronts along the fronts considered. Resulting delamination growth rates, $\Delta a/\Delta N$ (N = number of load cycles), have been plotted versus the computed total energy release rates of the corresponding front and included in Fig. 17. Computational results lie well between the experimentally determined straight lines for Mode I and Mode II failure, the points with small mixed-mode r ratio (according to Fig. 15), being closer to the failure line for Mode II. The influence of the mixed-mode r ratio on the location of points between the two Paris law lines has to be investigated further.

Conclusions and Outlook

The application of elastic fracture mechanics criteria has been tested as a possible tool for predicting delamination growth under tension-compression and tension-tension fatigue loading.

For tension-tension, experiments with Configuration B specimens show that measured delamination progression in a 0°/0° interface fits well into a Paris law as obtained experimentally using simple specimens for material characterization. These results suggest that growth prediction based on Paris law is an approach to be considered for further investigations. Configuration C specimens designed to simulate a structural part with ply drops, will serve for additional verification tests. Due to the observed mixed-mode behavior along the fronts, the knowledge of the conditions for Mode I and Mode II fatigue failure alone is not sufficient. Mixed-mode bending (MMB) tests become necessary to determine Paris law for mode mix.

In general, delamination growth occurs between plies of dissimilar orientations that is simulated, for example, by Configuration A specimens. Fracture toughness characterization of angle ply interfaces is in progress [27], data for fatigue failure, however, are currently not available. A quantitative comparison of the experimental behavior of the Configuration A specimens with predictions is therefore not yet possible. This will be the focus of future investigations.

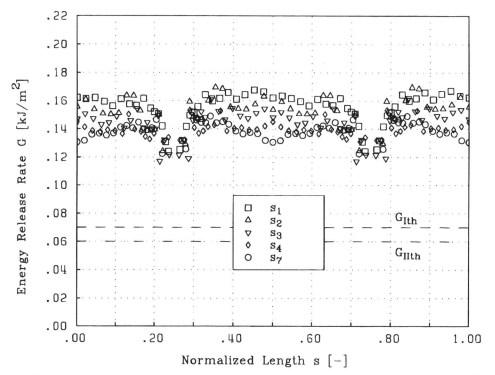

FIG. 15—*Computed total energy release rates along propagating* s_1, s_2, s_3, s_4, *and* s_7 *fronts of Specimen B.*

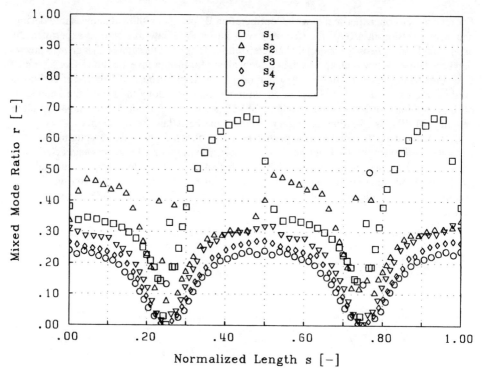

FIG. 16—*Computed mixed-mode ratio,* $r = G_I/(G_{II} + G_{III})$, *along propagating* s_1, s_2, s_3, s_4, *and* s_7 *Fronts of Specimen B.*

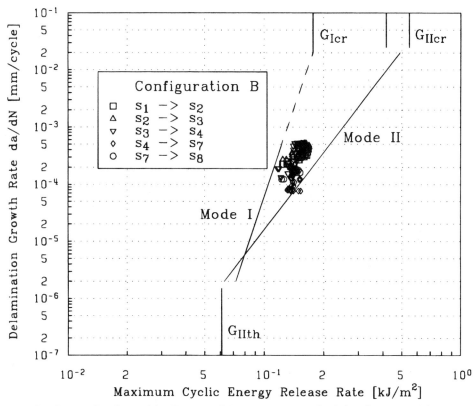

FIG. 17—*Results obtained for Specimen B in comparison to Paris law from material tests.*

References

[1] Miller, A. G., Lovell, D. T., and Seferis, J. C., "The Evolution of an Aerospace Material: Influence of Design, Manufacturing and In-Service Performance," *Composite Structures*, Vol. 27, 1994, pp. 193–206.

[2] Hart, W. G. J. and Frijns, R. H. W. M., "Delamination Growth in Improved Carbon Composites Under Constant Amplitude Fatigue Loading," Technical Report NLR TP 89008 U, National Aerospace Laboratory NLR, The Netherlands, 1989.

[3] Martin, R. H., "Local Fracture Mechanics Analysis of Stringer Pull-Off and Delamination in a Post-Buckled Compression Panel," *The Tenth International Conference on Composite Materials, Vol. I*, A. Poursartip and K. Street, Eds., Woodhead Publishing Ltd., 1995, pp. 253–260.

[4] Parisch, H., "A Continuum-Based Shell Theory for Non-Linear Applications," *International Journal for Numerical Methods in Engineering*, Vol. 38, 1995, pp. 1855–1883.

[5] Parisch, H., "A Consistent Tangent Stiffness Matrix for Three-Dimensional Non-Linear Contact Analysis," *International Journal for Numerical Methods in Engineering*, Vol. 28, 1989, pp. 1803–1812.

[6] Rybicki, E. F. and Kanninen, M. F., "A Finite Element Calculation of Stress Intensity Factors by a Modified Crack Closure Integral," *Engineering Fracture Mechanics*, Vol. 9, 1977, pp. 931–938.

[7] Raju, I. S., Shivakumar, K. N., and Crews, J. H., Jr., "Three-Dimensional Elastic Analysis of a Composite Double Cantilever Beam Specimen," *AIAA Journal*, Vol. 26, 1988, pp. 1493–1498.

[8] Buchholz, F. G., Grebner, H., Dreyer, K. H., and Krome, H., "2D- and 3D-Applications of the Improved and Generalized Modified Crack Closure Integral Method," *Computational Mechanics '88*, S. N. Atluri and G. Yagawa, Eds., Springer Verlag, New York, 1988.

[9] Shivakumar, K. N., Tan, P. W., and Newman, J. C., Jr., "A Virtual Crack-Closure Technique for Calculating Stress Intensity Factors for Cracked Three Dimensional Bodies," *International Journal of Fracture*, Vol. 36, 1988, pp. R43–R50.

[10] Krüger, R., König, M., and Schneider, T., "Computation of Local Energy Release Rates Along Straight and Curved Delamination Fronts of Unidirectionally Laminated DCB- and ENF-Specimens," *Proceedings*, The Thirty-fourth AIAA/ASME/ASCE/AHS/ASC Structures, Structural Dynamics and Materials Conference, La Jolla, CA, AIAA-93-1457-CP, 1993, pp. 1332–1342.

[11] Davidson, B. D. and Sundararaman, V., "A Single Leg Bending Test for Interfacial Fracture Toughness Determination," *International Journal of Fracture*, accepted for publication.

[12] Davidson, B. D., Krüger, R., and König, M., "Three Dimensional Analysis of Center Delaminated Unidirectional and Multidirectional Single Leg Bending Specimens," *Composites Science and Technology*, Vol. 54, No. 4, 1995, pp. 385–394.

[13] Krüger, R., "Three Dimensional Finite Element Analysis of Multidirectional Composite DCB, SLB and ENF Specimens," ISD-Report No. 94/2, Institute for Statics and Dynamics of Aerospace Structures, University of Stuttgart, Germany, 1994.

[14] Raju, I. S., Crews, J. H., and Aminpour, M. A., "Convergence of Strain Energy Release Rate Components for Edge-Delaminated Composite Laminates," *Engineering Fracture Mechanics*, Vol. 30, No. 3, 1988, pp. 383–396.

[15] Hwu, C. and Hu, J., "Stress Intensity Factors and Energy Release Rates of Delaminations in Composite Laminates," *Engineering Fracture Mechanics*, Vol. 42, No. 6, 1992, pp. 977–988.

[16] Gao, H., Abbudi, M., and Barnett, D. M., "Interfacial Crack-Tip Field in Anisotropic Elastic Solids," *Journal of the Mechanics and Physics of Solids*, Vol. 40, No. 2, 1992, pp. 393–416.

[17] Hänsel, C. and Eberle, K., "Measuring Propagation of Delaminations in CFRP-Laminates by Moiré Technique," *Composites Testing and Standardisation, ECCM-CTS, Amsterdam*, EACM, European Association for Composite Materials, 1992, pp. 417–424.

[18] Parisch, H., "*NOVA User Manual*, Institute for Statics and Dynamics of Aerospace Structures, University of Stuttgart, Germany, 1991.

[19] Parisch, H., "An Investigation of a Finite Rotation Four Node Assumed Strain Shell Element," *International Journal for Numerical Methods in Engineering*, Vol. 31, 1991, pp. 127–150.

[20] König, M., Albinger, J., and Hänsel, C., "Delamination Buckling: Numerical Simulation of Experiments," *Proceedings*, ICCM-9, Madrid, Vol. VI, 1993, pp. 535–542.

[21] Krüger, R., König, M., Albinger, J., and Hänsel, C., "Combined Experimental-Numerical Approach for the Determination of Mixed-Mode Energy Release Rates at Delamination Growth," *Proceedings*, The Thirty-fifth AIAA/ASME/ASCE/AHS/ASC Structures, Structural Dynamics and Materials Conference, Hilton Head, SC, AIAA-94-1460-CP, 1994, pp. 1212–1222.

[22] Nilsson, K.-F. and Storåkers, B., "On Interface Crack Growth in Composite Plates," *Journal of Applied Mechanics*, Vol. 59, 1992, pp. 530–538.

[23] Whitcomb, J. D., "Three-Dimensional Analysis of a Postbuckled Embedded Delamination," *Journal of Composite Materials*, Vol. 23, Sept. 1989, pp. 862–889.

[24] Krüger, R., Rinderknecht, S., Hänsel, C., and König, M., "Computational Structural Analysis and Testing: An Approach to Understand Delamination Growth," E. A. Armanios, Ed., *Fracture of Composites, Key Engineering Materials*, Vols, 120–121, Transtec Publications Ltd., Switzerland, 1996, pp. 181–202.

[25] Krüger, R. and König, M., "Computation of Energy Release Rates: A Tool for Predicting Delamination Growth Under Cyclic Loading?" *Proceedings*, International Conference on Composites Engineering, ICCE/1, D. Hui, Ed., New Orleans, 1994, pp. 805–806.

[26] Prinz, R. and Gädke, M., "Characterization of Interlaminar Mode I and Mode II Fracture in CFRP Laminates," *Proceedings*, International Conference: Spacecraft Structures and Mechanical Testing, ESA SP-321, 1991, pp. 97–102.

[27] Davidson, B. D., Krüger, R., and König, M., "Three Dimensional Analysis and Resulting Design Recommendations for Unidirectional and Multidirectional End-Notched Flexure Tests," *Journal of Composite Materials*, Vol. 29, No. 16, 1995, pp. 2108–2133.

Jeffery R. Schaff[1] and Barry D. Davidson[2]

A Strength-Based Wearout Model for Predicting the Life of Composite Structures

REFERENCE: Schaff, J. R. and Davidson, B. D., **"A Strength-Based Wearout Model for Predicting the Life of Composite Structures,"** *Composite Materials: Fatigue and Fracture (Sixth Volume), ASTM STP 1285*, E. A. Armanios, Ed., American Society for Testing and Materials, 1997, pp. 179–200.

ABSTRACT: A model to predict the residual strength and life of polymeric composite structures subjected to spectrum fatigue loadings is described. The model is based on the fundamental assumptions that the structure undergoes proportional loading, that the residual strength is a monotonically decreasing function of the number of fatigue cycles, and that both the life distribution due to continuous constant amplitude cycling and the residual strength distribution after an arbitrary load history may be represented by two parameter Weibull functions. The model also incorporates a "cycle mix factor" to account for the drastic reduction of fatigue life that may be caused by a large number of changes in the stress amplitude of the loading. The model's predictions are compared to experimentally determined fatigue life distributions for uniaxial loadings of a number of laminates comprised of different materials and layups. Constant-amplitude, two-stress level, and spectrum fatigue loadings, including the FALSTAFF (Fighter Aircraft Loading STAndard For Fatigue) spectrum, are considered. The theoretical fatigue life distributions are shown to correlate well with the experimental results. Moreover, excellent correlation of theory and experiment is obtained for an "average fatigue life" that is based on the 63.2% probability of failure.

KEYWORDS: polymer-matrix composites, fatigue (materials), spectrum loading, life prediction, fracture (materials), composite materials

Fiber-reinforced composites are used frequently in secondary structures for aerospace, marine, and automotive applications. In most instances, these existing structures are stiffness-critical, and meeting the stiffness requirements has simultaneously alleviated any concerns about structural durability. However, as composite materials are used in strength-critical applications, structural durability can become a primary driver of design. This is of particular concern, as current fatigue life prediction methodologies generally do not yield accurate results under realistic loading conditions. Consequently, the design process for a strength-critical structure is forced to depend on expensive experimental programs that include a large amount of validation testing. This presents two dilemmas for industry: first, the design cannot easily be optimized in terms of cost and performance, and second, an "updated" life cannot be predicted if the in-service loading is found to be significantly different from the assumed design condi-

[1]Associate research engineer, United Technologies Research Center, East Hartford, CT 06108.

[2]Associate professor, Department of Mechanical, Aerospace, and Manufacturing Engineering, Syracuse University, Syracuse, NY 13244.

tions. This latter issue is of particular importance to military aircraft, whose missions change frequently over the aircraft's service life.

In response to these concerns, a strength-based wearout model has been developed recently for predicting the residual strength and life of laminated, continuous fiber polymeric composites subjected to spectrum fatigue loadings [1–3]. This model is phenomenological and semi-empirical; however, in contrast to previous phenomenological models, only a limited amount of experimental input is required. Furthermore, it has been demonstrated that the model provides excellent predictive capability for two-stress level fatigue [2] and a variety of spectrum fatigue loadings [1,3].

In this work, a brief review of the model is first presented. This is followed by an assessment of its predictive capability for two-stress level fatigue and random-ordered loading spectra. The model's ability to account for sequencing effects is verified by comparing theoretical fatigue life predictions to experimental results for low-high and high-low two-stress level loadings. The predictive capabilities of the model for spectrum fatigue are demonstrated by comparison to experimental results from a number of different laminates that were subjected to various tension and compression-dominated load spectrums. These include the FALSTAFF (Fighter Aircraft Loading STAndard For Fatigue) spectrum, for which no previous fatigue model has provided reliable results. Predictions for fatigue life distributions are shown to correlate quite well with experiment for all laminates and load spectrums considered.

Theory

In this section, the fundamental assumptions used in the model's derivation are presented, as well as a brief review of the derivation itself. The model is applicable to any structure or structural feature that undergoes proportional loading, such that strength may be defined in terms of a single parameter. It is assumed that strength is a monotonically decreasing function of the number of cycles. Further, the distribution of structural lives that would be obtained from a set of geometrically "identical" structures subjected to "identical" loadings, as well as the distribution of structural strengths after an arbitrary load history, are assumed to be statistically variable quantities that may be represented by two parameter Weibull density functions [4]. The two parameters that describe the Weibull functions are defined as the scale which defines the 63.2 percentile of the distribution, and the shape which describes the degree of scatter in the data.

Constant-Amplitude Fatigue

Figure 1 presents the residual strength relation that is used by the model for constant amplitude fatigue. The solid line in the figure "tracks" the Weibull scale parameter for residual strength, that is, the 63.2 percentile of the strength distribution, as the scale parameter for residual strength decreases with increasing cycling. The equation of this line is given by

$$R(n) = R_o - (R_o - S_p)\left(\frac{n}{N}\right)^{\nu} \tag{1}$$

where

$R(n)$ = residual strength scale parameter after n cycles,
R_o = static strength scale parameter,
S_p = peak stress *magnitude* of the constant-amplitude loading,
N = scale parameter for the fatigue life distribution, and
ν = strength degradation parameter.

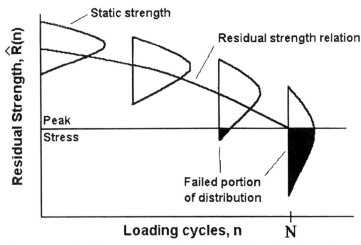

FIG. 1—*Strength distributions associated with a specific residual strength relation.*

For constant-amplitude loading with a positive mean stress, S_p is the maximum stress, and for constant-amplitude loading with a negative mean stress, S_p is the magnitude of the minimum stress. The conventional definition of mean stress is used, that is $S_{mn} = (S_{max} + S_{min})/2$. Also, R_o is always taken as positive. For a positive mean stress, R_o is the scale parameter for the distribution of static strength when the structural geometry of interest is subjected to a positive stress. Similarly, for a negative mean stress, R_o is the static strength scale parameter for the structure when subjected to a negative stress. The value of R_o is obtained from static strength test data. The scale parameter for the fatigue life distribution, N, is found by constant-amplitude fatigue tests at the peak stress and stress ratio for which Eq 1 is applied. The strength degradation parameter, v, controls the overall shape of the residual strength relation. It is obtained by comparing the predicted and observed life distributions for constant-amplitude fatigue loadings [2]. When $v \gg 1$, "sudden death" behavior [5,6] is simulated; when $v = 1$, a linear strength degradation is simulated; and when $v < 1$, the behavior of those laminates that experience an early, sudden loss in strength is simulated. The values of R_o, N, and v are dependent on laminate layup, material system, geometry, and the type of loading (that is, uniaxial, biaxial, shear, etc.); therefore, model characterization tests must be performed for each variation. Furthermore, the fatigue life, N, and the strength degradation parameter, v, depend on the stress amplitude and stress ratio of the fatigue loading. However, as will be discussed in a subsequent section, these latter relationships may be approximated from a relatively modest amount of static and constant amplitude fatigue test data.

Also shown in Fig. 1 are the residual strength distributions that would be displayed by a set of geometrically similar structural configurations subjected to the same loadings. To avoid subsequent confusion, the symbols $R(n)$, R_o, and N will continue to be used to denote the scale parameters described by Eq 1, and the symbols $\hat{R}(n)$, $\hat{R}_o$, and $\hat{N}$ will be used to denote the residual strength, static strength, and life distributions, respectively. For each distribution shown in Fig. 1, stronger structures are in the upper portion and weaker structures are in the lower portion. Initially, the strength distribution is equal to the static strength distribution. During fatigue loading, the entire distribution experiences a degradation in strength. As cycling continues and the mean strength degrades, the residual strength of the weaker structures will fall below the maximum applied stress and failure will occur. The percentage of failures is shown by the shaded areas. The probability of failure during constant-amplitude fatigue loading, that is, the probability that the residual strength is less than the peak stress, S_p, may also be expressed in the form of a Weibull distribution as

$$P[\hat{R}(n) \leq S_p] = 1 - \exp[-(S_p/R(n))^{B_f(n)}] \tag{2}$$

where $B_f(n)$ is a yet-to-be-determined Weibull shape parameter for residual strength. Substituting the residual strength relation, Eq 1, into Eq 2 gives

$$P[\hat{R}(n) \leq S_p] = 1 - \exp\left[-\left(\frac{S_p}{R_o - (R_o - S_p)\left(\frac{n}{N}\right)^v} \right)^{B_f(n)} \right] \tag{3}$$

A review of the fatigue data in the open literature indicates that strength data becomes increasingly scattered with increasing fatigue loading prior to static testing [for example, Refs 7 and 8]. In Eq 3, this is controlled by the Weibull shape parameter, B_f; smaller values of B_f correspond to broader (more disperse) distributions. This is also demonstrated schematically in Fig. 1: as the number of loading cycles increases, the range of residual strength values that the distribution covers expands. In our model, B_f is assumed to initially equal the static strength shape parameter, B_s, and to linearly degrade to the limiting value of the fatigue life shape parameter, B_l. The equation for B_f is given by

$$B_f(n) = B_s - (B_s - B_l)\frac{n}{N} \qquad n < N$$
$$B_f(n) = B_l \qquad\qquad\qquad n \geq N \tag{4}$$

Thus, characterization of the model (that is, experimental determination of all necessary parameters) for constant-amplitude loading requires a sufficient number of static tests to determine a Weibull strength distribution, described by R_o and B_s, and a sufficient number of constant-amplitude fatigue tests to determine a life distribution, described by N and B_l. The strength degradation parameter, v, is obtained by comparing the observed fatigue life distribution to that predicted by Eq 3. Additional details on this procedure are presented in Ref 2.

Multistress Level Fatigue

The basic assumptions and relationships that were derived for constant-amplitude fatigue may be used, with minor modifications, to predict the residual strength and probability of failure for multistress level loading. To illustrate, consider a two-stress amplitude fatigue sequence, that is, two separate constant-amplitude loading segments applied sequentially. Assume that the first stress segment consists of n_1 cycles at a peak stress, S_1, and that the second segment consists of n_2 cycles at a peak stress, S_2. For this example, assume $S_1 > S_2$. For the constant-amplitude loading defined by Stress Level, S_1, let the strength degradation and fatigue life scale parameters be denoted by v_1 and N_1, respectively. Similarly, for the constant-amplitude loading defined by Stress Level, S_2, let the strength degradation and fatigue life scale parameters be denoted by v_2 and N_2, respectively. The scale parameter for residual strength at the end of the first stress segment and, therefore, at the beginning of the second stress segment is $R_1(n_1)$. This strength is given by

$$R_1(n_1) = R_o - (R_o - S_1)\left(\frac{n_1}{N_1}\right)^{v_1} \tag{5}$$

In determining the residual strength after n_2 cycles of the second loading segment, we desire to use the constant-amplitude residual strength relation characterized by S_2, N_2, v_2, and n_2, that is

$$R_2(n_2) = R_o - (R_o - S_2)\left(\frac{n_2}{N_2}\right)^{v_2} \tag{6}$$

However, the loss-of-strength for the second segment of a two-stress level loading must be calculated beginning from $R_1(n_1)$, and the rate of strength loss must be that for the current state of the laminate. That is, except for the special case of a linear degradation law ($v = 1$), the rate at which the strength degrades depends on the amount of previous cycling. Physically, this corresponds to a damage-dependent rate and has been observed experimentally [9].

The way in which the two-stress loading sequence is handled by the model may be most easily described with the aid of Fig. 2. In this figure, Curve AB illustrates the strength degradation for laminates subjected to constant-amplitude loading at Stress Level S_1, and Curve ACD illustrates the strength degradation for laminates subjected to constant-amplitude loading at Stress Level S_2. Both of these curves assume that the specimens are of virgin material and therefore begin at the static strength. Point B represents the strength after n_1 cycles at Stress Level S_1. Point C represents a location of equivalent strength as that of Point B on the constant-amplitude loading curve corresponding to Stress Level S_2. Our model assumes that a laminate that arrived at Point B, along Path AB, has the same residual strength and essentially the same state of damage as a laminate that arrived at Point C along Path AC. Applying this idea to the two-stress level sequence, the beginning of the second segment should correspond to Point C, and the strength degradation for the second segment itself should

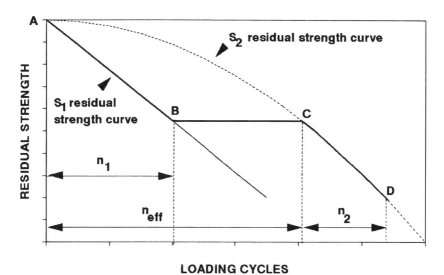

LOADING CYCLES

FIG. 2—*Demonstration of the residual strength relationship for two-stress level fatigue.*

occur along Path CD. Notice that Curve ACD exhibits nonlinear strength degradation, thus, simply replacing R_o with $R_1(n_1)$ in Eq 6 will *not* produce the correct result.

To achieve the requisite shift from Point B to Point C, an "effective" number of cycles, n_{eff}, is introduced. The value of n_{eff} is defined as the equivalent number of loading cycles necessary to produce the same strength loss within the second segment as that predicted to occur within the first segment. Thus, the effective number of cycles may be determined from Eqs 5 and 6 by setting $R_1(n_1) = R_2(n_{eff})$ and solving for n_{eff}, that is, by determining the number of cycles that defines the location of Point C in Fig. 2. This gives

$$n_{eff} = \left[\frac{R_o - R_1(n_1)}{R_o - S_2} \right]^{1/v_2} N_2 \tag{7}$$

Referring to Fig. 2, Eq 6 may now be used to predict the strength at the end of Segment 2 by replacing n_2 by $n_2 + n_{eff}$. Thus, the scale parameter for residual strength after n_1 cycles at S_1 and n_2 cycles at S_2 is given by

$$R_2(n_1 + n_2) = R_o - (R_o - S_2) \left[\frac{(n_{eff} + n_2)}{N_2} \right]^{v_2} \tag{8}$$

Notice that n_{eff} is also the mechanism through which sequencing effects are incorporated into the model. This technique can easily be extended to two-stress level repeating block loadings, as well as to multistress level loadings [3].

Probability of Failure

For the two-stress level sequence described earlier, the probability of failure during the first segment is given by

$$P[\hat{R}_1(n) \leq S_1] = 1 - \exp[-(S_1/R_1(n))^{Bf_1(n)}] \tag{9}$$

where $R_1(n)$ is given by Eq 5, with n replacing n_1 and $n \leq n_1$. The probability of failure during the second segment is given by

$$P[\hat{R}_2(n_1 + n) \leq S_2] = 1 - \exp[-(S_2/R_2(n_1 + n))^{Bf_2(n)}] \tag{10}$$

where $R_2(n_1 + n)$ is given by Eq 8, with n replacing n_2, and n is interpreted as the current number of cycles within the second segment.

To determine the probability of failure after an arbitrary number of cycles, a "tracking" technique is adopted by the model. That is, suppose in the current example that $S_2 << S_1$ and $N_2 >> N_1$. Under these circumstances, it is possible that the probability of failure at the beginning of Segment 2, as predicted by Eq 10, will be *less* than at the end of Segment 1 as predicted by Eq 9. When this occurs, the model stores the probability of failure at the end of Segment 1. The probability of failure during Segment 2 is taken to be the greater of that predicted at the end of Segment 1 or that predicted by Eq 10. This technique is also readily extended to two-stress level repeating block or multistress level loadings [3].

Cycle Mix Effects

The residual strength and fatigue life of composite laminates have been observed to decrease more rapidly when the loading sequence is repeatedly changed after only a few loading cycles [10]. This was classified by Farrow [10] as a "cycle mix effect." For illustration, consider Fig. 3, where large block and small block loading sequences are compared. In this figure, the large block loading consists of 10 000 cycles at the low stress level followed by 100 cycles at the high stress level, while the small block loading contains 1000 cycles at the low stress followed by 10 cycles at the high stress. Note that after 10 100 cycles, both the large block and small block loadings contain the same number of cycles at each stress level. However, it has been observed [10] that laminates that experience small block loadings have reduced average fatigue lives as compared to laminates that are subjected to large block loadings.

It is our belief that the cycle mix effect is a result of the damage that occurs during those "transition cycles," between different constant-amplitude segments, where the magnitude of the mean stress increases. This effect is accounted for in the model through application of a "cycle mix factor." The cycle mix factor is applied only when the magnitude of the mean stress increases from one segment to the next. During these transition cycles, the scale parameter for residual strength is degraded according to the relation

$$R(n) \Rightarrow R(n) - CM \qquad (11)$$

where *CM* is the cycle mix factor given by

$$CM = C_m R_o \left[\frac{\Delta S_{mn}}{R(n)} \right]^{(\Delta S_p / \Delta S_{mn})^2} \qquad (12)$$

In this equation, ΔS_p and ΔS_{mn} are the change in the peak stress magnitude and mean stress magnitude, respectively, during the transition between stress levels, and C_m is a nondimensional

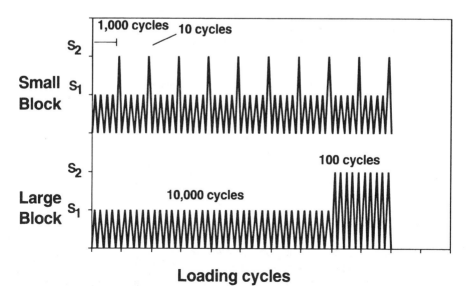

Loading cycles

FIG. 3—*Illustration of loading sequence with small cycle blocks (top) and large blocks (bottom).*

"cycle mix constant." The cycle mix constant is determined by comparing fatigue life data from small block and large block two-stress level fatigue loadings to predicted results. In general, it is preferable to choose two cycle mix constants: one for transitions to a negative mean stress and one for transitions to a positive mean stress. The details of this procedure are presented in a subsequent section. We have observed that the cycle mix constants can be negligibly small for certain laminates [1,3]. Our results also suggest that cycle mix is a significant contributor to strength loss only when the loading segments are such that the ratio of cycles in the segment, n_i, to cyclic life for that segment, N_i, is less than 0.001. In instances where n_i/N_i is greater than 0.001, the strength loss due to the cycle mix effect is generally small; thus, C_m is small and the cycle mix factor is likely not required. Of course, the cycle mix factor may be included for all loadings; exclusion of this effect for $n_i/N_i > 0.001$ is suggested only to reduce the required experimental input to the model.

Finally, we point out that Eq 12 was obtained through evaluation of a number of possible relationships comprised of nondimensional groupings of suitable governing parameters. Of those expressions evaluated, Eq 12 gave the best results in terms of the model's predictive capability for a wide variety of laminates and loadings. We are currently performing a more extensive investigation of the effect of cycle mix for a range of materials, layups, and geometries.

Model Characterization

This section provides a brief description of the procedures that are used to determine the fatigue life scale parameters, N, the fatigue life shape parameters, B_l, and the strength degradation parameters, v, that characterize each segment of a random-ordered load spectrum. These parameters are obtained from a relatively modest amount of static strength, constant-amplitude fatigue, and two-stress level fatigue test data through the use of a series of master diagrams. A more complete description is provided in Refs 1 and 3.

Experimental Database

Monotonic Tests—Monotonic tests are required to determine the structure's static strength and to obtain its associated Weibull distribution. If the model is to be used to predict the life of a structure that is subjected to mean stresses of only one sign, it is recommended that at least five monotonic tests be performed and the data reduced by the maximum likelihood method [11]. For structures subjected to loadings that contain mean stresses of both signs, at least five tension and five compression tests are recommended for the determination of the static tension and compression strength distributions.

Cyclic Tests—The results of constant-amplitude tests are used to determine those model parameters that are necessary to predict the strength degradation during a loading segment, and two-stress level test results are used to determine the one or two required cycle mix constants. Depending upon the loading spectrum of interest, constant-amplitude tests are required at either two or three stress ratios; here, the stress ratio, R_s is defined in the conventional manner as S_{min}/S_{max}. For convenience, denote the three stress ratios as R_{s0}, R_{s1}, and R_{s2}. In our notation, R_{s0} is always used in the test program, R_{s1} is only used when there are load cycles in the spectrum with a positive mean stress, and R_{s2} is only used when there are cycles with a negative mean stress. The way in which R_{s0}, R_{s1}, and R_{s2} are chosen is discussed below, along with the aid of the partial master diagram of Fig. 4. This diagram is for a material or structure with equal strengths in tension and compression. It contains no experimental data, but simply plots lines of constant stress ratio versus mean stress and stress amplitude.

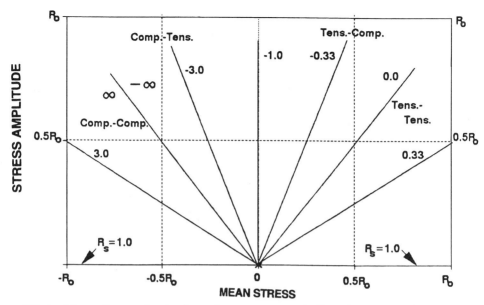

FIG. 4—*Master diagram showing lines of constant stress ratio. R_0 is the tension and compression strength. Numbers are values of R_s.*

Suppose that the spectrum of interest contains stress ratios only in the range $0 \leq R_s < 1$, denoted in Fig. 4 as tension-tension fatigue. With reference to the figure, for spectrums that contain mean stresses of only one sign, R_{s0} is always chosen as an "inner bound." In this example, R_{s0} would be chosen equal to the smallest stress ratio in the spectrum or else equal to zero. R_{s1} is chosen to "best characterize" the remaining tension dominated loads in the spectrum. It is always at positive mean stress, and will lie on the tension-dominated side of the master diagram between R_{s0} and $R_s = 1$. Typically, R_{s1} is chosen relatively close to the most commonly occurring tension-dominated stress ratio in the spectrum. If the loading spectrum contains stress ratios in the range $-1 \leq R_s < 1$, that is, tension dominated fatigue, then R_{s0} can be chosen equal to the smallest (most negative) stress ratio in the spectrum or else it can be taken equal to -1. R_{s1} is chosen in the same manner as before.

Similarly, for compression-compression fatigue, choosing R_{s0} as an "inner bound" means it is equal to or greater than the largest stress ratio in the spectrum; $R_{s0} = \infty$ is the upper limit for this case. Similar to R_{s1}, R_{s2} is chosen relatively close to the most commonly occurring compression-dominated stress ratio in the spectrum. For compression-dominated fatigue, defined by $|R_s| \geq 1$, R_{s0} is chosen equal to -1 or equal to the largest stress ratio of all compression-tension cycles ($-\infty \leq R_s < -1$) in the spectrum, and the procedure for choosing R_{s2} is unchanged. Finally, for loading spectrums containing mean stresses of both signs, $R_{s0} = -1$ is used, R_{s1} is chosen to best characterize the tension-dominated cycles, and R_{s2} is chosen to best characterize the compression-dominated cycles.

For each stress ratio chosen, individual S-N curves are plotted using Weibull fatigue life scale parameters. To this end, at a given stress ratio, at least five tests are performed at each of three stress amplitudes. Each set of five or more tests are reduced to obtain a fatigue life scale parameter, N, and a fatigue life shape parameter, B_l. The individual S-N curves therefore represent the behavior of the 63.2 percentile of the life distributions.

To determine the one or two cycle mix constants, two-stress level fatigue testing is conducted. The cycle mix constant for transitions to a positive mean stress is obtained by comparing the model's predictions to experimentally determined life distributions for small and large block tension-tension tests at Stress Ratio R_{s1} and choosing that value of C_m that best correlates the 63.2 percentile of the distributions. Similarly, the cycle mix constant for transitions to a negative mean stress is obtained by comparison of theoretical and experimental results for small and large block compression-compression testing at Stress Ratio R_{s2}. The stress amplitudes in these blocks should be typical of those in the loading spectrum of interest.

The type and number of tests that are required to characterize the model for an arbitrary loading spectrum are summarized in Table 1. In the table, "+" refers to a spectrum that contains only tension-dominated cycles (positive mean stress), and "−" refers to a spectrum that contains only compression-dominated load cycles. We also point out that this table presents the *minimum* number of tests that must be performed for each type of spectrum. More tests are of course preferable, as they provide a more statistically significant sample size.

Determination of N, B_l, and v for Arbitrary Loadings

For characterization of a structure's behavior under a fully reversed loading spectrum (used here to denote one that contains both positive and negative mean stress), the results of the previous experiments and associated data reduction procedures will provide the scale and shape parameters for static strength, the scale and shape parameters for fatigue life for constant-amplitude loadings at three stress amplitudes at each of three stress ratios, and the two cycle mix constants. The critical step in model characterization is to generalize these results to provide the key model parameters, N, B_l, and v, for each stress amplitude and stress ratio that occurs in the loading spectrum. For the fatigue life scale parameter, N, this is accomplished through a "standard" master diagram [for example, Ref *12*] and modified Goodman relations [*13*]. Arbitrary values of the fatigue life scale parameter, B_l, and the strength degradation parameter, v, are obtained through "modified" master diagrams.

Consider the construction of a master diagram that will be used to determine the fatigue life scale parameter, N, corresponding to an arbitrary mean stress and stress amplitude. First, the previously developed S-N curves for the 63.2% probability of failure may be expressed as

$$S_a = A_k \log N + B_k \tag{13}$$

TABLE 1—*Type and number of tests required for model characterization.*

Type of Test	Mean Stress in Spectrum		
	"+" only	"−" only	"+" and "−"
Static tension	5	0	5
Static compression	0	5	5
S-N CURVE AT STRESS RATIO			
R_{s0} (5 replicates at 3 stress amplitudes)	15	15	15
R_{s1} (5 replicates at 3 stress amplitudes)	15	0	15
R_{s2} (5 replicates at 3 stress amplitudes)	0	15	15
LARGE & SMALL BLOCK AT STRESS RATIO			
R_{s1} (5 tests of each type)	10	0	10
R_{s2} (5 tests of each type)	0	10	10
Total number of tests required	45	45	75

where

S_a = stress amplitude = $(S_{max} - S_{min})/2$;
A_k = slope of S-N curve for stress ratio R_{sk};
N = scale parameter for life; and
B_k = stress amplitude intercept of S-N curve for stress ratio R_{sk}.

If the structure's behavior under a fully reversed loading spectrum is being characterized, then there will be three values of A_k and three values of B_k. These constants are readily obtained by a least-squares curve fitting procedure.

Next, the two or three stress ratios that have been characterized using Eq 13 are plotted on a master diagram. A partial diagram is illustrated in Fig. 5 for the case where constant-amplitude fatigue test data has been obtained at stress ratios of $R_{s0} = -1.0$, $R_{s1} = 0.1$, and $R_{s2} = -10.0$. Note that Eq 13 may be used to determine the fatigue life scale parameter, N, for any arbitrary stress amplitude at the stress ratios tested. For other stress ratios, modified Goodman relations are used to obtain the fatigue life scale parameters. One modified Goodman relation is used for each of the constant life lines connecting R_0^c and R_{s0}, where R_0^c is the scale parameter for the compression strength, and one relation is used for each of the constant life lines connecting R_0^t and R_{s0}, where R_0^t is the scale parameter for the tension strength. The general form of the relationship is given by

$$S_a = S_{a0}\left[1 - \left(\frac{S_{mn} - S_{m0}}{R_o - S_{m0}}\right)^m\right] \qquad (14)$$

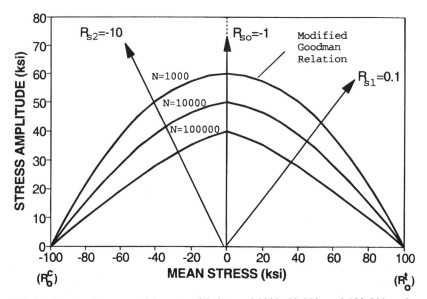

FIG. 5—*Master diagram with constant life lines of 1000, 10 000, and 100 000 cycles.*

where

$$m = \log\left(\frac{1 - S_{ai}}{S_{a0}}\right)\bigg/\log\left(\frac{S_{mi} - S_{m0}}{R_o - S_{m0}}\right) \tag{15}$$

In Eq 14, S_a and S_{mn} are the stress amplitude and mean stress, respectively, at which the scale parameter for life is desired; S_{a0} and S_{m0} are the stress amplitude and mean stress, respectively, obtained from Eq 13 for $R_s = R_{s0}$; and N is the unknown fatigue life scale parameter of interest. We point out that $S_{m0} = 0$ for the case where $R_{s0} = -1$, and Eqs 14 and 15 reduce to those used by Farrow [10]; however, $S_{m0} \neq 0$ for all other values of R_{s0}. For loadings that fall between R_0^t and R_{s0}, the value of R_o that appears in Eqs 14 and 15 is taken to be R_0^t; otherwise, R_0 is taken to R_0^c. Equation 15 defines the Goodman curve-fitting parameter, m. In this equation, S_{ai} is the stress amplitude on the "intermediate" stress ratio line through which the Goodman relation passes. For loadings defined by stress ratios between R_0^t and R_{s0}, S_{ai} is the stress amplitude corresponding to the life of interest (N) along R_{s1}; and for loadings defined by stress ratios between R_0^c and R_{s0}, S_{ai} is the stress amplitude corresponding to the life of interest along R_{s2}. Similarly, S_{mi} is the mean stress corresponding to the life of interest on the "intermediate" stress ratio line through which the Goodman relation passes.

Equations 13 through 15 mathematically define a master diagram for life, and the life corresponding to any arbitrary stress amplitude and stress ratio may be obtained by an iterative procedure. For example, for a given S_a and S_{mn} of interest, one "estimates" a corresponding scale parameter for life, N_e. Equation 13 is first applied at R_{s0} to obtain the stress amplitude and mean stress corresponding to N_e. These are denoted in Eqs 14 and 15 as S_{a0} and S_{m0}, respectively. Equation 13 is then applied at R_{si} to obtain the stress amplitude and mean stress corresponding to N_e; these are denoted in Eq 15 as S_{ai} and S_{mi}, respectively. These results give m (Eq 15) and S_a (Eq 14). If this value of S_a agrees with the value of interest, then the estimate of life was correct. Otherwise, N_e is adjusted and the process repeats until convergence is achieved.

The strength degradation parameter and fatigue life shape parameter are determined from "modified" master diagrams that display lines of constant v or B_l. First, values of v are determined for each stress amplitude and stress ratio at which constant-amplitude fatigue tests were performed. This is accomplished using the methodology described in Ref 2. The resulting values of v are used, as are the previously determined values of B_l, to develop linear S-v and S-B_l relationships for each stress ratio. This step is essentially the same as was done for the S-N data. Using linear least-squares curve-fitting procedures, the constants C_k, D_k, E_k, and F_k in the following equations are obtained

$$S_a = C_k v + D_k \tag{16}$$

$$S_a = E_k B_l + F_k \tag{17}$$

The parameters in Eqs 16 and 17 are defined similarly as for Eq 13; C_k and E_k are the slopes of the curve-fits (that is, corresponding to each stress ratio) and D_k and F_k are their intercepts. Master diagrams for v and B_l are constructed by linearly connecting like values. As for the case for the standard master diagram, the diagrams themselves are primarily for interpretation. Actual values of v and B_l for a given loading may be found from Eqs 16 and 17, and an additional set of equations that provide the slopes of the curves on the modified master diagrams. Additional details are presented in Refs 1 and 3.

Results

Two-Stress Level Fatigue

The model's accuracy for predicting sequencing effects was evaluated by comparison to two-stress level test results for cross-ply, glass/epoxy laminates [2]. All experimental results were taken from the work of Broutman and Sahu [14], who tested these laminates under 22 different tension-tension fatigue conditions. All tests were two-stress level run-out fatigue, that is, the second loading block was continued until all laminates failed. The 22 different conditions were created by varying the size of the first loading block and the magnitudes of the two stress levels. Representative results are presented in Figs. 6 through 8. In these figures, "model prediction" refers to those by the model described herein, and "P-M rule" refers to predictions that were obtained by the Palmgren-Miner rule [16,17] for the 50th percentile of the life distribution. Results are also shown for loading cases in which predictions were made by Yang and Jones [15], who also used this data for verification of a different wearout model. All of the experimental data is presented using log-normal cumulative distributions based on the mean lives and standard deviations presented in Ref 14. For our model, additional constant-amplitude results that were presented in Ref 14 were used for the determination of N, B_l, and v for each test stress level. Complete details of this procedure are presented in Ref 2.

Figures 6 and 7 present results for high-low loadings. Figure 6 is for the case where the first loading block continued for approximately 16% of the expected life, and Fig. 7 presents results for the case where the first loading block continued for approximately 40% of the expected life. Excellent correlation between the present model and the experimental results is observed in both instances. Also, for these two loading cases, better correlation is observed by this model than by the other two approaches. There were, however, a few loadings when

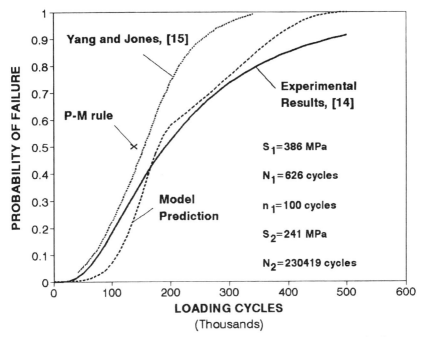

FIG. 6—*Comparison of predicted and observed results for a high-low loading.*

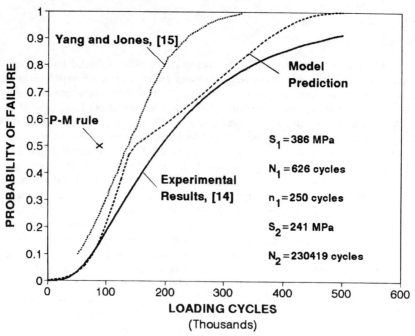

FIG. 7—*Comparison of predicted and observed results for a high-low loading.*

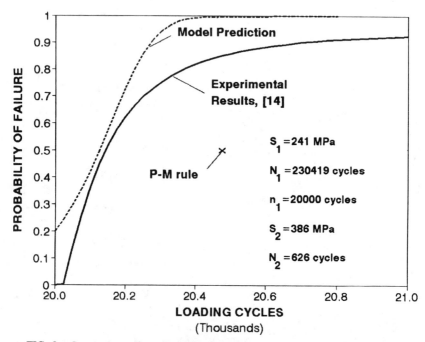

FIG. 8—*Comparison of predicted and observed results for a low-high loading.*

this was not the case, but overall the present model shows reasonable correlation to experiment for all two-stress level loadings evaluated [2]. The discontinuity in slope in the model's predictions occurs when $B_f(n)$ reaches its limiting value of B_l.

Figure 8 compares predicted and observed results for a low-high loading using the same two stress levels as those of Figs. 6 and 7. This is a relatively small first block loading, as the first loading block continued for less than 9% of the expected life. Note that the number of cycles on the abscissa begins at the second stress level. Once again, relatively good correlation is observed. This loading case was not evaluated by Yang and Jones [15]. In contrast to the high-low results, the Palmgren-Miner rule compared unfavorably to the model's predictions and to the experimental results. This over-prediction of the mean fatigue life by the Palmgren-Miner rule occurred in approximately 50% of all of the two-stress level loadings that were studied [2].

Multistress Level Fatigue: SGF Spectrum

In this section, life predictions by the model are compared to experimental results reported by Schutz and Gerharz [18] for uniaxial loadings of $[0_2/\pm45/0_2/\pm45/\overline{90}]_s$ graphite/epoxy test specimens. All specimens contained a gage section approximately 10 mm wide by 20 mm long housed in "Teflon coated anti-buckling supports." Comparisons are made for two Schutz and Gerharz Fighter (SGF) loading spectrums. Spectrum "SGF1" contains a peak compressive stress of 725 MPa, and "SGF2" contains a peak compressive stress of 652.5 MPa. Both spectrums simulate the stress history on the upper wing skin at the wing root of a fighter aircraft, and consist of a series of 16 compression loading segments. The first segment is at the peak compressive stress, and the next ten segments are at successively smaller stress magnitudes. The 11th segment contains the smallest stress magnitude, and the final five segments are at successively increasing stress magnitudes. These spectrums are comprised of 14 031 cycles, corresponding to approximately 200 flights or a typical year of service. Schutz and Gerharz reported that only three test replicates were performed for each spectrum. Although this is not a statistically significant sampling, we choose to present this data as it has also been used for comparison by Yang and Du [19]. We also compare these results to predictions made by Schutz and Gerharz [18] using the Palmgren-Miner rule [16,17].

Referring to Table 1, since both SGF1 and SGF2 contain stresses of compressive mean stress only, 45 tests are recommended to characterize the model. However, in view of the fact that the magnitude of the mean stress decreases in 11 out of the 16 load transitions, and that the loading is relatively large block, ignoring the cycle mix effect should not significantly alter the model's predictions. This may be accomplished by simply taking $C_m = 0$, thereby reducing the number of recommended tests to 35. The results of these 35 tests are used to obtain the static compression strength distribution and the fatigue life distributions for the stress ratios R_{s0} and R_{s2}. Schutz and Gerharz generated fatigue life data at three stress ratios: -1.0, -1.66, and -5.0. Data from these tests were presented graphically, and at least four replicates were performed at each stress amplitude and stress ratio. In what follows, we have used $R_{s0} = -1.0$ and $R_{s2} = -5.0$ to determine the model's parameters. These data were reduced to obtain N, B_l, v, and the linear curve fitting constants for the S-N, S-B_l, and S-v curves as defined by Eqs 13, 16, and 17. The static strength scale and shape parameters were based on 31 static tests and were reported to be 854.65 MPa and 28.1, respectively [18]. Additional details on the model characterization procedure, as well as all experimentally obtained parameters for this spectrum are presented in Ref 3.

Experimental and theoretical fatigue life results are shown in Fig. 9 for the SGF1 load spectrum. In the figure, 200 flights correspond to one time through the spectrum; that is, the 14 031 cycles comprising the 16 load levels. For this loading, the model demonstrates good

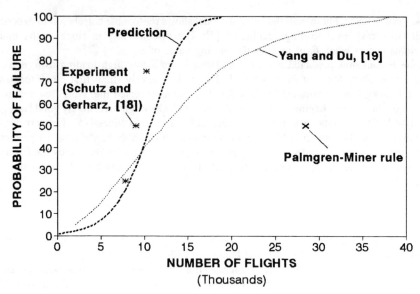

FIG. 9—*Comparison of prediction to the model of Yang and Du, the Palmgren-Miner rule, and to experiment for SGF1.*

correlation with experiment and provides somewhat better agreement than does the model of Yang and Du [19]. The Palmgren-Miner rule predicted the mean fatigue life, or 50th percentile, to be 28 490 flights. This is more than three times larger than the observed value.

Figure 10 compares the model's predictions with experiment and with predictions by the model of Ref 19 for SGF2. Again, the best predictions are obtained using the present approach.

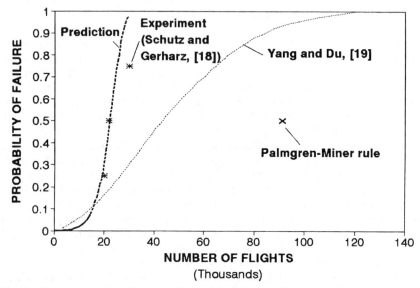

FIG. 10—*Comparison of prediction to the model of Yang and Du, the Palmgren-Miner rule, and to experiment for SGF2.*

The Palmgren-Miner rule predicted a mean fatigue life of 90 090 flights, which exceeded the experimental results by a factor of four.

Multistress Level Fatigue: FALSTAFF Spectrum

In this section, life predictions by the model are compared to experimentally obtained fatigue life distributions for graphite/epoxy laminates subjected to the FALSTAFF spectrum. This is a standardized random-ordered loading spectrum that simulates the in-flight load-time history of fighter aircraft. One spectrum "block" contains a series of 17 983 cycles that simulates the loading on the wing skin of a fighter aircraft over a period of 200 flights [20,21]. Depending on the FALSTAFF "mission" that is desired, a single flight may contain from 70 to 560 different loading segments comprised of up to 32 different stress levels. Individual loading segments may only be a few cycles in length. The stress level for any given segment is proportional to a user-defined test limit stress, that may be positive or negative. Because of its authenticity, FALSTAFF is commonly used in the production of fatigue life and crack growth data [20,21]. However, due to its complexity, no model of which the authors' are aware has yet been able to predict the life of structural composites subjected to this loading.

In what follows, all experimental results for the FALSTAFF spectrum are taken from Ref 10, where two different layups of XAS/914C graphite/epoxy were considered: (1) $[\pm45/90/(\pm45)_4]_s$, hereafter referred to as the angle-ply laminate; and (2) $[\pm45/0/90]_{3s}$, referred to subsequently as quasi-isotropic. All tests consisted of uniaxially loaded, 270-mm long by 30-mm wide, "fastener type" specimens. These specimens contained a 4.83-mm diameter drilled and countersunk hole, into which a 4.81-mm diameter countersunk fastener was placed and torqued to 5.0 Nm. All data were reported in the form of Weibull distributions; it is stated that Weibull scale and shape parameters were calculated using the maximum likelihood technique [11] and a minimum of six data points. All reported stresses are based on "net section area," that is, the specimen width, less the hole diameter, multiplied by the thickness.

For the FALSTAFF spectrum, the model's required characterization testing was performed under monotonic loadings to determine the static strength scale and shape parameters, and under constant amplitude fatigue loadings at stress ratios of 0.1, −1.0, and 10.0 to determine the fatigue life scale and shape parameters. For the angle-ply laminates, two-stress level testing was also performed at a stress ratio of 0.1 to determine the tension-dominated cycle mix constant. Due to the absence of data in Ref 10, this cycle mix constant is also used for the angle-ply laminates under compression-dominated loadings. Cycle mix testing was not performed for the quasi-isotropic laminates, so all subsequent predictions for this laminate use $C_m = 0$. It is shown that the correlation is still rather good for this case, which may indicate that the cycle mix effect is less important for the more fiber-dominated layups [1,3]. However, additional testing is required to verify this issue. Complete details of the model characterization procedure, as well as all experimentally obtained parameters for the FALSTAFF spectrum are presented in Ref 1.

Angle-Ply Laminates—The experimental data on the angle-ply laminates was compared to the model's predictions for FALSTAFF load spectrums containing test limit stresses of 208.9, 178.8, −259.9, and −228.1 MPa [1]. For reference, the static strength scale parameters for these laminates were 250.9 MPa for tension and −303.4 MPa for compression. Two representative cases are presented here. The model's predictions that are presented in this section used $C_m = 5.38 \times 10^{-7}$; this value was determined according to the procedure described in Ref 2. In order to demonstrate the effect of the cycle mix factor on the predicted life distributions, the "fatigue life," used subsequently without quotes to denote the 63.2 percentile of the failure distributions will be presented for both $C_m = 0$ and for $C_m = 5.38 \times 10^{-7}$.

Figure 11 shows the probability of failure as predicted by the model and as obtained by experiment for a test limit stress of 208.9 MPa. For this load case, the model's predictions show excellent correlation to experiment. The theoretical and experimental failure curves agree within 5% throughout the entire range of the probability of failure values. For the fatigue life (63.2 percentile), the model's prediction and the observed results are 31 and 29 blocks, respectively. For comparison, if cycle mix is neglected in the model (that is, $C_m = 0$), the fatigue life is over-predicted by a factor of two. Thus, by accounting for cycle mix, the predictive capability of the model is significantly improved. For the same load case, the Palmgren-Miner rule predicted the mean fatigue life, or 50th percentile, to be 185 blocks. This exceeds the experimental mean of 28 blocks by over a factor of six.

The FALSTAFF load case defined by a -228.1 MPa test limit stress is presented in Fig. 12. In this figure, reasonable correlation is obtained for the failure distributions. The model's predictions for fatigue life are approximately 11% less than the experimental results. As in the previous case, the predictions from the model without the cycle mix factor, and by the Palmgren-Miner rule, provided unconservative estimates of fatigue life.

Quasi-Isotropic Laminates—Figures 13 and 14 present the model's predicted probability of failure versus the experimental fatigue life distributions for quasi-isotropic laminates subjected to FALSTAFF spectrums with test limit stresses of -478.7 and -451.6 MPa, respectively. The static strength scale parameter for compression for this laminate equals -540.9 MPa. As described earlier, the model's results presented in this section do not account for cycle mix effects, and all predictions use $C_m = 0$. For the data of Fig. 13, the model under-predicts fatigue life by 10%; for comparison, the Palmgren-Miner rule predicts the average fatigue life to be 913 blocks, whereas the experimental value was 75 blocks. In Fig. 14, the model over-predicts the experimental fatigue life by 38%. The under-prediction of life for the spectrum of Fig. 13 and the over-prediction of life for that of Fig. 14 suggests that the degradation of strength and fatigue life due to cycle mix

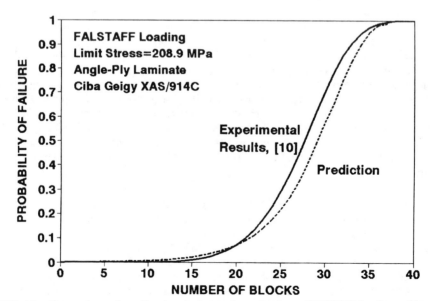

FIG. 11—*Comparison of predicted and observed results for a FALSTAFF loading with a test limit stress of 208.9 MPa.*

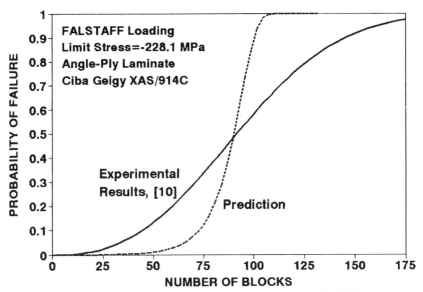

FIG. 12—*Comparison of predicted and observed results for a FALSTAFF loading with a test limit stress of −228.1 MPa.*

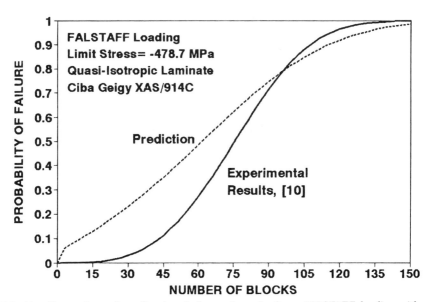

FIG. 13—*Comparison of predicted and observed results for a FALSTAFF loading with a test limit stress of −478.7 MPa.*

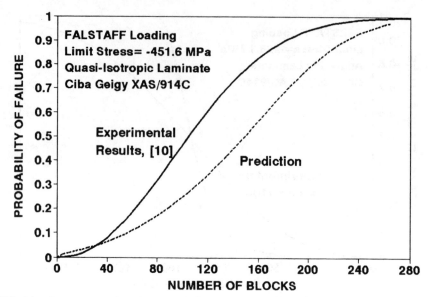

FIG. 14—*Comparison of predicted and observed results for a FALSTAFF loading with a test limit stress of −451.6 MPa.*

effects are not significant for the quasi-isotropic laminate in compression-dominated loadings. Further testing is required to ascertain whether this result can be extended to other materials, layups, and load spectrums.

Summary

Tables 2 and 3 summarize the results for all laminates subjected to spectrum loadings that have been considered to date. Due to the relatively small number of experiments performed for the SGF1 and SGF2 spectrums, the results in Table 2 are presented in terms of the mean, or 50th percentile, of the life distribution. In Table 3, results are presented for fatigue life, that is, for the 63.2 percentile of the predicted life distributions, both with ($C_m \neq 0$) and without ($C_m = 0$) inclusion of the cycle mix effect. In all cases, the model's fatigue life predictions with the cycle mix factor incorporated are quite good. This is particularly impressive considering the complexity of the various loadings and the relative simplicity of the model. In all cases, the Palmgren-Miner rule, which predicts the 50th percentile of the life distribution, is highly unconservative. Also, for all cases considered to-date, this model shows better correlation to experiment than all other fatigue models of which we are aware.

TABLE 2—*Summary of fatigue life predictions for SGF loadings (in flights).*

Limit Stress, MPa	Load Type	Model $C_m =$ 0, 50%	Observed Fatigue Life, 50%	P-M Prediction, 50%	Yang's Prediction, 50%
−725.0	SGF1	10 600	9 000	28 490	12 375
−652.5	SGF2	22 400	26 933	90 090	43 287

TABLE 3—*Summary of fatigue life predictions for FALSTAFF loadings (in blocks).*

Limit Stress, MPa	Laminate[a]	Model $C_m = 0$, 63.2%	Model $C_m \neq 0$, 63.2%	Observed Fatigue Life, 63.2%	P-M Prediction, 50%	Observed Fatigue Life, 50%
208.9	angle	59	31	29	185	28
178.8	angle	236	120	129	790	122
−259.9	angle	53	26	38	240	36
−228.1	angle	251	94	106	790	92
−478.7	quasi	76	NA[b]	84	913	75
−451.6	quasi	171	NA	124	1291	104

[a]Angle = predominately angle-ply and quasi = quasi-isotropic.
[b]NA = not available.

Conclusion

A residual strength-based wearout model has been presented for predicting the life of composite laminates subjected to randomly-ordered load spectra. The model is phenomenological and semi-empirical and requires a limited amount of fatigue data for its characterization. It has been shown that the model provides good accuracy for predicting the fatigue life of composite structures subjected to a wide variety of loadings. This is particularly true if fatigue life comparisons are based on the 63.2 percentile of the probability of failure distribution.

In terms of its general applicability and predictive capability, the model presented herein appears to be a significant improvement over currently used approaches. The primary differences between this model and traditional wearout models (see Ref 2 for an extensive literature review) are the inclusion of a cycle mix factor, an accounting for the dispersion of the strength distribution during fatigue loading, and an extensive amount of experimental verification for spectrum loadings. In an effort to further increase the model's accuracy, we are currently performing a more comprehensive investigation of the effects of cycle mix. Various other modifications to the model are also being evaluated to obtain better predictions for fatigue life at the lower end of the probability of failure curve, that is, in order to predict a statistically significant "minimum life." Primarily, these modifications involve the way in which the strength degradation parameter, v, and residual strength shape parameter, B_f, are defined or evaluated or both. However, it is our belief that the current model is sufficiently accurate to be incorporated into the present-day design process, thereby providing a valuable tool in the cost and performance optimization procedures used for composite structures. Incorporating the model into the design process will also offer significant benefit for cases where it is found that the loading spectra are different than those used for the structure's design. That is, rather then embarking on expensive test programs, this model may be used to determine an updated life based upon the true service usage.

References

[1] Schaff, J. R. and Davidson B. D., "Life Prediction for Composite Laminates Subjected to Spectrum Fatigue Loading," *Durability of Composite Materials*, MD-Vol. 51, R. C. Wetherhold, Ed., The American Society of Mechanical Engineers, New York, 1994, pp. 89–109.

[2] Schaff, J. R. and Davidson, B. D., "Life Prediction Methodology for Composite Structures. Part I—Constant Amplitude and Two-Stress Level Fatigue," *Journal of Composite Materials*, Vol. 31, No. 2, 1997, pp. 128–157.

[3] Schaff, J. R. and Davidson, B. D., "Life Prediction Methodology for Composite Structures. Part II—Spectrum Fatigue," *Journal of Composite Materials*, Vol. 31, No. 2, 1997, pp. 158–181.

[4]   Weibull, W. and Weibull, G. W., "New Aspects and Methods of Statistical Analysis of Test Data with Special Reference to the Normal, the Lognormal, and the Weibull Distributions," FOA Report D 20045-DB, Defense Research Institute, Stockholm, 1977.

[5]   Chou, P. C. and Croman, R., "Residual Strength in Fatigue Based on the Strength-Life Equal Rank Assumption," *Journal of Composite Materials*, Vol. 12, 1978, pp. 177–195.

[6]   Chou, P. C. and Croman, R., "Degradation and Sudden-Death Models of Fatigue of Graphite/Epoxy Composites," *Composite Materials, Testing and Design (Fifth Conference), ASTM STP 674*, S. W. Tsai, Ed., American Society for Testing and Materials, Philadelphia, 1979, pp. 431–454.

[7]   Yang, J. N. and Jones D. L., "Load Sequence Effects on Graphite/Epoxy [+−35]$_{2s}$ Laminates," *Long-Term Behavior of Composites, ASTM STP 813*, T. K. O'Brien, Ed., American Society for Testing and Materials, Philadelphia, 1983, pp. 246–262.

[8]   Yang, J. N., "Fatigue and Residual Strength Degradation for Graphite/Epoxy Composites Under Tension-Compression Cyclic Loadings," *Journal of Composite Materials*, Vol. 12, 1978, pp. 19–41.

[9]   Poursartip, A. and Beaumont, P. W. R., "The Fatigue Damage Mechanics of a Carbon Fibre Composite Laminate: II—Life Prediction," *Composites Science and Technology*, Vol. 25, 1986, pp. 283–299.

[10]   Farrow, I. R., "Damage Accumulation and Degradation of Composite Laminates Under Aircraft Service Loading: Assessment and Prediction, Volumes I and II," Ph. D. thesis, Cranfield Institute of Technology, Cranfield, UK, 1989.

[11]   Talreja, R., "Estimation of Weibull Parameters for Composite Materials Strength and Fatigue Life Data," *Fatigue of Fibrous Composite Materials, ASTM STP 723*, K. N. Lauraitis, Ed., American Society for Testing and Materials, Philadelphia, 1981, pp. 291–311.

[12]   Tsai, S. W., *Composites Design*, 4th ed., Think Composites, Dayton, OH, 1988.

[13]   Goodman, J., *Mechanics Applied to Engineering*, Vol. 1, 9th ed., Longmans Green, London, 1930.

[14]   Broutman, L. J. and Sahu, S., "A New Theory to Predict Cumulative Fatigue Damage in Fiberglass Reinforced Plastics," *Composite Materials: Testing and Design (Second Conference) ASTM STP 497*, American Society for Testing and Materials, Philadelphia, 1972, pp. 170–188.

[15]   Yang, J. N. and Jones D. L., "Load Sequence Effects on the Fatigue of Unnotched Composites Materials," *Fatigue of Fibrous Composite Materials, ASTM STP 723*, K. N. Lauraitis, Ed., American Society for Testing and Materials, Philadelphia, 1981, pp. 213–232.

[16]   Palmgren, A., "Die Lebensdauer von Kugellagern," *Zeitschrift des Vereins Deutscher Ingenieure*, Vol. 68, 1924, pp. 339–341.

[17]   Miner, M. A., "Cumulative Damage in Fatigue," *Journal of Applied Mechanics*, Vol. 12, 1945, pp. 159–164.

[18]   Schutz, D. and Gerharz, J. J., "Fatigue Strength of a Fibre-Reinforced Material," *Composites*, Vol. 8, No. 4, 1977, pp. 245–250.

[19]   Yang, J. N. and Du, S., "An Exploratory Study Into the Fatigue of Composites Under Spectrum Loading," *Journal of Composite Materials*, Vol. 17, 1983, pp. 511–526.

[20]   Van Dijk, G. M. and de Jonge, J. B., "Introduction to FALSTAFF," *Proceedings*, Eighth ICAF Symposium, ICAF Document No. 801, International Committee of Aeronautical Fatigue, 1975.

[21]   ten Have, A. A., "European Approaches in Standard Spectrum Development," *Development of Fatigue Loading Spectra, ASTM STP 1006*, J. M. Potter and R. T. Watanabe, Eds., American Society for Testing and Materials, Philadelphia, 1989, pp. 17–35.

Strength and Residual Properties

Sailendra N. Chatterjee,[1] Chian-Fong Yen,[1] and Donald W. Oplinger[2]

On the Determination of Tensile and Compressive Strengths of Unidirectional Fiber Composites

REFERENCE: Chatterjee, S. N., Yen, C.-F., and Oplinger, D. W., **"On the Determination of Tensile and Compressive Strengths of Unidirectional Fiber Composites,"** *Composite Materials: Fatigue and Fracture (Sixth Volume), ASTM STP 1285*, E. A. Armanios, Ed., American Society for Testing and Materials 1997, pp. 203–224.

ABSTRACT: Stress fields in tabbed unidirectional composite coupons and in cross-ply specimens are examined with a goal towards improving the methods for determining the axial strengths of the unidirectional material. Results of parametric studies for evaluation of the influence of tab materials and geometries as well as adhesive properties on the stress peaks in unidirectional tension coupons are presented. Use of ductile (but tough) adhesives, soft tabs, and low taper angles is recommended to reduce failures near tab ends.

Data reduction schemes for evaluation of cross-ply test data are critically examined with due consideration to subcritical damages (such as ply cracks) and expected failure modes. Test results from cross-ply and unidirectional tension and compression specimens of carbon and glass-fiber composites are compared. Some recommendations are made based on the results reported. Tests and data correlations for other composites are suggested for selecting a data reduction scheme acceptable to the composites community.

KEYWORDS: unidirectional fiber composites, tensile strength, tab effects, adhesive nonlinearity, cross-ply tension, ply cracks, cross-ply compression, data analysis, fatigue (materials), fracture (materials), composite materials

Test methods that worked well for first-generation composites are often found to be inadequate for new composites with fibers of increasing strength and stiffness. Various attempts have been and are being made to improve the reliability of existing test methods and to develop new specimens and test techniques suitable for composites now in use and for materials of the next generation that are under development. Since the unidirectional material is the building block in structural laminates, its properties must be well characterized for use in lamination analyses. A recent survey of test methods for composites [1] indicates that various stiffnesses of highly anisotropic unidirectional composites can be obtained using most of the available test methods. However, determination of the high axial tensile (or compressive) strengths of such composites is a challenge to experimentalists. The straight-sided coupon with bonded tabs, according to ASTM Test Method for Tensile Properties of Fiber-Resin Composites (D 3039-76), is widely used and has been found to perform better than others with varying widths such as ASTM Test Method for Tensile Properties of Plastics (D 638-91) Type I and linear tapered [2] specimens. The streamline [3] specimen has been used with some success, but the

[1]Staff scientist and research engineer, respectively, Materials Sciences Corporation, Fort Washington, PA 19034.

[2]Federal Aviation Administration, FAA Technical Center, Atlantic City International Airport, NJ 08405.

results are not yet conclusive. The straight-sided coupon, however, is still not an ideal one. Loads must be transmitted from the grips of the testing machine to the coupon via shear, and the shear strength of a unidirectional composite is typically at least an order of magnitude lower than its axial tensile strength. Shear failure near the gripped and tabbed region or damage to fibers under the grips are the most common problems. Other difficulties that arise in imperfect specimens [4,5] (such as splitting due to fiber misalignment) can be avoided by careful specimen preparation methods.

One interesting approach, which has been taken in recent years, involves testing of cross-reinforced specimens and inferring the tensile or compressive strength of the unidirectional material [4–6]. The strength is calculated usually by determining the stress in the 0° layers from the average stress in the laminate coupon using the elastic laminate theory. This approach yields the following value of the back-out (multiplication) factor, F, which is the ratio of the strength of the 0° layer to that of the laminate [7].

$$F = \frac{E_1[nE_2 + (1 - n)E_1] - (v_{12}E_2)^2}{[nE_1 + (1 - n)E_2][nE_2 + (1 - n)E_1] - (v_{12}E_2)^2} \tag{1}$$

where E_1, E_2 are the axial and transverse moduli, v_{12} is the axial Poisson's ratio of each lamina, and n is the fraction of 0° plies. The approach has been found to work well for some high-stiffness carbon-fiber cross-ply (0/90) composites. In the opinion of some investigators, "backing-out" lamina properties from laminate tests is unnecessary and fraught with uncertainties and there is no substitute for a good direct test. Proponents of the approach point out that laminate tests yield a good estimate of the "in situ" ply strength, since they are less influenced by splitting and other damages observed in unidirectional specimens. There is, however, still no agreement in the composites community on any standard data analysis scheme for backing out the required property.

The next section deals with the results of some analytical parametric studies on the stress and strain fields near the tab ends in the ASTM D 3039 specimen. Some suggestions are made to improve the specimen performance based on the results obtained. Results of some analytical, experimental, and data correlation studies employing alternative data analysis methods for cross-ply tension and compression tests are presented in the sections that follow. Effects of 90° plycracks as well as stiffening or softening of 0° layers are studied. The results should be of use in resolving the problem of selecting appropriate data reduction procedures.

Tabbed Tension Coupon

As discussed earlier, high stress peaks near tab ends are the reasons for early failure near such locations. Many interesting analytical and experimental studies dealing with the stress fields and observed failure patterns as well as other issues related to tabbing and gripping, such as clamping pressure, are reported in the literature [3,8–14]. A review of these studies [1] indicates that use of compliant tabs and low-taper angles as well as tapering of the tab end to zero thickness lowers the stress peaks and often yields higher tensile strength. Although the inelastic response of the adhesive material is expected to have a significant influence on the peak values of stresses, the effect of this factor has not been addressed in any analytical study. Limited test data [6] indicate that resilient adhesives, such as those needed for high temperature testing, may yield slightly higher strengths. The results of analytical parametric studies and discussions reported in this section deal with the effects of tab taper angle, tab material, and adhesive inelasticity with emphasis on the last two parameters.

Tab Taper Angle

Elastic plane-strain finite element studies were conducted to study effects of taper angles of glass/epoxy fabric tabs (7, 10, and 20°). The material properties for the carbon/epoxy coupon, adhesive, fabric tab (0/90), and steel grips used for calculations are given in Table 1. The laminate was considered to be 8 plies thick (1.016 mm = 0.04 in.), since this thickness is commonly used in tests. Because of symmetry, only a quarter of the geometry needs to be modeled. The geometry for a 10° taper angle for the tab is shown in Fig. 1. The total lengths were different for other taper angles, since the lengths of the straight portion of the tab and the gage section of the specimen were kept the same. The axial stress in the test section of the laminate is 69 MPa (10 ksi) for all the results reported. It may be noted that the steel grips with 11.2° tapers on the machine sides were included in finite element modeling. The grips were not allowed to have any deformation normal to the tapered faces, but could slide freely along those planes. All other boundaries were assumed to be stress free except the center of the test section, where a uniform axial displacement was imposed.

Stresses in the laminate close to the adhesive layer are of interest. Stress distributions are similar to those reported by Kural and Flaggs [10]. There are, however, two differences. The through-thickness normal stress (compressive) in the flat portion of the tab are higher in the present study, possibly because the tapers in the grips are different. Also, the peak stresses at the tab end are higher in the present study. The finer mesh used in the present study may be the reason for these higher values. The elements used near the tab end are 0.1 by 0.1 mm in size, which are much smaller than those used in Ref 10. Peak values of stresses for the 10° taper are compared in Table 2. It should be mentioned that because of the presence of the stress singularity, the peak values are dependent on the mesh size. However, they can be used for comparing the effects of geometry, as well as tab and adhesive properties. Maximum stresses (average over an element near the tab end) for different taper angles are given in Table 3. It is clear that lower taper angles yield smaller peak values since the stress singularity is less severe. Obviously, soft tabs and the inelastic response of the adhesive will yield lower

TABLE 1—*Properties of materials used in tabbed carbon/epoxy tensile coupon.*

Material Property[a]	Specimen T300/5208	Adhesive Layer[b]	(0/90) Fabric Tab Glass/Epoxy	Grip Steel
E_x, GPa (Msi)	139 (20.1)	2.2 (0.32)	32.5 (4.72)	200 (29.0)
E_y, GPa (Msi)	10.3 (1.5)	2.2 (0.32)	6.9 (1.0)	200 (29.0)
E_z, GPa (Msi)	10.3 (1.5)	2.2 (0.32)	32.5 (4.72)	200 (29.0)
ν_{xy}	0.3	0.42	0.3	0.3
ν_{yz}	0.3	0.42	0.064	0.3
ν_{zx}	0.022	0.42	0.077	0.3
G_{xy}, GPa (Msi)	4.6 (0.66)	0.76 (0.11)	4.1 (0.6)	77 (11.2)
G_{yz}, GPa (Msi)	3.2 (0.46)	0.76 (0.11)	4.1 (0.6)	77 (11.2)
G_{zx}, GPa (Msi)	4.6 (0.66)	0.76 (0.11)	6.9 (1.0)	77 (11.2)

NOTE: Steel tab properties are the same as those of grip materials. Aluminum tab—Young's modulus = 68.95 GPa (10 Msi), $\nu = 0.33$.

[a] x, y, z axes are in the specimen length-wise direction, thickness direction, and through-the-width (plane strain is assumed) direction, respectively. E and G denote Young's and shear moduli, respectively.

[b] Elastoplastic adhesive: yield stress (tension and compression) = 5 ksi and Tangent modulus beyond yield point = 19 ksi.

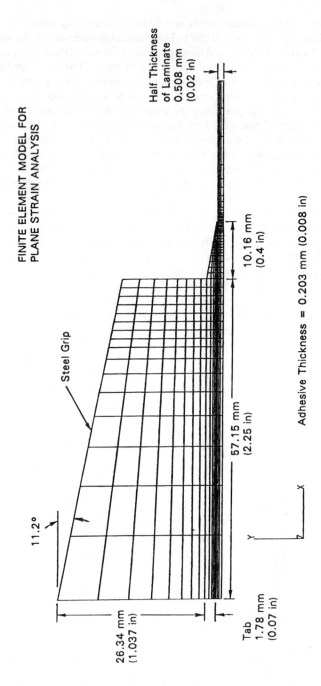

FINITE ELEMENT MODEL FOR
PLANE STRAIN ANALYSIS

Half Thickness
of Laminate
0.508 mm
(0.02 in)

Steel Grip

11.2°

10.16 mm
(0.4 in)

57.15 mm
(2.25 in)

26.34 mm
(1.037 in)

Tab
1.78 mm
(0.07 in)

X

Y

Z

Adhesive Thickness = 0.203 mm (0.008 in)

FIG. 1—*Geometry of a tensile specimen with 10° tapered tabs (quarter of the specimen is shown).*

TABLE 2—*Maximum stresses near tab end in carbon/epoxy tensile coupon (test section $\sigma_x = 69$ MPa = 10 ksi), 10° fabric tab.*

	Perfectly Bonded Case (Near Tab End)	
Stress	Present Study	Kural and Flaggs [Ref *12*]
σ_x, MPa (ksi)	89.3 (12.95)[a]	···
	81.8 (11.9)	71.0 (10.3)
σ_y, MPa (ksi)	2.8 (0.40)[a]	···
	2.5 (0.36)	0.76 (0.11)
σ_z, MPa (ksi)	2.5 (0.36)[a]	···
	2.5 (0.36)	1.7 (0.24)
τ_{xy}, MPa (ksi)	−2.8 (−0.41)[a]	···
	−2.5 (−0.36)	−1.8 (−0.26)

[a]Maximum value at an integration point. All other values reported are element averages.

TABLE 3—*Peak laminate stresses[a] near tab end (gage section stress = 69 MPa = 10 ksi) for various taper angles, (0/90) glass/epoxy tab.*

	Stresses,[a] MPa (ksi)			
Taper	Axial (σ_x)	Peel (σ_y)	Transverse (σ_z)	Shear (τ_{xy})
7°	78.6 (11.4)	1.8 (0.26)	2.2 (0.32)	−2.3 (−0.34)
10°	81.8 (11.9)	2.5 (0.36)	2.5 (0.36)	−2.5 (−0.36)
20°	90.3 (13.1)	4.8 (0.70)	3.3 (0.48)	−4.1 (−0.60)

[a]Element averages.

peak stresses, since the stress singularity at the tab end will be significantly weaker. Results of parametric studies on tab materials and adhesive inelasticity are discussed next.

Tab Material and Adhesive Inelasticity

The following cases were investigated for a fixed tab taper of 10° and zero cut-off thickness (Fig. 1):

1. (0/90) and (±45) glass/epoxy tabs with elastic adhesive; properties of the (±45) tab were evaluated via appropriate rotation of axes; and
2. steel, aluminum, and (0/90) fabric tabs with elastic and inelastic adhesives; properties are listed in Table 1.

Stresses in the elements next to the elastic adhesive for the two fabric tab orientations (0/ 90 versus ±45) are almost identical except for the peak values listed in Table 4. Peak values of all the stresses are lowered to some extent when the tab is softer (in the axial direction) except for the shear stress that is slightly increased. It is clear that the tab taper angle plays a more critical role than the axial stiffness of the tab.

Effects of the tab material properties were investigated further by performing stress analyses modeling elastic as well as elastoplastic behavior of the adhesive layer. Steel, aluminum, and (0/90) glass/epoxy tabs with the geometry shown in Fig. 1 were considered. Since the inelastic

TABLE 4—*Maximum stresses[a] in tabbed carbon/epoxy coupon for two fabric tabs, 10° taper, elastic adhesive (gage section stress = 69 MPa = 10 ksi).*

Fabric Tab Orientation	Stresses, MPa (ksi)			
	Axial, σ_x	Peel,[a] σ_y	Transverse,[a] σ_z	Shear, τ_{xy}
0/90	81.8 (11.9)	2.5 (0.36)	2.5 (0.36)	2.5 (−0.36)
±45	79.5 (11.5)	2.0 (0.29)	2.3 (0.33)	2.7 (−0.39)

[a]Element averages.

behavior of the adhesive makes the problem nonlinear, all results were computed for an average gage section stress of 1379 MPa (200 ksi), which is close to the ultimate tensile strength of common unidirectional carbon/epoxy composites such as T300/5208. Some representative stress distributions (axial, σ_x, and peel, σ_y, stresses in the laminate as well as the shear, τ_{xy}, stress in the adhesive in the elements near the laminate-adhesive interface are plotted in Figs. 2 through 4 that show the effects of adhesive nonlinearity for the case of the fabric tab. Similar effects are observed for other tab materials (see Ref 15 for details).

Examination of the results for various tab materials indicates that severe stress peaks exist for all tabs near the tab end when the adhesive is elastic. The peak values are higher for stiffer tabs. Stresses also show some oscillatory behavior near the tab ends. However, it may be noted that all stress peaks are reduced significantly when the adhesive is considered elastoplastic (Tables 5 and 6). For example, the axial stress peak is 2024 MPa (294 ksi) for steel tabs with elastic adhesive, but it reduces to 1544 MPa (224 ksi), which is only 12% higher than the average gage section stress when the adhesive is elastoplastic. Drastic reductions also occur

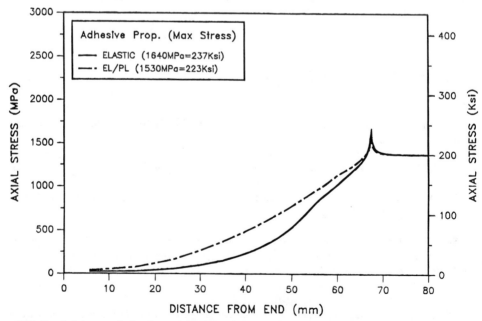

FIG. 2—*Axial stress in laminate for (0/90) glass/epoxy tabs, gage section stress = 1379 MPa (200 ksi).*

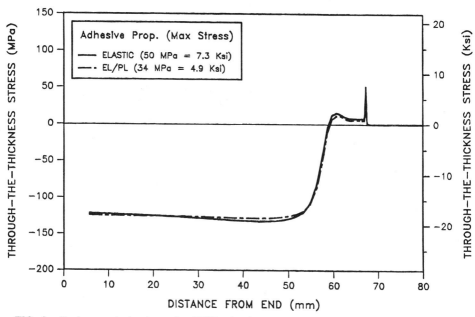

FIG. 3—*Peel stress in laminate for (0/90) glass/epoxy tabs, gage section stress = 1379 MPa (200 ksi).*

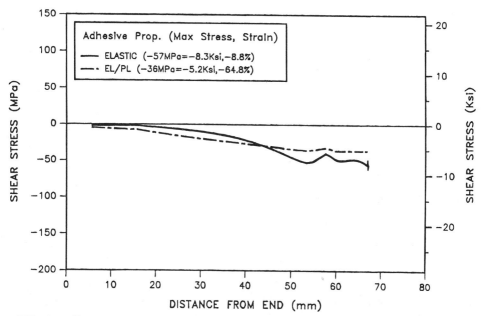

FIG. 4—*Adhesive shear stress for (0/90) glass/epoxy tabs, gage section stress = 1379 MPa (200 ksi).*

TABLE 5—*Comparison of peak values[a] (gage section stress = 1379 MPa = 200 ksi), various tabs, elastic adhesive.*

Tab	Axial	Peel	Transverse	Shear
		LAMINATE STRESSES, MPa (ksi)		
Steel	2024 (294)	128 (18.5)	79 (11.5)	−118 (−17.1)
Aluminum	1815 (263)	87 (12.6)	63 (9.2)	−80 (−11.6)
(0/90) glass fabric	1637 (237)	50 (7.3)	49 (7.1)	−50 (−7.2)
		ADHESIVE STRESSES, MPa (ksi)		
Steel	146 (21.2)	193 (28.0)	143 (20.7)	−156 (−22.6)
Aluminum	100 (14.5)	134 (19.4)	99 (14.3)	−108 (−15.6)
(0/90)glass fabric	60.7 (8.8)	79 (11.5)	59 (8.6)	−57 (−8.3)

[a]Element averages.

TABLE 6—*Comparison of peak values[a] (gage section stress = 1379 MPa = 200 ksi), various tabs, inelastic adhesive.*

Tab	Axial	Peel	Transverse	Shear
		LAMINATE STRESSES, MPa (ksi)		
Steel	1544 (224)	27 (3.9)	41 (5.9)	−34 (−5.0)
Aluminum	1536 (223)	30 (4.4)	41 (6.0)	−34 (−5.0)
(0/90) glass fabric	1534 (223)	34 (4.9)	43 (6.2)	−34 (−5.0)
		ADHESIVE STRESSES, MPa (ksi)		
Steel	48 (6.9)	50 (7.2)	49 (7.1)	−36 (−5.2)
Aluminum	49 (7.1)	55 (8.0)	52 (7.6)	−36 (−5.2)
(0/90) glass fabric	45 (6.5)	62 (9.0)	53 (7.7)	−36 (−5.2)

[a]Element averages.

in peak normal and shear stresses (shown in Tables 5 and 6), when the adhesive response is changed from elastic to elastoplastic. It may also be noted that the stresses (in the laminate and the adhesive) for all types of tabs become similar in nature when the adhesive is inelastic (Table 6). Furthermore, the shear stress distributions are more uniform (Fig. 4) for the inelastic case as compared to the case when the adhesive is elastic. Peel stress distributions in the laminate for different tabs are compared in Fig. 5 to demonstrate the similarity of the stress distributions. It appears that the tab geometry, the adhesive yield stress, and the force equilibrium are the major factors influencing the stress distribution for the case of inelastic adhesive. Elastic properties of the tab and other materials do not have any significant effect on the stress field. However, it may be noted that both the peel strain and the shear strain in the adhesive show large peaks at the tab ends (Fig. 6 shows the variations of the plastic shear strains in the adhesive for the three tab materials), the peak shear strain being extremely high. Although the peel component (Table 7) is highest for the (0/90) glass/epoxy tabs (8.1%), the shear component is highest for steel tabs (137.4%). Peak values of the inelastic shear strains for the three tab materials are compared in Table 7.

Peak stresses for inelastic adhesive layers are reduced significantly from those for the elastic case for all tabs (steel, aluminum, and 0/90 glass fabric), and the magnitudes do not differ appreciably for different tab materials. The peak laminate axial stress is about 12% higher than gage section values. However, since the stress drops off inside the specimen (away from the surface), the effect of this peak on strength may be lower. However, it is likely that

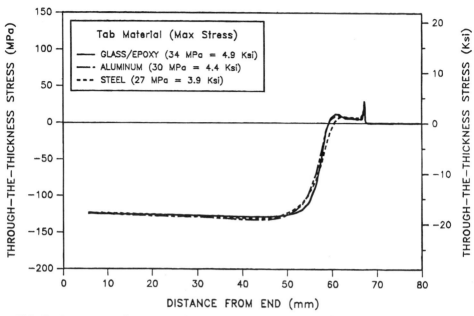

FIG. 5—*Laminate peel stress for different tabs with inelastic adhesive, gage section stress =* 1379 MPa (200 ksi).

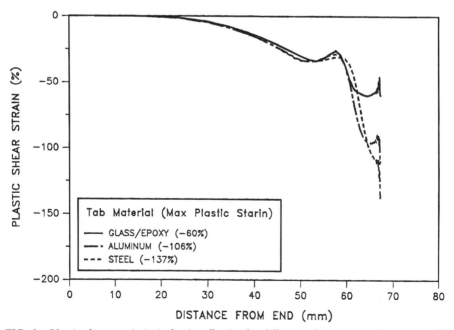

FIG. 6—*Plastic shear strain in inelastic adhesive for different tabs, gage section stress = 1379 MPa (200 ksi).*

TABLE 7—*Peak adhesive plastic strains[a] (percent) in inelastic adhesive for various tabs, laminate stress ≈ 200 ksi.*

Tab	Peel	Shear	Effective
Steel	6.5	−137	80
Aluminum	8.5	−106	62
(0/90) glass fabric	8.1	−60	36

[a]Element averages.

debonding may occur due to high peel and shear strains in the adhesive and in the laminate. Peel stress and strain are high, but they peak over a small region. Shear strains, however, are high over a large region even for the case of inelastic adhesive. Peak plastic shear strains are lowest for (0/90) glass fabric tabs (much less than that for steel and aluminum tabs). Therefore, the use of soft tabs and low taper angles are still desirable. Debonding may occur even with the use of soft tabs and strengths obtained may be affected to some extent by such debonding. Elastic analyses of a specimen with delaminations in the adhesive layers reported elsewhere [15] indicate that all stress components near the tips of delaminations are expected to be higher than those near the tab ends, and delaminations once initiated may propagate towards the grips. Therefore, the use of ductile but tough adhesives and soft (and tough) tab materials may improve the performance of tabbed tension coupons. However, the matrix rich area of the composite remains the weak link of the system. Delamination growth is expected to be shear-dominated aided by peel stresses. It is likely, however, that the influence of this weak link may be reduced by careful surface preparation (sanding without fiber damage) and bonding. Composites with high toughness (tough resins) may perform better than standard epoxy-matrix systems.

For compression coupons, which are designed primarily to avoid buckling failure, peel stresses are compressive and friction may prevent delamination growth in shear. However, the use of ductile and tough adhesives as well as soft tabs may still be desirable even in such specimens. It appears worthwhile to perform suitably designed tension as well as compression tests using the adhesives, tabs, and specimens (composites) mentioned earlier. Careful observation of delamination growth near tab ends through real-time observations (after magnification) as well as other nondestructive evaluation (NDE) techniques (such as acoustic emission monitoring) will also be useful.

Cross-Ply Tension Tests

As discussed earlier, cross-reinforced specimens are being considered for inferring the axial strengths of unidirectional composites [4–6]. It is known, however, that severe subcritical damages, such as ply cracks parallel to fibers, occur in all the off-axis plies of cross-ply (0/90) or other (such as (0/±45)) layups used in the tension tests before the failure of the 0° plies. Some delamination growth from these ply-cracks may also occur. The role of such damages on specimen performance and accuracy of data analysis schemes to infer 0° strength has not been studied in much detail. In what follows, some analytical estimates of the stresses in the 0° layers at the point of tensile failure of cross-ply (0/90) carbon and glass fiber-reinforced laminates with ply cracks are given. In the next subsection, some data analysis schemes, including the use of the elastic lamination theory, are examined based on these estimates. Correlations with test data from cross-ply tension specimens and comparisons with measured strengths of unidirectional specimens are reported in the subsections that follow. Results of cross-ply compression specimens are examined in the next section.

Analytical Estimates for Cross-Ply Tension Test

Various mechanics-based approaches [16–22] have been suggested to determine the laminate stiffness loss and average ply stresses due to the existence of ply cracks of known densities ($\omega = a/\ell$, $2a$, and ℓ being the thickness of and the average crack spacing in an inner cracked 90° layer, respectively). In all of these approaches, the 0° layers are considered to be linearly elastic in the fiber direction, and the ply cracks are assumed to extend through the entire width of the test coupon (an approximation that appears to be pretty accurate at the point of failure of such coupons). The estimates reported here are based on simple approximate formula for evaluation of the change in strain energy density in a representative volume element due to the ply cracks [22]. These formulas have been found to yield stiffnesses (secant moduli for the tension coupon) that are higher than lower bound solutions [16,19] as well as test data for some carbon and glass/epoxy systems. Also, effects of delaminations are not considered. For these reasons, it is expected that elastic laminate theory solutions should yield average stresses in the 0° layers that are closer to the estimates reported here than those obtained from lower bound [16,19] or other accurate solutions.

The calculations are based on assumed 90° ply-crack densities (ω_{max}) of 0.4 and 0.5 for carbon/epoxy and glass/epoxy systems, respectively, with the plies as dispersed as possible. These values are representative of those observed in tests [20,21] before final failure. Ply properties used for calculations are given in Table 8. Calculations were performed without and with due consideration to thermal residual stresses due to cool-down from processing temperature to room temperature (change in temperature, $\Delta T = -111°C = -200°F$) and the results are given in Figs. 7 and 8. The average stress-strain responses (schematic except for low strains and at the point of failure) of cross-ply carbon and glass/epoxy coupons are shown in the figures. Axial and Poisson (transverse) strains at the point of failure (maximum stresses of 952 MPa (138 ksi) for carbon/epoxy and 552 MPa (80 ksi) for glass/epoxy are typical of such systems) shown in Figs. 7 and 8, and 0° layer stresses at that point (shown in the tables below the figures) are obtained from the calculations performed. Figure 7 shows that the ply cracks have a significant influence on the Poisson strain, but the stiffness loss and the nonlinearity in the axial stress-strain response of the carbon-epoxy system are not significant. Since the effect of Poisson strain on the 0° layer stress is small, the 0° strength (or the back-out factor) estimated from elastic laminate theory differs from that obtained with due consideration to ply cracks by only -5% (shown in the table in Fig. 7). Residual stress does not have much influence on the results. On the other hand, for glass/epoxy systems, the stiffness loss due to ply cracks is quite significant (Fig. 8), since the contribution of 90° layers to the initial laminate stiffness is high (as reported in many studies [16,19–22]). For this reason the laminate theory estimate of 0° strength (or the back-out factor) is 22% lower than that obtained when effects

TABLE 8—*Layer properties used for calculation for cross-ply specimens.*

Property[a]	Carbon/Epoxy	Glass/Epoxy
E_1, GPa (Msi)	130 (18.85)	41.7 (6.05)
E_2, GPa (Msi)	10.5 (1.52)	13.0 (1.89)
ν_{12}	0.35	0.3
$\alpha_1 10^{-6}/°C(°F)$	−0.77 (−0.43)	8.6 (4.8)
$\alpha_2 10^{-6}/°C(°F)$	24.5 (13.6)	22.1 (12.3)
ω_{max}	0.4	0.5

[a]1 denotes the fiber direction (axial) and 2 the transverse. E denotes the modulus and α the thermal expansion. ν_{12} is the Poisson strain due to unit axial strain. ω_{max} = assumed maximum crack density in 90° layers. Temperature change for calculation of residual stresses = $-111°C$ ($-200°F$).

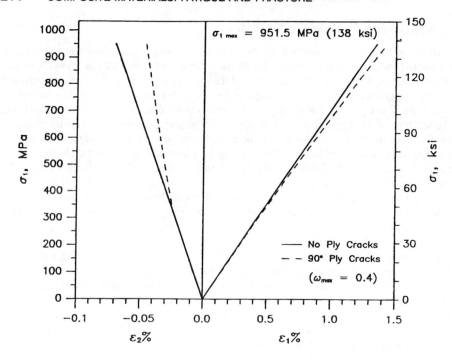

	No Ply Cracks		With Ply Cracks
	No Residual Stress	With Residual Stress	With Residual Stress
$\sigma^{0^\circ}_{1\,max}$ MPa (ksi)	1761.7 (255.5)	1735.5 (251.7)	1843.7 (267.4)
Factor: $\sigma^{0^\circ}_{1\,max} / \sigma_{1\,max}$	1.851	1.824	1.938

FIG. 7—*Effects of 90° ply cracks in (0/90) carbon/epoxy system loaded in 0° direction. Note: standard theory yields 5% underestimation of 0° strength with ply cracks, but the calculation based on strains is* $\sigma^{0^\circ}_{1max} \approx 1843.7$ MPa (267.4 ksi).

of ply cracks are considered. Average transverse stresses in the 90° layers (in the loading direction) at the point of coupon failure are 269 MPa (39 ksi) for no ply cracks and 69 MPa (10 ksi) in the cracked ($\omega_{max} = 0.5$) condition. This large change, which is transferred to 0° layers, causes the severe underestimation of 0° strength by the use of elastic laminate theory. This underestimation, the reasons for which are quite obvious, is expected but has never been addressed in connection with cross-ply tests, since they are not commonly employed for the determination of the tensile strength of unidirectional glass/epoxy composites. It should, how-ever, be given due consideration when any standardization of the cross-ply test and associated data reduction schemes is attempted.

Another approach to back-out the 0° strength is to use the measured axial and Poisson strains and the liner elastic properties of the 0° layer (these properties can be determined very accurately from tests on unidirectional specimens), that is

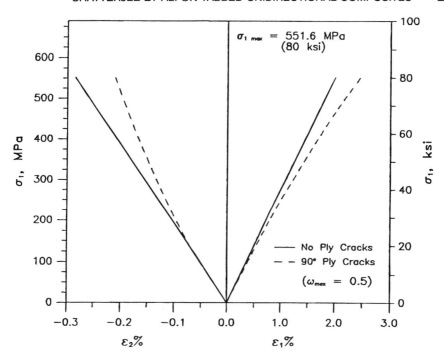

	No Ply Cracks		With Ply Cracks
	No Residual Stress	With Residual Stress	With Residual Stress
$\sigma_{1\,max}^{0^\circ}$ MPa (ksi)	846.7 (122.8)	834.3 (121.0)	1031.5 (149.6)
Factor: $\sigma_{1\,max}^{0^\circ} / \sigma_{1\,max}$	1.535	1.512	1.870

FIG. 8—*Effects of 90° ply cracks in (0/90) glass/epoxy system loaded in 0° direction. Note: standard theory yields 22% underestimation of 0° strength with ply cracks, but the calculation based on strains is $\sigma_{1max}^{0^\circ} \approx 1031.5$ MPa (149.6 ksi).*

$$\sigma_1 = E_1(\epsilon_1 + \nu_{21}\epsilon_2)/(1 - \nu_{12}\nu_{21}) \approx E_1\epsilon_1 \qquad (2)$$

where $\nu_{21} = \nu_{12}E_2/E_1$. Results of calculations for the tests on the two systems simulated in this section are shown in the notes below tables in Figs. 7 and 8 and they are identical to those computed with due consideration to ply cracks, since the 0° layers are assumed to be linear elastic in the fiber direction.

Before concluding this subsection, we note that factors other than ply cracks may be of significance in cross-ply tension tests. Therefore, the data reduction schemes are re-examined next in conjunction with test data from cross-ply and unidirectional specimens. The elastic laminate theory has also been used for analysis of data from cross-ply compression tests. For this reason, test data from compression specimens are analyzed in the next section to examine some simple data reduction schemes for backing out the 0° compressive strength with the objective of determining how these schemes perform.

Materials and Test Specimens

The cross-ply test data utilized here were generated by the Composite Materials Research Group, University of Wyoming, as a part of a comprehensive test program described elsewhere [15] involving tests on unidirectional and cross-ply coupon specimens.

Tension and compression tests were conducted on three material systems listed below:

1. AS4/3501-6,
2. Glass/L-20, and
3. IM7/8551-7A.

The first material is a standard epoxy-matrix composite used widely in the industry. The second one is commonly used in fabric composites made by European suppliers, whereas the third one is a thermoplastic composite (IM7/8551-7A, a product under development at the time of procurement) with tensile strengths slightly higher than that of the first. It is worth noting that the final product (IM7/8551-7) has a tensile strength higher than that of IM7/8551-7A and a compressive strength comparable to that of AS4/3501-6. Flat rectangular coupons as well as thickness-tapered unidirectional specimens with different taper geometries were tested. Thickness tapering was chosen because its effect is expected to be less severe than width tapering that is known to be detrimental for unidirectional composites. Although performances of thickness-tapered unidirectional specimens are not addressed here in much detail, the results discussed elsewhere [15] indicate that properly designed tapering can yield strengths higher than those from tabbed straight-sided coupons (ASTM D 3039). In addition, tabbed and untabbed straight-sided cross-ply coupons were tested since they are being considered by many investigators. Some test data from unidirectional coupons with tapered and constant cross-sections are reported for comparison with strengths backed out from cross-ply tests.

ASTM D 3039 specimens for all materials were 12.7-mm (0.5 in.) wide, 305-mm (12 in.) long, rectangular coupons with 12° tapered tabs with a gage length of 127 mm (5 in.) between tabs. Unidirectional coupons were eight plies thick, whereas the cross-ply specimens had nine plies (five 0° in load direction, and four 90°). Untabbed AS4/3501-6 cross-ply specimens were also tested. Nominal ply thickness was 0.14 mm (0.0055 in.). Fiber volume fractions were close to 0.62. Standard serrated grips as well as flame-sprayed grips were used for ASTM D 3039 specimens (unidirectional and cross-ply) made from AS4/3501-6. Standard grips were employed for the other two systems. Comparisons of results obtained from cross-ply tests with those measured from unidirectional specimens are shown in Table 9. Material properties used for analysis of cross-ply data are listed below the table.

Data Correlation for AS4/3501-6 Material

The average strength from untabbed cross-ply specimens was found to be very low with a high standard deviation (obtained by using the back-out factor of 1.7 in this case) indicating possible damages at the grips due to the absence of tabs. However, strengths backed out from tabbed cross-ply specimens tested with serrated and flame-sprayed grips were also lower than those from all the unidirectional specimens tested. Although it is expected (as discussed in the subsection dealing with analytical estimates) that strengths of a typical unidirectional carbon/epoxy material backed out from cross-ply data using elastic theory may be about 5%

TABLE 9—Analysis of cross-ply tension data.

Material	Mean of Cross-Ply Data ϵ_{ult}, %	σ_{ult}, MPa (ksi)	Estimated 0° Strength,[a] MPa (ksi) from Cross-Ply Data, using Back-Out Factor[c]	from Strains,[d] Initial E_1	from Strains,[d] Secant E_1	Mean Values from Unidirectional Specimens[b] 0° Strength,[a] MPa (ksi)	ϵ_{ult}, %	Specimen Type[d]
AS4/3501-6[e] Serrated grips Flame-sprayed grips	1.31	1069(155)	1792(260)	1793(260)	1931(280)	2013(292)	1.39	tapered geometry ASTM D 3039 (flame sprayed grips)
Glass/L20[f] Serrated grips	1.28	1014(147)	1692(246)	1682(244)	1884(273)	1793(260)	1.34	tapered geometry
	3.16	800(116)	1139(165)	1456(211)	...	1296(188)	3.10	ASTM D 3039
						1269(184)	2.88	
IM7/8551-7A[g] Serrated grips	1.40	1393(202)	2315(336)	2284(331)	...	2337(339)	1.43	tapered geometry
						2172(315)	1.33	ASTM D 3039

[a]Normalized to 60% fiber volume.
[b]Highest and lowest mean values.
[c]Material properties and back-out factors (B.F.) are listed below in e, f, and g.
[d]Values are also close to $E_1 \epsilon_{ult}$.
[e]AS4/3501-6: Initial $E_1 \approx$ serrated grip = 138 GPa (20 Msi), flame-sprayed grip = 132 GPa (19.2 Msi), B.F. = 1.7, Secant E_1 = 149 GPa (21.6 Msi), E_2 = 10.3 GPa (1.5 Msi), and ν_{12} = 0.28.
[f]Glass/L20: E_1 = 47.6 GPa (6.9 Msi), E_2 = 12.4 GPa (1.8 Msi), ν_{12} = 0.28, and B.F. = 1.5.
[g]IM7/8551-7A: E_1 = 170 GPa (24.6 Msi), E_2 = 9.0 GPa (1.3 Msi), ν_{12} = 0.31, and B.F. = 1.73.

lower than the "true" strength, the values in Column 4 of Table 9 are much lower (about 10%) than the highest mean strength 2013 MPa (292 ksi) obtained from a tapered geometry specimen.

It was suggested in the subsection on analytical estimates that a better estimate may be obtained by utilizing measured axial and transverse strains at the point of failure of the cross-ply coupon to compute the stress (in situ strength) in the 0° layer, but the results shown in Column 5 of Table 9 indicate that the use of initial moduli yields strength values comparable to those obtained using the elastic back-out factor of 1.7. However, the strengths obtained using the secant modulus at the point of failure of the unidirectional material (stress-strain plots showing some stiffening are reported elsewhere [15]) are much higher than both the previous estimates and are only slightly lower than the highest mean strength obtained from a tapered geometry specimen. However, it should be emphasized that use of the elastic back-out factor utilizing the secant modulus at failure will yield values comparable to results obtained using a back-out factor of 1.7 since the back-out factor is not a strong function of the axial modulus. The value of the back-out factor (obtained using the secant modulus of 149 GPa = 21.6 Msi) is 1.706, which differs very slightly from 1.7, calculated using the initial modulus. In addition, it should be pointed out that, since $\sigma_{11} \approx E_1 \epsilon_1$, the coefficient of variation of unidirectional strengths obtained using measured strains in the cross-ply layup and unidirectional moduli may be considered to be the same as those for ultimate strains of cross-ply specimens that are 4 and 6%, respectively, for standard and flame-sprayed grips. Unidirectional tapered geometry specimens show a coefficient of variation of 7%. It may be noted that results shown in Columns 5 and 6 of Table 9 are not significantly influenced by the transverse strains that are small. A good estimate of 0° strength may also be obtained by multiplying the axial failure strain by the 0° axial modulus (using initial or secant modulus as appropriate).

Data Correlation for Glass/L20 Material

Use of the elastic back-out factor (its value is 1.50 in this case) and measured cross-ply coupon strengths yields a low mean value of in situ 0° strength (Table 9) as compared to that from unidirectional specimens. This trend is consistent with the results discussed in the subsection dealing with analytical estimates. Calculations based on the elastic back-out factor do not consider the influence of ply cracks in the 90° layers and the load transfer from 90° to 0° layers. As suggested in the previous subsection, we obtained estimates of the 0° stress at failure based on measured axial and transverse strains in the laminate and elastic properties of the unidirectional material listed below the table.

It may be noted that this unidirectional material does not show any stiffening, which is observed in AS4/3501-6 material. The result shown in Column 5 of Table 9 is not influenced significantly by the transverse strain. A good estimate of 0° strength may also be obtained by multiplying the axial failure strain by the 0° axial modulus. This estimate is only 2% lower than that in Column 5. The result shows that the mean strength estimated from strains is higher than those obtained from unidirectional specimens by about 12%.

IM7/8551-7A Material

In contrast to the results for AS4/3501-6 and Glass/L20, the mean 0° strength backed out from cross-ply tests (using the back-out factor) is higher than that obtained from unidirectional ASTM D 3039 specimens and is comparable to the highest mean strength measured from specimens with tapered geometries (Table 9). Since the 90° layers do not contribute much to the initial laminate stiffness and there is no significant stiffening effect (similar to that observed in AS4/3501-6), it is expected that the use of an elastic solution (which yields a back-out factor of 1.73) will underestimate strengths by less than 5%. It appears, therefore, that the elastic back-out factor should work well for materials with similar or higher axial strength and stiffness (as compared to transverse stiffness), provided there is no stiffening in the unidirectional material that causes the secant modulus at failure to be higher than the initial modulus. For completeness, the estimate of strength based on measured strains (similar to those obtained for AS4/3501-6 and Glass/L20) is shown in Column 5 of Table 9. The unidirectional properties used to obtain these estimates are listed below the table.

The result shows that the estimate is very close to that obtained using the elastic back-out factor. It may be noted that a simple estimate obtained by multiplying the axial failure strain by the corresponding modulus is also very close to the computed 0° strength shown in Table 9.

Analyses of Cross-Ply Compression Test Data

Under compression load, transverse ply cracks perpendicular to lamination planes are not expected. Other kinds of damage, such as cracks parallel to fibers at an angle close to 45° to loading direction may develop due to shear. However, such cracks will possibly allow compressive load transfer across crack planes (due to contact). In addition, initiation of such cracks is also expected to occur at comparatively high transverse compressive stresses. In any event, some minor stiffness loss may occur due to such transverse cracks that may introduce some nonlinearity in the cross-ply response. In addition, major stiffness loss may be expected due to localized failure or softening (as opposed to stiffening in tension) of the fibers in the 0° layers at higher loads. Some secondary bending effects were noticed in most of the cross-ply specimens [15]. They varied from specimen to specimen, but no significant differences were observed in the three materials studied. Such bending may cause additional stiffness loss. Although a detailed stress analysis considering all such nonlinearities may be performed, it will be more complicated than that required to consider ply cracking under tensile loads (discussed earlier), since micro- and mini-level mechanics under compression are strongly influenced by the presence of minor imperfections. For this reason, an attempt was made to obtain some bounds on the 0° strengths based on some simple but realistic assumptions on failure modes as discussed here.

1. Elastic solution to calculate the load carried by 0° layers—Elastic back-out factor yields a good estimate of this load and it yields an approximate lower bound ($\leq \sigma_{ult}$) on the strength of 0° layers, since the back-out factor is not strongly influenced by the axial stiffness of 0° layers.
2. Calculation of stress in the 0° layer from measured strains (axial and transverse) at failure assuming the 0° layers to behave elastically—This approach possibly yields an upper bound ($\geq \sigma_{ult}$) to the 0° strength. This estimate will be exact if the 0° layers behave linearly (as discussed in the case of tension). Since the transverse strains were not measured and their effects on axial stress is small, a simpler estimate is obtained here by multiplying the failure strain of the laminate by the axial modulus of the 0° layer ($= E_1 \epsilon_{ult}$).
3. Another upper-bound estimate is obtained by dividing the average stress in the laminate by the volume fraction of 0° layers ($= 9/17$ in the layup tested). Contribution of 90° layers to laminate strength is neglected in this case.

The lowest of the two upper bounds listed in Estimates 2 and 3 yields the best upper bound of the two. The properties used for obtaining the bounds and the results obtained from tabbed (straight tabs) cross-ply IITRI specimens are given in Table 10. All of the specimens tested had nine 0° plies and eight 90° plies.

Results in Table 10 indicate that Estimate 3 neglecting 90° layers yields the best upper bound for carbon fiber composites since the contribution of 90° layers is small and the 0° layers show significant nonlinearity at failure. Estimate 3 is much higher especially for AS4/3501-6 (as compared to that for IM7/8551-7A) possibly because of significant softening of the 0° layers (see Fig. 9). On the other hand, for Glass/L20, the contribution of 90° layers is significant and the 0° layers do not show much nonlinearity up to the point of failure (Fig. 10), and Estimate 2 based on strain yields the best upper bound. Lower bounds for all three materials are within 5 to 7% of the best upper bounds and are higher than the highest strengths obtained from unidirectional specimens. A tapered geometry specimen for unidirectional AS4/3501-6 yields values close to the lower bound. For Glass/L20 and IM7/8551-7A, no tapered geometry was used and results from standard IITRI specimens are quite low. Hand-sanded, modified ASTM Test Method for Compressive Properties of Rigid Plastics (D 695-89) specimens of IM7/8551-7A yielded unidirectional strengths slightly lower than Estimate 1.

Stiffness loss in compression may be expected due to damage in 90° and 0° plies as well as softening of some types of graphite fibers. However, a good lower-bound-type estimate of 0° strength can be obtained for all three materials studied, using the elastic back-out factor, possibly because the damage in the 90° layers is not as severe as in tension and the back-out factor changes very slightly due to change (or loss) in the stiffness of the 0° layers. The better of the two simple upper-bound-type estimates (one based on calculated stress in the 0° layer using measured strain and Hooke's law and the other neglecting 90° contribution to the load carried by the laminate) is also found to be close (within 5 to 7%) to the estimate using the elastic back-out factor. Strengths measured from unidirectional specimens are found to be lower than or comparable to that obtained using the factor.

Concluding Remarks

Parametric studies of stress fields near tab ends show that the inelastic behavior of adhesives lowers the stress peaks considerably. Strains in the adhesive, however, are high because of inelastic strains. Tests with various ductile but tough adhesives, softer tabs, and low taper angles are recommended.

Subcritical damages such as 90° ply cracks and stiffening or softening of 0° layers play an important role in the performance of cross-ply specimens and they should be given due consideration for data reduction in some cases. For the three materials studied, the elastic back-out factor works well for compression tests and appears to yield good lower-bound results that are close to the lower of the two upper bounds reported herein. For cross-ply tension tests, use of this factor is found to be appropriate only for composites with high axial modulus (compared to transverse modulus) provided the fibers (and the unidirectional material) do not show much stiffening. In this study, such stiffening is noticed in AS4, but not in IM7. It is known that some other types of carbon fibers exhibit stiffening under tension. On the other hand, computation of 0° stress based on elastic 0° layer properties and measured strains (at failure) in the cross-ply specimens yields results that appear to be acceptable for all three materials studied. However, use of the secant axial modulus at the point of failure of the 0° layers is required for materials that show stiffening (such as AS4/3501-6). Additional cross-ply tests on other commonly used composites of varying axial moduli and strengths and correlation of backed-out 0° strengths with those from carefully conducted and well-designed unidirectional specimens (thickness-tapered specimens as well as straight-sided ones with proper adhesives and tabs) are needed to decide on data reduction schemes, which can be accepted as standards.

TABLE 10—Analysis of cross-ply compression data.

Material	Mean of Cross-Ply Data		Estimated 0° Strength,[a] MPa (ksi) from Cross-Ply Data			Mean Values from Unidirectional Specimens		
	ϵ_{ult}, %	σ_{ult}, MPa (ksi)	Elastic Solution Lower Bound[a] (1)[b]	Based on Strain, Upper Bound[a] (2)	Neglecting 90° Layers, Upper Bound[a] (3)	0° Strength,[a] MPa (ksi)	ϵ_{ult}, %	Specimen Type[c]
AS4/3501-6	1.68	965(140)	1689(245)	2065(300)	1805(262)	1585(230)[d]	1.83	A
						1386(201)	1.33	B
Glass/L20	2.43	710(103)	1075(156)	1142(166)	1310(190)	972(141)[d]	1.93	C
						848(123)	1.83	B (10 ply)
IM7/8551-7A	1.16	855(124)	1482(215)	1664(241)	1564(227)	1462(212)[d]	...	D (8 ply)
						1227(178)	0.87	B

NOTE: Lamina properties used:

AS4/3501-6: E_1 = 124 GPa (18 Msi), E_2 = 10.3 GPa (1.5 Msi), and ν_{12} = 0.28.

Glass/L20: E_1 = 48.3 GPa (7 Msi), E_2 = 12.4 GPa (1.8 Msi), and ν_{12} = 0.28.

IM7/8551-7A: E_1 = 148 GPa (21.5 Msi), E_2 = 9.0 GPa (1.3 Msi), and ν_{12} = 0.31.

[a]Values normalized to 60% fiber volume.

[b]Using back-out factors ~1.76 (AS4/3501-6), 1.55 (Glass/L20), and 1.79 (IM7/8551-7A).

[c]Specimen Type:

A = tapered test section, IITRI fixture, no tabs.

B = IITRI, straight tabs (bonded to hand-sanded specimens for AS4/3501-6 and IM7/8551-7A).

C = IITRI, pair of tapered at one end (failure occurred at the non-tapered tab ends).

D = modified D 695, tabs bonded to hand-sanded specimens.

[d]Highest mean value from unidirectional specimens.

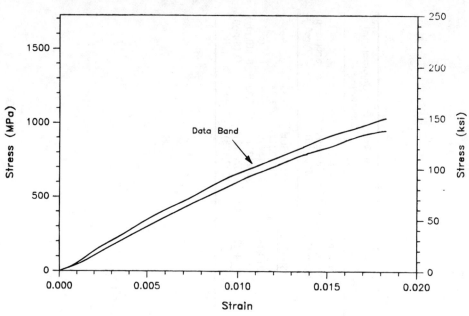

FIG. 9—*Stress-strain plots for cross ply AS4/3501-6 compression specimens.*

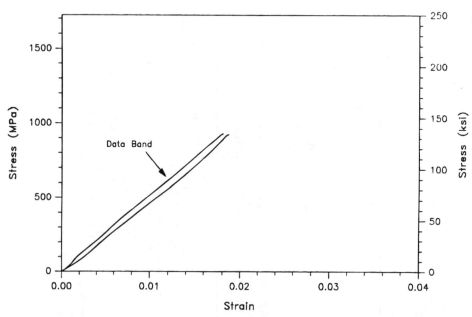

FIG. 10—*Stress-strain plots for cross ply Glass/L20 compression specimens.*

Acknowledgments

The results reported were a part of the work sponsored by the FAA Technical Center, Atlantic City, New Jersey, under ARL Contract No. DAAL04-89-C-0023. Sincere appreciations are due to Mr. Robert Pasternak, technical monitor at ARL, Watertown, Massachusetts; Mr. Lawrence Neri of the FAA Technical Center; and Mr. Joseph Soderquist of FAA, Washington, DC. Thanks are also due to Mr. Jeff Kessler and Prof. Donald Adams of the Composites Research Group, University of Wyoming, for generation of the test data used for correlation.

References

[1] Chatterjee, S., Adams, D., and Oplinger, D. W., *Test Methods for Composites: A Status Report, Volumes I, II, and III*, DOT/FAA/CT-93/17, FAA Technical Center, Atlantic City, NJ, June 1993.
[2] Dastin, S., Lubin, G., Munyak, J., and Slobodzinski, A., "Mechanical Properties and Test Techniques for Reinforced Plastic Laminates," *Composite Materials: Testing and Design, ASTM STP 460*, American Society for Testing and Materials, Philadelphia, 1969, pp. 13–26.
[3] Oplinger, D. W., Parker, B. S., Gandhi, K. R., Lamothe, R., and Foley, G., "On the Streamline Specimen for Tension Testing of Composite Materials," *Recent Advances in Composites in the United States and Japan, ASTM STP 864*, J. R. Vinson and M. Taya, Eds., American Society for Testing and Materials, Philadelphia, 1985, pp. 532–555.
[4] Hart-Smith, L. J., "Some Observations About Test Specimens and Structural Analysis for Fibrous Composites," *Composite Materials: Testing and Design (Ninth Volume), ASTM STP 1059*, S. P. Garbo, Ed., American Society for Testing and Materials, Philadelphia, 1988, pp. 86–120.
[5] Hart-Smith, L. J., "Generation of Higher Composite Material Allowables Using Improved Test Coupons," *Proceedings*, Thirty-sixth International SAMPE Symposium, 1991, pp. 1029–1044.
[6] Rawlinson, R. A., "The Use of Crossply and Angleply Test Specimens to Generate Improved Property Data," *Proceedings*, Thirty-sixth International SAMPE Symposium, 1991, pp. 1050–1067; see also, *Proceedings* on crossply test results, MIL-HDBK-17 Coordination Group Meeting by G. Hansen (compression tests) and R. A. Rawlinson (tension tests), April 1989.
[7] Camponeschi, E. T., Jr., and Hoyns, D., "Determination of Effective 0° Properties from [0/90] Laminate Theory," *Proceedings*, ASTM D30-04 Subcommittee Meeting, April 1991.
[8] Oplinger, D. W., Gandhi, K. R., and Parker, S., "Studies of Tension Specimens for Composite Material Testing," AMMRC-TR82-17, Army Materials and Mechanics Research Center, Watertown, MA, April 1982.
[9] Cunningham, M. E., Schoultz, S. V., and Toth, J. M., Jr., "Effect of End-Tab Design on Tension Specimen Stress Concentrations," *Recent Advances in Composites in the United States and Japan, ASTM STP 864*, I. R. Vinson and M. Taya, Eds., American Society for Testing and Materials, Philadelphia, 1985, pp. 253–262.
[10] Kural, M. H. and Flaggs, D. E., "A Finite Element Analysis of Composite Tension Specimens," *Composite Technology Review*, Spring 1983, pp. 11–17.
[11] Cernosek, J. and Sims, D., "The Effect of Tab Geometry on the Fatigue Life of Fibrous Composites," *Experimental Techniques*, Dec. 1982, pp. 5–11.
[12] Abdallah, M. G. and Westberg, R. L., "Effect of End Tab Design of the ASTM 3039 Tension Specimen on Delivered Strength of HMS1/3501-6 Graphite/Epoxy Materials," *Proceedings*, 1987 SEM Spring Conference in Experimental Mechanics, 1987, pp. 362–366.
[13] Lenoe, E. M., Knight, M., and Schoene, C., "Preliminary Evaluation of Test Standards for Boron/Epoxy Laminates," *Composite Materials: Testing and Design, ASTM STP 460*, American Society for Testing and Materials, Philadelphia, 1969, pp. 122–139.
[14] Abdallah, M. G. and Muller, C. S., "An Experimental Study of the Effect of Clamping Pressure on the Tensile Properties of Unidirectional Graphite/Epoxy Composite Materials," Hercules Inc., Magna, UT, March 1987.
[15] Chatterjee, S. N., Yen, C. F., Kessler, J. A., and Adams, D. F., "Standardization of Test Methods for Laminated Composites, Vol. I and II," MSC TFR 3313/1706-002, Final Report, ARL Contract No. DAAL04-89-C-0023, ARL, Watertown, MA, Oct. 1993.
[16] Hashin, Z., "Analysis of Cracked Laminates: A Variational Approach," *Mechanics of Materials*, Vol. 4, 1985, pp. 121–136.
[17] Dvorak, G. J. and Laws, N., "Progressive Transverse Cracking in Composite Laminates," *Journal of Composite Materials*, Vol. 23, 1988, pp. 900–916.

[18] Tan, S. C. and Nuismer, R. J., "A Theory of Matrix Cracking in Composite Laminates," *Journal of Composite Materials*, Vol. 23, 1989, pp. 1029–1047.

[19] Nairn, J. A., "The Strain Energy Release Rate of Composite Microcracking: A Variational Approach," *Journal of Composite Materials*, Vol. 23, 1989, pp. 1106–1129.

[20] Lee, J., Allen, D. H., and Harris, C. E., "Internal State Variable Approach for Predicting Stiffness Reductions in Fibrous Laminated Composites with Matrix Cracks," *Journal of Composite Materials*, Vol. 23, 1989, pp. 1273–1290.

[21] Highsmith, A. L. and Reifsnider, K. L., "Stiffness Reduction Mechanisms in Composite Laminates," *Damage in Composite Materials, ASTM STP 775*, K. L. Reifsnider, Ed., American Society for Testing and Materials, Philadelphia, 1982, pp. 103–117.

[22] Chatterjee, S. N. and Yen, C.-F., "Ply Cracks and Load Redistribution in Laminated Composites," *Proceedings*, Twelfth Army Symposium in Solid Mechanics, Plymouth, MA, Nov. 1991, pp. 619–633.

Adam Sawicki[1] and Nhien Nguyen[1]

The Influence of Transverse Bearing Loads Upon the Bypass Strength of Composite Bolted Joints

REFERENCE: Sawicki, A. and Nguyen, N., **"The Influence of Transverse Bearing Loads Upon the Bypass Strength of Composite Bolted Joints,"** *Composite Materials: Fatigue and Fracture (Sixth Volume), ASTM STP 1285*, E. A. Armanios, Ed., American Society for Testing and Materials, 1997, pp. 225–243.

ABSTRACT: The performance of biaxially loaded composite joints was examined at the coupon level using a simple, low-cost test apparatus designed to apply transverse bearing loads independent of longitudinal bypass loads. The apparatus consisted of clevises, load links, and attachments to test machine supports. Bending and extensional strain gage bridges were placed on the load links to measure transverse bearing stress levels as longitudinal bypass loads were applied and used to determine when bearing damage initiated. Two laminate configurations were examined to compare transverse bearing effects in relatively high and low stiffness laminates. At low bearing stress levels, transverse bearing was found to be a less damaging load condition than longitudinal on-axis bearing; however, bearing-related failure modes became prominent at lower stress levels in the transverse loading case than in the on-axis case. High stiffness laminates exhibited greater sensitivity to transverse hole deformation, laminate damage, and fastener torque than low stiffness laminates. Experimental data were compared with failure predictions generated with the Boeing-developed BEARBY analysis code. Predictions for the high stiffness laminate were in general agreement with test results, but were conservative for the low stiffness laminate as bearing damage relieved the stress concentration.

KEYWORDS: fatigue (materials), fracture (materials), composite materials, bolted composite joints, biaxial loading, bearing-bypass interaction, bolt torque, hole elongation

Historically, structural design allowables for mechanically fastened composite joints have been developed using uniaxially loaded test specimens, simulating various degrees of load transfer between adherends using a small number of fasteners. Extensive research has been performed examining test and analysis techniques for uniaxial bearing-bypass load cases and subsequent design allowables methodology (notably, Crews and Naik [*1,2*], Ramkumar [*3*], Garbo [*4*], and Grant and Sawicki [*5*]). The performance of joints subjected to biaxial states of stress typically has been examined using larger and more complex multi-fastener specimens due to the added complexity of applying multi-axial loads. The failure modes of joints subjected to biaxial loading have not yet been rigorously examined experimentally using simple specimens with predefined, measurable bearing-bypass load ratios. A notable exception is the recent work of Hoa [*6*], who examined bearing-dominated failure of biaxially loaded joints using a cruciform specimen. He demonstrated significant reductions (~50%) in damage initiation bearing load for specimens under relatively low transverse strains.

---

[1]Technical specialists, Structures Technology, Boeing Defense & Space Group, Helicopters Division, Philadelphia, PA 19142–0858.

The objective of this investigation was to examine at the coupon level the longitudinal tension bypass strength of joints subjected to transverse bearing loads and damage, using a simple, low-cost mechanism for applying biaxial loads. Of particular importance was a comparison of bypass-dominated failure modes for laminates subjected to longitudinal, or on-axis, and transverse bearing loads. As indicated in Fig. 1, on-axis bearing loads ($\alpha = 0°$) primarily increase the stress concentration at the location of peak longitudinal bypass stress; bearing-induced hole damage is generally created away from this location. In the case of transverse bearing ($\alpha = 90°$), bearing loads create fiber and matrix damage at the location of peak bypass stress. In transverse bearing-load cases, bypass strength is reduced by bearing-induced damage as well as through an increase in the stress concentration.

Two laminates, one fabric and one tape-fabric hybrid, were tested to examine the influence of laminate longitudinal stiffness upon bearing-bypass failure. Specimens were tested at normal

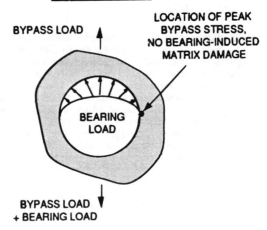

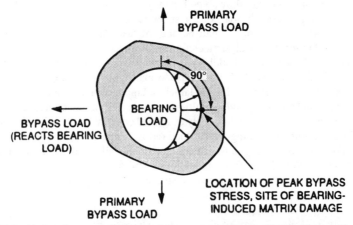

FIG. 1—*Load cases for on-axis and transverse bearing-bypass interaction.*

and reduced fastener installation torque to compare levels of hole deformation and subsequent failure loads. Experimental results were compared with predictions made using a Boeing-developed bolted-joint analysis code.

Experimental Approach

Loading Apparatus

The test configuration used in this investigation consisted of a test specimen, a mechanical apparatus designed to apply transverse bearing loads, and a uniaxial hydraulic test machine. The use of this configuration permitted direct examination of the effects of transverse bearing stresses and hole elongation upon longitudinal bypass strength, as the applied bearing loads did not induce global bending or shear stresses in the specimen. Development of the transverse loading apparatus, which can carry up to a 11.1-kN (2500-lb) transverse load, eliminated the need for a biaxial hydraulic test machine.

Longitudinal bypass loads were applied using an MTS hydraulic test machine. A 100-kN (22-kip) MTS load cell and linear variable differential transformer, along with a Pegasus controller and Hewlett-Packard X-Y plotter, were used to control the application of bypass loads and record longitudinal load-deflection behavior. All transducers and instrumentation were maintained and calibrated per MIL-STS-45662.

As shown in Fig. 2, the transverse loading apparatus consisted of two L-shaped clevises, an adapter, a load link, and attachments to test machine supports for each loaded hole. The

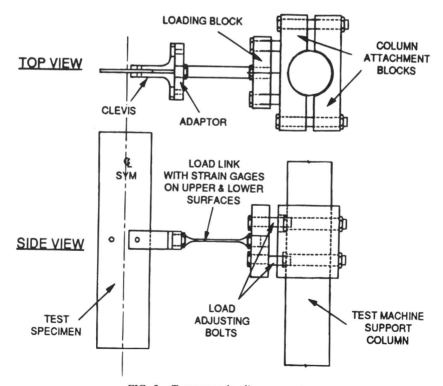

FIG. 2—*Transverse loading apparatus.*

clevises were aluminum, 6.35 mm (0.25 in.) thick, with 6.35-mm-diameter holes for the bolted attachment to the test specimen. Slots for attachment of the clevises to the adapters permitted adjustment of clevis separation to match the specimen thickness. The adapters were steel, with two 6.35-mm-diameter holes for attachment of clevises and one 15.88-mm (0.625-in.) diameter threaded hole for load link attachment. Load links were steel, 15.88 mm in diameter and threaded at the ends, necked down to a 15.9 by 2.54 mm (0.63 by 0.1 in.) square section at the center.

Each support attachment consisted of two steel column attachment blocks, one steel loading block, and eight 12.7-mm (0.5-in.) diameter steel bolts. The column attachment blocks were secured to a 85.85-mm (2.6-in.) diameter test machine crosshead support column using four steel bolts. Teflon film was used to protect the surface of the column. Loading blocks were bolted to the column attachments and then connected to the load links to complete the load path. Four load-adjusting bolts were used to secure each loading block to the column attachments to ensure a purely axial load was applied through the load links to the specimen.

Six strain gages were installed and wired into one four-arm bending bridge and one two-arm tension bridge (for temperature compensation) on the flat surfaces of each load link. This permitted measurement of extensional and bending strains in the load links, and subsequently the transverse bearing load applied to the specimen during the tests. As shown in Fig. 2, the links were positioned so that bending bridges measured vertical displacements caused by longitudinal bypass loading.

Test Specimens

Test specimens were designed to be similar in configuration to previously tested Boeing bearing-bypass interaction test specimens. As shown in Fig. 3, specimens were 305 mm (12 in.) long, 76.2 mm (3-in.) wide with 6.35-mm holes. This provided a specimen width-to-diameter (*w/D*) ratio = 6 and an edge margin-to-diameter (*e/D*) ratio = 3. Steel bolts (6.35 mm in diameter) were used to attach the specimen to the transverse loading apparatus.

The two laminates used in this investigation, shown in Table 1, were similar to laminates previously used in generating a failure database for the IM6/3501-6 tape and AS4/3501-6 plain-weave fabric material systems [5]. The configurations chosen had ply orientations equivalent to those of the laminates previously used, but were thinner to permit testing in a 100-kN machine. The use of two laminate configurations allowed comparison of transverse bearing effects in relatively low stiffness fabric laminates (typical of a basic fuselage skin) and high stiffness tape-fabric hybrid laminates (typical of a skin padup or edgeband).

Previously obtained data indicated that bolted joint failures can be classified as bearing-dominated and bypass-dominated, and that the failure of each type is highly dependent upon fastener clamp-up provided through installation torque [5]. High fastener clamp-up appears to increase fastener-laminate friction and inhibit delamination, and tends to increase bearing strength and decrease tension bypass strength. As the transverse bearing tests performed were anticipated to be bypass-dominated, the majority of specimens tested were fastened at a relatively high clamp-up level, namely, the normal installation torque of 9.6 N-m (85 in.-lb) for a 6.35-mm bolt. For the highest transverse bearing load case, specimens were tested at torque levels of 9.6 N-m and 3.9 N-m (35 in.-lb), or 40% of normal installation torque. It is believed that this torque level simulates the reduction in fastener clampup over a lifetime of service due to relaxation of laminate through-the-thickness stiffness, as estimated in work by Crews and Shivakumar [7,8].

Test Procedure

Prior to biaxial testing, load link strain gages were calibrated by loading each link under axial tension load in the MTS machine. The load-strain data obtained were used to develop

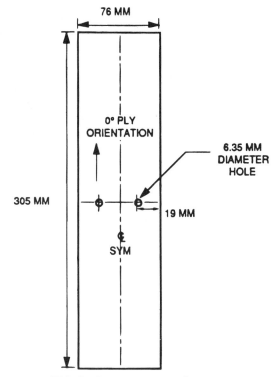

FIG. 3—*Test specimen configuration.*

TABLE 1—*Laminate configurations.*

Laminate Type	Stacking Sequence[a]	Nominal Thickness	% 0/45/90	Nominal Longitudinal Modulus
Fabric	$(45_f/0_f)_{3s}$	2.286 mm (0.090 in.)	25/50/25	46.5 GPa (6.74 msi)
Tape/Fabric Hybrid	$[45_f/0_f/0_t]_{2s}$	2.276 mm (0.0896 in.)	50/33/17	78.9 GPa (11.45 msi)

[a]*f* indicates an AS4/3501-6 3K-70-PW fabric ply and *t* indicates an IM6/3601-6 Grade 190 tape ply.

relationships between link extensional and bending strains and applied bearing loads. This permitted accurate measurement of transverse bearing loads throughout the bearing-bypass tests.

The transverse load apparatus was then assembled on the MTS machine. First, support attachments were secured to the test machine column supports, using dial indicators to ensure that both attachments were secured at a level position. Load links, adapters, and clevises were attached to the support attachments, with lock washers placed on the load links to prevent rotation. Strain gage bridges were balanced under zero applied load. A specimen was then secured to the clevises by inserting steel fasteners. Load-adjusting bolts were torqued until extensional link strains were observed. The fasteners securing the specimen to the clevises were then torqued to required levels per Table 2. Specimens with no applied transverse load per Table 2 were secured to clevises and adapters disconnected from the load links and support attachments so no transverse loads were induced by specimen deformation during the test.

The specimen was then loaded to the initial bearing load levels shown in Table 2 by tightening the load-adjusting bolts. By tightening the bolts in sequence and minimizing measured bending strains, purely extensional transverse loads were applied to the specimen. Care was taken to ensure the specimen was not bent or twisted during application of loads, and that bolt tightening centered the specimen relative to the hydraulic grips. Once transverse load levels were within 110 N (±25 lb) of the desired test level, the specimen was gripped in the test machine hydraulic grips, at a pressure of 3.45 to 6.90 MPa (500 to 1000 psi) as necessary to prevent specimen slippage. A photograph of a specimen installed in the apparatus prior to application of longitudinal bypass loads is shown in Fig. 4.

The specimen was then loaded in longitudinal tension at a rate of 1.27 mm (0.05 in.)/min in displacement control. Load versus crosshead travel and load link strain data were recorded at a rate of two readings per second. All tests were performed in room temperature, ambient humidity conditions. Typically, 1 h was required to install a specimen in the apparatus, apply transverse bearing loads, apply longitudinal bypass load, and remove the failed specimen from the apparatus.

Twenty-nine specimens were tested until they could no longer carry load. All failures were longitudinal net tension emanating from the fastener holes. Two specimens were tested to 95% of the lowest recorded longitudinal bypass failure load for 8.9-kN (2000-lb) transverse loading, 3.9-N-m (35-in.-lb) fastener clamp-up. These specimens underwent ultrasonic inspection to identify areas containing hole damage or delaminations.

Analytical Approach

Bearing-bypass test results were compared with predictions made using the BEARBY analysis code, which consists of pre- and post-processors to the Bolted Joint Stress Field Model (BJSFM) developed by McDonnell Douglas [9]. BJSFM determines the stress field around a loaded hole in a homogeneous anisotropic plate under plane-stress conditions using a closed-form analytic solution based upon the anisotropic theory of elasticity and laminated plate theory. It assumes a cosine radial stress distribution for bearing loads and uses the principle of superposition for combined bearing and bypass loads.

The BEARBY methodology predicts bearing-bypass failure loads by applying a point stress failure criterion at the lamina level for the tangential fiber tension, tangential fiber compression, radial fiber tension, and radial fiber compression failure modes. For each failure mode, a characteristic dimension is calculated semi-empirically using unnotched uniaxial strength and

TABLE 2—*Test matrix.*

Laminate Type	Fastener Torque	Number of Tests			
		0 kN Off-Axis Load	2.9 kN (650 lb) Off-Axis Load	5.9 kN (1325 lb) Off-Axis Load	8.9 kN (2000 lb) Off-Axis Load
Fabric	9.6 N-m (85 in.-lb)	2	3	3	3
	3.9 N-m (35 in.-lb)	···	···	···	4[a]
Tape/Fabric Hybrid	9.6 N-m (85 in.-lb)	3	3	3	3
	3.9 N-m (35 in.-lb)	···	···	···	4[a]

[a]Fourth specimen loaded to 95% of lowest previous failure load to inspect for delaminations.

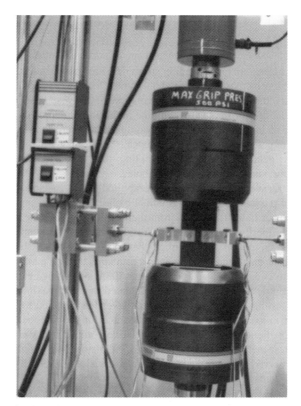

FIG. 4—*Photograph of test configuration prior to application of bypass loads.*

notched strength data (pure bypass data for tangential fiber failure and bearing data for radial fiber failure). Lamina failure is assumed to occur when the fiber-direction stress at a characteristic dimension away from the fastener hole, produced by a far-field load equivalent to that at which notched laminate failure occurs, equals the unnotched uniaxial material strength. Functional relationships between the characteristic dimension, the elastic stress concentration factor along the principal axis at the hole boundary, and the laminate configuration parameter AML (angle minus loaded plus, defined as the percentage of ±45° plies minus percentage of 0° plies) are used to relate notched laminate strength to layup.

Once characteristic dimensions have been calculated for each failure mode along each fiber orientation, BEARBY calculates the stress field about a loaded hole using BJSFM for multiple bearing-bypass ratios. The point stress criterion is then reapplied for the tangential and radial fiber failure modes for each lamina, and used to calculate the bearing and bypass load levels at which lamina failure occurs. Any inaccuracies created by the neglect of matrix failure modes are accounted for by the semi-empirical nature of the calculation of the characteristic dimension.

Previously obtained pure bypass and bearing data [10] were used in BEARBY characteristic dimension calculations for transverse bearing-longitudinal bypass analysis. The data were obtained using fasteners installed at critical clamp-up levels for the failure modes (normal installation torque for bypass and reduced torque for bearing).

Results

Transverse Load Measurements

Typical load link extensional and bending strain data are shown in Fig. 5. Link extensional strains were observed to increase as longitudinal bypass loads were applied. It is believed this was caused primarily through Poisson's deformation of the test specimen induced by the longitudinal bypass loads. For load cases with the highest initial applied transverse loads (8.9 kN), link extensional strains (and the slope of bending strain-bypass load curves) were observed to decrease at high bypass load levels. This was caused by a combination of hole elongation and damage, such as delamination or fiber buckling, that would locally reduce the stiffness of the specimen in the vicinity of the fastener hole.

Initially, as only transverse bearing loads are applied to the specimen, no significant link bending strains were recorded. When longitudinal bypass loads were applied, bending strains increased as the fastener holes translated from their initial position (the upper hydraulic grip remained fixed while the lower grip was displaced). Therefore, the angle between the load link and the longitudinal axis of the specimen changed from $\alpha = 90°$. Utilizing test machine longitudinal load versus displacement data, it was estimated that the maximum vertical translation of the fastener holes during all tests was 2.3 mm (0.09 in.). Using beam theory to represent the load link as a cantilevered beam, it was calculated that the maximum change in α was 1.2° throughout all testing and the maximum cantilever load applied to the links was 6.5 N (1.5 lb). The change in loading angle was not accounted for in subsequent calculations, as its effect upon failure predictions was found to be relatively minor, and should be considered as a limitation of the test method. As shown in Fig. 5, the slope the longitudinal load-bending strain curve was observed to decrease at high transverse load levels, most probably due to hole elongation and damage.

Bearing-bypass load paths at various initial transverse load levels are shown for the fabric and hybrid laminate types in Figs. 6 and 7, respectively. The data are plotted to show the relationship between measured transverse bearing stress versus longitudinal bypass strain, in a format typically used in bolted joint bearing-bypass strength analysis. At all initial bearing stress levels, both laminate types exhibited similar increases in bearing stress due to specimen Poisson's deformation. The greater the initial bearing stress, however, the greater was the observed reduction caused by hole elongation or damage at high bypass strain levels. Specimens with approximately 415 MPa (60 ksi) initial bearing stress showed slight decreases in bearing stress near failure, while specimens with approximately 620 MPa (90 ksi) initial bearing stress exhibited significant decreases in bearing stress near failure.

The fabric and hybrid laminates exhibited differing transverse loading behavior for low and high installation torques. For the fabric laminate, transverse bearing stresses increased for both high and low torque specimens until approximately 2800 $\mu\epsilon$ longitudinal bypass strain was achieved, at which time they decreased for low-torque specimens while continuing to increase for the high-torque specimens. For the hybrid laminate, transverse bearing stresses began to decrease at approximately 2800 $\mu\epsilon$ bypass strain for specimens of both torque levels, but initial stress reductions were more severe for low-torque than high-torque specimens. Bypass failure of the stiff hybrid laminate appeared to be influenced by transverse hole elongation, damage, and fastener installation torque to a greater degree than the softer fabric laminate.

Examination of Failure Modes

Typical fabric and hybrid laminate failures at 0 and 8.9 kN transverse loading are shown in Figs. 8 and 9, respectively. All specimens ultimately failed in a net tension bypass mode.

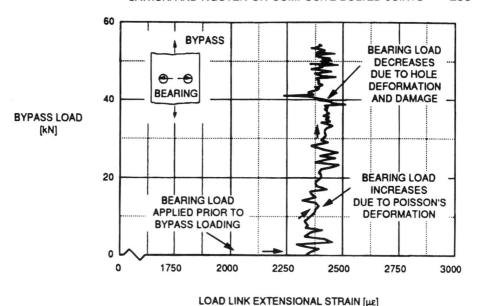

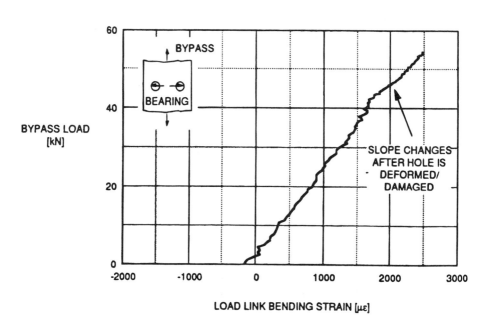

FIG. 5—*Typical load link extensional and bending strain measurements (hybrid laminate).*

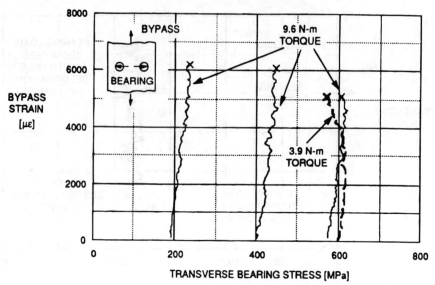

FIG. 6—*Bearing-bypass load paths for fabric laminate.*

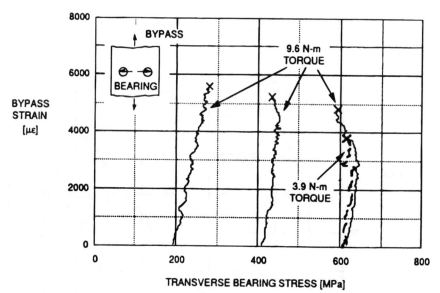

FIG. 7—*Bearing-bypass load paths for hybrid laminate.*

However, specimens with transverse loads applied exhibited significant elongation of the fastener holes (in the transverse direction) prior to net tension failure.

One specimen for each laminate configuration was tested to 95% of the lowest recorded longitudinal bypass failure load at a 8.9-kN initial transverse loading, 3.9-N-m fastener torque. For both laminate configurations, ultrasonic inspection identified delaminations extending up to 6.35 mm from the hole edges where bearing loads were applied. Hole diameters typically elongated 2% as measured with calipers.

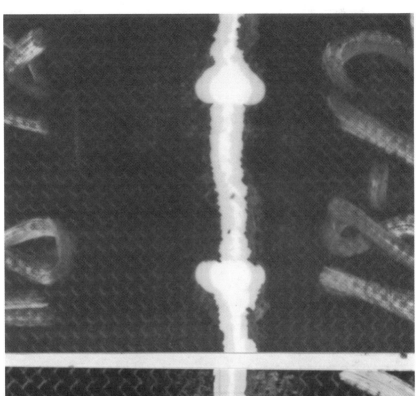
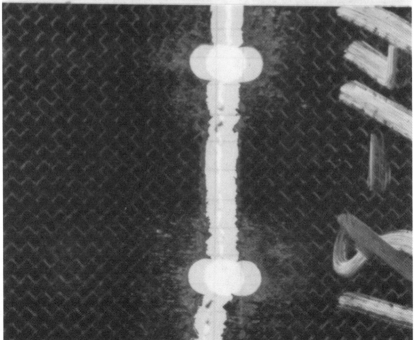

FIG. 8—*Transverse hole elongation for fabric laminate:* (left) *0 kN and* (right) *8.9 kN transverse loads.*

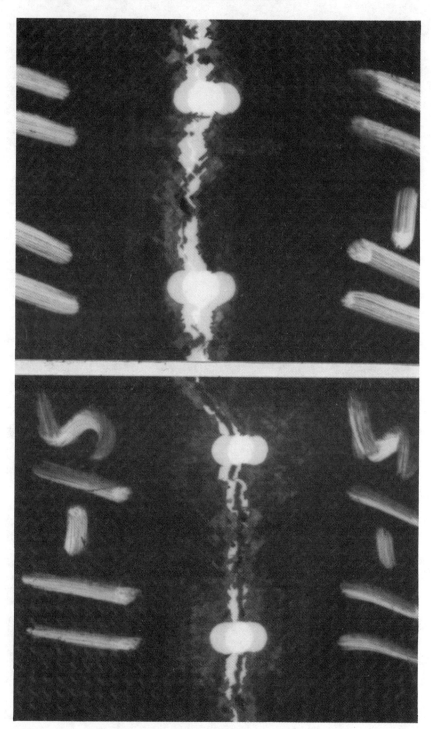

FIG. 9—Transverse hole elongation for hybrid laminate: (left) 0 kN and (right) 8.9 kN transverse loads.

Bearing-Bypass Analysis

Bearing-bypass interaction results are shown in Fig. 10, where bypass failure strains are plotted versus applied transverse bearing stress. For both laminate types, the longitudinal bypass failure strain of the two laminates was observed to decrease in a near-linear relationship with applied transverse bearing stress. Bypass failure strains were observed to be more sensitive to transverse bearing loads for the stiff hybrid laminate than for the softer fabric laminate. This seems logical in that the bypass strength of the hybrid laminate is highly dependent upon the strength of the tape ply fibers local to the stress concentration, which are broken as the hole is elongated and damaged.

As demonstrated in Fig. 11, the ratio of transverse bearing stress to longitudinal bypass strain changed throughout the tests. This differs from typical on-axis tests in which bearing-bypass ratios remain constant until failure, which is more representative of actual loading in aircraft, automotive, and marine structures. The transverse bearing-bypass test results must be assumed to be path dependent due to the changing bearing-bypass ratio. This is acceptable in that ultimate failure was driven by bypass-dominated modes, which are the modes of primary interest for design purposes.

The influence of fastener installation torque upon transverse bearing-bypass strength was observed to be more significant for the hybrid laminate than the fabric laminate. This result seems logical in that low torque increases the likelihood of bearing damage at a given stress level, and the hybrid laminate configuration is more susceptible to delamination due to its less homogeneous nature.

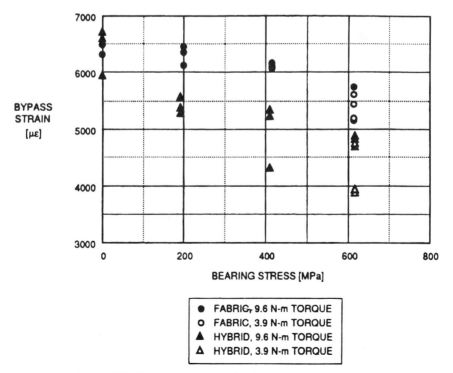

FIG. 10—*Transverse bearing-bypass test results.*

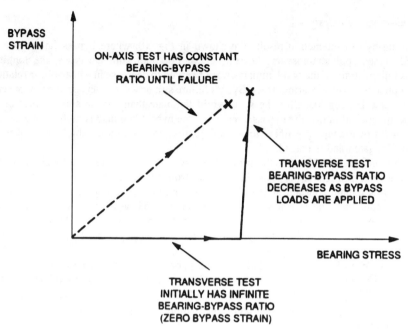

FIG. 11—*Differences between on-axis and transverse bearing-bypass ratios during test.*

Figures 12 and 13 compare transverse bearing-bypass results to previously obtained on-axis results [*10*]. High clamp-up on-axis data were generated using lockbolt fasteners, which provide a clamp-up load greater than or equal to 9.6 N-m for a 6.35-mm-diameter fastener. All specimens except those under pure bearing loads failed in bypass-dominated modes. At the bearing stress levels tested, transverse bearing was observed to be a less severe load condition that on-axis bearing. This occurs because the increase in stress concentration created by transverse bearing loads is less severe than that produced by on-axis bearing loads.

In the 600 to 800 MPa bearing stress range, transverse bearing-bypass strength was generally lower with 3.9-N-m fastener installation torque than with 9.6-N-m installation torque. This contrasts with on-axis bearing-bypass strength, which was generally higher for 3.9 N-m installation torque. Therefore, transverse bearing-bypass strength appears to become low-torque critical at lower bearing stress levels than does on-axis strength, indicating that bearing-related failure modes become prominent at lower stress levels for transverse bearing than for on-axis bearing. For this reason, the use of a conservative bearing allowable stress, which minimizes hole deformation and local damage, becomes even more of a necessity for transverse bearing load cases than in the on-axis case.

Experimental bearing-bypass results are compared to BEARBY predictions in Figs. 14 and 15. It should be noted that the pure bypass data vary from BEARBY analysis predictions which were calculated using failure strain-AML regression relationships rather than actual failure strain results for the particular layups [*10*]. The pure bypass data obtained in this investigation also varied slightly from that previously obtained, due to differences in fastener configuration and subsequent clamp-up force imparted by normal installation.

The comparisons demonstrate that the BEARBY analysis code can be used to predict trends for bypass-dominated failure of the hybrid laminate under both on-axis and transverse bearing-load cases. Failure trends for the fabric laminate under on-axis bearing load cases were also

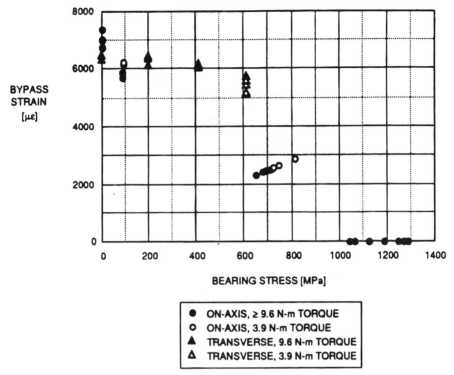

FIG. 12—*On-axis and transverse bearing-bypass results for fabric laminate.*

predicted by the BEARBY analysis. However, for the transverse bearing load cases, BEARBY failure predictions were overly conservative. Using a linear regression of bypass strain versus transverse bearing stress test data, it was estimated that the sensitivity of bypass failure strain to transverse bearing stresses (as measured by the slope of the regression line) was 15% of the sensitivity predicted using BEARBY. This difference is most likely caused by bearing-induced local failures at the hole and the ability of the fabric material form to redistribute loads away from damaged tows.

Conclusions

A simple, low-cost method for applying biaxial loads to bolted-joint test specimens has been developed. Use of this method provided information on hole deformation and strength for several transverse bearing-longitudinal bypass interaction load cases, and permitted a comparison of joint performance for different fastener torque levels.

Based upon the bearing-bypass interaction behavior observed, it appears that increases in stress concentration caused by transverse bearing loads are less severe than for on-axis bearing loads. Consequently, at bearing stress levels below 620 MPa (90 ksi), transverse bearing-bypass interaction strengths were higher than on-axis strengths. As transverse bearing stress levels increased, factors such as hole deformation and fastener torque began to significantly influence bearing-bypass strength. Such effects appear to become important at lower bearing stress levels for the transverse bearing case than in the on-axis case.

The fabric and hybrid laminates exhibited different behavior under transverse bearing-bypass interaction. The stiff hybrid laminate exhibited differences in hole deformation and bearing-

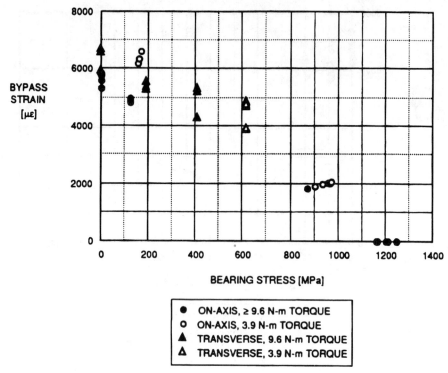

FIG. 13—*On-axis and transverse bearing-bypass results for hybrid laminate.*

bypass strength depending upon fastener installation torque, whereas the softer fabric laminate exhibited little difference in strength due to fastener torque. The greater sensitivity of the hybrid laminate to bearing-induced failure modes and its dependence upon longitudinal fibers to provide bypass strength seem to be the major factors associated with this behavior.

The results of this investigation show the importance of using conservative bearing strengths in developing bolted-joint allowables. As transverse bearing-bypass load cases were demonstrated to be affected by low fastener torque and bearing-induced local failures at low bearing stress levels, they should be particularly sensitive to fatigue-induced hole elongation and damage. The selection of allowable bearing strengths should consider the potential effects of fatigue damage upon structures subjected to such bearing-bypass load conditions. Therefore, it is recommended that bearing allowables be developed with consideration of a maximum permissible hole elongation caused by fatigue loading over an expected operating lifetime.

Future Work

Testing completed thus far focused on the tension bypass side of the transverse bearing-bypass interaction strength envelope. Recently, a stabilization fixture has been developed to permit longitudinal compression bypass loading of the specimens. Future tests will examine the compression bypass side of the strength envelopes.

Greater insight into damage and failure mechanisms associated with transverse bearing could be gained through additional testing of tape, fabric, and hybrid laminates of differing configurations.

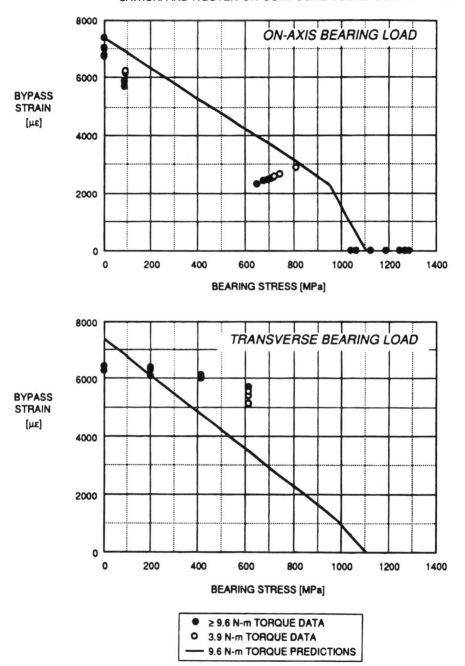

FIG. 14—*Comparison of fabric laminate bearing-bypass test results and analysis predictions.*

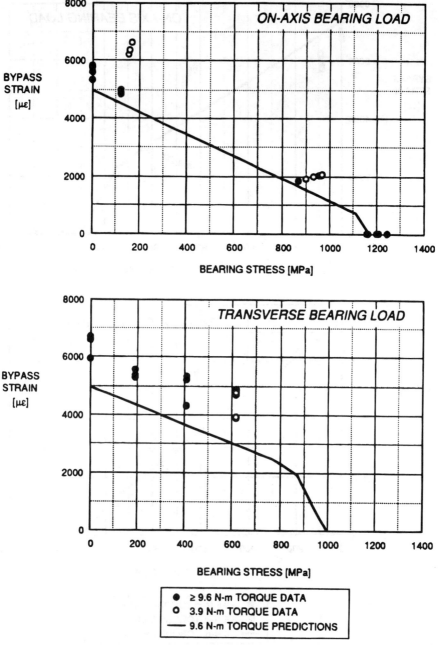

FIG. 15—*Comparison of hybrid laminate bearing-bypass test results and analysis predictions.*

Acknowledgments

The authors gratefully acknowledge the efforts of Mr. M. Kelleher in the development of test procedures, as well as improving the design of test specimens and the loading apparatus. The work of Mr. P. Grant and Mr. G. Mabson regarding data interpretation and the BEARBY analysis code was invaluable. The contributions of Dr. C. K. Gunther (manager, Structures Research and Development), Mr. D. MacKenzie (manager, Engineering Laboratories), and Dr. P. Minguet are also greatly appreciated.

References

[1] Crews, J. and Naik, R., "Combined Bearing and Bypass Loading on a Graphite/Epoxy Laminate," *Composite Structures*, Vol. 6, 1986, pp. 21–40.

[2] Naik, R. and Crews, J., "Ply Level Failure Analysis of Graphite/Epoxy Laminates Under Bearing-Bypass Loading," *Composite Materials: Testing and Design (Ninth Volume), ASTM STP 1059*, S. P. Garbo, Ed. American Society for Testing and Material, Philadelphia, 1990, pp. 191–211.

[3] Ramkumar, R., "Bolted Joint Design, Test Methods and Design Analysis for Fibrous Composites," *Test Methods and Design Allowables for Fibrous Composites, ASTM STP 734*, C. Chamis, Ed., American Society for Testing and Materials, Philadelphia, 1981, pp. 376–395.

[4] Garbo, S., "Effects of Bearing/Bypass Load Interaction on Laminate Strength," AFWAL–TR-81-3114, Air Force Wright Aeronautical Laboratories, OH, Sept. 1981.

[5] Grant, P. and Sawicki, A., "Development of Design and Analysis Methodology for Composite Bolted Joints," *Proceedings*, American Helicopter Society Rotorcraft Structures Specialist's Meeting, Williamsburg, VA, Oct. 1991.

[6] Hoa, S., "Biaxial Bearing/Bypass Testing of Graphite/Epoxy Plates," *Journal of Composites Technology and Research*, Vol. 17, No. 2, April 1995, pp. 125–133.

[7] Shivakumar, N. and Crews, J., "Bolt Clampup Relaxation in a Graphite/Epoxy Laminate," *Long Term Behavior of Composites, ASTM STP 813*, T. K. O'Brien, Eds., American Society for Testing and Materials, Philadelphia, 1983, pp. 5–22.

[8] Shivakumar, N. and Crews, J., "An Equation for Bolt Clampup Relaxation in Transient Environments," NASA Technical Memorandum 84480, Hampton, VA, April 1982.

[9] Garbo, S. and Ogonowski, J., "Effect of Variances and Manufacturing Tolerances on the Design Strength and Life of Mechanically Fastened Composite Joints. Volume 1: Methodology Development and Data Evaluation," Report AFWAL-TR-81-3041, Air Force Wright Aeronautical Laboratories, OH, April 1981.

[10] Kesack, W., et al., "V-22 Material Substantiating Data and Analysis Report," Contract No. N00019-85-C-0145, Bell-Boeing Report No. 901-930-022, Revision A, Bell Helicopter Textran, Inc. and Boeing Helicopters, Nov. 1991.

Ming Wu[1] and Dale Wilson[2]

Residual Strength of Metal-Matrix Laminated Panels

REFERENCE: Wu, M. and Wilson, D., **"Residual Strength of Metal-Matrix Laminated Panels,"** *Composite Materials: Fatigue and Fracture (Sixth Volume), ASTM STP 1285*, E. A. Armanios, Ed., American Society for Testing and Materials, 1997, pp. 244–259.

ABSTRACT: The primary objective of this study was to investigate the residual strength of ARALL-3 and GLARE-2 center-notched panels without stiffeners. The *R*-curve approach in linear elastic fracture mechanics (LEFM) was used for the residual strength predictions.

The applicability of LEFM was verified through a series of tests of ARALL-3 and GLARE-2 center-notched panels with different layups. They demonstrated limited crack-tip plastic deformation. The *R*-curves calculated from the tests of different size panels with various initial crack extensions showed that they were independent of initial crack length and specimen width, which is true for the monolithic aluminum alloy. Polynomial curve fitting was used to obtain the *R*-curves for each laminate and laminate layup to be used for the *R*-curve residual strength predictions. The predictions were made by superimposing the crack driving force curves onto these *R*-curves to locate the tangent points. The results of prediction of unidirectional fiber/metal laminates proved that the *R*-curve approach was not only a suitable but simple method that has a great potential in the damage tolerance characterization of certain unstiffened and stiffened laminate materials.

KEYWORDS: fiber/metal laminates, linear elastic fracture mechanics, *R*-curve, stress intensity factor, slow crack growth, residual strength, fatigue (materials), fracture (materials), composite materials

Since various components of the aircraft structure are loaded in different ways, the failure modes can also be different. If components are predominantly loaded in compression, they are designed usually for stability as a limiting criterion. A high elastic modulus and a high yield stress are required. If cyclic loading in tension is predominant, components have to be designed against fatigue. The fatigue resistance and fracture toughness are significant material properties. An example of the first category of structures is the upper skin of an aircraft wing that is predominantly loaded into compression due to the wing bending moment. Examples for the second category are the lower wing skin and the fuselage pressure cabin. The lower wing skin is loaded by a cyclic stress in tension due to bending of the wing as caused by gust loads and maneuvers. The skin of a pressurized fuselage carries a biaxial cyclic tensile stress during each flight. ARALL laminate is a highly fatigue-resistant material with a high tensile strength. Its superior fatigue properties are a great advantage for fatigue critical sheet structures like the pressurized fuselage and the lower wing skin.

[1]Engineer, Mechanics and Materials Department, Failure Analysis Associates, Inc., Menlo Park, CA 94025.

[2]Professor, Mechanical Engineering Department, Tennessee Technological University, Cookeville, TN 38505.

Extensive research has been conducted to study every aspect of this new type of fiber/metal laminated material, especially its fatigue and fracture properties. The goal of this study was to investigate the residual strength characteristics of this material and to explore the possibility of predicting it by using conventional fracture mechanics theories for the isotropic metallic materials.

ARALL laminate, developed by Delft University, is an adhesive bonded laminate that combines the advantages of high-strength isotropic aluminum sheet with the fatigue and fracture resistance of aramid or glass fibers. The material is built up as laminated sheet material with thin high-strength aluminum alloy sheets; strong unidirectional or woven aramid or glass fibers, impregnated with a thermoset or thermoplastic adhesive. Post-stretch of the material after curing (if desired) will produce a compressive residual stress in the metal sheets. ARALL laminates are a new family of structural composite materials. The final properties are highly dependent on configurations of the material. ARALL laminates can be tailored for many different applications by varying fiber-resin systems, aluminum alloys and sheet gages, stacking sequences, fiber orientations (such as uniaxial and cross-ply), surface preparation techniques, and by the degree of post-cure stretching or rolling. Four products, ARALL-1 through ARALL-4, have been standardized for commercial availability by ALCOA. A variety of layup sequences is available. A 3/2 (three plies of aluminum and two plies of fiber/adhesive) ARALL-1 laminate layup is shown in Fig. 1. The ARALL-3 and GLARE-2 are the two laminates used in this research for the investigation of residual strength and the size effect on the residual strength. The ARALL-3 consists of thin 7475-T761 sheets (0.31 mm thick) alternating with aramid fiber layers (0.22 mm thick) embedded in a special resin matrix. Similarly, GLARE-2 is made up of 2024-T3 aluminum layers (0.31 mm thick) alternating with unidirectional glass fiber prepreg layers (0.25 mm thick).

ARALL laminate tensile ultimate strengths in the reinforcement direction are significantly better than those of the respective aluminum counterparts; also they are competitive with those of graphite/epoxy composites. This is especially true for the ARALL laminates with high-strength glass fiber. Due to the ineffectiveness of the fibers in bearing, the ratio of bearing

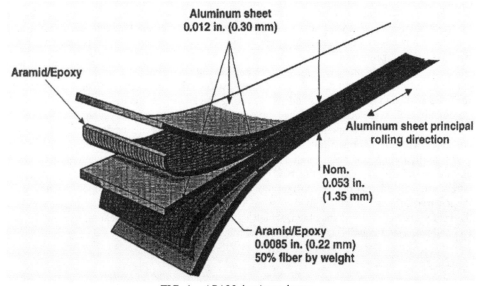

FIG. 1—*ARALL laminate layup.*

strength to tensile strength is lower than that experienced by aluminum. Some of the static mechanical properties of ARALL laminates are shown in Table 1.

Residual Strength

For more than three decades, there has been extensive interest in characterizing the residual strength of center-cracked tension panels. There have been different approaches to this problem area [1,2]. The Feddersen theory [3] and the R-curve approach are two of the approaches. The most important drawback of most of these engineering methods is the fact that slow crack growth is not incorporated as an essential part of plane-stress behavior. The R-curve concept is based on the energy balance concept, and it reveals the slow crack growth behavior information under plane-stress condition. There is a continuous balance between released and consumed energy during slow stable crack growth. Therefore, this characterization is obtainable from R-curves. The R-curves characterize the resistance to fracture of materials during incremental slow stable crack extension and result from growth of the plastic zone as the crack extends from a sharp crack. They provide a record of the toughness development as a crack is driven stably under increasing crack extension forces. This approach shows that a modified linear elastic methodology can be extended to handle elastic-plastic crack-tip conditions. This methodology incorporates a plasticity-corrected stress intensity factor, K_R, based on an appropriate effective crack size. The K_R-values are equivalent to values of deformation theory (J-integral based on deformation theory of plasticity). The appropriate effective crack size, a_e, is determined by compliance methods and the equivalence between K_R and J_R is completely maintained up to 90% of net section yield load in tension configurations [4]. K_R is used to generate the R-curves in this research.

Material property characteristics can be incorporated completely into the R-curve by plotting K_R versus the effective crack growth, Δa_e. To predict instability, the crack driving force can be calculated using a pseudo-linear elastic R-curve methodology. This elimination of plastic deformation analysis from the crack driving force (necessary with J-Δa_p R-curve analysis) also provides a much simplified method for predicting instability.

This K_R-curve approach is of particular value for complex geometries or stress conditions for which elastic-plastic J-solutions may not exist. Cases of local stress field excursions caused by residual stresses, reinforcement straps, or neighboring holes are handled more easily, and only the knowledge of the elastic K_1 stress field behavior of the component is needed for the residual strength prediction.

For an untested geometry, the K_R-curve can be matched with the crack extension force curves to estimate the load necessary to cause unstable crack propagation. The K_R-curves are regarded as though they are independent of starting crack size, a_0, and the specimen configuration in which they are developed. They are only a function of crack extension, Δa. To predict crack instability in a component, the K_R-curve is positioned as in Fig. 2, so that the origin coincides with the assumed initial crack length, a_0. The crack extension force curve for a given configuration can be generated by assuming applied loads or stresses and calculating crack extension force, K_R, as a function of crack size using the appropriate expression for K_1 (stress intensity factor) of the configuration. The unique curve that develops tangency with the R-curve defines the critical load or stress that will cause onset of unstable fracturing. For the center-cracked tension specimen, which is the type used in this research, use one of the two following and equally appropriate expressions

$$K_R = \sigma\sqrt{\pi a_e}\left(\sec\frac{\pi a_e}{W}\right)^{1/2} \tag{1}$$

TABLE 1—Mechanical properties of aircraft materials [7].

		GLARE-1[a]	GLARE-2[a]	GLARE-3[a]	GLARE-4[b]	ARALL-1[c]	ARALL-2[c]	2024-T3	7075-T6
Tensile ultimate strength, MPa (ksi)	L	1300 (188.7)	1229 (178.5)	755 (109.6)	1040 (151)	799 (116.1)	717 (104.1)	440 (63.9)	538 (78.1)
	LT	360 (52.3)	319 (46.4)	755 (109.6)	618 (89.7)	385 (56)	316 (46)	434 (63.1)	538 (78.1)
Tensile yield strength, MPa (ksi)	L	549 (79.8)	400 (58.1)	319 (46.4)	359 (52.2)	640 (93)	358 (52.1)	323 (47)	482 (70.1)
	LT	339 (49.3)	230 (33.4)	319 (46.4)	259 (37.7)	330 (48)	228 (33.1)	290 (42.1)	482 (68.1)
Elastic tensile strength, GPa (msi)	L	64 (9.4)	65 (9.5)	57 (8.4)	57 (8.3)	67 (48)	64 (9.3)	72 (10.5)	70 (10.3)
	LT	48 (7.1)	50 (7.3)	57 (8.4)	50 (7.3)	48 (7)	48 (7.1)	72 (10.5)	70 (10.3)
Ultimate strain, %	L	4.6	5.1	5.1	5.1	1.9	2.5	13.6	8
	LT	7.7	13.6	5.1	5.1	7.7	12.7	13.6	8
Bearing ultimate strength, GPa (msi)	L	770 (111.8)	704 (102.2)	690 (100.2)	697 (01.2)	737 (107.1)	564 (82)	890 (129.2)	1076 (156.2)
Blunt notch strength, GPa (msi) (circular hole)	L	804 (116.8)	775 (112.5)	500 (72.7)	604 (87.8)	494 (71.8)	405 (58.8)	420 (61)	549 (79.8)
	LT	360 (52.3)	290 (42.1)	500 (72.7)	420 (61)	385 (56)	310 (45.1)	420 (61)	549 (79.8)
Sharp notch strength, GPa (msi) (sawcut)	L	709 (103)	649 (94.3)	409 (59.4)	529 (76.9)	376 (54.7)	361 (52.50)	319 (46.4)	350 (50.8)
	LT	230 (33.4)	230 (33.4)	409 (59.4)	319 (46.4)	250 (36.3)	250 (36.3)	319 (46.4)	350 (50.8)
Density, kg/cm³ (lb/in.³)		0.002519 (0.091)	0.002512 (0.091)	0.00252 (0.091)	0.0025 (0.089)	0.00233 (0.084)	0.00233 (0.084)	0.0028 (0.1)	0.0028 (0.1)

[a]3/2 lay-up; thickness is 1.4 mm (with 0.3-mm-thick aluminum).
[b]3/2 lay-up; thickness is 1.65 mm (with 0.3-mm-thick aluminum).
[c]3/2 lay-up; thickness is 1.3 mm (with 0.3-mm-thick aluminum).

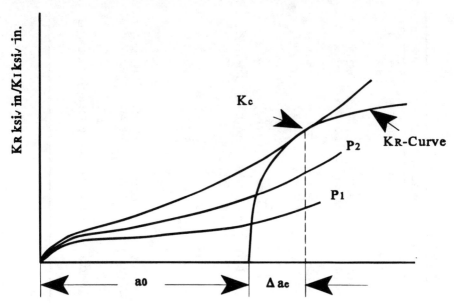

FIG. 2—*Crack growth resistance and crack extension force for a load-controlled test.*

or

$$K_R = \sigma\sqrt{a_e}\left[1.77 - 0.177\left(\frac{2a_e}{W}\right) + 1.77\left(\frac{2a_e}{W}\right)^2\right] \qquad (2)$$

where $a_e = a_0 + \Delta a_p + r_p$ is the effective crack length decided by compliance technique; r_p is the radius of crack-tip plastic zone.

Test Program

Materials used in this investigation are 3/2 and 5/4 ARALL-3 and 5/4 GLARE-2 laminates. These two materials differ by both aluminum alloy and prepreg system: the 7475-T761 is the present metallic constituent and aramid/epoxy is the prepreg system of commercial ARALL-3. The 2024-T3 and glass/epoxy are the metallic constituent and prepreg system of GLARE-2.

Specimen Preparation

The specimens used in this investigation were middle-cracked tension, M(T), panels shown in Fig. 3. For 3/2 ARALL-3 panels, both outside surfaces are bare (not clad) and phosphoric acid anodized. For one of the 5/4 ARALL-3 panels, one side was clad and one side was bare, while both sides were chrome acid anodized. For the other panel, both sides were bare and chrome acid anodized. For the 5/4 GLARE-2 panel, both sides were bare, chrome acid anodized, and primed. All these surface treatments are for the corrosion protection, and they have very little effect on the static properties of the laminates. Specimens were made 76.2 mm (3 in.), 152.4 mm (6 in.), 254 mm (10 in.), 304.8 mm (12 in.), 406.4 mm (16 in.), and 457.2 mm (18 in.) wide for the purpose of evaluating the width effect on the *R*-curve. Only the 3/2

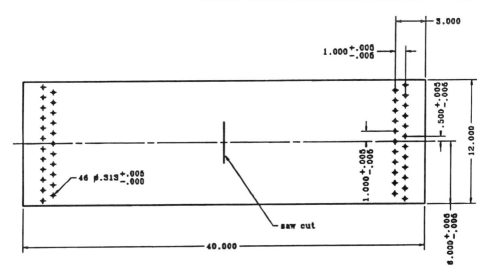

FIG. 3—*Unstiffened ARALL test specimen.*

Note: **A d = 0.062 (in) hole drilled at the center of the saw cut. All units are in inches.**

1 in = 25.4 mm

ARALL-3 laminate was used for 76.2 mm (3 in.) and 152.4 mm (6 in.) wide specimens. Different initial crack lengths, a_0, were cut for each specimen width to evaluate the initial crack length effect on the R-curve.

The different size specimens made for the tests are shown in Table 2. Two 4.45 mm (0.175 in.) thick, 19.05 by 6.35 mm (0.75 by 0.25 in.) aluminum knife edges were bonded above and below the notch for use with a crack opening displacement (COD) gage. The gage could be placed between the two knife edges and, when the specimen was loaded, a measure of compliance could be obtained. The compliance (which is the reciprocal of the load-displacement slope normalized for elastic modulus and specimen thickness) can then be used to calculate the effective crack lengths for the R-curves.

Fixture Design and Testing Machine

In these center-cracked tension tests, the grip fixtures were designed to develop uniform load distribution on the specimen. Since the largest panel to be tested was wider than 304.8 mm (12 in.), a multiple pin grip design was used. To ensure uniform stress entering the crack plane, the length between the loading pins was chosen to be two to three times of specimen widths. As the stress level for the tests in this program would all be above zero, a typical pin and clevis grip based on design recommendation in ASTM Practice for R-Curve Determination (E 561-92a) [5] was adopted. This design is easy to manufacture, simple to adjust, align, and maintain.

Four Z-shaped clamping rails were designed to be used to prevent the larger panels from out-of-plane (plane of the panel surface) buckling during testing. The buckling could cause the knife edges to debond and, even if these knife edges did not debond, it could greatly affect

TABLE 2—*Test specimens.*

Laminates	Layup	Width, W, mm (in.)	Length, L, mm, (in.)	Crack Length, $2a_0$, mm (in.)
ARALL-3	3/2	76.2 (3)	279.4 (11)	12.1 (0.4788)
ARALL-3	3/2	76.2 (3)	279.4 (11)	24.9 (0.9816)
ARALL-3	3/2	76.2 (3)	279.4 (11)	23.0 (0.909)
ARALL-3	3/2	76.2 (3)	279.4 (11)	24.9 (0.9813)
ARALL-3	3/2	76.2 (3)	279.4 (11)	24.8 (0.9771)
ARALL-3	3/2	76.2 (3)	279.4 (11)	37.8 (1.489)
ARALL-3	3/2	152.4 (6)	609.6 (24)	24.8 (0.979)
ARALL-3	3/2	152.4 (6)	609.6 (24)	37.5 (1.479)
ARALL-3	3/2	152.4 (6)	609.6 (24)	37.3 (1.47)
ARALL-3	3/2	152.4 (6)	609.6 (24)	49.8 (1.9628)
ARALL-3	3/2	152.4 (6)	609.6 (24)	50.1 (1.974)
ARALL-3	3/2	152.4 (6)	609.6 (24)	62.5 (2.4632)
ARALL-3	3/2	254 (10)	914.4 (36)	78.8 (3.103)
ARALL-3	3/2	304.8 (12)	914.4 (36)	98.5 (3.878)
ARALL-3	3/2	304.8 (12)	914.4 (36)	99.6 (3.924)
ARALL-3	3/2	406.4 (16)	914.4 (36)	132.9 (5.233)
ARALL-3	3/2	304.8 (12)	914.4 (36)	96.2 (3.789)
ARALL-3	5/4	457.2 (18)	1143 (45)	150.7 (5.935)
ARALL-3	5/4	457.2 (18)	1143 (45)	151.4 (.9634)
GLARE-2	5/4	254 (10)	1143 (45)	83.1 (3.274)
GLARE-2	5/4	304.8 (12)	1143 (45)	98.5 (3.88)
GLARE-2	5/4	406.4 (16)	1143 (45)	132.6 (5.223)

the accurate reading from the COD gage. These clamping rails can be seen in Fig. 4, above and below the initial crack.

The tests were conducted on an Instron Servohydraulic Model 1332 test system shown in Fig. 5, which includes a Hewlett Packard computer for the test control and data acquisition. An Instron COD gage with ±0.085 mm/mm (in./in.) range was used for the crack opening displacement readings.

General Testing Procedures

A total of 22 tests were performed. In general, these tests were in accordance with ASTM E 561-92a [5]. Each specimen tested was numbered according to ARALL version type, its layup sequence, specimen width, and initial crack length. The 3/2 ARALL-3 specimens with 76.2 mm (3 in.) widths were numbered as 3/2 ARALL-3-W3-*aij*, where W3 represents 76.2 mm (3 in.) width, i is the initial crack length, and j is the test number. The 3/2 ARALL-3 specimens with 152.4 mm (6 in.) widths were numbered as 3/2 ARALL-W6-*aij*. The additional specimens were numbered accordingly.

Each specimen was handsawed with a jeweler's saw to achieve the desired initial crack length. The surface around the notch was cleaned with alcohol and two knife edges were bonded above and below the notch with super glue. Specimens were cured for 6 h, before any loads were applied to them. For specimens larger than 152.4 mm (6 in.), the knife edges were screwed to the panels, because the knife edges would debond as the panel started to buckle with the increase of the load. The buckling was due to the compressive stress, $-\sigma$, in the x direction (parallel to the initial crack direction). For smaller panels, the applied stress was sufficiently lower to prevent the debonding.

FIG. 4—*Test specimen and fixture assembly.*

Test Results and Analysis

Only 18 *R*-curve tests were performed on ARALL-3 and GLARE-2 specimens of different widths and initial crack lengths as shown in Table 2. The purpose of these was to show that the validity of the linear elastic fracture mechanics (LEFM) *R*-curve concept could be applied equally to ARALL and GLARE unidirectional laminates. The *R*-curve of a high-strength thin metallic sheet with enough width ($2a < W/3$) and initial crack extension is independent of initial crack size [*3,4*] and the plane-stress toughness could be decided by finding the tangent point between crack driving force curves and *R*-curve. The crack driving force curve of a unstiffened thin sheet M(T) specimen is given in Eqs 1 or 2. The failed test panel is shown in Fig. 6.

FIG. 5—*Instron servohydraulic Model 1332 test system.*

Comparison of Residual Strength Prediction with the Test Data

For high-strength isotropic metal notched plates, the *R*-curves are assumed to be independent of the initial crack lengths. Although unidirectional fiber-reinforced ARALL-3 and GLARE-2 laminates used in this study satisfy the conditions specified by Wu [6] to permit the direct application of linear isotropic fracture mechanics to anisotropic plates, the verification was done through a series of tests of ARALL-3 and GLARE-2 panels with different widths and initial crack sizes.

The residual strengths for 3/2 and 5/4 ARALL-3 and 5/4 GLARE-2 panels were predicted by first generating a fitting *R*-curve to several *R*-curves calculated from the tests (Figs. 7, 9, and 11), and then finding the crack driving force curve that would be tangent to the fitted *R*-curve (Figs. 8, 10, and 12). The critical load was then determined from the load used to calculate the crack driving force curve (see ASTM E 561–92a).

The *R*-curves for 152.4 mm (6 in.), 304.8 mm (12 in.), and 406.4 mm (16 in.) wide 3/2 ARALL-3 panels were fitted with an eighth order polynomial (Eq 3) that was generated through a statistical program (SAS)

$$y = -401.716x^8 + 7830.121x^7 - 60834x^6 + 241778x^5 - 532144x^4$$
$$+ 682473x^3 - 554946x^2 + 270889x + 22726. \qquad (3)$$

The selection of polynomial was done through a "stepwise" program routine in SAS that stepped through polynomials of different orders to select the one that best fit the *R*-curves. The fitting of the *R*-curves for 3/2 ARALL-3 is shown in Fig. 7 and examples of the predictions are shown in Fig. 8. (Part of the *R*-curve beyond the maximum value is due to the final unstable fracture of the specimen.)

FIG. 6—*Failed 457.2 mm (18 in.) 5/4 ARALL-3 panel with 152.4 mm (6 in.) initial crack.*

The R-curves for 304.8 mm (12 in.) and 457.2 mm (18 in.) wide 5/4 ARALL-3 notched panels were fitted with a sixth order polynomial in SAS (Eq 4). The fitting is shown in Fig. 9. This fitted R-curve is

$$y = -9166.153x^6 + 111029x^5 - 507583x^4$$
$$+ 1078088x^3 - 1050146x^2$$
$$+ 402898x + 17159. \tag{4}$$

It will also be used later for the residual strength predictions of stiffened panels. The diagram of the R-curve prediction of the 5/4 ARALL-3 notched panels is shown in Fig. 10.

The R-curves for the 254 mm (10 in.), 304.8 mm (12 in.), and 406.4 mm (16 in.) 5/4 GLARE-2 notched panels were also fitted with an eighth order polynomial in SAS (Eq 5).

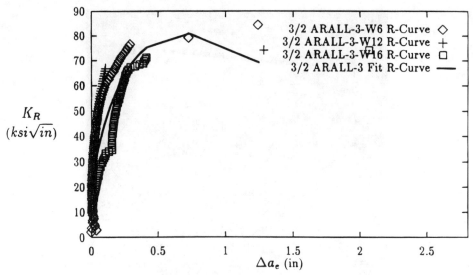

FIG. 7—*Polynomial curve fit for 3/2 ARALL-3 R-curves (1 in. = 25.4 mm and 1 ksi = 6.89 MPa).*

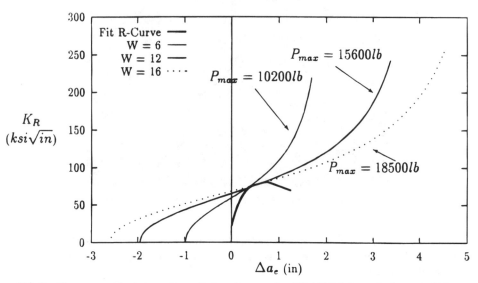

FIG. 8—*R-curve residual strength prediction diagram for 3/2 ARALL-3 notched panels (1 in. = 25.4 mm, 1 ksi = 6.89 MPa, and 1 lb = 4.45 N).*

Since the COD gage used for the test and the 50-kip load capacity testing machine both ran out of range when testing the GLARE-2 panels, only part of the load versus COD data were able to be collected for all three GLARE-2 panel tests. An approximated extension curve was added to the fitted *R*-curve to enable the predictions of the residual strengths. The fitted *R*-curve is

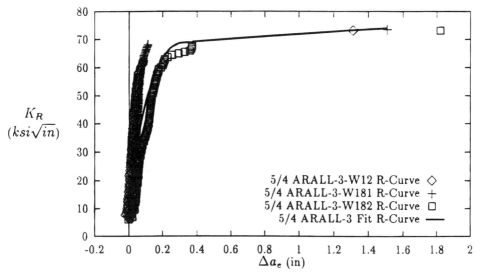

FIG. 9—*Polynomial curve fit for 5/4 ARALL-3 R-curves (1 in. = 25.4 mm and 1 ksi = 6.89 MPa).*

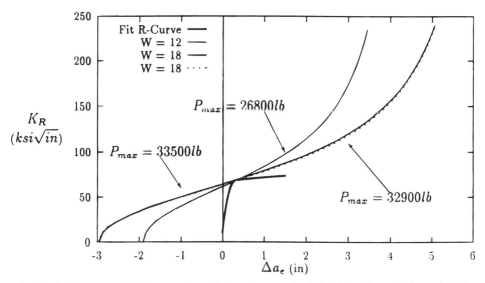

FIG. 10—*R-curve residual strength prediction diagram for 5/4 ARALL-3 notched panels (1 in. = 25.4 mm, 1 ksi = 6.89 MPa, and 1 lb = 4.45 N).*

$$y = -2014021x^8 + 11359226^7 - 26913600x^6 + 34811475x^5 - 26811848x^4$$
$$+ 12647175x^3 - 3659902x^2 + 727637x + 9413.235. \qquad (5)$$

The *R*-curve for the 254 mm (10 in.) panel starts to deviate from the *R*-curves for the 304.8 mm (12 in.) and 406.4 mm (16 in.) panels as the load is increased. This explains that the testing panel width for the GLARE-2 laminate needs to be much larger than that required for

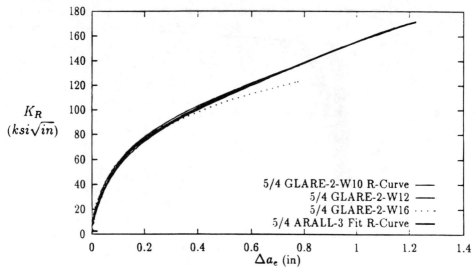

FIG. 11—*Polynomial curve fit for 5/4 GLARE-2 R-curves (1 in. = 25.4 mm and 1 ksi = 6.89 MPa).*

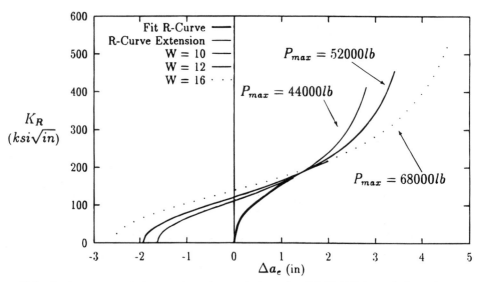

FIG. 12—*R-curve residual strength prediction diagram for 5/4 GLARE-2 notched panels (1 in. = 25.4 mm, 1 ksi = 6.89 MPa, and 1 lb = 4.45 N).*

the ARALL-3 laminate to produce a valid *R*-curve. This is due to the lower yield strength of 2024 aluminum used in GLARE-2 that causes a bigger plastic deformation zone around the crack tip. The fitted *R*-curve and prediction diagram for GLARE-2 testing panels are shown in Figs. 11 and 12.

The residual strengths for all the panels tested except the 76.2 mm (3 in.) wide panels were predicted using the fitted *R*-curves. The results from the predictions were in good agreement with those from the tests as can be seen in Fig. 13 and Table 3. The *R*-curve approach was not only able to reveal the slow crack growth behavior, but also able to measure the size influence to the residual strength of the center-notched unstiffened laminate panels. The small differences between the results could have been caused mainly by the random errors in the testing processes, such as from the measuring and alignment of the fixture. The limited number of test specimens used to generate the fitted *R*-curves could have also been the reason. The comparison of these results indicated that the LEFM *R*-curve approach was a suitable analytical tool for these fiber/metal laminates.

Conclusions

Tests on unstiffened center-notched 3/2 and 5/4 ARALL-3 and 5/4 GLARE-2 panels showed that these laminates, like monolithic aluminum alloys, exhibit slow stable tearing prior to rapid fracture. For wide panels, there was visible buckling around the crack tip due to the compressive stress in the *x* direction (parallel to the initial crack direction) but the buckling has very small effects on the Model I stress intensity factor calculations. The plastic deformation around the crack tips was small compared to the crack tip plastic zone of center-cracked monolithic 7074 and 2024 aluminum panels. This is due to the high yield strength of prepreg layers in the laminates. Therefore, the linear analysis approach selected for this study was proper. This greatly simplified the calculation. The effective crack extension concept and ASTM E561-92a was demonstrated to be applicable to the *R*-curve and residual strength determination of

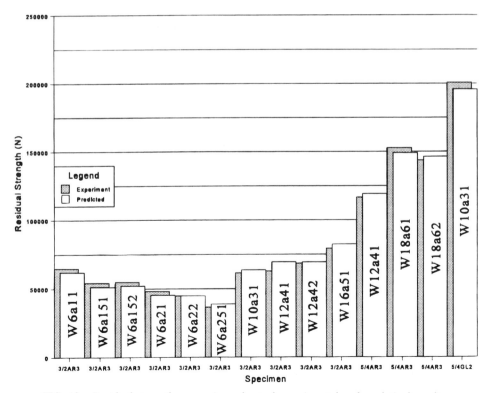

FIG. 13—*Residual strength comparison chart of experimental and analytical results.*

TABLE 3—*Residual strengths from tests and predictions.*

Specimen	$2a_0$, mm (in.)	P_{exp}, kN (kip)	P_{pred}, kN (kip)
3/2 ARALL–3–W6a11	24.8 (0.979)	64.5 (14.5)	61.8 (13.9)
3/2 ARALL–3–W6a151	37.5 (1.479)	54.2 (12.2)	51.1 (11.5)
3/2 ARALL–3–W6a152	37.3 (1.47)	53.8 (12.1)	52.0 (11.7)
3/2 ARALL–3–W6a21	49.8 (1.963)	47.6 (10.7)	45.3 (10.2)
3/2 ARALL–3–W6a22	50.1 (1.974)	44.9 (10.1)	44.9 (10.1)
3/2 ARALL–3–W6a251	62.5 (2.463)	36.4 (8.2)	37.8 (8.5)
3/2 ARALL–3–W10a31	78.8 (3.103)	61.4 (13.8)	63.6 (14.3)
3/2 ARALL–3–W12a41	98.5 (3.878)	62.7 (14.1)	69.4 (15.6)
3/2 ARALL–3–W12a42	99.6 (3.924)	68.5 (15.4)	69.4 (15.6)
3/2 ARALL–3–W16a51	132.9 (5.233)	79.2 (17.8)	82.3 (18.5)
5/4 ARALL–3–W12a41	96.2 (3.789)	116.5 (26.2)	119.2 (26.8)
5/4 ARALL–3–W18a61	150.7 (5.935)	152.6 (34.3)	149.0 (33.5)
5/4 ARALL–3–W18a62	151.4 (5.963)	143.7 (32.3)	146.4 (32.9)
5/4 ARALL–2–W10a31	83.1 (3.274)	200.6 (45.1)	195.8 (44)
5/4 ARALL–3–W12a41	98.5 (3.88)	not found	231.4 (52)
5/4 ARALL–2–W16a51	132.6 (5.223)	not found	302.6 (68)

test panels. The *R*-curves developed from wide, center-notched ARALL-3 and GLARE-2 laminate panels were independent of initial crack extensions. The *R*-curves from the 76.2 mm (3 in.) 3/2 ARALL-3 panels were not valid to be used to determine the residual strengths of wide panels through the *R*-curve approach due to the large-scale yielding that happens for a panel with its width smaller than the minimum width requirement when the load reached the critical level. This verified the panel size requirements by Feddersen theory [3] of residual strength prediction for the center-cracked tension panels. The *R*-curves developed from the 5/4 GLARE-2 center-notched laminate panels show slow crack growth over a much larger range of loading than those for the 3/2 and 5/4 ARALL-3 laminate panels. This is due to the higher ductility and lower yield strength of the 2024-T3 aluminum in GLARE-2 contrasted to the higher strength 7475-T76 aluminum in ARALL-3. This also causes the panel width requirement for GLARE-2 center-notched laminate panels to yield a valid *R*-curve higher than the ARALL-3 laminate panels. The *R*-curve approach revealed the effect of slow crack growth on the residual strength of center-notched laminate panels, and it predicted the size influence on the residual strength of center-notched laminate panels. The results of predictions showed less than 10% difference with respect to the experimental results, and this indicated that the conventional fracture mechanics methodology is suitable for metal laminates in these cases.

Also, the fabrication of test panels used for the tests show that the machinability of these laminates was comparable to monolithic aluminum even though certain procedures had to be taken to prevent the laminates from delamination. Tool wear was slightly high.

References

[1] Wu, Ming., "Residual Strength Prediction of Center Notched Unstiffened and Bonded Stringer Stiffened ARALL-3 and GLARE-2 Panels through The R-Curve Approach," Ph.D. thesis, Tennessee Technological University, Cookeville, 1994.
[2] Feddersen, C. E., "Evaluation and Prediction of the Residual Strength of Center Crack Tension Panels," *Damage Tolerance in Aircraft Structures, ASTM STP 486*, American Society for Testing and Materials, Philadelphia, 1970, pp. 50–78.
[3] Mccabe, D. E. and Schwalbe, K. H., "Prediction of Instability Using the K_R-Curve Approach," *Elastic-Plastic Fracture Mechanics Technology, ASTM STP 896*, Newman and Loss, Eds., American Society for Testing and Materials, Philadelphia, 1985, pp. 99–113.

[4]   Wu, E. M., "Fracture Mechanics of Anisotropic Plates," *Composite Materials Workshop*, S. W. Tsai, J. C. Halpin, and N. J. Pagano, Eds., Technomic Publishing Co., Inc., Lancaster, PA, 1968, pp. 20–43.
[5]   "GLARE-Aerospace ARALL," Technical Information, AkZO Nobel, Arnhem, The Netherlands, 1991.
[6]   Christensen, R. H. and Denke, P. H., "Crack Strength and Crack Propagation Characteristics of High Strength Mentals," ASD-TR-61-207, Douglas Aircraft Company, Long Beach, CA, 1961.
[7]   Kuhn, P., "Residual Strength in the Presence of Fatigue Cracks," *Proceedings*, Structures and Materials Panel, Advisory Group for Aeronautical Research and Development, Turin, Italy, April 1967.

Jennifer L. Miller,[1] *Marc A. Portanova,*[2] *and W. Steven Johnson*[3]

Impact Damage Resistance and Residual Property Assessment of [0/±45/90]$_s$ SCS-6/TIMETAL 21S

REFERENCE: Miller, J. L., Portanova, M. A., and Johnson, W. S., **"Impact Damage Resistance and Residual Property Assessment of [0/±45/90]$_s$ SCS-6/TIMETAL 21S,"** *Composite Materials: Fatigue and Fracture (Sixth Volume), ASTM STP 1285*, E. A. Armanios, Ed., American Society for Testing and Materials, 1997, pp. 260–280.

ABSTRACT: The impact damage resistance and residual mechanical properties of [0/±45/90]$_s$ SCS-6/TIMETAL 21S composites were evaluated. Both quasi-static indentation and drop-weight impact tests were used to investigate the impact behavior at two nominal energy levels (5.5 and 8.4 J) and to determine the onset of internal damage. Through X-ray inspection, the extent of internal damage was characterized nondestructively. The composite strength and constant-amplitude fatigue response were evaluated to assess the effects of the sustained damage. Scanning electron microscopy was used to characterize internal damage from impact in comparison to damage that occurs during mechanical loading alone. The effect of stacking sequence was examined by using specimens with the long dimension of the specimen both parallel (longitudinal) and perpendicular (transverse) to the 0° fiber direction. Damage in the form of longitudinal and transverse cracking occurred in all longitudinal specimens tested at energies greater than 6.3 J. Similar results occurred in the transverse specimens tested above 5.4 J. Initial load drop, characteristic of the onset of damage, occurred on average at 6.3 J in longitudinal specimens and at 5.0 J in transverse specimens. X-ray analysis showed broken fibers in the impacted region in specimens tested at the higher impact energies. At low impact energies, visible matrix cracking may occur, but broken fibers may not. Matrix cracking was noted along fiber swims, and it appeared to depend on the surface quality of the composite. At low impact energies, little damage had been incurred by the composite and the residual strength and residual life was not greatly reduced as compared to an undamaged composite. At higher impact energies, more damage occurred and a greater effect of the impact damage was observed.

KEYWORDS: metal-matrix composites, impact damage, residual strength, residual fatigue, fatigue (materials), fracture (materials), composite materials

Titanium-matrix composites (TMCs) are candidate materials for high-temperature structural applications, such as gas turbine engines, where their high specific strength at elevated temperatures and good general corrosion resistance are beneficial. These materials provide a strong, lightweight alternative to conventional structural alloys due to their ability to maintain mechanical integrity at elevated temperatures [1]. Much research has been conducted on the mechanical behavior of TMCs under various types of thermal, mechanical, and combined thermomechanical loadings as well as on the various influences of notches and holes [2–7]. However, another

---

[1]Resident research associate, NASA Langley Research Center, Hampton, VA 23681–0001.
[2]Senior materials engineer, Lockheed Matrin Corporation, Hampton, VA 23681–0001.
[3]Professor of Materials Science and Engineering, Georgia Institute of Technology, Atlanta, GA 30332–0150.

critical aspect of the service conditions has received little attention: low-energy impact damage. Considerable damage may result from a seemingly innocuous event such as a dropped tool. Characterizing a material's residual properties after impact should be considered in the component design process.

Although several studies have been conducted on polymeric composites [8–11], few studies exist on the impact behavior of continuous-fiber metal-matrix composites (MMCs). Those studies that do exist are primarily focused on boron fiber-reinforced aluminum composites for turbine blade applications. Impact tests on unidirectional boron-aluminum composites have shown that a considerable reduction in residual strength can occur from low-velocity (low-energy) impact with a hard object. Carlisle et al. noted a 25% reduction in residual strength at the lowest impact velocity used in the study [12]. When residual fatigue of boron/aluminum composites was considered, at the slowest test velocity, Gray found the fatigue life was reduced by an order of magnitude [13]. In both studies, the residual properties continued to decrease as impact velocity, and consequently impact energy, increased. Comparisons to unreinforced titanium alloys in the previously mentioned studies showed the boron-aluminum composites to be less damage resistant and less damage tolerant than the monolithic material.

Similar impact studies conducted on polymer-matrix composites (PMCs) show distinctive differences in the damage mechanisms occurring in these materials as compared to MMCs (delamination versus fiber/matrix cracking); however, some general trends of the PMC behavior may apply to MMCs. Greszczuk found cross-ply laminates resist impact damage better than unidirectional or pseudoisotropic laminates [11] when investigating PMCs. Another trend noted in Greszczuk's study was that damage resistance increased when a stronger matrix material is used. Since the strength of a metal can be varied easily through heat treatment, this effect could apply to MMCs. The differences in damage resistance occurring in PMCs due to variations in laminate layup and constituent elastic properties may also apply to MMCs.

In this study, the impact resistance and residual mechanical properties of quasi-isotropic SCS-6/TIMETAL 21S[4] composites is evaluated. The onset of internal damage is described in terms of impact energy. Residual strength and residual fatigue tests were conducted at room temperature, and the results are compared to those of nonimpacted materials. The influence of prior impact on fracture behavior and damage acculumation is also examined.

Materials and Processing

The SCS-6/TIMETAL 21S composites tested were manufactured into [0/±45/90]$_s$ quasi-isotropic laminates by hot isostatically pressing thin foils of Ti-15Mo-3Nb-3Al-0.2Si (TIMETAL 21S) between unidirectional tapes of SCS-6 silicon-carbide fibers. The 0.14-mm-diameter fibers were held in place by crosswoven titanium-niobium (Ti-Nb) wires. Several laminates of varying thicknesses between 1.70 to 1.88 mm were used in this study. Several sections were examined to determine any variation in fiber spacing and the average fiber volume fraction. Of the laminates examined, those with the smaller nominal thickness showed a greater variation in fiber spacing. The average fiber volume fraction for the laminates ranged from 0.348 to 0.357. Figure 1 displays photomicrographs of the polished cross sections of two laminates. As shown, the average fiber spacing (0.216 mm) did not vary greatly in the thicker panels (Fig. 1a), whereas there is a larger variation in thin panels (Fig. 1b).

Of the various laminated sheets from which specimens were manufactured, many variations in quality occur, both internally and on the exterior surfaces of the sheets. The photograph and radiograph of two different specimens in Fig. 2 illustrate some of these anomalies. Fiber

---

[4]TIMETAL 21S is a registered trademark of Timet Corporation, Denver, CO.

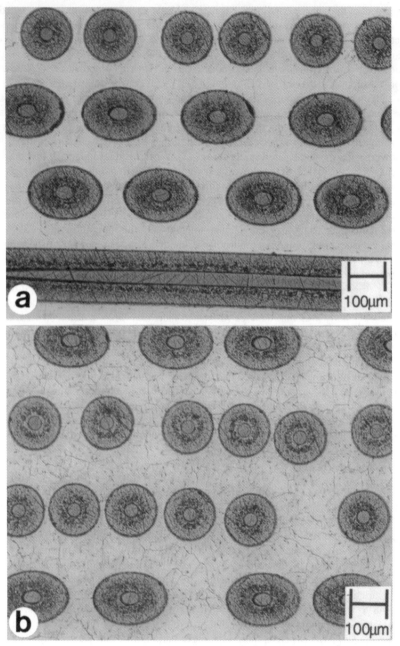

FIG. 1—*Optical micrographs of [0/±45/90]ₛ SCS-6/TIMETAL 21S laminates: (a) typical cross section representing specimens having thickness ranging from 1.80 to 1.87 mm and (b) typical cross section representing specimens of 1.70-mm thickness.*

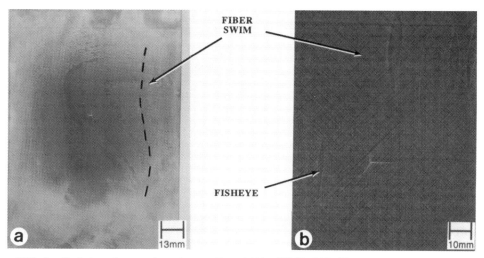

FIG. 2—*Variations in manufacturing quality of [0/±45/90]ₛ SCS-6/TIMETAL 21S composites: (a) photograph of an impact panel showing surface roughness and fiber swim and (b) radiograph of an impact panel showing fisheyes and fiber swim.*

swim describes the waviness in the fibers as indicated in the figure. Fish eyes are areas where the fibers separate and rejoin, forming a gap. Both of these anomalies contribute to nonuniform fiber spacing throughout the composite. The laminates were examined prior to machining to determine the best arrangement to use to machine the impact specimens to avoid placing theses anomalies in the center of a specimen. However, the anomalies could not be avoided altogether.

The 152 by 102 mm impact specimens were machined using a diamond-impregnated abrasive cutting wheel. The long dimension of the panel was oriented both parallel and perpendicular to the 0° fiber direction, yielding two different panel designations and layups: the original [0/±45/90]ₛ are the longitudinal specimens and the 90° rotated orientation yields a [90/±45/0]ₛ layup for the transverse specimens. By varying the panel orientation, the effects of stacking sequence on mechanical behavior could be examined. Prior to heat treating, specimens were degreased and chemically cleaned using a diluted mixture of hydrofluic and nitric acid followed by a dilute hydrochloric acid wash. The specimens were then subjected to a β-stabilization heat treatment consisting of an 8-h soak in vacuum at 620°C to prevent α precipitation during future elevated temperature testing [14]. All specimens were examined radiographically and by ultrasonic C-scan both prior to and after impact testing to assess the damage state of the specimens. The results will be discussed in a later section.

After the impact tests were completed, 152 by 25.4 mm specimens were machined from the damaged panels for residual property evaluation. The entire impact area was contained within the cross-sectional area of these specimens. Some permanent bending deformation may have occurred in some specimens due to the impact event. Strain gages were applied to the back and front surface of these specimens along the centerline to determine the magnitude of the initial bending stress applied when the specimens straighten during placement in the grips of the testing machine. Several other 152 by 12.5 mm specimens were machined from the edges of the panels for use in baseline tension and constant-amplitude fatigue studies. End tabs were applied to all specimens with cyanoacrylate adhesive to reduce the gripping stress and prevent specimen failure in the grip sections. Table 1 describes the residual property test matrix used in the study.

TABLE 1—*Residual property test matrix.*

Specimen	Laminate Orientation	Nominal Impact Energy, J	Residual Property Test Method
91L01D	longitudinal	5.4	tension
99L01D	longitudinal	5.4	tension
90L04D	longitudinal	8.4	tension
92L01D	longitudinal	8.4	tension
118L01D	longitudinal	8.4	tension
95T03D	transverse	5.4	tension
96T01D	transverse	5.4	tension
48T01D	transverse	8.4	tension
99L03D	longitudinal	5.4	345 MPa CAF[a]
92L03D	longitudinal	5.4	414 MPa CAF
90L05D	longitudinal	8.4	345 MPa CAF
92L05D	longitudinal	8.4	414 MPa CAF
95T04D	transverse	5.4	345 MPa CAF
98T02D	transverse	5.4	414 MPa CAF
93T04D[b]	transverse	8.4	345 MPa CAF
95T02D	transverse	8.4	414 MPa CAF

[a]Constant amplitude fatigue.
[b]12.7-mm-wide specimen.

Experimental Procedure

Impact Tests

Two different test methods were employed to assess the damage resistance of the TMCs: quasi-static indentation (QSI) and drop-weight impact (DWI) [8]. Two impact energies, 5.4 J (4.0 ft-lbf) and 8.4 J (6.2 ft-lbf), were recommended by industry as energies typical of tool drops. The QSI tests were performed using a servohydraulic test frame at a constant displacement rate of 0.51 mm/min. During testing, the specimens were clamped firmly in an aluminum test fixture that contained a 127 by 76.2 mm opening with corner radii of 12.7 mm. An instrumented tup attached to a 12.7-mm-diameter hemispherical indenter was used to measure load. The tup was mounted in the grips of the load frame such that the indenter traveled normal to the plane of the specimen. The indentation load and stroke output were recorded at a rate of one data point per second throughout the loading history using a digital storage oscilloscope.

Drop-weight impact tests were conducted using the same instrumented tup and test fixture as in QSI tests, resulting in the same plate boundary conditions. The total free-falling mass of the indentor, tup, and steel weight was 3.03 kg. The impactor was centered above the panel at the required height to impart the desired impact energy. After the impactor struck the specimen, a dummy panel was quickly moved between the fixture and specimen to prevent multiple impacts. The impact force-time history was then recorded in real time using a digital storage oscilloscope.

Residual Property Tests

The room-temperature tension and constant-amplitude fatigue tests were conducted in a 100 kN closed-loop servohydraulic test frame equipped with hydraulic grips. A 7 MPa gripping pressure was used in all tests. Tension tests were conducted in stroke control at a rate of 1.27

mm/min. Constant-amplitude fatigue tests were conducted in load control using a sinesoidal waveform at a frequency of 1 Hz and an *R*-ratio of 0.1. The bending stress induced when gripping the damaged specimens was measured using 350-Ω electrical resistance strain gages mounted to the front and back face of the specimens, oriented longitudinally and transversely with respect to the gage length, and positioned 25.4 mm above and below the damage area. The measured strains were small in comparison to the applied loads and did not appear to influence the results greatly. Axial strain during loading was measured using a strain gage extensometer with a 25.4-mm gage section. The damage area was centered in the gage section. Baseline data for both tension tests and constant-amplitude fatigue tests were generated by testing the undamaged coupons cut from the edges of the impact panels.

Results and Discussion

Impact Damage Resistance

The impact damage resistance of the TMCs was evaluated by examining the force versus displacement response of the panels when subjected to both QSI and DWI. The energy applied during loading was calculated by integrating the force versus displacement curves. Two nominal impact energies, 5.4 J and 8.4 J, were sought throughout the study when comparing results since slight variations occur in the impact energy for each individual panel tested. Figures 3 and 4 compare the TMCs response to QSI and DWI tests at 8.4 J for longitudinal and transverse specimens, respectively. The oscillation in the force-displacement response of the DWI test is due to vibrations that occur as the incident wave reflects off the clamped plate boundaries. The vibration is inherent in the test method [8]. The response of the TMCs to both types of tests was similar: as the contact force on the panel increases, the displacement of the panel increases, and subsequently, the applied energy increases. If the contact force is increased enough, strain will accumulate in the composite until reaching the fiber failure strain wherein

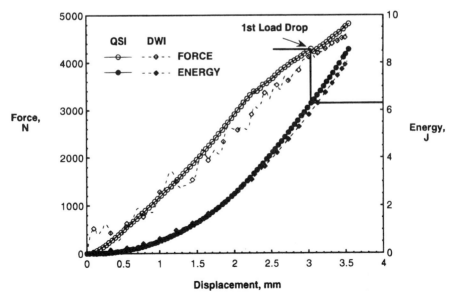

FIG. 3—*Comparison of the response of longitudinal [0/±45/90]$_s$ SCS-6/TIMETAL 21S to QSI and DWI tests.*

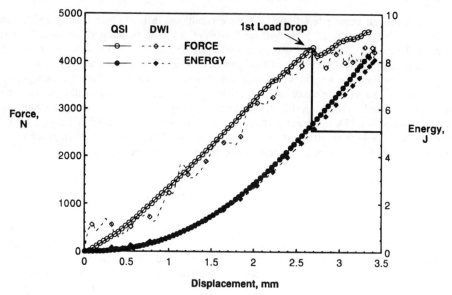

FIG. 4—*Comparison of the response of transverse [90/±45/0]ₛ SCS-6/TIMETAL 21S to QSI and DWI tests.*

the fibers break. When this occurs, the contact force decreases rapidly since the dominant load-carrying component of the composite is damaged. Matrix cracking usually precedes fiber failure as shown in past studies on the mechanical response of TMCs [2–4]. Since there was no significant difference between the force-displacement response of the panels subjected to QSI or DWI, the QSI was determined to be the best method of testing the impact resistance. This method provided a repeatable test that allowed the contact force to be increased slowly, thereby permitting the test to be interrupted periodically to examine the specimen to determine if any damage was visible.

The impact results for all the longitudinal and transverse specimens show the mean value of the first load drop occurs at 4.5 kN. The first load drop indicates that damage has occurred and is identified on Figs. 3 through 5. The mean applied energy corresponding to this mean load drop is 6.3 J and 5.0 J for the longitudinal and transverse specimens, respectively. The difference in applied energy at initial load drop may be due in part to the variation in the bending stiffnesses of the two different stacking sequences. Figure 5 compares a typical response for a longitudinal and a transverse specimen subjected to a nominal 8.4-J QSI. The transverse specimen is stiffer in bending since the fibers in the outer ply span the short dimension of the rectangular plate during loading. As shown in Fig. 4, in order to produce the same amount of deflection of the plate, a greater force must be applied to the transverse specimens. Although there is a difference in the energy at the first load drop, it is within the statistical variation of the test results. Figure 6 is a histogram displaying the mean energy associated with the first load drop as well as the maximum and minimum values for each specimen orientation. The numbers above the bars indicate the number of specimens used to determine the mean while the error bars represent one standard deviation above and below the mean. The range of energies for both orientations overlap. Since there were several composite sheets from which specimens were made, all having slight differences in the surface quality, degree of fiber swim, and in other manufacturing anomalies, the resulting variations in mechanical response would be expected.

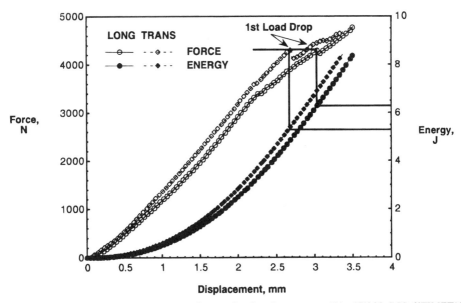

FIG. 5—*Comparison of the response of longitudinal and transverse [0/±45/90]ₛ SCS-6/TIMETAL 21S to QSI tests to a nominal 8.4-J impact.*

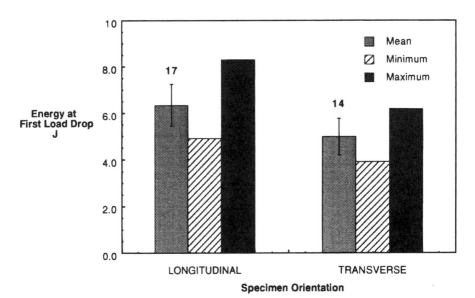

FIG. 6—*Energy associated with the first load drop occurring in the force-displacement response of longitudinal and transverse specimens of [0/±45/90]ₛ SCS-6/TIMETAL 21S.*

When examining the panels tested at the lower energy levels, a few displayed matrix cracks; however, X-rays did not show internal fiber breaks occurring. All the longitudinal panels tested showed internal damage when subjected to a force greater than 4.1 kN or at energies above 5.6 J. Similar results are obtained for the transverse panels for forces greater than 3.6 kN or at 3.9 J. The matrix cracking on the backface of the panels tested at low energies (lower than the mean energy associated with the first load drop) indicates that the appearance of visible damage does not give a clear indication of the true damage state of the material. Matrix cracking on the surface does not imply fiber breakage in the interior. Although, at higher energies, when fibers are broken, matrix cracking also occurs. The development of matrix cracks prior to the first load drop in the force-displacement response suggests that the first load drop is characteristic of fiber damage occurring within the composite and not due to the matrix cracking.

Damage Assessment

Damage varied greatly in both specimen orientations depending on the impact energy. Longitudinal and transverse matrix cracks and broken fibers were found at the higher impact energies. Crack lengths were measured on the surface of the specimens and from radiographs. The exterior surface cracks were in general longer than those shown on the radiographs. Table 2 shows the surface crack measurements and the interior (X-ray) crack measurements for cracks running in the longitudinal and transverse plate directions as a function of impact energy. There is a considerable amount of scatter in the data. As mentioned previously, in some of the specimens tested at low energies, no fiber breaks were found by radiography. The extent of the damage incurred by the TMCs appears to be sensitive to material variations in the laminate due to stress concentrations produced by nonuniform fiber distributions. This sensitivity to manufacturing anomalies was suspected when specimens undergoing residual

TABLE 2—*Comparison of crack length measurements for impacted specimens.*

Specimen	Nominal Impact Energy, J	Surface Measurements		X-ray Measurements	
		Longitudinal Crack, mm	Transverse Crack, mm	Longitudinal Crack, mm	Transverse Crack, mm
91L01D	5.4	7.14	3.13	7.14	...
99L01D	5.4	1.63	2.77	...	...
90L04D	8.4	7.95	4.78	7.94	4.78
92L01D	8.4	7.69	4.51	7.54	3.18
118L01D	8.4	5.94	5.56	4.37	4.76
95T03D	5.4	8.26	4.11	7.94	3.18
96T01D	5.4	7.95	17.85	7.94	17.86
48T01D	8.4	11.93	6.20	11.91	5.94
99L03D	5.4	6.47	1.28	a	...
92L03D	5.4	4.81	...	...	...
90L05D	8.4	8.78	4.80	7.94	3.57
92L05D	8.4	3.77	0.617	2.78	...
95T04D	5.4	2.84	3.43	...	...
98T02D	5.4	...	...	...	...
93T04D[b]	8.4	7.37	5.59	7.13	3.96
95T02D	8.4	16.28	4.91	16.28	3.18

[a]Not available.
[b]12.7-mm-wide specimen.

fatigue tests did not fail at the impact site. These results will be discussed in the next section. Ultrasonic C-scan inspections did not provide any further insight into the extent of damage due to the local permanent deformation at the contact site compounded by the bending deformation of the panels. The bending displacement caused a change in signal attenuation that could not be discerned from the attenuation due to internal damage. Unlike polymer-matrix composites, where large delaminations occur due to impact [*10*], the C-scan does not provide a method of quantifying damage in the TMC.

Residual Property Assessment

Residual Strength—A comparison of the results from selected tension tests on both nonimpacted and impacted longitudinal specimens are shown in Fig. 7. During the residual tension tests, failures occurred in the damage area for all but one specimen that failed in the grip area. Results for the transverse specimens were similar to those of the longitudinal specimens. From the stress-strain response, the initial elastic modulus, the 0.2% offset yield stress, the ultimate strength, and the failure strain of the composites were determined and are given in Table 3. Figures 8 through 11 are histograms representing the mean values of these properties as a function of nominal impact energy. The error bars displayed on each of these figures represent one standard deviation above and below the mean for each property, while the numbers above each bar give the number of tests.

Figure 8 compares the mean initial elastic moduli (E_I) of nonimpacted and impacted materials for both specimen orientations. The elastic modulus was determined from the initial loading portion of the curve prior to the knee that occurs at approximately 200 MPa. The elastic response for the longitudinal and transverse specimens were similar. For impacted specimens, the E_Is fall within the range shown by the error bars for the undamaged materials. The impact event does not appear to have caused considerable fiber-matrix debonding that would have otherwise resulted in a reduced elastic modulus. The presence of local matrix cracks would

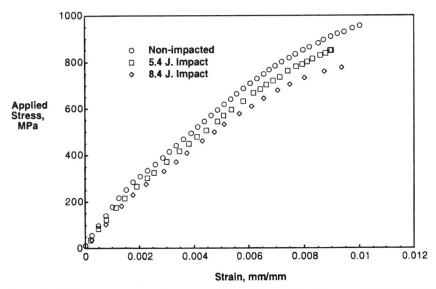

FIG. 7—*Applied stress versus strain response for nonimpacted and impacted [0/±45/90]$_s$ SCS-6/TIMETAL 21S. The legend describes the nominal impact energy.*

TABLE 3—*Residual tension results.*

Specimen	Nominal Impact Energy, J	Initial Elastic Modulus, GPa	0.2% Offset Yield Stress, MPa	Ultimate Strength, MPa	Failure Strain
91L01D	5.4	123.8	a	1010.7	0.00916
99L01D	5.4	146.2	720.5	850.2	0.00902
90L04D	8.4	128.2	a	678.1	0.00706
92L01D	8.4	157.1	752.0	908.7	0.00923
118L01D	8.4	133.1	682.9	777.0	0.00936
95T03D	5.4	147.1	748.5	988.5	0.1092
96T01D	5.4	122.1	799.8	820.4	0.00894
48T01D	8.4	149.1	686.1	776.6	0.00799

[a]Insufficient yielding to calculate the 0.2% offset yield stress.

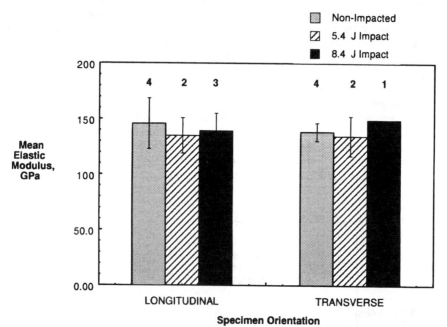

FIG. 8—*Mean elastic modulus from tension tests of nonimpacted and impacted [0/±45/90]$_s$ SCS-6/TIMETAL 21S. The legend describes the nominal impact energy.*

not be expected to change a global property like E_l. Similarly, the prior impact does not seem to influence the 0.2% offset yield stress (σ_y) as shown in Fig. 9. During a few of the tests, insufficient yielding occurred and σ_y could not be determined. Table 3 displays the test conditions during which the insufficient yielding occurred. The bars of histogram in Fig. 9 are labeled with the number of tests used to calculate the mean 0.2% offset yield stress. The lower failure strain for the impacted specimens can be attributed to a local effect of the impact damage on the fracture behavior of the material. As shown in Figs. 10 and 11, the mean ultimate strength (σ_u) and the mean failure strain (ϵ_f) are reduced for the impacted specimens. The premature failures are due to the local effect of the impact damage, not simply a net

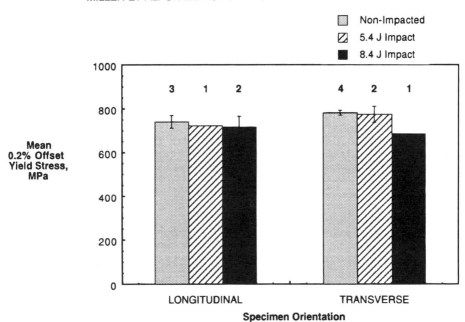

FIG. 9—*Mean 0.2% offset yield stress from tension tests of nonimpacted and impacted [0/±45/90]ₛ SCS-6/TIMETAL 21S. The legend describes the nominal impact energy.*

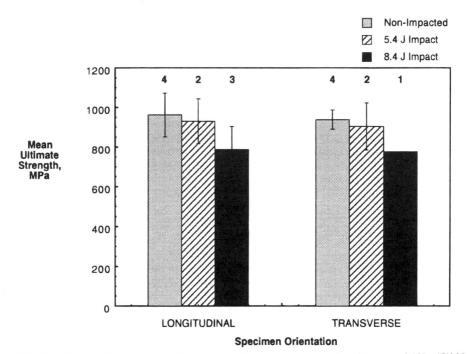

FIG. 10—*Mean ultimate strength from tension tests of nonimpacted and impacted [0/±45/90]ₛ SCS-6/TIMETAL 21S. The legend describes the nominal impact energy.*

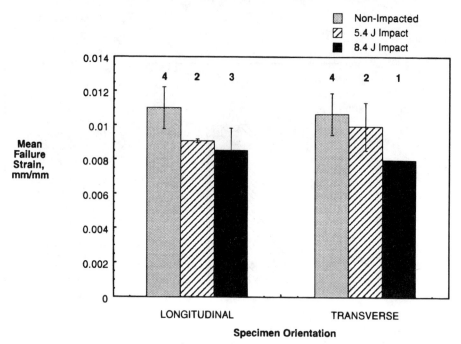

FIG. 11—*Mean failure strain from tension tests of nonimpacted and impacted* $[0/\pm45/90]_s$ *SCS-6/TIMETAL 21S. The legend describes the nominal impact energy.*

section effect. The damage increases the fiber stresses locally, causing fibers to begin to fail at lower applied (global) stresses. This results in a reduction in the global (applied) ultimate stress. The reduction in ϵ_f (measured globally) is also attributed to the local stress increase in the fibers. Both the longitudinal and transverse specimens tested at the 8.4-J impact energy showed a greater decrease in σ_u and ϵ_f than those tested at 5.4 J. These specimens also suffered the most severe damage, fiber breaks, and matrix cracks. The tension test results for the longitudinal and transverse specimens were very similar; the variation in the stacking sequence did not seem to influence the tensile response of the TMCs examined.

Since residual properties are a concern of this study, the mean ultimate strength for the impacted materials was normalized with respect to the mean ultimate strength for the undamaged materials. This will result in a relative measure of the material's damage tolerance. Figure 12 shows the normalized residual strengths as a function of impact energy for both specimen orientations. The results show that low-energy impacts, where little damage is incurred by the composite, did not greatly effect the strength of the material. In particular, when only matrix cracking occurred, the residual strength was within the statistical variation of the undamaged material strengths. The mean residual strength for a 5.4-J impact is 95% of the mean ultimate for undamaged materials. As the amount of fiber damage increased, the retention of composite strength decreased. For the 8.4-J impact, the residual strength is effectively reduced on average by 20%.

Baseline Fatigue Study—Constant-amplitude fatigue tests were conducted on undamaged specimens to establish a baseline for assessing the residual life of impact-damaged materials. Both longitudinal and transverse specimens were tested to determine if the laminate layup

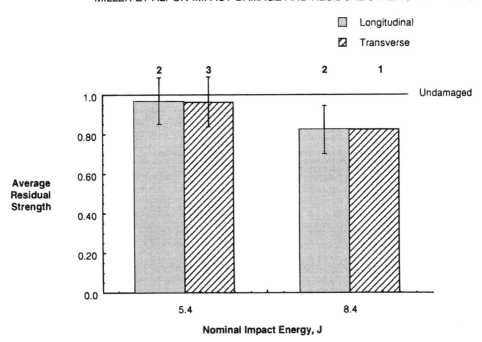

FIG. 12—*Average residual tensile strength of impacted [0/±45/90]ₛ SCS-6/TIMETAL 21S. The legend describes the specimen orientation with respect to the 0° fiber direction.*

affected the fatigue life of the materials. Figure 13 displays the results of the baseline tests. Each data point represents one specimen. A run out criterion of 10^6 cycles was used to set an endurance limit for the material and is indicated by the arrows shown in the figure. Both specimen orientations showed similar fatigue lives at the applied stress levels tested with the transverse specimens typically having a longer life. In terms of overall fatigue life, little effect of laminate layup is shown.

A longitudinal specimen tested at 310 MPa and a transverse specimen tested at 276 MPa failed at much lower fatigue lives than the other tests. Both of these specimens were from the thinner panels (1.7 mm). X-rays showed nonuniform fiber distributions and a considerable amount of fiber swim in comparison to the other panels tested. Figure 1 showed a typical cross section of the material from which the transverse specimen was machined. As discussed previously, the fiber spacing varied greatly through the thickness of this material. Stress concentrations due to the higher fiber density may have increased matrix cracking and produced a higher net section stress, increasing strain accumulation leading to composite failure. Although the thinner composites had similar strengths to other composites in tension, the local effect of fiber spacing would be more significant in fatigue where crack propagation is greatly influenced by local stress fields in the material.

Residual Fatigue Life—The results of the fatigue tests on impacted specimens are shown in Fig. 14 along with the baseline fatigue results for two applied stresses. There is considerable variation in the residual fatigue lives of the impacted specimens. For the longitudinal specimens tested at 345 MPa, the trend is as expected—the higher the impact energy, the more initial damage and the shorter residual fatigue life. However, the longitudinal specimen impacted at 5.4 J did not have any fiber breaks. The specimen did have substantial fiber swim, particularly

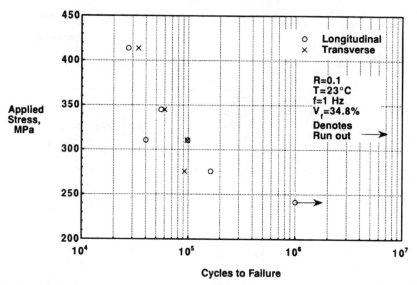

FIG. 13—*Applied stress versus number of cycles to failure for unnotched [0/±45/90]$_s$ SCS-6/ TIMETAL 21S at room temperature. The legend describes the orientation of the specimen with regard to the 0° fiber direction.*

in the 0° surface plies. Since no fiber breaks occurred, a fatigue life similar to an undamaged specimen would be expected (for the same test conditions). As the results show, the fatigue life was much lower than the undamaged composite. The specimen also failed outside of the impacted region. X-rays show a large gap separating 0° fibers where only matrix is found. By examining the fracture surface, it was determined that in this area where only the matrix exists, there should be approximately 15 fibers. The fiber gap essentially reduces the total number of 0° fibers in the composite by approximately 5%. Since the 0° fibers are the dominant load-carrying component in the composite, reducing their number may have contributed to the reduced fatigue life. The 0° fibers also bridge fatigue cracks occurring in the composite resulting in slower fatigue crack growth [3–5]. Similarly, the undamaged specimen tested at 414 MPa had a shorter life than the impacted specimens. Again, X-rays show substantial fiber swim and several fisheyes occurring along the length of the specimen.

When comparing the results for the specimens impacted at 8.4 J, little difference is shown for the two applied stress levels. The specimen tested at 345 MPa had a longer transverse crack length and may have suffered more internal damage initially, causing a reduction in fatigue life. The amount of impact damage from the 8.4-J tests varied a great deal, so it would be expected that the residual fatigues lives would also vary considerably.

Little difference is shown in the fatigue results at 345 MPa between the undamaged transverse specimen and the one impacted at 5.4 J, where no fiber breaks occurred. The longer life may be typical of the statistical variation of this material's properties. The 8.4-J impact had a very short life in comparison, but may be due to a finite width effect. The specimen had a large transverse crack with respect to its width and did not give a true indication of the material's damage tolerance due to the greatly reduced cross section. The results for the 414-MPa tests showed the expected trend, as discussed previously. The 5.4-J impacted specimen had a similar life to the undamaged specimen. No fiber breaks occurred in this specimen. The 8.4-J impact specimen had a small transverse crack and, as shown, its residual fatigue life was reduced.

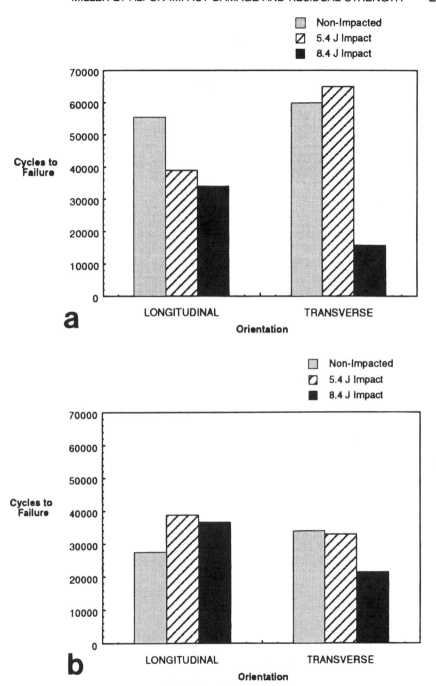

FIG. 14—*Comparison of fatigue lives of nonimpacted and impacted SCS-6/TIMETAL 21S [0/±45/90]$_s$ subjected to constant-amplitude fatigue at two applied stress levels: (a) 345 MPa and (b) 414 MPa. The legends describe the nominal impact energy.*

The variation in the fatigue lives and the location of failures with respect to manufacturing anomalies, seems to indicate variations in mechanical properties are dependent on the quality of the laminate. These anomalies are more damaging to the composite than low-energy impacts. Although there is considerable variation in the test results, the general trend of reduced life with an increasing amount of initial damage has been shown. Figure 15 shows the initial crack lengths (longitudinal and transverse) measured on X-rays compared to the fatigue life for each specimen tested. The solid vertical lines represent the average fatigue lives for the two applied stresses shown. The longitudinal crack length does not appear to influence fatigue life; however, as the initial transverse crack length gets longer, the residual fatigue life decreases.

Fractography—Fracture surfaces of the specimens from the baseline constant-amplitude fatigue (CAF) tests and the residual fatigue tests were examined using a scanning electron microscope (SEM). Micrographs of a nonimpacted specimen subjected to CAF at 345 MPa are shown in Fig. 16. A step-like fracture surface occurs (Fig. 16a), typical of this type of TMC [15], indicating fatigue crack initiation at multiple sites within the material. Fatigue crack growth is controlled by crack initiation at debonded fiber/matrix interfaces on off-axis plies. Figure 16b shows multiple initiation sites occurring along a 45° fiber. Final fracture occurs via ductile rupture as indicated by the equiaxed dimples shown in the matrix of the 0° ply adjacent to this 45° ply. This type of fracture behavior has been identified in other angle ply TMCs [15]. None of the specimens tested that were previously subjected to the nominal 5.4-J impact showed any initial fiber breakage in the X-rays. Of these four specimens, three did not fail in the impacted area. The fracture surfaces of those specimens were no different from those that were not impacted. As discussed earlier, they did fracture along areas where fiber swim can be seen on the outside surface of the specimen.

Figure 17 shows a longitudinal specimen impacted at the nominal 8.4-J energy. The fracture surface was tilted to show the longitudinal matrix crack running in the 0° ply. The initial

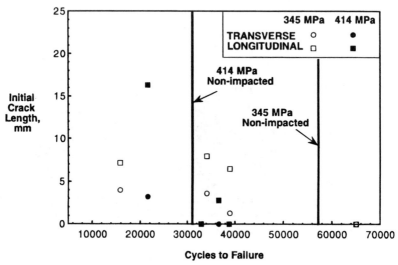

FIG. 15—*Longitudinal and transverse crack lengths versus number of cycles to failure for impacted [0/±45/90]ₛ SCS-6/TIMETAL 21S. Solid lines represent average fatigue lives for nonimpacted specimens at given applied stresses. The legend describes the crack orientation with respect to the 0° fiber direction.*

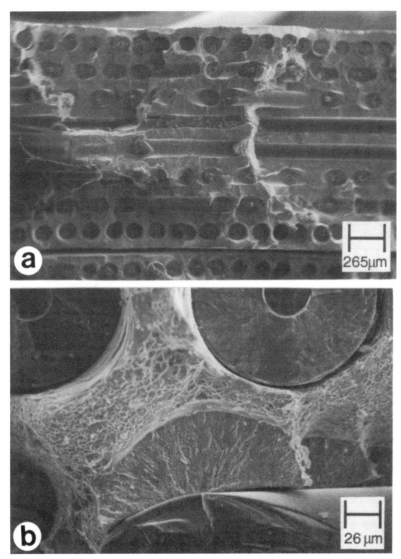

FIG. 16—*Fracture surface of a nonimpacted [0/±45/90]ₛ SCS-6/TIMETAL 21S subjected to constant-amplitude fatigue at room temperature: (a) step-like fracture surface showing fatigue crack initiation on multiple planes and (b) higher magnification of a region between a 0° and 45° ply showing multiple fatigue crack initiation sites along the 45° fiber and ductile rupture in the matrix around the 0° fiber.*

transverse crack that appears on X-rays traverses the entire thickness of the specimen, breaking the off-axis fibers as seen in Fig. 17a, but runs around the 0° fibers, causing debonding. Figure 17b is a magnified view of the cross section showing the crack running around the 0° fiber and propagating into the 45° ply. Note that the 0° fiber is broken in a different plane than the matrix, indicating fiber pullout during final fracture. The matrix around the 0° fiber also shows

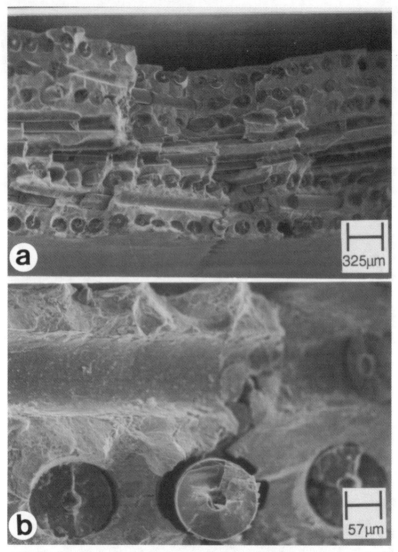

FIG. 17—*Fracture surface of a [0/±45/90]$_s$ SCS-6/TIMETAL 21S subjected to a nominal 8.4-J impact and constant-amplitude fatigue at room temperature: (a) impacted region showing longitudinal crack on outside surface and through-thickness crack and (b) magnification of the through thickness crack.*

ductile rupture. Away from the damage area, the fracture surface is similar to the undamaged material. In reviewing the residual fatigue results, the presence of transverse cracks in the TMCs do not appear to alter the mechanism of crack growth, but provide a larger initial damage area for crack propagation. The initial crack adds to the numerous small fatigue cracks growing from debonded fibers to accumulate sufficient strain to fail the composite.

Conclusions

The impact damage resistance of [0/±45/90]$_s$ SCS-6/TIMETAL 21S composites was evaluated experimentally using both quasi-static indentation and drop-weight impact tests. Longitudinal and transverse specimens were tested to examine the effect of stacking sequence. Results showed that the quasi-isotropic TMCs were able to resist impact damage when subjected to a contact force of 4.5 kN corresponding to impact energies of 6.3 J and 5.0 J for the longitudinal and transverse specimen orientations, respectively. The difference in the impact energy associated with the onset of damage is due to the greater plate bending stiffness for the transverse specimen orientation. The extent of the damage incurred by the TMCs was evaluated nondestructively through X-ray inspection. At higher impact energies, fibers were broken and residual properties were affected. Both the residual tensile strength and residual fatigue life as a function of impact energy were evaluated. The composites were able to withstand 5.4-J impacts without a substantial loss of tensile strength or fatigue crack growth resistance. At higher impact energies, the initial impact damage affects these properties more greatly. Results showed that matrix cracking alone is not sufficient to reduce tensile strength or fatigue life. Only when fibers are broken are the TMCs tensile strengths and failure strains reduced. The initial elastic modulus and 0.2% offset yield stress are not affected by the impact damage. The TMCs impacted nominally at 5.4 J had a residual tensile strength of 95% of the undamaged strength, whereas those impacted at 8.4 J had 80% of the nonimpacted strength. The variation in fatigue life and the location of failure with respect to manufacturing anomalies seems to indicate that variations in mechanical properties are dependent on the quality of the laminate; the anomalies are more damaging than the low-energy impacts. Although there is considerable variation in the test results, the general trend of reduced life with an increasing amount of initial damage has been shown. The presence of initial longitudinal cracks does not appear to influence fatigue life; however, as the initial transverse crack length gets longer, the residual fatigue life decreases. From examination of the fracture surfaces, the presence of transverse cracks in the TMCs appears not to alter the mechanism of crack growth, but provides a larger initial damage area for crack propagation. The initial crack adds to the numerous small fatigue cracks growing from debonded fibers to accumulate sufficient strain to fail the composite.

Acknowledgment

Ms. Miller would like to acknowledge the support extended by the National Research Council (NRC), Washington, DC, through its associateship program.

References

[1]   Smith, P. R. and Froes, F. H., "Developments in Titanium Metal Matrix Composites," *Titanium Technology: Present Status and Futures Trends*, Titanium Development Association, Dayton, 1985, pp. 157–164.
[2]   Mirdamadi, M., Johnson, W. S., Bahei-El-Din, Y. A., and Castelli, M. G. "Analysis of Thermomechanical Fatigue of Unidirectional Titanium Metal Matrix Composites," *Composite Materials: Fatigue and Fracture, Fourth Volume, ASTM STP 1156*, W. W. Stinchcomb and N. E. Ashbaugh, Eds., American Society for Testing and Materials, Philadelphia, 1993, pp. 591–607.
[3]   Bakuckas, J. G., Jr., Johnson, W. S., and Bigelow, C. A., "Fatigue Damage in Cross-Ply Titanium Metal Matrix Composites Containing Center Holes," NASA TM-104197, NASA Langley Research Center, Hampton, VA, 1992.
[4]   Herrmann, D. J., Ward, G. T., Lawson, E. J., and Hillberry, B. M., "Prediction of Matrix Fatigue Crack Initiation From Notches in Titanium Matrix Composites," *Life Prediction Methodology for Titanium Matrix Composites, ASTM STP 1253*, W. S. Johnson, J. M. Larsen, and B. N. Cox, Eds., American Society for Testing and Materials, West Conshohocken, PA, 1996, pp. 359–376.
[5]   Larsen, J. M., Jira, J. R., John, R., and Ashbaugh, N. E., "Crack Bridging Effects in Notch Fatigue of SCS-6/TIMETAL 21S Composite Laminates," *Life Prediction Methodology for Titanium Matrix*

Composites, ASTM STP 1253, W. S. Johnson, J. M. Larsen, and B. N. Cox, Eds., American Society for Testing and Materials, West Conshohocken, PA, 1996, pp. 114–136.

[6]   Castelli, M. G., Bartolotta, P. A., and Ellis, J. R., "Thermomechanical Fatigue Behavior of SiC (SCS-6)/Ti-15-3," *Composite Materials: Testing and Design (Tenth Volume), ASTM STP 1120,* G. C. Grimes, Ed., American Society for Testing and Materials, Philadelphia, 1991, pp. 70–86.

[7]   Russ, S. M., Nicholas, T., Bates, M., and Mall, S., "Thermomechanical Fatigue of SCS-6/Ti-24Al-11Nb Metal Matrix Composite," *Failure Mechanisms in High Temperature Composite Materials,* ASME, AD-Vol. 22/AMD Vol. 122, American Society of Mechanical Engineers, New York, 1991, pp. 37–43.

[8]   Poe, C. C., Jr., Portanova, M. A., Masters, J. W., Sankar, B. V., and Jackson, W. C., "Comparison of Impact Results for Several Polymeric Composites over a Wide Range of Low Impact Velocities," NASA CP-3104, NASA Langley Research Center, Hampton, VA, 1990.

[9]   Jackson, W. C. and Poe, C. C., Jr., "The Use of Impact Force as a Scale Parameter for the Impact Response of Composite Laminates," *Journal of Composites Technology & Research,* Vol. 15, No. 4, 1993, pp. 282–289.

[10] Portanova, M. A., Poe, C. C., Jr., and Whitcomb, J. D., "Open Hole and Post-Impact Compression Fatigue of Stitched and Unstitched Carbon/Epoxy Composites," *Composite Materials: Testing and Design (Tenth Volume,) ASTM STP 1120,* G. C. Grimes, Ed., American Society for Testing and Materials, Philadelphia, 1992, pp. 37–53.

[11] Greszczuk, L. B., "Damage in Composite Materials Due to Low Velocity Impact," *Impact Dynamics,* Wiley, New York, 1982, pp. 55–94.

[12] Carlisle, J. C., Crane, R. L., Jaques, W. J., and Montulli, L. T., "Impact Damage Effects on Boron-Aluminum Composites," *Composite Reliability, ASTM STP 580,* American Society for Testing and Materials, Philadelphia, 1975, pp. 458–470.

[13] Gray, T. D., "Foreign Object Damage and Fatigue Interaction in Unidirectional Boron/Aluminum-6061," *Fatigue of Composite Materials, ASTM STP 569,* American Society for Testing and Materials, Philadelphia, 1975, pp. 262–279.

[14] "Data Sheet for TIMETAL 21S (Ti-15Mo-3Nb-3Al-0.2Si) High Strength, Oxidation Resistant Strip Alloy," Timet Corp., Denver, CO.

[15] Johnson, W. S., Miller, J. L., and Mirdamadi, M., "Fractographic Interpretation of Failure Mechanisms in Titanium Matrix Composites," TMS/ASM Symposium on Mechanisms and Mechanics of MMC Fatigue, Rosemont, II, 2–6, Oct. 1994.

Mode Mixity and Delamination

Konstantinos Trakas[1] and Mark T. Kortschot[1]

The Relationship Between Critical Strain Energy Release Rate and Fracture Mode in Multidirectional Carbon-Fiber/Epoxy Laminates

REFERENCE: Trakas, K. and Kortschot, M. T., "**The Relationship Between Critical Strain Energy Release Rate and Fracture Mode in Multidirectional Carbon-Fiber/Epoxy Laminates,**" *Composite Materials: Fatigue and Fracture (Sixth Volume), ASTM STP 1285*, E. A. Armanios, Ed., American Society for Testing and Materials, 1997, pp. 283–304.

ABSTRACT: It is proposed that the fracture surface of delaminated specimens, and hence the critical strain energy release rate, is dependent on both the mode of fracture and the orientation of the plies on either side of the delamination with respect to the propagation direction. Recent fractographs of Mode III delamination surfaces obtained by the authors have reinforced the idea that the properties, G_{IIc} and G_{IIIc}, are structural rather than material properties for composite laminates. In this study, the relationship between the mode of fracture, the ply orientation, and the apparent interlaminar toughness has been explored. Standard double-cantilever-beam and end-notched flexure tests have been used, as has the newly developed Mode III modified split-cantilever beam test. Delaminations between plies of various orientations have been constrained to the desired plane using Teflon inserts running along the entire length of the specimen. As well, scanning electron microscopy (SEM) fractography has been extensively used so that measured energies can be correlated to the surface deformation. While fractographs show that Modes II and III share common fractographic features, corresponding values of G_c do not correlate, and it is shown that the large plastic zone of fractured Mode II specimens eliminates any comparison between the two. In contrast, Mode I delamination is found to be independent of the orientation of the delaminating plies.

KEYWORDS: delamination (materials), fractography, interlaminar fracture, fracture modes, multidirectional laminates, plasticity, plastic zone size, strain energy release rate, fracture (materials), composite materials

The delamination of laminated composites has been studied extensively over the past 15 years. To date, many aspects of delamination have been studied, including Mode I [1–4], Mode II [5–8], Mode III [9–12], and mixed-mode loading [13–16], as well as the effects of loading rate and specimen geometry [17,18], temperature and moisture effects [19,20], and the effect of matrix toughening [21,22]. Fractographic studies have provided a useful tool in the search for why composite laminates behave as they do. Scanning electron micrographs (SEMs) of delaminations have been used with the intent of verifying "qualitatively" why the strain energy release rate, G_c, varies under the different conditions just mentioned.

Little work has addressed the "physical" origin of the "material properties" G_{Ic}, G_{IIc}, and G_{IIIc}. Studies of mixed-mode fractures and delaminations between off-axis plies have been

[1]Graduate student and associate professor, respectively, Department of Chemical Engineering and Applied Chemistry, University of Toronto, Toronto, Ontario, Canada M5S 1A4.

largely phenomenological to date. The existing tests allow for the ranking of various materials, but are of little use to designers who do not have the appropriate toughness values for cracks running at an angle to the fibers or between plies of differing orientations. For example, many mixed-mode fracture studies have resulted in empirical laws of the form

$$(G_I/G_{Ic})^m + (G_{II}/G_{IIc})^n = 1 \tag{1}$$

without an explanation for the physical basis of such laws. Several authors have proposed values for m and n, including Refs 23, 24, and 25, but agreement has not been achieved.

Mode III tests have existed for quite some time, but obtaining "pure" Mode III at the crack tip (particularly in the split-cantilever-beam (SCB) specimen) has been elusive. Use of other specimen geometries has been successful [9,11], but these specimens are difficult to fabricate and test, and conventional beam theory is useless in these cases. A new loading scheme designed to produce almost pure Mode III delamination in a standard SCB specimen has recently been introduced by Sharif, Kortschot, and Martin [12]. Their work on Mode III led to some interesting observations about the relationship between delamination surfaces and the value of the apparent critical strain energy release rate. Whereas Mode II surfaces are typically characterized by shear hackles, Mode III surfaces were found to be dominated by "shear crevices," that is, matrix cracks running at 45° to the fracture surface but parallel to the fibers.

In this study, the physical origin of toughness is examined by looking at G_c for Mode I, Mode II, and Mode III loaded 0°/0°, 90°/90°, and 0°/90° specimens, and fractography is used in an attempt to elucidate the similarities and differences between the various modes of fracture.

Materials Aspects

Griffith proposed that the energy required to propagate a crack (dW) would be equal to the amount of new surface produced (dA), multiplied by the energy absorbed per unit of crack face created (2γ), or

$$dW = dA \cdot 2\gamma \tag{2}$$

In reality, the term 2γ is inappropriate because most materials exhibit considerable plasticity around the advancing crack tip. Thus, the 2γ term (which may be on the order of 0.01 to 1 J/m^2) is replaced by G_c (usually 0.1 to 100 kJ/m^2), where G_c is the total energy absorption (both plastic and surface) associated with crack growth.

Plasticity accompanying composite delamination has been studied by Hunston et al. [26]. In order to correlate the Mode I laminate and matrix G_c values, they tested numerous brittle and toughened laminate systems. Brittle matrix-based composites showed a threefold increase in G_{Ic}, while toughened matrix-based composites showed a 1 J/m^2 laminate G_c increase for every 3 J/m^2 increase in the matrix G_c. They attributed this effect to the fibers restricting the size of the plastic zone, thus reducing the expected toughness of the material.

Brittle matrices also exhibit considerable plasticity. Evidence of shear yielding of the (brittle) matrix of AS4/3501-6 fracture surfaces was detected by Bascom et al. [27]. After fracture, heating to a temperature above the matrix, T_g, caused a strain relaxation on the order of 30 to 50%.

Previous Fractographic Studies

SEM studies of delaminated specimens by Morris [28] showed that characteristic features (which he referred to as "hackles") existed at a tilt angle of 45° that was correlated with the fracture direction. Richards-Frandsen and Naerheim [29] noted that the hackle direction coincided with the direction of crack propagation for graphite/epoxy three-point-bend fatigue specimens, and that the spacing and size of hackles were related to the distance between graphite fibers (the closer the fibers, the denser the hackles). This was confirmed by Corleto and Bradley [6] with real-time observations of Mode II delamination showing the formation of hackles and the subsequent tilting process.

Garg [30] examined Mode I and Mode II fracture (using double-cantilever beam (DCB), compact tension (CT), center-notched tension (CNT), and cracked-lap shear (CLS) tests) in graphite/epoxy laminates. Mode I delamination was associated with resin-rich zones and matrix-coated fibers, while Mode II delamination was dominated by clean fibers and raised epoxy between fibers. For a brittle epoxy laminate, Arcan et al. [31] showed that fiber debonding and bridging (broken fibers) along with matrix-cracking-characterized Mode I failure. On the other hand, hackles were prevalent for Mode II failure. A toughened epoxy exhibited more ductile behavior.

Johannesson and Blikstad [32] studied the effects of mixed-mode loading on angle-ply composites. They found that G_c was strongly dependent on the ratio between G_I and G_{II}, with the variation caused by different levels of shear hackling.

Fractographic studies of Mode III specimens have been rare, and most of the Mode III studies have focused on a quantitative assessment of G_{IIIc}. Some examples should be noted. Becht and Gillespie [9] witnessed hackle-like formation, but not of the same type as in a Mode II specimen. Lee [11] stated that the fracture surfaces of Mode III specimens were rough, and exhibited "hackle-like" formations, as well as river markings. Finally, Sharif et al. [12] noted unique "shear crevices" in their Mode III specimens.

Delamination Between Off-Axis Plies

It is suggested that the standard methods of measuring delamination resistance are really only useful as ranking methods, since cracks running at an angle other than 0° with respect to the fiber direction may result in a different fracture surface, and hence might be expected to result in a different value of G_c. It should be noted that a recent study by Chou et al. [33] showed that G_{IIc} did not change for delaminations propagating between off-axis plies, although G_{Ic} increased slightly as the orientation of the fibers inclined towards the delamination direction.

Figure 1 illustrates a delamination propagating between two 0° plies. In this case, shear hackles are created if the delamination is propagating in Mode II; for Mode III propagation, shear crevices are generated. For a delamination propagating between two 90° plies, Mode II delamination would be expected to produce shear crevices, and Mode III would produce shear hackles. The initial premise of this study was that the values of G_{IIc} and G_{IIIc} should be reversed for delaminations propagating between 90° plies, since the surface features are essentially reversed. In contrast, observations of Mode I delamination surfaces show no effect of ply orientation, and hence it might be expected that G_{Ic} would be independent of ply orientation.

In this study, these ideas have been examined for delaminations propagating at 0°/0°, 90°/90°, and 0°/90° interfaces in all three modes.

Experimental

Specimen Fabrication

Twenty-four ply samples were fabricated from overaged AS4/3501-6 prepreg (originally supplied by Hercules) with a fiber content of 62%. Samples were fabricated from properly stored prepreg that was between 4 and 18 months past its expiry date. A previous study on

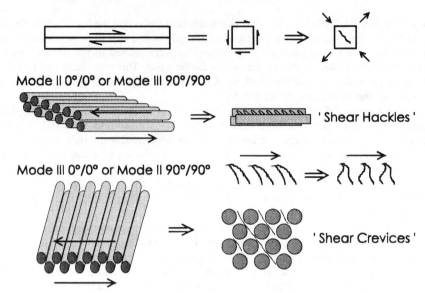

FIG. 1—*Formation of shear hackles and shear crevices in laminates.*

overaged prepreg by Ginty and Chamis [34] showed that no degradation in performance should be expected for this degree of overaging.

The laminate was debulked during layup after every 6 to 8 plies. Samples were cured in accordance with the manufacturer's recommended curing cycle and were cut using a diamond-tipped rotary saw.

Artificial delaminations were created by placing a 12.7-μm thick Teflon film between the 12th and 13th plies. The shape of the insert varied depending on the delamination interface and the mode of fracture for which the laminate was to be used. The 0°/0° laminates had conventional inserts, however, the 90°/90° and 0°/90° laminates incorporated the edge delamination developed by Robinson and Song [35] (see Fig. 2) in order to constrain the delamination to the plane of the insert.

Unfortunately, it was not possible to create both symmetric laminates and symmetric sublaminates for tests in which the delamination was constrained to a single 0°/90° interface. To avoid the difficulties inherent in tracking the advance of two delaminations, an attempt was made to balance each sublaminate (delaminated half of the specimen thickness) and to balance the overall laminate as closely as possible. The layup chosen was [0/90/0$_8$/90/0//90/0$_{10}$/90], which is slightly unsymmetric. The strain energy release rate generated by residual stresses for this laminate is expected to be a small fraction of the overall strain energy release rate. All layups used are shown in Table 1.

Mode I Tests

Mode I tests were performed using the DCB specimen. The dimensions of each specimen were 125 by 20 by 3 mm as shown in Fig. 3. Specimens were loaded at a constant crosshead speed of 0.5 mm/min, with the load being applied via hinges that were bonded onto the delaminated end of each arm of the specimen with a 5-min cure epoxy. Specimen edges were painted with a white typewriter correction fluid, so that the extent of crack propagation could be readily determined. Load and displacement were recorded at standard amounts of crack

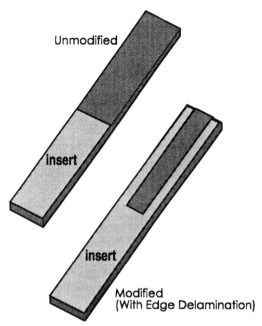

FIG. 2—*Modified (with edge delamination) and unmodified specimen insert configurations.*

TABLE 1—*Specimen layup.*

	$0°/0°$ Interface	$90°/90°$ Interface	$0°/90°$ Interface
Mode I	$[0_{12}//0_{12}]$	$[90/0_{10}/90//90/0_{10}/90]$	$[0/90/0_8/90/0//90/0_{10}/90]$
Mode II	$[0_{12}//0_{12}]$	$[90/0_{10}/90//90/0_{10}/90]$	$[0/90/0_8/90/0//90/0_{10}/90]$
Mode III	$[0_{12}//0_{12}]$	$[90/0_{10}/90//90/0_{10}/90]$	$[0/90/0_8/90/0//90/0_{10}/90]$

advance (0, 1, 2, 3, 4, 5, 10, 15, 20, and 25 mm), and the initiation point was defined as the point of deviation of the load-displacement curve from linearity.

Mode II Tests

End-notched flexure (ENF) tests were used to characterize Mode II delamination. The dimensions of each specimen were 150 by 20 by 3 mm (see Fig. 4). Once again, specimens were loaded at a constant crosshead speed of 0.5 mm/min until a delamination occurred. For each specimen, a compliance calibration was performed by shifting the specimen to vary the crack length, a, in the fixture (where a = 0, 15, 20, 25, 30, 35, and 40 mm). Load and displacement were recorded at the onset of delamination using the peak load and the corresponding displacement as the initiation point.

Mode III Tests

Mode III tests were accomplished using the modified split-cantilever-beam (MSCB) m of Sharif et al. [*12,36*] and Cicci et al. [*37*]. Specimen dimensions are shown in Fig. 5.

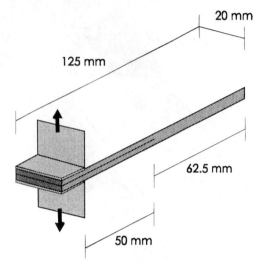

FIG. 3—*Double-cantilever-beam specimen geometry.*

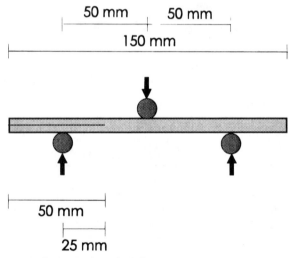

FIG. 4—*End-notched flexure specimen geometry.*

the Mode II tests, a compliance calibration was performed for each specimen by shifting the specimen in the fixture to vary the crack length (a = 35, 40, 45, 50, 55, 60, 65, and 70 mm), after which the specimen was loaded until the delamination propagated. The initiation point was defined as the point corresponding to the maximum load and the corresponding displacement.

Fractography

Delaminated specimens of interest were sputter coated with a 10 Å thick layer of gold and examined in a scanning electron microscope at 20 kV.

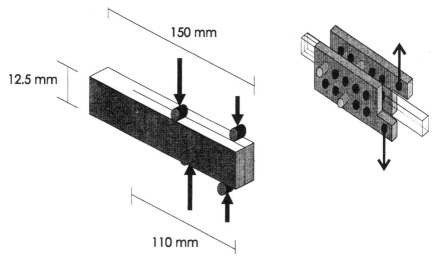

FIG. 5—*Modified split-cantilever-beam specimen geometry.*

Data Reduction

Mode I data were reduced using the modified-beam theory (MBT) method according to ASTM Test Method for Mode I Interlaminar Fracture Toughness of Unidirectional Fiber-Reinforced Polymer Matrix Composites (D 5528-94A). This method is considered to be the most conservative of the Mode I data reduction methods and is therefore favored by most researchers. The MBT method utilizes the load (P), displacement (d), specimen width (b), crack length (a), and a special parameter, χh, the x-intercept of a plot of the cube root of the specimen compliance versus crack length. The extra parameter is designed to account for rotation at the delamination front by simulating a slightly larger delamination than actually exists. For the purposes of this study, initiation values of G_{Ic} were used in order to minimize the effects of fiber bridging. The MBT equation is

$$G_{Ic} = (3P\delta)/[2b(a + \chi h)] \tag{3}$$

Mode II data were reduced using the experimental compliance calibration (ECC) method. This method is more direct than the direct beam theory and corrected beam theory methods, and was therefore favored for this study. The ECC method uses the load (P), specimen width (b), crack length (a), and m, the slope of a plot of the specimen compliance versus the cube root of the crack length. The value of m is obtained by shifting the specimen in the grips and obtaining load-displacement plots in the elastic range. The ECC equation is

$$G_{IIc} = (3ma^2P^2)/(2b) \tag{4}$$

Finally, Mode III data were reduced using the beam theory (BT) method. This method was favored over experimental compliance calibration-based methods that were found to be inadequate in yielding accurate G_{IIIc} values. The MSCB specimen's very low compliance compared to that of the DCB and the ENF specimens resulted in a great deal of scatter in the data of compliance-based methods. Cicci et al. [37] found that a high compliance cannot be obtained for the MSCB specimen since the specimen arms are deformed in the plane of the laminate. Also, dC/da plots obtained using compliance-based methods were nonlinear, invalidating the main assumption associated with these methods.

The BT method used in this study (data reduction methodology is described in more detail by Cicci et al. [37]) uses the specimen load (P), specimen width (b), and the rate of chance of specimen compliance versus crack length (dC/da) that is obtained directly from BT equations and not from experimental data. The BT equation is

$$G_{IIIc} = (P^2/2b)(dC/da) \qquad (5)$$

For their edge delamination specimens, Robinson and Song [35] used b values in their data reduction formula that were simply taken to be one half their normal value. So, for a specimen with a width of 20 mm and an edge delamination of 10 mm (5 mm on either side), the value of b would equal 10 mm. This was done when reducing data for these experiments as well.

Results and Discussion

Calculated results for G_c (along with standard deviations) are summarized in Table 2 and are presented graphically in Fig. 6. For all three modes, the initiation value of G_c was reported to avoid the possible effect of fiber bridging.

TABLE 2—*Numerical results.*

	0°/0° Interface		0°/90° Interface		90°/90° Interface	
	G_c, J/m²ᵃ	No. of Tests	G_c, J/m²ᵃ	No. of Tests	G_c, J/m²ᵃ	No. of Tests
Mode I	128 ± 40.2	8	93 ± 40.6	12	102 ± 28.7	12
Mode II	653 ± 69.1	12	520 ± 35.0	12	388 ± 44.0	12
Mode III	676 ± 174.2	12	277 ± 111.6	12	356 ± 55.0	12

ᵃ± limits are standard deviations.

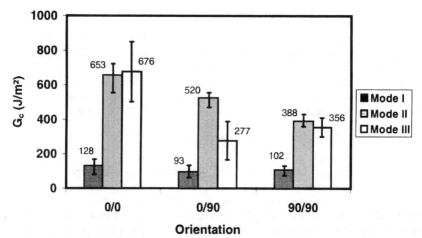

FIG. 6—G_c *distribution: mode and orientation.*

The value of G_{Ic} for the 0°/0° specimens (128 ± 28.1 J/m^2) was slightly higher than the 0°/90° and 90°/90° specimens (103 ± 39.7 and 93 ± 39.4 J/m^2, respectively). This value was within the range normally reported in the literature for AS4/3501-6. Student t-tests showed that the difference between the 0°/0° specimen results and the 90°/90° specimen results was significant at the 95% confidence level but not significant at the 96% confidence level. Figure 7, which is a plot of the Mode I values with 95% confidence bars around the mean, shows that the difference between the G_{Ic} values is small, and therefore they may be taken to be identical.

Mode II G_c values ranged from 653 ± 69.1 J/m^2 for the unidirectional 0°/0° specimens to 388 ± 44.0 J/m^2 for the 90°/90° specimens. The unidirectional value is similar to the value of 605 J/m^2 reported by Russell and Street [8] for the similar material, AS1/3501-6. Interestingly, the value of G_{IIc} for the 0°/90° specimen was exactly halfway between those of the 0°/0° and 90°/90° specimens. This would be expected if the energy absorption was confined to the plies on either side of the interface. Some crack jumping was observed for both the 0°/90° and the 90°/90° specimens; however, close examination of the delaminated specimens revealed that the first part of delamination growth for all specimens was in the plane of the starter crack. The t-test results show that all three values were distinct, with virtually 100% certainty.

Values for G_{IIIc} exhibited neither of the trends that the Mode I and Mode II values did. While G_{IIIc} was high for the 0°/0° specimens (676 ± 174.2 J/m^2), it was low for the 90°/90° specimens (356 ± 55.0 J/m^2). The initial hypothesis was that the value of G_{IIIc} for the 90°/90° specimen would be equivalent to the value of G_{IIc} for a 0°/0° specimen, since fractography reveals that the dominant surface feature is shear hackles in both cases. This was clearly not the case, and calculations of plastic zone radii discussed later may account for the discrepancy. The corresponding comparison of G_{IIIc} for a 0°/0° specimen and G_{IIc} for the 90°/90° specimen also failed, although both delamination surfaces displayed the shear crevices first documented by Sharif et al. [12]. The t-test results for these "critical combinations" of mode and orientation showed no statistically significant correlation between the numerical results.

The value of G_{IIIc} for the 0°/90° specimen was actually lower than those of "both" the 0°/0° and 90°/90° specimens, indicating that either poor bonding between the 0° and the 90° plies had occurred, or that some more complex mechanism of energy absorption was involved.

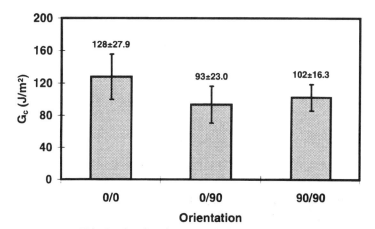

FIG. 7—G_{Ic} distribution and 95% confidence.

Fractographic Results

Fractographs of Mode I specimens are shown in Figs. 8 and 9. Both fractographs, one of which is a 0°/0° specimen (Fig. 8), and the other being a 90°/90° specimen (Fig. 9), show general features associated with Mode I delamination, including lifted fibers, serrations on the fibers indicative of tensile cracks, and cleaved matrix. The similarity of the two surfaces suggests that the Mode I interlaminar fracture toughness is independent of interface ply orientation. This is consistent with the experimental values of G_{Ic} that show little variation at the various ply orientations.

A fractograph of a Mode II 0°/0° delamination specimen (Fig. 10) exhibits the well-documented shear hackle and matrix damage features associated with delamination. When the Mode II 90°/90° fractograph is examined (also shown in Fig. 11), the features, which are visible from the edge, consist of lifted fibers, general matrix damage, and "shear crevices." Since the G_{IIc} value is greater for the 0°/0° specimen, it appears that the formation of the smaller but higher density hackles is the dominant energy absorbing process, while the formation of the less dense but longer shear crevices requires less energy. Figures 12 and 13 are fractographs of Mode II 0°/90° specimens, and clearly show shear crevices penetrating into the matrix. Fortunately, the matrix of the upper ply (which has been hackled) still remains, offering a striking contrast between the geometry of hackles (top of fractograph) and crevices. As stated earlier, the value of G_c for the 0°/90° specimens lies halfway between that obtained for the 0°/0° and the 90°/90° specimens. Since the 0°/90° specimen fractographs exhibit both hackles and crevices, this result would be expected, if the assumption is made that fractographs of specimen surfaces can be tied to the corresponding values of G_c.

Fractographs of the fracture surfaces of Mode III specimens are shown in Figs. 14 and 15. For the Mode III 0°/0° specimen (Fig. 14), shear crevices are evident across the edge of the specimen (similar to the Mode II 90°/90° specimens), which validates the assumption that

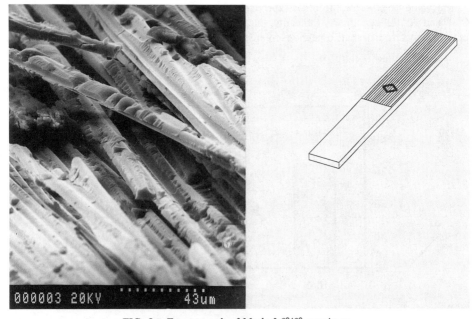

FIG. 8—*Fractograph of Mode I 0°/0° specimen.*

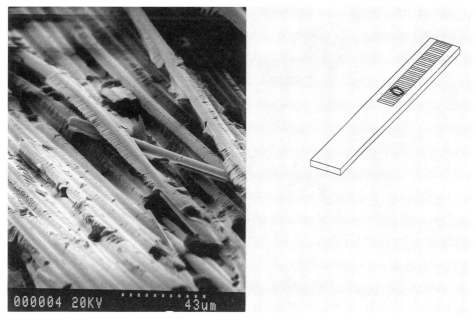

FIG. 9—*Fractograph of Mode I 90°/90° specimen.*

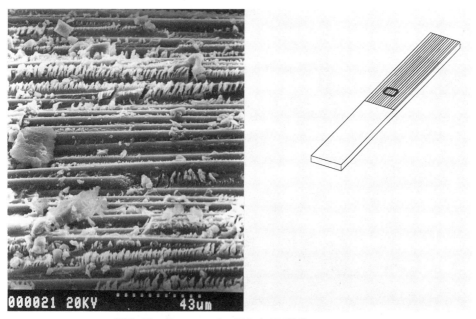

FIG. 10—*Fractograph of Mode II 0°/0° specimen.*

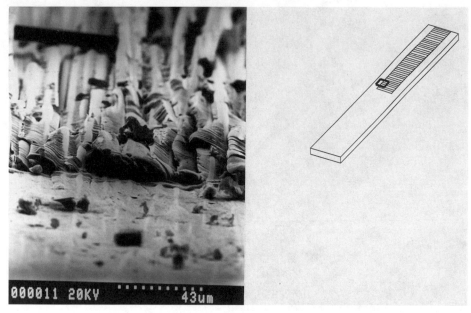

FIG. 11—*Fractograph of Mode II 90°/90° specimen.*

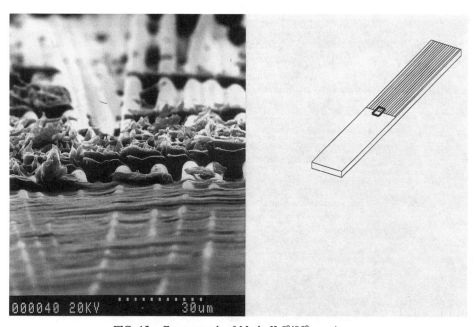

FIG. 12—*Fractograph of Mode II 0°/90° specimen.*

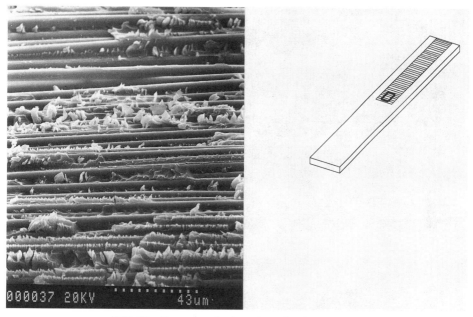

FIG. 13—*Fractograph of Mode II 0°/90° specimen.*

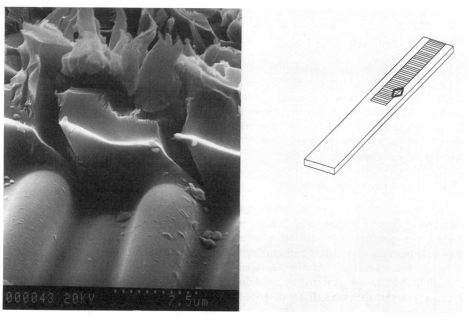

FIG. 14—*Fractograph of Mode III 0°/0° specimen.*

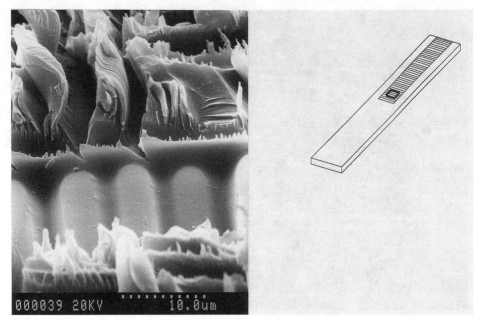

FIG. 15—*Fractograph of Mode III 90°/90° specimen.*

Mode II 90°/90° delamination and mode III 0°/0° delamination share a similar fracture mechanism. Figure 15 shows the fracture surface of a Mode III 90°/90° specimen, which exhibits shear hackles similar to the Mode II 0°/0° specimen. Unfortunately, the corresponding G_{IIIc} values do not correlate with the values obtained for the Mode II specimens, even though the fracture surfaces do.

Finally, in Fig. 16, the Mode III 90°/90° specimen fracture surface near one of the edge delaminations is illustrated. It is known that the G_{III} component decays to zero near this edge and that the G_{II} component dominates [12]. This fractograph shows a clear transition from the Mode III 90°/90° morphology to a smooth morphology. It is not clear from this fractograph, but the dominance of Mode II can be confirmed in this specimen by the presence of shear crevices in this region. For a Mode II 0°/0° specimen, the surface features in the center and near the edge are reversed.

Plasticity and the Plastic Zone Size

The observations of fracture surface morphology seem to be unable to fully account for the variation of G_c with fracture mode and interface ply orientation. The evidence suggests that there is a fair amount of subsurface deformation contributing to the total energy absorption. This deformation may consist of both crazing and plastic yielding. Although epoxy is considered to be brittle in tension and has a tensile failure strain of approximately 1.7%, there is actually a much higher strain represented by the deformation of the shear hackles in Figs. 10 and 13. It is assumed that the carbon fibers do not absorb any energy through deformation or friction with the epoxy.

As a preliminary calculation, assume that all of the epoxy between the layer of fibers on either side of the delamination is deformed to failure. Using the fiber matrix arrangement shown in Fig. 17, the area represented by the triangle is

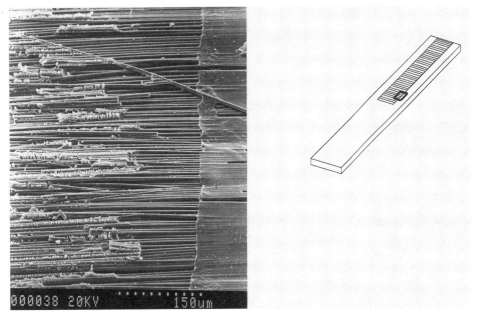

FIG. 16—*Fractograph of Mode III 90°/90° specimen at edge.*

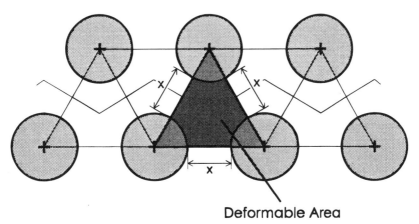

Deformable Area

FIG. 17—*Approximation of fiber and matrix areas.*

$$A_T = 0.5(2r + x)^2\cos 30° \tag{6}$$

where x is the spacing between the fibers, while the area represented by the fibers themselves is

$$A_f = 0.5\pi r^2 \tag{7}$$

If the fiber volume fraction, defined as A_f/A_T, is assumed to be 0.62 (from the manufacturer's data sheet), and the radius of a fiber is 3 μm (from fractographs), then a value for x can be obtained. This value is 1.26 μm, although it could be larger if there is a resin-rich region between the plies. If G_c is defined as the energy absorbed per volume of material multiplied by the volume of material deformed per unit area of new crack (assuming failure occurs only in the matrix), then

$$G_c = (W_{epoxy}/V)(V/A) = (W_{epoxy}/V) \cdot t \tag{8}$$

where t is the thickness of the deformed layer, and W_{epoxy}/V is the area under the stress-strain curve

$$t = (2r + x)\cos 30° \tag{9}$$

$$W_{epoxy}/V = 0.5 \ \sigma_y \epsilon_y \tag{10}$$

Using the values of $\epsilon_f = 0.017$ and $\sigma_f = 69$ MPa, a value of $G_c = 1.87$ J/m^2 is obtained. This number represents the value of G_c if the epoxy exhibited virtually no plasticity. This compares to a value of $G_c = 128$ J/m^2 for a Mode I 0°/0° laminate, indicating that there must be considerable plasticity occurring around the crack tip even for a "simple" delamination in Mode I. The surface energy of the epoxy is on the order of a few mJ/m^2 and could not account for the discrepancy, even if a large ratio of actual-to-nominal surface area is taken into account. Unfortunately, accurate values for the shear yield stress and strain for the 3501-6 matrix are not available in the literature and are rather difficult to measure.

A second calculation that may prove to be of some value is a calculation of the radius of the plastic zone (r_p), obtained from the stress intensity factor. Unfortunately, a conventional approach to obtaining r_p is not suitable because of the anisotropic nature of the laminate. For anisotropic plates, Sih et al. [38] presented equations relating the critical strain energy release rate to the critical stress intensity factor, and in turn relating the critical stress intensity factor to the stresses in the material around the crack tip (refer to the Appendix). Using the values of the crack-tip stresses together with a suitable failure criterion, a crude estimate of the radii of the plastic zones may be determined for all three modes.

Table 3 shows the values of r_p (in terms of the number of fiber layers obtained from Eq 9) using a simple Tresca criterion, equating the maximum shear stress *in the epoxy* to a shear yield stress of either 50 or 100 MPa. The values of r_p are on the order of one fiber layer for Mode I deformation, and hence it would be reasonable to tie a fractograph of delamination surface morphology directly to the value of G_c. These values are in accordance with those obtained by Crews et al. [39], who found that their thickness of the plastic zone was less than the thickness of the interlaminar resin-rich layer. On the other hand, Parker and Yee [40] noted a plastic zone that was several fiber layers for their rubber modified polycarbonate.

TABLE 3—*Number of fiber layers of the plastic zone.*

Fail	100 MPa			50 MPa		
	Mode I	Mode II	Mode III	Mode I	Mode II	Mode III
0°/0°	1	6	8	4	25	32
90°/90°	1	35	4	3	140	17

For the 0°/0° interface, the calculated values of r_p for Mode II and Mode III are much higher than those calculated for Mode I, and this combined with the higher values of G_{IIc} and G_{IIIc} leads to the conclusion that a large amount of the energy absorption is occurring below the surface. There is also surface friction and other effects to consider, but it seems likely that subsurface deformation must play a significant role in Modes II and III.

The value of r_p for the Mode II 90°/90° interface was 5 to 6 times higher than that of the Mode II 0°/0° interface. For the 0°/0° interface, the value of τ_{xy} was used in the Tresca criterion since this represents the maximum shear stress in the matrix for this layup. The maximum resolved shear stress in the x-y plane was used for the laminates with the 90°/90° interface. The elevated values of r_p were not reflected in the values of G_c for the 90°/90° laminates. These calculations, however, are extremely crude, since they use continuum mechanics to predict stresses at a scale where the continuum approximation is clearly invalid. They have been used to explore the potential significance of plastic deformation, but could not be used in any sort of quantitative modeling work.

Conclusions

The original premise of the work—that the fracture surface morphology could be quantitatively linked to the critical strain energy release rate—must be modified. Subsurface deformation and crazing is necessary to account for the relatively large values of G_c in Modes II and III, and hence some means of quantifying these mechanisms is needed if the dependence of G_c on fracture mode is to be explained from a fundamental perspective. From an engineering design point of view, this type of modeling is unnecessary: it is sufficient to measure G_c and to use it as a material property, but this can only be done if both the interfacial ply orientations and the mode of fracture is identical to the end use situation. Since real failures often occur under the influence of mixed modes, a complete data set for a particular material would become rather large. A fundamental understanding of the origin of toughness would greatly reduce the experimental effort needed to characterize a material. This approach should be pursued, in spite of the limited success of this preliminary study.

Acknowledgments

The authors wish to acknowledge the financial support of the National Sciences and Engineering Research Council. We would also like to thank Mr. Wei Ding for his comments on the manuscript.

APPENDIX

Calculation of Plastic Zone Radii

From Sih et al. [38], K_c for an anisotropic plate is defined as

$$K_c^2 = E_{\text{eff}}G_c \tag{11}$$

where E_{eff} is the effective modulus. E_{eff} is defined as

$$E_{eff} \text{ (Mode I)} = ((a_{11}a_{22})/2)^{-0.5}[(a_{22}/a_{11})^{0.5} + ((2a_{12} + a_{66})/2a_{11})]^{-0.5}$$

$$E_{eff} \text{ (Mode II)} = (2^{0.5}/a_{11})[(a_{22}/a_{11})^{0.5} + ((2a_{12} + a_{66})/2a_{11})]^{-0.5} \qquad (12)$$

$$E_{eff} \text{ (Mode III)} = 2(c_{44}c_{55})^{0.5}$$

where the a_{ij} values are values of the compliance matrix, and the c_{ij} values are defined in Ref 38. For plane strain, the a_{ij} values may be replaced by b_{ij} values, where b_{ij} is defined as

$$b_{ij} = a_{ij} - ((a_{i3}a_{j3})/(a_{33})) \qquad (13)$$

For unidirectional laminates, the a_{ij} values are defined as

$$a_{11} = 1/E_{11}, \qquad a_{22} = 1/E_{22}, \qquad a_{33} = 1/E_{33}, \qquad a_{44} = 1/G_{23}, \qquad a_{55} = 1/G_{31}, \quad (14)$$

$$a_{66} = 1/G_{12}, \qquad a_{12} = -v_{12}/E_{11}, \qquad a_{13} = -v_{13}/E_{11}, \qquad a_{23} = -v_{23}/E_{22}$$

These values may also be used for the 0°/90° and 90°/90° specimens, since they are composed of predominately unidirectional plies.

Substituting the material properties allows for a determination of K_c. Next, the stresses may be solved. The stresses are defined as [41]

Mode I

$$\sigma_x = (K_I/(2\pi r)^{0.5}) \, \text{Re}[(s_1 s_2)/(s_1 - s_2) * [(s_2/(\cos\theta + s_2\sin\theta)^{0.5}) - (s_1/(\cos\theta + s_1\sin\theta)^{0.5})]]$$

$$\sigma_y = (K_I/(2\pi r)^{0.5}) \, \text{Re}[(1)/(s_1 - s_2) * [(s_1/(\cos\theta + s_2\sin\theta)^{0.5}) - (s_2/(\cos\theta + s_1\sin\theta)^{0.5})]]$$

$$\tau_{xy} = (K_I/(2\pi r)^{0.5}) \, \text{Re}[(s_1 s_2)/(s_1 - s_2) * [(1/(\cos\theta + s_1\sin\theta)^{0.5}) - (1/(\cos\theta + s_2\sin\theta)^{0.5})]]$$

Mode II

$$\sigma_x = (K_{II}/(2\pi r)^{0.5}) \, \text{Re}[(1)/(s_1 - s_2) * [(s_2^2/(\cos\theta + s_2\sin\theta)^{0.5}) - (s_1^2/(\cos\theta + s_1\sin\theta)^{0.5})]]$$

$$\sigma_y = (K_{II}/(2\pi r)^{0.5}) \, \text{Re}[(1)/(s_1 - s_2) * [(1/(\cos\theta + s_2\sin\theta)^{0.5}) - (1/(\cos\theta + s_1\sin\theta)^{0.5})]]$$

$$\tau_{xy} = (K_{II}/(2\pi r)^{0.5}) \, \text{Re}[(1)/(s_1 - s_2) * [(s_1/(\cos\theta + s_1\sin\theta)^{0.5}) - (s_2/(\cos\theta + s_2\sin\theta)^{0.5})]]$$

$$(15)$$

Mode III

$$\tau_{xz} = -(K_{III}/(2\pi r)^{0.5}) \, \text{Re}[(s_3)/(\cos\theta + s_3\sin\theta)^{0.5}]$$

$$\tau_{yz} = (K_{III}/(2\pi r)^{0.5}) \, \text{Re}[(1)/(\cos\theta + s_3\sin\theta)^{0.5}]$$

where the values of s_1, s_2, and s_3 are the roots of the characteristic equation

$$a_{11}\mu^4 - 2a_{16}\mu^3 + (2a_{12} + a_{66})\mu^2 - 2a_{26}\mu + a_{22} = 0 \qquad (16)$$

For orthotropic materials, $a_{16} = a_{26} = 0$, and Eq 16 now becomes

$$a_{11}\mu^4 + (2a_{12} + a_{66})\mu^2 + a_{22} = 0 \qquad (16a)$$

Lekhnitskii showed that the roots of the equation (μ_i) would either be complex or purely imaginary and would have the form

$$\mu_x = s_x = \alpha_x + i\beta_x \qquad (17)$$

Before solving Eq 16a, the real component in the stress equations (Eq 15) may be extracted by substitution of Eq 17, expanding, and simplifying. The result (with $\alpha_x = 0$, Eq 19) is Mode I

$$\sigma_x = -(K_I/(4\pi r)^{0.5})[(\beta_1\beta_2)/(\beta_1 - \beta_2) * (\beta_2^{0.5} - \beta_1^{0.5})]$$

$$\sigma_y = (K_I/(4\pi r)^{0.5})[(1)/(\beta_1 - \beta_2) * (\beta_1/\beta_2^{0.5} - \beta_2/\beta_1^{0.5})]$$

$$\tau_{xy} = (K_I/(4\pi r)^{0.5})[(\beta_1\beta_2)/(\beta_1 - \beta_2) * (\beta_1^{-0.5} - \beta_2^{-0.5})]$$

Mode II

$$\sigma_x = (K_{II}/(4\pi r)^{0.5})[(1)/(\beta_1 - \beta_2) * (\beta_2^{1.5} - \beta_1^{1.5})]$$

$$\sigma_y = -(K_{II}/(4\pi r)^{0.5})[(1)/(\beta_1 - \beta_2) * (\beta_2^{-0.5} - \beta_1^{-0.5})]$$

$$\tau_{xy} = (K_{II}/(4\pi r)^{0.5})[(1)/(\beta_1 - \beta_2) * (\beta_1^{0.5} - \beta_2^{0.5})] \qquad (17a)$$

Mode III

$$\tau_{xz} = (K_{III}/(4\pi r)^{0.5})[\beta_1^{0.5}]$$

$$\tau_{yz} = -(K_{III}/(4\pi r)^{0.5})[\beta_1^{-0.5}]$$

Solving Eq 16a gives

$$\mu_1 = -[-2a_{12} - a_{66} + (4a_{12}^2 - 4a_{11}a_{22} + 4a_{12}a_{66} + a_{66}^2)^{0.5}]^{0.5}/(2a_{11})^{0.5}$$

$$\mu_2 = -[-2a_{12} - a_{66} - (4a_{12}^2 - 4a_{11}a_{22} + 4a_{12}a_{66} + a_{66}^2)^{0.5}]^{0.5}/(2a_{11})^{0.5} \qquad (18)$$

$$\mu_3 = -\mu_1$$

$$\mu_4 = -\mu_2$$

Substituting the values of a from Eqs 13 and 14 gives

$$\mu_1 = 0 + 0.925462i$$

$$\mu_2 = 0 + 4.2513i$$

$$\mu_3 = 0 - 0.925462i$$

$$\mu_4 = 0 - 4.2513i \qquad (19)$$

The values of μ may now be substituted directly into Eq 15a to obtain the stresses in all three modes. These stresses may now be used to calculate the principal stresses, σ_{max}, σ_{min}, and

τ_{max}. Once a failure criterion has been established (say, for example, when the maximum shear stress is 100 MPa), a value of r $(=r_p)$ may be calculated. The values of σ_{max}, σ_{min}, and τ_{max} can be found using the equations

$$\sigma_{max} = \sigma_1 = (\sigma_x + \sigma_y)/2 + [((\sigma_x - \sigma_y)/2)^2 + \tau_{xy}^2]^{0.5} \qquad (20)$$

$$\sigma_{min} = \sigma_2 = (\sigma_x + \sigma_y)/2 - [((\sigma_x - \sigma_y)/2)^2 + \tau_{xy}^2]^{0.5} \qquad (21)$$

$$\tau_{max} = [((\sigma_x - \sigma_y)/2)^2 + \tau_{xy}^2]^{0.5} \qquad (22)$$

References

[1]   Chai, H., "The Characterization of Mode I Delamination Failure in Non-Woven Multidirectional Laminates," *Composites*, Vol. 15, No. 4, Oct. 1984, pp. 277–290.
[2]   Devitt, D. F., Schapery, R. A., and Bradley, W. L., "A Method for Determining the Mode I Delamination Fracture Toughness of Elastic and Viscoelastic Composite Materials," *Journal of Composite Materials*, Vol. 14, Oct. 1980, pp. 270–285.
[3]   Lee, S., Gaudert, P. C., Dainty, R. C., and Scott, R. F., "Characterization of the Fracture Toughness Property (G$_{Ic}$) of Composite Laminates Using the Double Cantilever Beam Specimen," *Polymer Composites*, Vol. 10, No. 5, Oct. 1989, pp. 305–312.
[4]   O'Brien, T. K. and Martin, R. H., "Round Robin Testing for Mode I Interlaminar Fracture Toughness of Composite Materials," *Journal of Composites Technology and Research*, Vol. 15, No. 4, Winter 1993, pp. 269–281.
[5]   Carlsson, L. A., Gillespie, J. W., Jr., and Pipes, R. B., "On the Analysis and Design of the End Notched Flexure (ENF) Specimen for Mode II Testing," *Journal of Composite Materials*, Vol. 20, Nov. 1986, pp. 594–604.
[6]   Corleto, C. R. and Bradley, W. L., "Mode II Delamination Fracture Toughness of Unidirectional Graphite/Epoxy Composites," *Composite Materials: Fatigue and Fracture, Second Volume, ASTM STP 1012*, P. Lagace, Ed., American Society for Testing and Materials, Philadelphia, 1989, pp. 201–221.
[7]   Zhou, J. and He, T., "On the Analysis of the End Notched Flexure Specimen for Measuring Mode II Fracture Toughness of Composite Materials," *Composites Science and Technology*, Vol. 50, 1994, pp. 209–213.
[8]   Russell, A. J. and Street, K. N., "The Effect of Matrix Toughness on Delamination: Static and Fatigue Fracture Under Mode II Shear Loading of Graphite Fiber Composites," *Toughened Composites, ASTM STP 937*, N. J. Johnson, Ed., American Society for Testing and Materials, Philadelphia, 1987, pp. 275–294.
[9]   Becht, G. and Gillespie, J. W., Jr. "Numerical and Experimental Evaluation of the Mode III Interlaminar Fracture Toughness of Composite Materials," *Polymer Composites*, Vol. 10, No. 5, 1989, pp. 293–304.
[10] Robinson, P. and Song, D. Q., "A New Mode III Delamination Test for Composites," *Advanced Composite Letters*, Vol. 1, No. 5, 1992, pp. 160–164.
[11] Lee, S. M., "An Edge Crack Torsion Method for Mode III Delamination Fracture Testing," *Journal of Composites Technology and Research*, Vol. 15, No. 3, Fall 1993, pp. 193–201.
[12] Sharif, F., Kortschot, M. T., and Martin, R. H., "Mode III Delamination Using a Split Cantilever Beam," *Composite Materials: Fatigue and Fracture (Fifth Volume), ASTM STP 1230*, R. H. Martin, Ed., American Society for Testing and Materials, Philadelphia, 1995, pp. 85–89.
[13] Hashemi, S., Kinloch, A. J., and Williams, G., "Mixed-Mode Fracture in Fiber-Polymer Composite Laminates," *Composite Materials: Fatigue and Fracture, 3rd Volume, ASTM STP 1110*, T. K. O'Brien, Ed., American Society for Testing and Materials, Philadelphia, 1991, pp. 143–168.
[14] Reeder, J. R. and Crews, J. H., Jr., "Mixed-Mode Bending Method for Delamination Testing," *AIAA Journal*, Vol. 28, No. 7, pp. 1270–1276.
[15] Benzeggagh, M. L., Davies, P., Gong, X. J., Roelandt, J. M., Mourin, M., and Prel, Y. J., "A Mixed Mode Specimen for Interlaminar Fracture Testing," *Composites Science and Technology*, Vol. 34, 1989, pp. 129–143.

[16] Reeder, J. R., "A Bilinear Failure Criterion for Mixed-Mode Delamination," *Composite Materials: Testing and Design, 11th Volume, ASTM STP 1206*, E. T. Camponeschi, Jr., American Society for Testing and Materials, Philadelphia, 1993, pp. 303–322.

[17] Smiley, A. J. and Pipes, R. B., "Rate Effects on Mode I Interlaminar Fracture Toughness in Composite Materials," *Journal of Composite Materials*, Vol. 21, July 1987, pp. 670–687.

[18] Rybicki, E. F., Hernandez, T. D., Jr., Deibler, J. E., Knight, R. C., and Vinson, S. S., "Mode I and Mixed Mode Energy Release Rate Values for Delamination of Graphite/Epoxy Test Specimens," *Journal of Composite Materials*, Vol. 21, Feb. 1987, pp. 105–123.

[19] Clark, G., Saunders, D. S., and Van Blaricum, T. J., "Moisture Absorption in Graphite/Epoxy Laminates," *Composites Science and Technology*, Vol. 39, 1990, pp. 355–375.

[20] De Wilde, W. P. and Frolkovic, P., "The Modelling of Moisture Absorption in Epoxies: Effects at the Boundaries," *Composites*, Vol. 25, No. 2, 1994, pp. 119–127.

[21] Martin, R. H. and Murri, G. B., "Characterization of Mode I and Mode II Delamination Growth and Thresholds in AS4/PEEK Composites," *Composite Materials: Testing and Design (Ninth Volume), ASTM STP 1059*, S. P. Garbo, American Society for Testing and Materials, Philadelphia, 1990, pp. 251–270.

[22] Sue, H.-J., Jones, R. E., and Garcia-Meitin, E. I., "Fracture Behaviour of Model Toughened Composites Under Mode I and Mode II Delaminations," *Journal of Materials Science*, Vol. 28, 1993, pp. 6381–6391.

[23] Yoon, S. H. and Hong, C. S., "Interlaminar Fracture Toughness of Graphite/Epoxy Composites Under Mixed-Mode Deformations," *Experimental Mechanics*, Sept. 1990, pp. 234–239.

[24] Law, G. E., "A Mixed-Mode Fracture Analysis of $(\pm 25/90_n)_s$ Graphite/Epoxy Composite Laminates," *Effects of Defects in Composite Materials, ASTM STP 836*, K. L. Reifsnider, Ed., American Society for Testing and Materials, Philadelphia, 1984, pp. 143–160.

[25] Jurf, R. A. and Pipes, R. B., "Interlaminar Fracture of Composite Materials," *Journal of Composite Materials*, Vol. 16, 1982, pp. 386–394.

[26] Hunston, D. L., Moulton, R. J., Johnston, N. J., and Bascom, W. D., "Matrix Resin Effects in Composite Delamination: Mode I Fracture Aspects," *Toughened Composites, ASTM STP 937*, N. J. Johnson, Ed., American Society for Testing and Materials, Philadelphia, 1987, pp. 74–94.

[27] Bascom, W. D., Boll, D. J., Fuller, B., and Phillips, P. J. "Fractography of the Interlaminar Fracture of Carbon-Fibre Epoxy Laminates," *Journal of Materials Science*, Vol. 20, 1985, pp. 3184–3190.

[28] Morris, G. E., "Determining Fracture Directions and Fracture Origins on Failed Graphite/Epoxy Surfaces," *Nondestructive Evaluation and Flow Criticality for Composite Materials, ASTM STP 696*, R. Pipes, Ed., American Society for Testing and Materials, Philadelphia, 1979, pp. 274–297.

[29] Richards-Frandsen, R. and Naerheim, Y., "Fracture Morphology of Graphite/Epoxy Composites," *Journal of Composite Materials*, Vol. 17, March 1983, pp. 105–113.

[30] Garg, A. C., "Interlaminar and Intralaminar Fracture Surface Morphology in Graphite/Epoxy Laminates," *Engineering Fracture Mechanics*, Vol. 23, No. 6, 1986, pp. 1031–1050.

[31] Arcan, L., Arcan, M., and Daniel, I. M., "SEM Fractography of Pure and Mixed-Mode Interlaminar Fractures in Graphite/Epoxy Composites," *Fractography of Modern Engineering Materials: Composites and Metals, ASTM STP 948*, Masters and Au, Eds., American Society for Testing and Materials, Philadelphia, 1987, pp. 41–67.

[32] Johannesson, T. and Blikstad, M., "Fractography and Fracture Criteria of the Delamination Process," *Delamination and Debonding of Materials, ASTM STP 876*, W. S. Johnson, Ed., American Society for Testing and Materials, Philadelphia, 1985, pp. 411–423.

[33] Chou, I., Kimpara, I., Kageyama, K., and Ohsawa, I., "Mode I and Mode II Fracture Toughness on Differently Oriented Interlaminae of Graphite/Epoxy Composites," *Composite Materials: Fatigue and Fracture (Fifth Volume), ASTM STP 1230*, R. H. Martin, Ed., American Society for Testing and Materials, Philadelphia, 1995, pp. 132–151.

[34] Ginty, C. A. and Chamis, C. C., "Fracture Characteristics of Angleplied Laminates Fabricated from Overaged Graphite/Epoxy Prepreg," *Fractography of Modern Engineering Materials: Composites and Metals, ASTM STP 948*, Masters and Au, Eds., American Society for Testing and Materials, Philadelphia, 1987, pp. 101–130.

[35] Robinson, P. and Song, D. Q., "A Modified DCB Specimen for Mode I Testing of Multidirectional Laminates," *Journal of Composite Materials*, Vol. 26, No. 11, 1992, pp. 1554–1577.

[36] Sharif, F., "A New Mode III Delamination Test for Composites," Master's thesis, University of Toronto, Toronto, Canada, 1994.

[37] Cicci, D., Sharif, F., and Kortschot, M. T., "Data Reduction for the Split Cantilever Beam Mode III Delamination Test," *Proceedings*, ACCM 10, Whistler, British Columbia, Canada, 14–18 Aug. 1995.

[38] Sih, G. C., Paris, P. C., and Irwin, G. R., "On Cracks in Rectilinearly Anisotropic Bodies," *International Journal of Fracture Mechanics*, Vol. 1, 1965, pp. 189–203.

[*39*] Crews, J. H., Jr., Shivakumar, K. N., and Raju, I. S. "A Fiber-Resin Micromechanics Analysis of the Delamination Front in a Double Cantilever Beam Specimen," *Phase Interaction in Composite Materials*, A. Paipetis and G. C. Papanicolaou, Eds., Omega Scientific, Oxon, UK, 1992, pp. 396–405.

[*40*] Parker, D. S. and Yee, A. F., "Factors Influencing the Mode I Interlaminar Fracture Toughness of a Rubber Toughened Thermoplastic Matrix Composite," *Journal of Thermoplastic Composite Materials*, Vol. 2, Jan. 1989, pp. 2–18.

[41] Gdoutos, E. E., *Fracture Mechanics Criteria and Applications*, Klower, Dordrecht, 1990.

Roderick H. Martin[1] and Peter L. Hansen[1]

Experimental Compliance Calibration for the Mixed-Mode Bending (MMB) Specimen

REFERENCE: Martin, R. H. and Hansen, P. L., "**Experimental Compliance Calibration for the Mixed-Mode Bending (MMB) Specimen,**" *Composite Materials: Fatigue and Fracture (Sixth Volume), ASTM STP 1285*, E. A. Armanios, Ed., American Society for Testing and Materials, 1997, pp. 305–323.

ABSTRACT: A novel method to control the mixed-mode bending (MMB) specimen is presented. By maintaining a constant opening, or Mode I, displacement rate, stable delamination growth is achieved for all mixed-mode ratios. A constant-opening displacement rate is achieved by attaching a second displacement transducer to the hinges of the specimen. The test machine is then controlled externally by the second displacement transducer. By achieving stable delamination growth and monitoring the opening displacement, an experimental compliance calibration may be derived for the Modes I and II parts. This new approach to control the test and determine the experimental compliance calibration, overcomes the potential inaccuracies of the previously used beam theory expressions to determine compliance and to separate the modal values of G.

KEYWORDS: fatigue (materials), fracture (materials), composite materials, mixed-mode delamination, interlaminar fracture, compliance calibration, end-notched flexure, double-cantilever beam

Nomenclature

a	Delamination length
A_0, A_1, A_2, A_3	Constants in ENF experimental compliance calibration expressions
b	Specimen width
c	Distance from center loading roller and attachment to test machine
C	Compliance
C_I, C_{II}	Modes I and II compliance, respectively
$E_{11}, E_{22}, G_{12}, \nu_{12}$	Elastic properties of the composite
G	Strain energy release rate
G_I, G_{II}, G_{tot}	Mode I, Mode II, and total values of strain energy release rate, respectively
G_{Ic}, G_{IIc}	Modes I and II interlaminar fracture toughness, respectively
h	Specimen half thickness
m	Slope of double-cantilever beam (DCB) compliance calibration curve
n	Exponent in Berry compliance calibration method
L	Semi span

[1]Head, Advanced Composites and senior test engineer, respectively, Materials Engineering Research Laboratory, Hertford, UK SG13 7DG.

P Load applied on lever arm by test machine
P_1, P_2 End and center point loads in finite element model, respectively
P_I, P_{II} Modes I and II components of load, respectively
δ Test machine displacement
δ_A, δ_B Center point displacements from beam rotation and flexure
δ_I, δ_{II} Modes I and II components of displacement, respectively
Δ_I, Δ_{II} Modes I and II modified beam theory correction factors, respectively

In structural configurations, delaminations are generally initiated and driven by multi-axial stresses resulting in mixed-mode delamination growth. The mixed-mode bending (MMB) [1–5] specimen allows the interlaminar fracture toughness to be evaluated for a wide variety of mixed-mode, I and II, ratios. Data may be reduced by a beam theory expression, Eqs 1 through 3 [1].

$$G_I = \frac{3a^2P^2}{4b^2h^3L^2E_{11}} (3c - L)^2 \tag{1}$$

$$G_{II} = \frac{9a^2P^2}{16b^2h^3L^2E_{11}} (c + L)^2 \tag{2}$$

$$G_{tot} = \frac{3a^2P^2}{16b^2h^3L^2E_{11}} [4(3c - L)^2 + 3(c + L)^2] \tag{3}$$

where the notation is defined in Fig. 1. These equations may be further extended to include the effects of shear deformation and beam rotation at the crack tip [1].

$$G_I = \frac{3P^2(3c - L)^2}{4b^2h^3L^2E_{11}} \left[a^2 + \frac{2a}{\lambda} + \frac{1}{\lambda^2} + \frac{h^2E_{11}}{10G_{13}} \right] \tag{4}$$

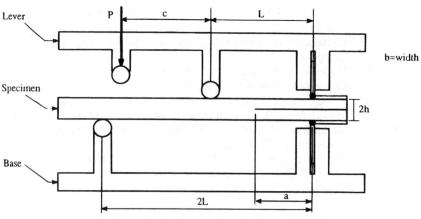

FIG. 1—*Notation for the mixed mode bending specimen.*

where

$$\lambda = \sqrt[4]{\frac{6E_{22}}{h^4E_{11}}} \tag{5}$$

and

$$G_{\text{II}} = \frac{9P^2(c + L)^2}{16b^2h^3L^2E_{11}}\left[a^2 + \frac{0.2h^2E_{11}}{G_{13}}\right] \tag{6}$$

The effects of geometric nonlinearities may also be included in these equations [3].

However, these theories may not always compare well with the actual deflections or compliances experienced in the experiments. To overcome the difference between beam theory expressions and experimental data in the double cantilever beam (DCB) specimen and the end-notched flexure (ENF) specimen, experimental compliance calibration expressions are determined by measuring the compliance, C, during the experiment for different delamination lengths, a. These experimental compliance expressions were then differentiated to determine G. For the DCB, two methods commonly used are (1) the Berry method where a curve fit is made to the experimental compliance data [6]

$$C = ma^n \tag{7}$$

and (2) the modified compliance calibration method. In this latter method, the measured delamination length is corrected by an amount, Δ_{I}, that accounts for end rotation and crack deflection. The value of Δ_{I} is the intercept on the x-axis of the $C^{1/3}$ versus a plot or [7]

$$C = \frac{8(a + \Delta_{\text{I}})^3}{bh^3E_{11}} = m(a + \Delta_{\text{I}})^3 \tag{8}$$

For the ENF test, a compliance calibration is obtained by measuring compliance of the specimen at different delamination lengths prior to testing to delamination growth. Different delamination lengths are achieved by moving the specimen in the fixture. This is necessary because delamination growth is unstable for the ENF and compliance cannot be measured at different delamination lengths during the test using the conventional machine control. Compliance for the ENF is usually expressed as a cubic polynomial of a, Eq 9

$$C = A_0 + A_1a + A_2a^2 + A_3a^3 \approx A_0 + A_3a^3 \tag{9}$$

The terms A_1 and A_2 are usually neglected. There is a need to develop a similar experimental compliance calibration expression for the Modes I and II part of the MMB test. This expression should take the form of the modified beam theory as for the DCB or the compliance calibration for the ENF such that the material properties are inherently accounted for as are deflection assumptions of the specimen.

For the MMB specimen, even under displacement control, delamination growth is stable only for certain delamination lengths that depend on the mixed-mode ratio, Fig. 2 [1]. For stable delamination growth at the higher Mode II ratios, the crack front must be close to the center loading roller making measurement of the delamination length difficult. Because the hinges are bonded onto the specimen, it is not possible to shift the specimen in the fixture to give different delamination lengths as in the ENF test. Removal and reapplication of the hinges is possible but is not a feasible practice. Hence, it is not possible to generate a compliance expression for all mixed-mode loading ratios using the conventional loading techniques.

In addition, using the conventional loading technique, it is not possible to extract the Mode I and II experimental compliances even for the mixed-mode ratios and delamination lengths that give stable delamination growth. One method of generating the compliance calibration techniques is to use the experimental compliance data from the DCB test for Mode I and the ENF or end-loaded shear (ELS) specimen for the Mode II [3]. The modified beam theory expressions, Eq 8, for the Mode I and II parts of G_{tot} may be expressed as [3]

$$G_I = \frac{3(a + \Delta_I)^2 P^2}{4b^2 h^3 L^2 E_{11}} (3c - L)^2 \tag{10}$$

$$G_{II} = \frac{9(a + \Delta_{II})^2 P^2}{16b^2 h^3 L^2 E_{11}} (c + L)^2 \tag{11}$$

$$G_{tot} = \frac{3[(a + \Delta_I)^2 + (a + \Delta_{II})^2]P^2}{16b^2 h^3 L^2 E_{11}} [4(3c - L)^2 + 3(c + L)^2] \tag{12}$$

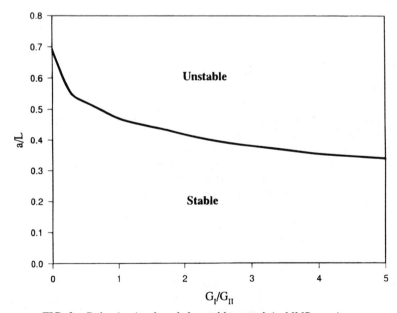

FIG. 2—*Delamination length for stable growth in MMB specimen.*

However, the values of Δ are generated as an average from separate DCB and ENF or ELS specimens and these values can vary from specimen to specimen. An alternate approach is to use a second displacement measurement, such as at the center point [5]. However, this will still not allow an experimental compliance calibration to be generated for situations where the delamination growth is unstable.

A method is required to determine the Mode I and II experimental compliance calibration expression for all mode ratios. This requires devising a control method that produces stable delamination growth for all modes and allows the individual compliances to be determined. This paper describes such a procedure. By maintaining a constant opening displacement rate rather than a constant machine displacement rate in the MMB test, delamination growth becomes stable as in the DCB specimen. By also measuring the opening displacement and the crosshead displacement, a compliance calibration for the Mode I and II component may be determined using the load equilibrium condition of the beam. This controlling method has analogies to the crack sliding displacement for the ENF specimen detailed in Ref 8.

Materials and Test Procedures

The material used in this work was HTA/R6376. The resin system is an improved toughness epoxy system generally used for the fabrication of aircraft primary structures. Examples of applications of this resin system with different fibers include flaps and spoilers on various Airbus aircraft, cowls and transcowels on GE, CASA, and Grumman aircraft; plus uses on the B2 and MD-11 aircraft. The material properties from the data sheets were

$$E_{11} = 145.0 \text{ GPa} \quad E_{22} = 10.5 \text{ GPa} \quad G_{12} = 4.16 \text{ GPa} \quad G_{23} = 3.55 \text{ GPa} \quad \nu_{12} = 0.293$$

Because the E_{11} modulus from the data sheets was the axial modulus, the flexural modulus of $E_{11} = 114$ GPa [9] was used in the analysis.

The DCB, MMB, and ENF specimens were manufactured as 32-ply unidirectional beams, with a 13-μm PTFE insert placed between the 16th and 17th ply during manufacture. The following autoclave cure cycle was used: (1) vacuum was applied to the panels; (2) a pressure of 0.7 MPa was applied; (3) temperature was raised to 175°C at 2°C/min; (4) the vacuum was vented when the pressure reached 0.14 MPa; (5) the temperature was then held for 2 h until; (6) allowed to cool and the pressure was released when the temperature fell below 70°C. Extruded aluminum hinges (Stock No. MS 20001-6) were bonded to the DCB and MMB specimen using Redux 420(A&B). The surfaces of the composite and aluminum to be bonded were lightly abraded using grit paper, rinsed in water, and then cleaned with a volatile solution prior to bonding.

The DCB tests were conducted according to ASTM Test Method for Mode I Interlaminar Fracture Toughness of Unidirectional Fiber-Reinforced Polymer Matrix Composites (D 5528), and three replicates were used. For all of the ENF tests, an experimental compliance calibration was conducted on each specimen prior to fracture testing. Because the exact length of the insert was not known during the compliance calibration or the fracture test, a compliance measurement was taken for known distances of overhang outside the span of the fixture. Following the fracture test, the specimen was broken open and the insert length measured. By subtracting the overhang during the compliance tests, the exact delamination length was determined. ENF specimens were tested so that delamination grew from the insert, from a shear pre-crack, and from a tension precrack. The shear pre-crack was created by using the ENF fixture to grow the delamination instability until it stopped approximately under the center roller. The specimen was then shifted in the fixture so that the crack front was between the loading and support rollers and retested. The tension precrack was created by lightly clamping the specimen 15 mm away from the insert end. The specimen was then pried open using a wedge until the crack grew to the clamped region. Three replicates of each configuration were used. To maintain a constant-opening displacement for the MMB tests, a direct-current displacement transducer (DCDT) was attached to the loading pins of the hinges, Fig. 3. This DCDT was connected to the external controls of the test system. Thus, the actuator of the machine was controlled from the external control. An opening displacement rate of 0.5 mm/min was used, resulting in a variable loading rate of the beam. For compliance calibration purposes, the crack length was monitored visually using a video microscope. Three replicates for each mode ratio were used.

Analysis

Stability of MMB from Control of Opening Rate

By controlling the loading on the lever such that the opening displacement rate of the MMB specimen is kept constant, the delamination growth will be stable. To demonstrate this, and to make the algebra easier, consider the MMB to remain linear and to displace according to beam theory with no shear or rotation corrections. Using the notation given in Fig. 4

$$C_I = \frac{\delta_I}{P_I} = \frac{2a^3}{3E_{11}I} \tag{13}$$

FIG. 3—*Experimental setup showing external DCDT.*

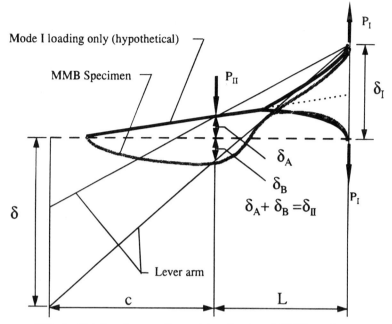

FIG. 4—*Notation for compliance determination of MMB specimen.*

where

$$P_I = P \frac{3c - L}{4L} \tag{14}$$

or

$$P = \frac{\delta_I}{\dfrac{2a^3}{3E_{11}I} \dfrac{3c - L}{4L}} \tag{15}$$

substituting Eq 15 into Eq 3

$$G_{\text{tot}} = \frac{3\delta_I^2[4(3c - L)^2 + 3(c + L)^2]}{\left[\dfrac{2}{3E_{11}I} \dfrac{3c - L}{4L}\right]^2 16b^2h^3L^2E_{11}} \left[\frac{1}{a^4}\right] \tag{16}$$

Therefore, G_{tot} is proportional to a^{-4}. Hence, $\partial G/\partial a$ is always negative indicating stable delamination growth, as for the DCB specimen.

Finite-Element Analysis

A two-dimensional, plane strain, nonlinear finite element analysis was conducted using FLEXPAC, a proprietary code at the Materials Engineering Research Laboratory (MERL).

The mesh consisted of 520 eight-noded plane-strain elements. The elements measured 1 by 0.5 mm in the regions modeled with a crack. The deformed mesh is shown in Fig. 5. Loads P_1 and P_2 were calculated assuming two different loading arm lengths, $c = 90$ mm and $c = 32.5$ mm. The P_1 and P_2 loads were found iteratively to result in a unit opening displacement, δ_I. The virtual crack closure technique (VCCT) [10] was used to determine the individual modal values as well as G_{tot}. Plots of normalized strain energy release rate versus delamination length from the finite element analysis (FEA) results, Figs. 6 and 7, confirm that delamination growth is stable, Eq 16. The Mode I, Mode II, and total values of G determined from Eqs 4 and 5 are also plotted in Figs. 6 and 7 where P_I and P_{II} were determined from the loads applied to the finite element analysis model using Eqs 17 and 18. The variation of mode ratio with delamination length determined from the FEA and from beam theory give similar results for longer delamination lengths at a mode ratio of 4:1 (circles) and 1:1 (squares), Fig. 8. The mode mix ratio diverges at smaller delamination lengths but these tests are not usually run at values of a less than 20 mm in a 50 mm semi-span.

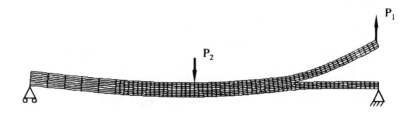

Note: Exagerated Scale

FIG. 5—*Deformed MMB finite element mesh.*

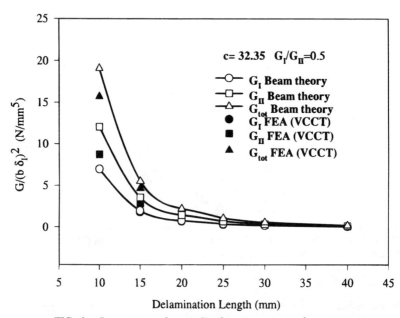

FIG. 6—*Comparison of normalized strain energy release rate.*

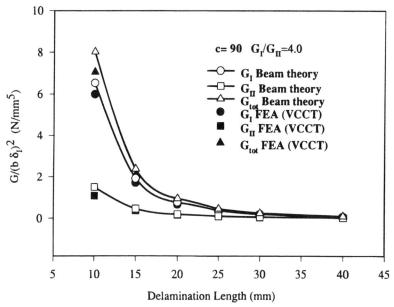

FIG. 7—*Comparison of normalized strain energy release rate.*

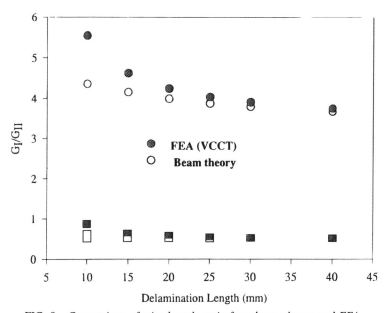

FIG. 8—*Comparison of mixed mode ratio from beam theory and FEA.*

Experimental Compliance Calibration

By obtaining stable delamination growth and a direct measurement of opening displacement, experimental compliance calibration equations may be derived. The force equilibrium of the MMB is given as [1]

$$P_I = \frac{(3c - L)}{4L} P \tag{17}$$

$$P_{II} = \frac{(c + L)}{L} P \tag{18}$$

Because a direct measurement of the opening displacement, δ_I, is obtained, an expression for compliance can be expressed as

$$C_I = \frac{\delta_I}{P_I} = \frac{4L\delta_I}{(3c - L)P} = f_I(a) \tag{19}$$

where $f_I(a)$ represents either Eq 7 or 8. Similarly, an expression for the Mode II compliance may be defined from Fig. 4 as

$$\delta_{II} = \delta_A + \delta_B \tag{20}$$

where

$$\delta_A = \frac{\delta_I}{4} \tag{21}$$

assuming small displacements

$$\delta_B = \delta - \left(\left(\frac{c}{c + L} \right)(\delta_I + \delta) \right) \tag{22}$$

Hence,

$$C_{II} = \frac{\delta_{II}}{P_{II}} = \frac{\delta - \left(\frac{c}{c + L}(\delta_I + \delta) \right) + \frac{\delta_I}{4}}{\frac{c + L}{L} P} = f_{II}(a) \tag{23}$$

where $f_{II}(a)$ may take the form of Eq 9. Thus, the Modes I and II values of G may be determined from

$$G_I = \frac{nP_I\delta_I}{2ba} = \frac{n(3c - L)P\delta_I}{8bLa} \qquad (24)$$

$$G_I = \frac{3P_Im(a + \Delta_I)^2}{2b} = \frac{3P_I\delta_I}{2b(a + \Delta_I)} = \frac{3(3c - L)P\delta_I}{8bL(a + \Delta_I)} \qquad (25)$$

$$G_{II} = \frac{P_{II}^2}{2b}[A_3a^2] = \frac{(c + L)^2P}{2bL^2}[A_3a^2] \qquad (26)$$

Equations 25 and 26 are used throughout this work.

Experimental Results and Discussion

Double-Cantilever Beam (DCB) Tests

At the onset of delamination from the insert, the load dropped sharply with no noticeable deviation from linearity, Fig. 9. Therefore, the maximum load value was used to determine the Mode I fracture toughness, G_{Ic}, at initiation. At the onset of delamination growth, the load dropped sharply but not sharply enough to indicate unstable delamination growth. The R-curves for all three specimens tested indicates a significantly higher value of G_{Ic} at the insert

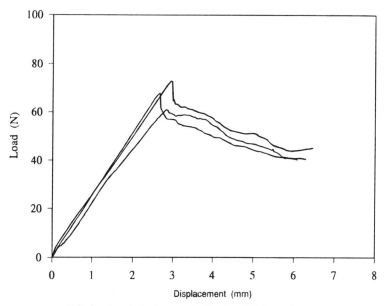

FIG. 9—Load-displacement plots for DCB specimens.

and consistent propagation values, Fig. 10. This indicates there was negligible fiber bridging, as also evident by examination of the failure surfaces. The insert obviously did have an effect on the toughness values.

End-Notched Flexure Tests

For the tests where fracture was grown from the insert, the load displacement curve showed the conventional sudden load drop with delamination growth, Fig. 11. There was also no significant deviation from linearity prior to delamination growth. Therefore, the maximum values of load were used to determine the Mode II fracture toughness, G_{IIc}. For the specimens where delamination growth was from a precrack, the load displacement curves did show some deviation from linearity prior to rapid delamination growth. Therefore, the maximum loads and the loads at the deviation from linearity were used to determine G_{IIc}. The mean results from the specimens tested are shown in Table 1 with the standard deviations in parentheses.

Similar to the DCB results, the values of toughness at the insert are significantly larger than those from a precrack. The lowest mean value was obtained from specimens with shear precracks.

The Mixed-Mode Bending (MMB) Specimen

Compliance Measurements and Calibration—Using the opening-displacement control mode, stable delamination was achieved, as shown by the load-displacement plots, Fig. 12. The load versus opening-displacement, $P - \delta_1$ plot appeared similar to the DCB plots, Fig. 9, in that there was linear loading followed by a decrease in load with delamination growth. The load versus machine-displacement plot, $P - \delta$, shows the machine displacement decreasing after the onset of delamination to maintain a constant opening displacement rate. The opening displacement rate was kept constant while the machine displacement rate varied, Fig. 13. In

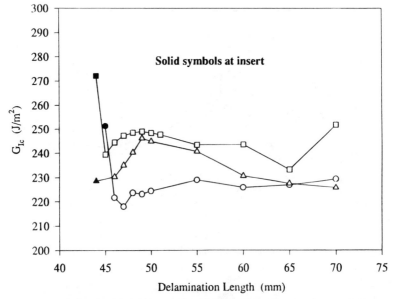

FIG. 10—*Mode I interlaminar fracture toughness values.*

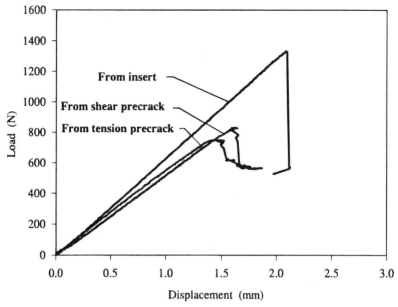

FIG. 11—*Load-displacement plots for ENF specimens.*

TABLE 1—*Mode II fracture toughness results.*

No Precrack		Shear Precrack		Tension Precrack	
G_{IIc}^{NL} (J/m^2)	G_{IIc}^{Max} (J/m^2)	G_{IIc}^{NL} (J/m^2)	G_{IIc}^{Max} (J/m^2)	G_{IIc}^{NL} (J/m^2)	G_{IIc}^{Max} (J/m^2)
1090 (200)	1090 (200)	640 (110)	700 (140)	820 (150)	860 (150)

one specimen, the delamination growth was unstable for a short amount of delamination growth from the insert until the machine could respond sufficiently to maintain the opening rate.

The Modes I and II compliances were determined for each specimen using Eqs 19 and 23. For the Mode I compliance calibration, the value of Δ_I to be used to calculate G_{Ic} from Eq 25 was determined for each specimen by plotting $C_I^{1/3}$ versus a. The value of A_3 to be used to calculate G_{IIc} from Eq 26 was determined for each specimen by plotting C_{II} versus a^3. Typical calibration plots are shown in Figs. 14 through 16. Included in the plots are all the individual compliance data from the DCB and ENF tests. In all of the specimens tested, the Mode I data from the MMB specimen compared very well with the DCB data resulting in similar values of Δ_I. For the Mode II data, the comparison between the compliance from the MMB specimen and ENF specimen differed from specimen to specimen; the comparison being very good in Fig. 14 and less good in Figs. 15 and 16. Because of these potential differences between the Mode II compliance calibration from the MMB to that of the ENF, it is necessary to determine the compliance calibration constants from the MMB test itself. For most specimens where there was a difference between the ENF and Mode II compliance from the MMB, the latter had a higher compliance. This increased compliance may be caused by compliance of the fixture or load train. However, any compliance from the machine or fixture would be

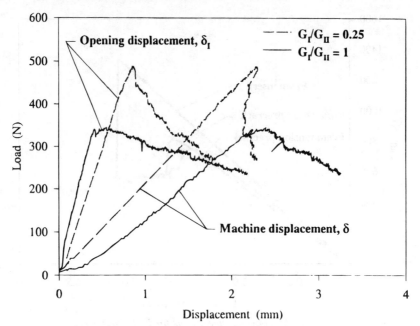

FIG. 12—*Load displacement plot for MMB specimen.*

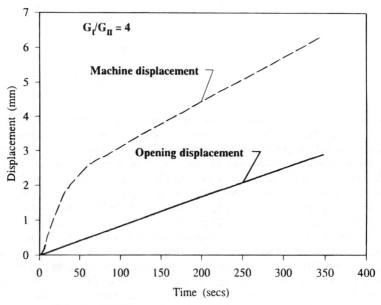

FIG. 13—*Machine and opening displacement versus time.*

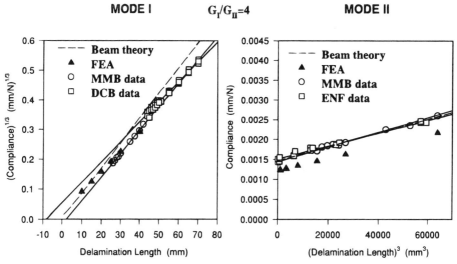

FIG. 14—*Compliance calibration* $G_I/G_{II} = 4$.

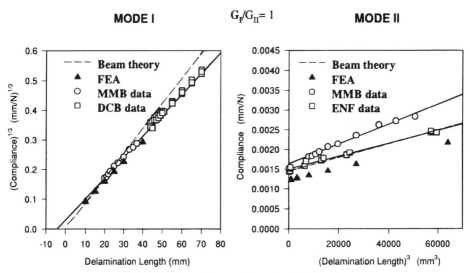

FIG. 15—*Compliance calibration* $G_I/G_{II} = 1$.

constant with delamination length resulting in a similar slope to the ENF. Another possibility is the fact that the MMB specimen is open and, hence, there is no friction between the delaminated surfaces, unlike the ENF specimen. In the ENF specimen, as the delamination gets longer, so do the effects of friction become larger, resulting in a greater difference in compliance at larger delamination lengths, Figs. 15 and 16.

Also plotted in Figs. 14 through 16 are the results from the beam theory and the results from the FEA. The beam theory results compare reasonably well with the experimental results for the DCB, but they underestimate the Mode II compliance from the MMB. The FEA results

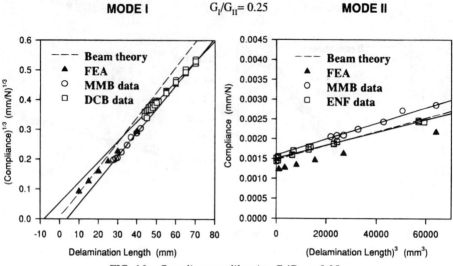

FIG. 16—*Compliance calibration* $G_I/G_{II} = 0.25$.

tend to underestimate the beam theory and experimental compliance values for Mode II but give a closer comparison for Mode I.

Mixed-Mode Fracture Toughness—The variation of fracture toughness with delamination length is shown in Figs. 17 through 19 for a single representative specimen at each of the three different mode ratios. The values of toughness were determined using the experimental

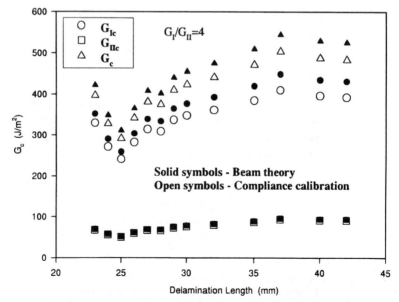

FIG. 17—*Resistance curve for MMB specimen,* $G_I/G_{II} = 4$.

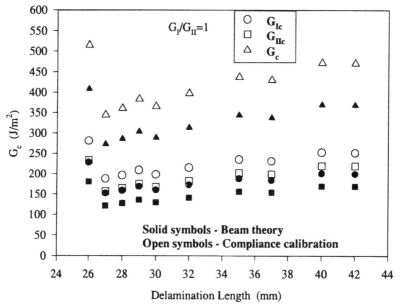

FIG. 18—*Resistance curve for MMB specimen,* $G_I/G_{II} = 1$.

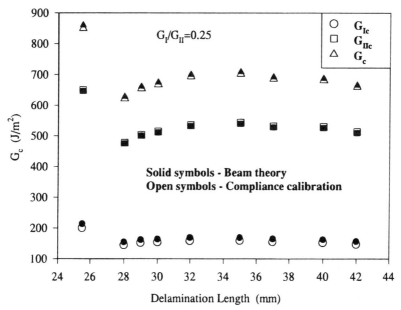

FIG. 19—*Resistance curve for MMB specimen,* $G_I/G_{II} = 0.25$.

compliance calibration expressions, Eqs 25 and 26, and the beam theory expressions, Eqs 4 and 6, using the flexural modulus. A difference between the two methods was apparent for some specimens, but this is largely down to experimental variations mainly in the Mode II values. Similar to the DCB and ENF tests, a larger value of interlaminar fracture toughness was observed at the insert than in the next few values taken as the delamination propagation. For the high Mode I ratios, a slight increase in G_{tot} or an R-curve effect was observed. Because the DCB tests gave no evidence of fiber bridging, it is thought this increase may be caused by rotation of the block used to attach the external DCDT. This effect may be minimized by increasing the thickness of the beam, or may be accounted for in the data reduction [3].

A plot of the G_{IIc} values versus the G_{Ic} values calculated using the compliance calibration technique for all specimens tested is given in Fig. 20. Two values for each specimen are given, one from the insert and the first value taken after the insert. In the majority of specimens, the first propagation values were lower than the values from the insert.

Summary and Conclusions

By maintaining a constant-opening, or Mode I, displacement rate, stable delamination growth was achieved for all mixed-mode ratios. A constant-opening displacement rate was achieved by attaching a second displacement transducer to the hinges of the specimen and externally controlling the test machine using this transducer. Compliance calibration expressions for the Mode I and Mode II parts of the MMB specimen were then developed. The Mode I calibration compared well to the experimental compliance calibration of the DCB but not to the beam theory solution. The Mode II compliance calibration was different from specimen to specimen where most compared well to the ENF tests. Because the beam theory expressions showed differences from experimental variations compared with the experimental data, they should not be used to reduce the MMB data. Also, because the MMB data does not compare consistently

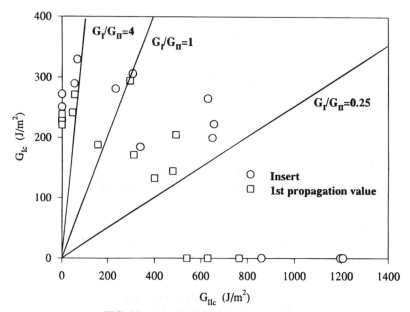

FIG. 20—*Mixed mode fracture toughness.*

with the ENF data, the constants in the experimental compliance calibration tests should not be conducted from separate tests.

In addition, the initiation values from the insert in this material was higher for all mode ratios than the propagation values. The insert was straight and did not stick and was believed to be well made. Therefore, the ASTM DCB test methods and standards may need to be modified to represent materials where this may occur.

Acknowledgments

The authors are grateful to Ian Gurnell of Hexcel (formerly CIBA Composites) for the supply of the HTA 6376 material.

References

[1] Reeder, J. R. and Crews, J. H., Jr., "Mixed-Mode Bending Method for Delamination Testing," *AIAA Journal*, Vol. 28, No. 7, July 1990, pp. 1270–1276.

[2] Reeder, J. R. and Crews, J. H., Jr., "Redesign of the Mixed-Mode Bending Delamination Test to Reduce Non-linear Effects," *Journal of Composites Technology and Research*, Vol. 14, 1992, p. 12.

[3] Kinloch, A. J., Wang, J., Williams, J. G., and Yayla, P., "The Mixed-Mode Delamination of Fibre Composite Materials," *Composites Science and Technology*, Vol. 47, 1993, pp. 225–237.

[4] Zhao, S., Gädke, and Prinz, R., "Mixed-Mode Interlaminar Fracture Toughness and Delamination Fatigue Growth of Carbon/Epoxy Composites," DLR Technical Report IB-131-93/24, Braunschweig, Germany, 1993.

[5] Theksen, J. C., Brandt, F., and Nilsson, S., "Investigations of Delamination Growth and Criticality Along Heterogeneous Interfaces," ICAS-94-9.3.2, 19th Congress of the International Council of Aeronautical Sciences, 18–23 Sept. 1994, pp. 995–1014.

[6] Berry, J. P., "Determination of Fracture Energies by the Cleavage Technique," *Journal of Applied Physics*, Vol. 34, No. 1, Jan. 1963, pp. 62–68.

[7] Hashemi, S., Kinloch, A. J., and Williams, J. G., "Corrections Needed in Double Cantilever Beam Tests for Assessing the Interlaminar Failure of Fibre Composites," *Journal of Materials Science Letters*, Vol. 8, 1989, pp. 125–129.

[8] Chang, W.-T., Kimpara, I., Kageyama, K., and Ohsawa, I., "New Data Reduction Schemes for the DCB and ENF Tests of Fracture-Resistant Composites," *Proceedings*, Fourth European Conference on Composite Materials, 25–28 Sept. 1990, Stuttgart, Germany, pp. 503–508.

[9] Bhashyam, S. and Davidson, B. D., "An Evaluation of Data Reduction Methods for the Mixed Mode Bending Test," *Proceedings*, Thirty-seventh AIAA/ASME/ASCE/AHS/ASC, AIAA-96-1419, Salt Lake City, 15–17 April 1996.

[10] Shivakumar, K. N., Tan, P. W., and Newman, J. C., Jr., "A Virtual Crack-Closure Technique for Calculating Stress Intensity Factors for Cracked Three Dimensional Bodies," *International Journal of Fracture*, Vol. 36, 1988, pp. R43–R50.

Jack L. Beuth[1] and Shri Hari Narayan[1]

Separation of Crack Extension Modes in Composite Delamination Problems

REFERENCE: Beuth, J. L. and Narayan, S. H., **"Separation of Crack Extension Modes in Composite Delamination Problems,"** *Composite Materials: Fatigue and Fracture (Sixth Volume), ASTM STP 1285*, E. A. Armanios, Ed., American Society for Testing and Materials, 1997, pp. 324–342.

ABSTRACT: In the analysis of composite delamination problems, the magnitudes of Mode I and Mode II stress intensity factors or energy release rates are typically not unique, due to oscillatory behavior of near-tip stresses and displacements. This behavior currently limits the ability to consistently apply interfacial fracture mechanics to predict composite delamination. The virtual crack closure technique (VCCT) is a method used to extract Mode I and Mode II components of energy release rates from finite element fracture models. Energy release rate components extracted from an oscillatory delamination model using the VCCT are dependent on the virtual crack extension length, Δ.

In this paper, a recently-developed modified VCCT from the literature is used to extract Δ-independent energy release rate quantities for composite delamination problems where the delamination occurs between two plies or ply groups modeled as in-plane orthotropic materials. Numerical cases studied are taken from existing work in the literature. Energy release rate ratios are compared to those obtained using other methods proposed in the literature for the analysis of oscillatory delamination problems. Unlike other methods, the method of mode separation used in this work does not involve altering the fracture model to eliminate its oscillatory behavior. Instead, it is argued that consistent, Δ-independent energy release rate quantities can be extracted directly from oscillatory solutions. For the cases studied, results show that mode mix values resulting from application of the modified VCCT are comparable to those obtained via existing methods. Some mode mix predictions obtained (in the literature) using existing methods lie outside the range of reasonable values for such problems, however. The Δ-independent energy release rate quantities extracted using the modified VCCT can serve as guides for testing the convergence of finite element models. This technique also has potential as a consistent method for extracting energy release rate quantities from numerical models of composite delamination.

KEYWORDS: orthotropic materials, interfacial fracture mechanics, mode mix, virtual crack closure technique, energy release rate, fatigue (materials), fracture (materials), composite materials

The problem of delamination in composite materials has been studied extensively in the literature. From this work, methods for analyzing delamination problems have been established, where the problem is treated within the framework of fracture mechanics. To predict delamination resistance, an existing interfacial delamination crack is modeled and an applied stress intensity factor, K, or energy release rate, G, is calculated. The applied K or G is then compared to a critical value of K or G determined from a toughness test. The work in Refs *1–3* was some of the earliest to use this type of approach.

---

[1]Assistant professor and graduate student, respectively, Department of Mechanical Engineering, Carnegie Mellon University, Pittsburgh, PA 15213.

Critical K or G values are typically a strong function of mode mix. This requires that an applied stress intensity factor be compared to an interfacial toughness measured using a specimen experiencing the same ratio of Mode I to Mode II stress intensity factors or energy release rates. Commonly used interfacial toughness tests include the double cantilever beam (DCB) test, which yields a pure Mode I interfacial toughness, and the end-notched flexure (ENF) test, which yields a toughness close to that for pure Mode II. In order to limit the number of required tests, composite interfacial toughness tests are typically performed on unidirectional $0°$ laminates. It is then assumed that toughnesses as a function of mode mix measured for delamination between $0°$ plies are comparable to those for delamination between plies of other orientations. Although there is some debate regarding the accuracy of this assumption, the alternative of obtaining toughnesses as a function of mode mix for all combinations of ply orientations is not practical.

Because the homogeneous elastic properties of adjacent plies are the same, the mode mix associated with tests on $0°$ laminates can be determined unambiguously. In contrast, in most applications, delaminations occur between plies of different orientation. In such cases, modeling the delamination as a crack along a distinct interface between plies typically results in near-tip stresses and displacements that oscillate as the tip is approached. As a result, in interfacial fracture models of most composite applications, the mode mix is not uniquely defined. Under such conditions, a match of mode mix between the application and toughness tests cannot readily be made. A method is needed for extracting a consistent measure of mode mix from delamination analyses that exhibit oscillatory behavior. In this study, a method recently proposed in the literature is outlined for consistently extracting energy release rate quantities from oscillatory delamination analyses where the plies or ply groups defining the cracked interface are modeled as in-plane orthotropic materials (with one principal axis aligned with the crack front). The method is then applied to two composite delamination problems analyzed in the literature. Results are compared to those of other methods proposed in the literature for extracting fracture quantities from oscillatory delamination models.

Background

Two methods have been proposed in the literature to explicitly address oscillatory composite delamination problems. Each of these existing methods involves altering the delamination model to eliminate its oscillatory behavior. Once this is done, unique Mode I and Mode II stress intensity factors and energy release rates exist that can be extracted from the model. The first of these methods is termed the resin interlayer method and was originally proposed in Ref 4 for analyzing isotropic interfacial fracture problems. The method has been applied more recently to composite delamination problems in Refs 5 and 6. It involves inserting a thin homogeneous layer (the resin interlayer) between the layers forming the cracked interface and placing the crack within it. Because the crack tip is fully embedded in a homogeneous material, non-oscillatory stresses and displacements result.

Because composites have a resin-rich region between adjacent plies, the resin interlayer method has appeal in modeling composite delamination. The method has some disadvantages, however. First, significantly more effort is required in finite element analyses to mesh the interlayer region with sufficient resolution to extract near-tip fracture quantities. Another disadvantage is the addition of the interlayer thickness and properties as variables in the problem. The sensitivity of energy release rate predictions to relatively small changes in the modeling of the interlayer is not clear. A final disadvantage of this method involves its true relationship to the physical delamination problem on an interlayer scale. As noted in Ref 4, placing the crack fully within the interlayer ignores the fact that a crack under mixed-mode conditions in a homogeneous material will not grow in a self-similar manner. In the general case of mixed-mode loading, a crack within an interlayer will not remain there as it grows, but will branch to an interface. Modeling interlayer cracking as a straight crack embedded fully within the interlayer ignores this. On an interlayer scale, modeling a delamination as the extension of a single crack may also be nonphysical. The work in Ref 7 suggests that macroscopically Mode II dominated crack extension in brittle interlayers occurs by the growth of pure Mode I microcracks ahead of the tip that then link up with the main crack tip.

Another approach for addressing oscillatory composite delamination problems, termed the $\beta = 0$ method, involves changing "physically insignificant" properties of one or both layers forming the interface to make the oscillatory exponent parameter, ϵ, equal zero (definitions for ϵ are given in Eq 3 in the next section). This method, proposed for orthotropic interfacial fracture problems in Refs 6 and 8, typically involves changing a Poisson's ratio of one of the layers forming the cracked interface. It is based on an analogous method proposed for isotropic bimaterial fracture problems (see Refs 9 and 10). The key to this method is the identification of "physically insignificant" properties that can be changed without significantly altering the physics of the model. For the isotropic bimaterial case, it is well-established that most interfacial fracture problems are very weakly dependent on ϵ or the related Dundurs parameter β. Thus, in most analyses of isotropic bimaterials, it is possible to change properties of one or more of the layers forming the interface to make $\epsilon = 0$ without significantly altering the modeled physics. An analogous role for the ϵ parameter for orthotropic and anisotropic interface problems has not been established. Although this method has potential, its standardized application requires formulation of consistent criteria for identifying which properties can be altered to make $\epsilon = 0$ without significantly changing the physical aspects of the model.

Each of the existing methods for approaching oscillatory delamination models currently has drawbacks related to the ease or consistency, or both, with which it can be applied by practitioners. In the current study, a third method from the literature [11] is outlined for extracting nonoscillatory fracture quantities from oscillatory composite delamination models. The goal of this method is to extract a consistent measure of mode mix directly from oscillatory models. This is done by using asymptotic near-tip stress and displacement relationships to define energy release rate quantities that are separated from oscillatory effects. In this work, ratios of these quantities are compared to ratios of energy release rates obtained using the β = 0 and resin interlayer methods for identical composite delamination problems. Comparisons with the method from Ref 11 allows the $\beta = 0$ and resin interlayer methods to be evaluated for their consistency in extracting mode mix values.

Theory

In this section, a description is given of a method outlined in Ref 11 for extracting Δ-independent energy release rate quantities from an oscillatory interfacial fracture solution, where the two materials forming the cracked interface are modeled as in-plane orthotropic materials. This method can be applied to composite delamination problems where the plies or groups of plies forming the cracked interface are modeled as in-plane orthotropic. Theory details are also given in Ref 11 and are provided here for clarity and for the convenience of the reader. In the first subsection, expressions are given from the literature for near-tip stress and displacement fields for orthotropic interfacial fracture problems. After a short subsection on energy release rate definitions, these near-tip displacement and stress fields are used to determine explicitly the oscillatory behavior of Mode I and Mode II crack-tip energy release rate quantities. Energy release rate quantities are subsequently defined that are independent of the oscillatory behavior of the solution. In the final two subsections, the virtual crack closure technique (VCCT) is described and a modified VCCT is presented that allows the defined Δ-independent energy release rate quantities to be extracted from a finite element model.

Orthotropic Interfacial Fracture

The interface crack problem to be considered is illustrated in Fig. 1, where Materials 1 and 2 are orthotropic with the *x-y* plane as a plane of material symmetry. Equations have been formulated in Ref *12* for the near-tip stresses and displacements for such problems. The stress field just ahead of the interfacial crack tip (along the positive *x*-axis) takes the form

$$\sigma_H \equiv \sqrt{\frac{H_{22}}{H_{11}}} \, \sigma_{yy} + i\sigma_{xy} = \frac{(K_1 + iK_2)x^{i\epsilon}}{\sqrt{2\pi x}} \tag{1}$$

The near-tip crack face opening and sliding displacements, δ_2 and δ_1 (along the negative *x*-axis), are given by

$$\delta_H \equiv \sqrt{\frac{H_{11}}{H_{22}}} \, \delta_2 + i\delta_1 = \frac{2H_{11}(K_1 + iK_2)|x|^{(1/2+i\epsilon)}}{\sqrt{2\pi}(1 + 2i\epsilon)\cosh(\pi\epsilon)} \tag{2}$$

In Eqs 1 and 2, the subscript, *H*, in the complex quantities, σ_H and δ_H, is used to emphasize that the normal stress, σ_{yy}, and crack face opening displacement, δ_2, are multiplied by H_{ij} quantities defined below. In Eqs 1 and 2, the oscillatory exponent, ϵ, is given by

$$\epsilon = \frac{1}{2\pi} \ln\left(\frac{1 - \beta}{1 + \beta}\right) \tag{3}$$

which is the same form as that used in isotropic interfacial fracture problems (see, for instance, Refs *13* and *14*). For the orthotropic case, the β parameter is given by

$$\beta = \frac{[\sqrt{S_{11}S_{22}} + S_{12}]_2 - [\sqrt{S_{11}S_{22}} + S_{12}]_1}{\sqrt{H_{11}H_{22}}} \tag{4}$$

where $[\ ]_2$ and $[\ ]_1$ denote the enclosed quantities evaluated for Materials 2 and 1, respectively. The H_{11} and H_{22} quantities are functions of the elastic properties of the layers forming the interface and are defined by

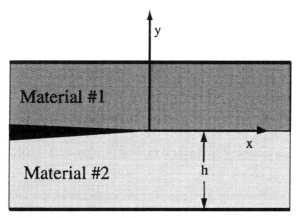

FIG. 1—*Interface crack between two orthotropic layers.*

$$H_{11} = [2n\lambda^{1/4}\sqrt{S_{11}S_{22}}]_1 + [2n\lambda^{1/4}\sqrt{S_{11}S_{22}}]_2$$

$$H_{22} = [2n\lambda^{-1/4}\sqrt{S_{11}S_{22}}]_1 + [2n\lambda^{-1/4}\sqrt{S_{11}S_{22}}]_2 \qquad (5)$$

Nondimensional parameters used to define H_{11} and H_{22} are

$$n = \sqrt{\frac{1}{2}(1 + \rho)} \qquad \lambda = \frac{S_{11}}{S_{22}} \qquad \rho = \frac{2S_{12} + S_{66}}{2\sqrt{S_{11}S_{22}}} \qquad (6)$$

Finally, for plane stress problems, the S_{ij} in Eqs 4 , 5, and 6 are the reduced compliances (i and j take on values of 1 through 6). For plane-strain problems, the S_{ij} are replaced by S'_{ij} defined by

$$S'_{ij} = S_{ij} - \frac{S_{i3}S_{j3}}{S_{33}} \qquad (7)$$

The oscillatory behavior of the near-tip stresses and displacements comes from the $x^{i\epsilon}$ terms in Eqs 1 and 2. Additionally, however, the H_{ij} quantities in the complex quantities, σ_H and δ_H, complicate the task of defining a measure of mode mix for the orthotropic problem. In Ref 12 a definition of mode mix in terms of the relative amounts of K_1 and K_2 is suggested for orthotropic interfacial fracture problems. This is analogous to methods used to define a mode mix for isotropic interfacial fracture problems. One problem associated with this definition (which is pointed out in Ref 12) is that it does not simplify to the classical definition of mode mix for cases where $\epsilon = 0$ but $H_{11} \neq H_{22}$. A related concern with such a definition is its lack of a direct relationship to physical quantities of interest, such as the ratio of normal stress to shear stress ahead of the crack tip. Even for nonoscillatory problems, the H_{ij} quantities multiplying σ_{yy} in Eq 1 and δ_2 in Eq 2, which are not present in the isotropic case, make the ratio of K_1 to K_2 different than the ratios of δ_2 to δ_1 and σ_{yy} to σ_{xy}. Finally, although this type of mode mix definition could perhaps still provide consistent comparisons between fracture problems where H_{11} and H_{22} are the same (in such cases it would still consistently relate the ratio of normal stress to shear stress ahead of the crack tip), this is not the typical case in composite delamination problems, where toughnesses are used to predict delamination between different combinations of ply orientations (with different values of H_{11} and H_{22}). In the next section, a definition of mode mix is introduced that is related to the relative amounts of energy released by an extending crack due to opening versus that due to sliding. A mode mix definition in terms of energy release rates is chosen because most existing work in the study of composite delamination is expressed in terms of energy release rates (as opposed to stress intensity factors).

Energy Release Rate Definitions

For the interface crack geometry shown in Fig. 1, the energy release rate for crack extension in the x-direction can be expressed as

$$G_{\text{total}} = \lim_{\Delta \to 0} \frac{1}{2\Delta} \int_0^\Delta [\sigma_{yy}(x)\delta_2(\Delta - x) + \sigma_{xy}(x)\delta_1(\Delta - x)] \, dx \qquad (8)$$

where arguments in Ref 15 have been used to express the energy per unit width released in propagating the crack a distance, Δ, along the x-axis as the energy of the stresses ahead of the tip acting through the displacements behind the tip. Dividing by Δ and taking the limit as

Δ approaches zero gives the energy release rate. In Eq 8, G_{total} is the total energy release rate. The first term in Eq 8 is commonly designated as the Mode I component, of G_{total}, corresponding to the energy released by normal stresses acting through crack face opening displacements. The second term in Eq 8 is commonly designated as the Mode II component of G_{total}, corresponding to the energy released by shear stresses acting through crack face sliding displacements. The oscillatory nature of orthotropic interfacial fracture solutions causes the Mode I and Mode II components of G_{total} to be functions of the crack extension length, Δ. No limiting value of mode mix is reached as Δ approaches zero. Although the individual components of G are a function of Δ, G_{total} is Δ independent. Details concerning this behavior can be found in Refs *16–18*.

Energy Release Rate Quantities for Composite Delamination Problems

In this section, the oscillatory nature of the Mode I and Mode II components of G_{total} (designated respectively as G_I and G_{II}) is determined explicitly for orthotropic interfacial fracture problems. Some of the procedures used follow those outlined in Ref 5 for the isotropic case. Modified energy release rate definitions are then presented that are separated from oscillatory quantities and are thus independent of the virtual crack extension length, Δ. A measure of mode mix based on these definitions corresponds to the ratio of energy release due to crack face sliding versus that due to crack face opening. In the analysis of composite delamination problems, mode mix is commonly expressed in terms of energy release rates (as opposed to stress intensity factors). Thus, a mode mix definition based on energy release rates is appropriate here.

Because the oscillatory behavior of σ_H and δ_H are known, the initial step in determining the oscillatory behavior of G_I and G_{II} is to express them in terms of σ_H and δ_H. First, define two real functions, Φ_1 and Φ_2, in terms of σ_H and δ_H

$$\Phi_1 \equiv \text{Re}\left[\int_0^\Delta \sigma_H(x)\overline{\delta_H(\Delta - x)}\, dx\right]$$

$$\Phi_2 \equiv \text{Re}\left[\int_0^\Delta \sigma_H(x)\delta_H(\Delta - x)\, dx\right] \tag{9}$$

where the overbar designates the complex conjugate of the quantity below it. Using the energy release rate definitions given in Eq 8, G_I, G_{II}, and G_{total} can be expressed in terms of Φ_1 and Φ_2 as follows

$$G_I = \lim_{\Delta \to 0} \frac{1}{4\Delta} [\Phi_1 + \Phi_2]$$

$$G_{II} = \lim_{\Delta \to 0} \frac{1}{4\Delta} [\Phi_1 - \Phi_2] \tag{10}$$

$$G_{total} = G_I + G_{II} = \lim_{\Delta \to 0} \frac{1}{2\Delta} [\Phi_1]$$

Through the definitions for Φ_1 and Φ_2, the energy release rates have now been expressed in terms of σ_H and δ_H. The oscillatory behavior of Φ_1 and Φ_2 can be determined by substituting the asymptotic stress and displacement Eqs 1 and 2 into Eq 9 and integrating. The forms of Φ_1 and Φ_2 to be integrated are

$$\Phi_1 = \frac{H_{11}}{\pi \cosh \pi\epsilon} K\bar{K} \, \mathrm{Re}\left[\frac{1}{1 - 2i\epsilon} \int_0^\Delta \left(\frac{\Delta - x}{x}\right)^{1/2 - i\epsilon} dx\right]$$

$$\Phi_2 = \frac{H_{11}}{\pi \cosh \pi\epsilon} \, \mathrm{Re}\left[\frac{K^2}{1 + 2i\epsilon} \int_0^\Delta \left(\frac{\Delta - x}{x}\right)^{1/2} ((\Delta - x)x)^{i\epsilon} \, dx\right] \qquad (11)$$

Use of the trigonometric substitution, $x = \Delta \sin^2(\beta)$, in the integral in the expression for Φ_1 gives an expression in terms of the beta function. Identities between the beta and gamma functions allow this expression to be simplified so that Φ_1 is given by

$$\Phi_1 = \frac{H_{11}}{2 \cosh^2 \pi\epsilon} K\bar{K}\Delta \qquad (12)$$

which is nonoscillatory. Use of the same trigonometric substitution in the integral contained in Φ_2 also results in an expression in terms of the beta function; however, the expression for Φ_2 cannot be simplified in the way that Φ_1 can. The function, Φ_2, is thus given by

$$\Phi_2 = \frac{2\Delta H_{11}}{\pi \cosh \pi\epsilon} \, \mathrm{Re}\left[\frac{(Kh^{i\epsilon})^2}{1 + 2i\epsilon} \left(\frac{\Delta}{h}\right)^{2i\epsilon} \int_0^{\pi/2} \cos^2(\beta)(\cos(\beta)\sin(\beta))^{2i\epsilon} \, d\beta\right] \qquad (13)$$

where the integral is equivalent to $1/2 \, B(z,w)$ with $z = 1/2 + i\epsilon$ and $w = 3/2 + i\epsilon$. In Eq 13, the crack extension length, Δ, has been normalized by the characteristic length, h. With respect to the near-crack-tip fields, h is the dominant length scale determining the size of the region in which the near-tip oscillations dominate. In a stress model of a long delamination crack, the modeled layer thickness (or the smallest layer thickness if the layers forming the interface have unequal thicknesses) is the characteristic length. It is important to make a distinction between the modeled layer thickness and the ply thickness in composite problems. In composite delamination models, properties of adjacent plies of the same orientation are typically smeared, allowing the ply group to be modeled as a single orthotropic layer. In such cases, h, the layer thickness in the model, is the thickness of the smeared ply group.

By determining the oscillatory behavior of Φ_2, the oscillatory behaviors of the energy release rate quantities for orthotropic interfacial fracture problems have been determined. The oscillatory behavior of Φ_2 can be eliminated by defining a Φ_2' as

$$\Phi_2' \equiv \mathrm{Re}\left[\left(\frac{\Delta}{h}\right)^{-2i\epsilon} \int_0^\Delta \sigma_H(x)\delta_H(\Delta - x) \, dx\right] \qquad (14)$$

which is Φ_2 with the oscillatory quantity $(\Delta/h)^{2i\epsilon}$ extracted. Now define G_1, G_2 (designated with Arabic subscripts to differentiate them from G_I and G_{II} in Eq 10, and G_{total} as

$$G_1 = \lim_{\Delta \to 0} \frac{1}{4\Delta} [\Phi_1 + \Phi_2']$$

$$G_2 = \lim_{\Delta \to 0} \frac{1}{4\Delta} [\Phi_1 - \Phi_2'] \qquad (15)$$

$$G_{total} = G_1 + G_2 = \lim_{\Delta \to 0} \frac{1}{2\Delta} [\Phi_1]$$

In these modified energy release rate definitions, G_{total} is still Δ independent, however, the G_1 and G_2 portions are also Δ independent. The relationships given in Eqs 14 and 15 define energy release rate quantities that are independent of the crack extension length, Δ, for orthotropic oscillatory interfacial fracture models. As a result, consistent values for G_1 and G_2 exist in the limit as Δ approaches zero. These quantities have been derived by first isolating the $(\Delta/h)^{2i\epsilon}$ oscillatory behavior of the Φ_2 quantity and then defining energy release rate components in terms of a Φ_2' quantity that has this oscillatory quantity extracted from it.

The definitions for energy release rate components derived in this section are analogous to a mode mix definition used for isotropic interfacial fracture problems (see Refs 13 or 14, for example). In the isotropic case, mode mix can be expressed in terms of the phase angle of the complex quantity, $Kh^{i\epsilon}$, that is related to the ratio of normal stresses to shear stresses ahead of the crack tip, separated from the oscillatory quantity, $(r/h)^{i\epsilon}$. The same approach has been taken in deriving Eqs 14 and 15, except that mode components are given in terms of energy release rates.

The basis for the mode mix definitions derived in this section is that the near-tip oscillatory behavior of interfacial fracture solutions is nonphysical. It is therefore argued that a ratio of energy release rate quantities defined to be separated from oscillatory quantities in the solution can serve as a reasonable and consistent definition for comparing mode mixes between problems. The normalization of Δ by h used in Eq 14 is chosen because h is the length scale determining the size of the region in which the near-tip oscillations dominate. It is important to note, however, that this normalization is not the only one that could be reasonably argued. For example, one could argue that the zone dominated by the near-crack-tip stresses and displacements (often referred to as the K field) will extend to a distance that is a fraction of h. Therefore, a normalization of Δ by some fraction of h in Eq 14 could be proposed. In contrast, normalization of Δ by some multiple of h would be difficult to justify.

The validity of any proposed normalization (including the one proposed in this paper) must be proven eventually by its ability to consistently relate interfacial fracture resistances. It is important to recognize, however, that because the magnitude of ϵ is very small for most physical problems (see Ref 12), the difference in the G_1 and G_2 components relative to G_{total} that would result from any physically reasonable change in normalization (over a range of $h/20$ to h, for instance) will generally be small. From a practical standpoint, determination of which normalization is the "correct" one within this range may be difficult or impossible. It would also likely have minimal bearing on the task at hand, which is to consistently compare delamination problems and toughness tests.

Virtual Crack Closure Technique

The virtual crack closure technique (VCCT) is a method used to extract Mode I and Mode II energy release rate components from finite element models. The method, which is detailed in Ref 19, is based on the Irwin representation of energy release rate for self-similar crack extension given in Eq 8. The method involves expressing the Mode I and Mode II portions of G_{total} in terms of nodal forces ahead of the crack tip and nodal displacements behind the tip. This can be represented by the expressions

$$G_I = \frac{1}{2\Delta} \sum_{j\text{nodes}} F_{22j}\delta_{2j} \qquad G_{II} = \frac{1}{2\Delta} \sum_{j\text{nodes}} F_{12j}\delta_{1j} \qquad (16)$$

where F_{22j} and F_{12j} are, respectively, the normal and shear nodal forces per unit thickness over distance, Δ, ahead of crack tip and δ_{2j} and δ_{1j} are "corresponding" nodal crack opening and sliding displacements over a distance, Δ, behind the crack tip. Corresponding nodes are designated in Fig. 2. The relationships in Eq 16 assume that nodes are equally spaced ahead of and behind the crack tip, as would be the case if non-singular elements are used in the analysis. The technique can also be used with singular crack-tip elements, however. In the VCCT, Δ is referred to as the virtual crack extension length. The VCCT is applied by using Eq 16 to calculate energy release rates for smaller and smaller values of Δ, until no significant changes in G-values occur. In applying the VCCT to oscillatory interface crack solutions, only G_{total} reaches a limiting value as Δ is decreased. The VCCT is typically applied using one element ahead of and behind the crack tip, with the virtual crack extension length, Δ, thus equal to the size of the first element from the tip. When this is done, changes in Δ are accompanied by changes in the near-crack-tip element resolution. Although it is not typically done, the VCCT can be applied over multiple elements ahead of the crack tip so that Δ can be varied without changing the near-tip element size. Such an approach is taken in this study so that changes in Δ are for a fixed level of near tip mesh refinement.

Modified Virtual Crack Closure Technique

A modified VCCT is formulated here to allow extraction of the energy release rates defined in Eq 15 from a numerical model that exhibits oscillatory behavior. To do this, the energy release rate quantities in Eq 15 are equivalently expressed in terms of nodal forces and displacements. Define

$$F_{H_j} \equiv \sqrt{\frac{H_{22}}{H_{11}}} \, F_{22j} + iF_{12j}$$

$$\delta_{H_j} \equiv \sqrt{\frac{H_{11}}{H_{22}}} \, \delta_{2j} + i\delta_{1j} \tag{17}$$

where F_{22j}, F_{12j}, δ_{2j}, and δ_{1j} are defined with respect to Fig. 2 as they are for the traditional VCCT. With these complex quantities defined, Φ_1 and Φ_2' can be expressed as

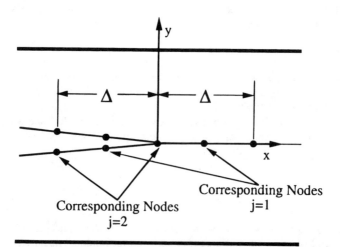

FIG. 2—*Virtual crack closure technique parameters.*

$$\Phi_1 = \mathrm{Re}\left[\sum_{j\text{nodes}} F_{H_j}\,\overline{\delta_{H_j}}\right]$$

$$\Phi_2' = \mathrm{Re}\left[\left(\frac{\Delta}{h}\right)^{-2i\epsilon}\sum_{j\text{nodes}} F_{H_j}\delta_{H_j}\right] \tag{18}$$

and the energy release rate (G_1, G_2, and G_{total}) definitions in terms of Φ_1 and Φ_2' are simply those given in Eq 15.

In applying this modified VCCT, the same quantities (nodal forces ahead of tip and nodal displacements behind the tip) are used as in the traditional VCCT. The quantities are simply re-packaged in the complex forms given in Eq 17. The only additional effort required is that needed to calculate the H_{11}, H_{22}, and ϵ quantities (defined in Eqs 3 through 7) and to perform the complex multiplication indicated in Eq 18. Use of the modified VCCT outlined in this section is similar to methods by which the traditional VCCT has been applied to oscillatory models in the literature (see Ref 2, for example), where a virtual crack extension length, Δ, is used that is large compared to the small zone dominated by ϵ effects. In a pragmatic way, use of large values of Δ takes advantage of the fact that outside of a region very close to the crack tip, mode mix is only weakly dependent on Δ. Use of large values of Δ avoids oscillatory effects by simply avoiding the near-tip region. The advantage of the modified VCCT presented in the current study is that small values of Δ can be used while still avoiding oscillatory effects because the oscillatory effects are extracted analytically. This allows limiting values of energy release rates to be obtained as Δ approaches zero.

A relationship exists between the mode mix obtained using the traditional and modified VCCT. This is due to the fact that the value of Δ used in a traditional VCCT is directly related to the normalization parameter chosen for the modified VCCT. Because of the normalization with respect to h used in the modified VCCT (see Eqs 13, 14, and 18), the mode mix obtained using the modified technique corresponds to that one obtained using $\Delta = h$ in the traditional VCCT, if the zone dominated by the near-crack-tip fields extended to a distance, h, from the crack tip. Because a second interface exists at a distance, h, from the crack tip, in practical problems, the near-tip zone will generally not extend to a distance, $\Delta = h$. As a result, the energy release rate components, G_1 and G_2, will not equal G_I and G_{II} obtained using a traditional VCCT with $\Delta = h$. This issue is addressed further near the end of the next section.

Application

A modified VCCT (from Ref *11*) for extracting the energy release rate quantities, G_1 and G_2, defined in Eq 15 has been presented in the previous section. In theory, application of this modified VCCT using nodal force and displacement results from a finite element model of an oscillatory delamination problem will result in Δ-independent mode mix values. This section demonstrates how well the methods outlined in this study perform in practice. Results are presented for application of the modified VCCT to finite element models of two plane strain composite delamination problems analyzed in the literature. Mode separation for these problems has also been attempted (in the literature) using the $\beta = 0$ and resin interlayer methods, and a comparison of available results of all three methods is given.

The plane strain problems analyzed are diagrammed in Fig. 3. In each problem studied, the left edge of the laminate has net forces and bending moments applied above and below the crack line. The right edge of each model is constrained so that x- and y-displacements equal zero (built-in conditions). A full model is studied for the $[0_8/90_4]_s$ case. The model for the $[0/\pm35/90]_s$ case is a half model with symmetric boundary conditions (designated by rollers)

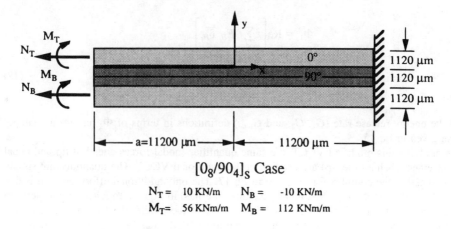

$[0_8/90_4]_s$ Case

$N_T =$ 10 KN/m $N_B =$ -10 KN/m

$M_T =$ 56 KNm/m $M_B =$ 112 KNm/m

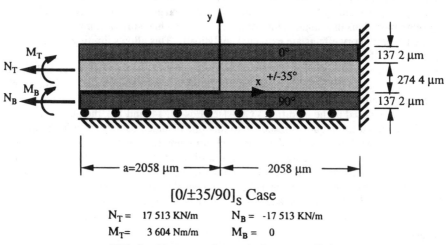

$[0/\pm35/90]_s$ Case

$N_T =$ 17 513 KN/m $N_B =$ -17 513 KN/m

$M_T =$ 3 604 Nm/m $M_B =$ 0

FIG. 3—*Diagrams of numerical cases studied.*

along its base. The $[0_8/90_4]_s$ and $[0/\pm35/90]_s$ cases have been studied, respectively, in Refs 6 and 8. Each problem in Fig. 3 represents a two-dimensional slice of a composite laminate that extends lengthwise in the z-direction. The coordinate system used in Fig. 3 is identical to the crack-tip coordinate system used in Fig. 1 and is different than the coordinate system traditionally used to describe composite laminates. Because of this, all ply angles are defined with respect to the z-axis so that a 0° ply corresponds to a ply with its fibers along the z-axis in Fig. 3. A 90° ply corresponds to a ply with its fibers along the x-axis in Fig. 3.

The finite element mesh used to model the $[0/\pm35/90]_s$ case analyzed in this study is shown in Fig. 4. The mesh used to model the $[0_8/90_4]_s$ case is similar. Each model is constructed out of eight-noded, plane strain quadrilateral interpolation elements using the finite element package ABAQUS. A minimum of four elements is used through the thickness of each modeled material layer. Refined non-singular elements have been used near the crack tip. Because they have

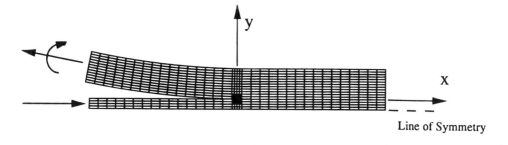

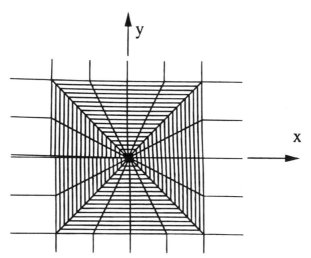

FIG. 4—*Finite element mesh with non-singular crack-tip elements for the* $[0/\pm35/90]_s$ *case.*

node locations at equal radial distances ahead of and behind the crack tip, non-singular near-tip elements allow straightforward application of the VCCT over multiple elements. Results given in Ref 5 and the authors' own experience indicate that accurate energy release rate values can be obtained using the VCCT without having to use quarter-point crack-tip elements to capture the $1/\sqrt{r}$ stress and strain singularity at the crack tip. For the results presented in this study, the near-tip meshes consist of 18 rings of elements meshed over a length equal to $h/2$, where h is the smaller thickness of the two modeled layers forming the interface. In the problems presented in this study, adjacent plies of the same orientation (a ply group) are modeled as a single orthotropic layer. The layer thickness, h, is thus the thickness of a modeled ply group.

The first problem studied is that of a $[0_8/90_4]_s$ graphite-epoxy laminate with a single delamination between the 0° and 90° plies. This problem is identical to a composite debonding problem studied in Ref 6 as part of an analysis of various delamination problems with identical geometries to that shown in Fig. 3, but with different applied loadings. Properties used (which are identical to those used in Ref 6) are given in Table 1. For this problem, Materials 1 and 2 as designated in Fig. 1 are the 0° and 90° ply groups, respectively, with $h = 1120$ μm. This material combination gives a value of the oscillatory exponent, ϵ, equal to 0.0479. For this problem, VCCT and J-integral values of G_{total} have been found to equal 13.2 J/m², agreeing with the value reported in Ref 6.

TABLE 1—*Graphite-epoxy properties used in the [$0_8/90_4$]$_s$ case (all moduli are in GPa).*[a]

	0° PLIES	
$E_{xx} = 14.5$	$\nu_{xy} = 0.25$	$\mu_{xy} = 5.80$
$E_{yy} = 14.5$	$\nu_{xz} = 0.0263$	$\mu_{xz} = 5.90$
$E_{zz} = 137.9$	$\nu_{yz} = 0.0263$	$\mu_{yz} = 5.80$
	90° PLIES	
$E_{xx} = 137.9$	$\nu_{xy} = 0.25$	$\mu_{xy} = 5.80$
$E_{yy} = 14.5$	$\nu_{xz} = 0.25$	$\mu_{xz} = 5.90$
$E_{zz} = 14.5$	$\nu_{yz} = 0.25$	$\mu_{yz} = 5.80$

[a]See Fig. 1 for x, y, and z-axis definitions. Ply thickness used was 140.0 μm. These properties are the same as those used in Ref 6.

Figure 5 gives a plot of the G_{II}/G_I and G_2/G_1 ratios as a function of Δ/h for the [$0_8/90_4$]$_s$ graphite-epoxy laminate case, where G_{II}/G_I relates to a traditional measure of mode mix obtained using the VCCT and G_2/G_1 is the mode mix calculated using the non-oscillatory energy release rate quantities of Eq 15, extracted using the modified VCCT. In application of the VCCT to this case, the near-tip element size was not changed as Δ was changed. In the plot, the four pairs of points correspond to application of the traditional or modified VCCT over distances of one, two, three, and four elements ahead of and behind the crack tip.

From the figure, it is clear that use of a traditional VCCT for small values of Δ/h results in significant changes in G_{II}/G_I with decreases in Δ. The change is exaggerated due to the fact that G_I is small relative to G_{II}; however, the first data point for $\Delta/h = 0.0278$ has not even been plotted because it would have disrupted the plotting scale. In contrast, use of the modified VCCT results in G_2/G_1 ratios that are essentially independent of Δ. Thus, the modified VCCT appears to work well in extracting a nonoscillatory measure of mode mix from the oscillatory solution. The energy release rate quantities obtained using the modified VCCT are

$$G_1 = 0.566 \text{ J/m}^2 \qquad G_2 = 12.6 \text{ J/m}^2 \qquad G_{total} = 13.2 \text{ J/m}^2 \qquad (19)$$

As mentioned in the previous section, the mode mix extracted using the modified VCCT

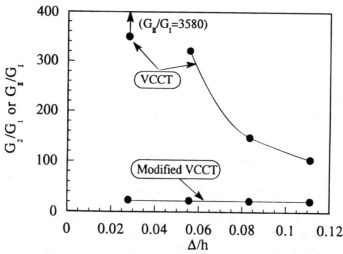

FIG. 5—*Energy release rate ratios versus normalized virtual crack extension length, Δ, for the [$0_8/90_4$]$_s$ case.*

corresponds to that from a traditional VCCT using $\Delta = h$. The trends in the results of Fig. 5 as Δ/h is increased are consistent with this fact. This issue is addressed further in the discussion of Fig. 8 later in this section.

Mode separation for this problem has also been attempted in Ref 6 using the $\beta = 0$ and resin-interlayer methods. The value for G_I obtained using the $\beta = 0$ method equals 0.4422 J/m². The value for G_I obtained using the resin-interlayer method equals 0.9503 J/m². The result from application of the modified VCCT lies between that from the $\beta = 0$ and resin-interlayer methods. By definition, the resin-interlayer method (with the highest of the G_I values) is providing a mode mix prediction for this case that corresponds to that from application of a traditional resin interlayer VCCT for a value of Δ greater than h (if the K field extended to such a distance). The resin interlayer prediction thus appears to be outside of the range of mode mix values that would be considered reasonable for this problem.

The second problem analyzed in this study is that of a $[0/\pm35/90]_s$ T300/5208 laminate with delaminations between each of the $-35°$ and $90°$ plies. The model used is similar to that of the $[0_8/90_4]_s$ case, but in this case symmetric boundary conditions are imposed on the base of a half-model to account for the symmetrically extending delamination at the second interface. This problem is identical to a composite debonding problem studied in Ref 8 as part of an analysis of a delamination problem studied in Ref 5 using the resin interlayer method. Properties used (which are identical to those used in Ref 8) are given in Table 2, where the properties of the individual $+35$ and -35 plies have been smeared into the properties of a ±35 ply group to make the analysis two dimensional. Because the $+35$ and -35 plies are modeled as a single orthotropic ply group, the methods outlined in this paper can be applied directly (even though the actual $+35$ and -35 plies do not exhibit orthotropic symmetry with respect to the x-y plane). For this second problem, Materials 1 and 2 (see Fig. 1) are the ±35 and $90°$ layers, respectively, with $h = 137.2$ μm. This material combination gives a value of the oscillatory exponent, ϵ, equal to 0.0365. For this problem, VCCT and J-integral values of G_{total} have been found to equal 74.2 J/m² that is close to the value of 72.7 J/m² reported in Ref 8.

Figure 6 gives a plot of the ratios G_{II}/G_I and G_2/G_1 as a function of Δ/h for the $[0/\pm35/90]_s$ T300/5208 laminate case, analogous to that given in Fig. 5. The plotted results are similar to those of Fig. 5. Use of a traditional VCCT for small values of Δ/h results in significant changes in G_{II}/G_I with decreases in Δ and, because it would have disrupted the plotting scale, the first data point for $\Delta/h = 0.0278$ is not plotted. Again, however, changes with Δ are exaggerated due to the fact that G_I is small relative to G_{II}. As in the $[0_8/90_4]_s$ case, use of the

TABLE 2—*T300/5208 properties used in the $[0/\pm35/90]_s$ case (all moduli are in GPa).*[a]

	0° PLY	
$E_{xx} = 10.2$	$\nu_{xy} = 0.49$	$\mu_{xy} = 3.43$
$E_{yy} = 10.2$	$\nu_{xz} = 0.0228$	$\mu_{xz} = 5.52$
$E_{zz} = 134.4$	$\nu_{yz} = 0.0228$	$\mu_{yz} = 5.52$
	±35° PLY GROUP	
$E_{xx} = 13.2$	$\nu_{xy} = 0.274$	$\mu_{xy} = 4.12$
$E_{yy} = 10.2$	$\nu_{xz} = 0.425$	$\mu_{xz} = 31.43$
$E_{zz} = 35.3$	$\nu_{yz} = -0.0358$	$\mu_{yz} = 4.83$
	90° PLY	
$E_{xx} = 134.4$	$\nu_{xy} = 0.30$	$\mu_{xy} = 5.52$
$E_{yy} = 10.2$	$\nu_{xz} = 0.30$	$\mu_{xz} = 5.52$
$E_{zz} = 10.2$	$\nu_{yz} = 0.49$	$\mu_{yz} = 3.43$

[a]See Fig. 1 for x, y, and z-axis definitions. Ply thickness used was 137.2 μm. These properties are the same as those used in Ref 8.

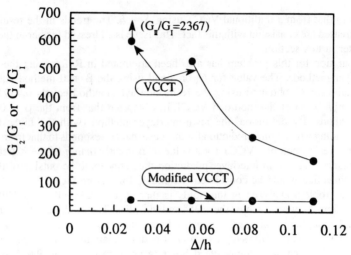

FIG. 6—*Energy release rate ratios versus normalized virtual crack extension length,* Δ, *for the [0/±35/90]ₛ case.*

modified VCCT results in G_2/G_1 ratios that are essentially independent of Δ. The energy release rate quantities obtained are

$$G_1 = 1.83 \text{ J/m}^2 \qquad G_2 = 72.4 \text{ J/m}^2 \qquad G_{\text{total}} = 74.2 \text{ J/m}^2 \qquad (20)$$

Figure 7 provides a plot of the G_1/G_{total} and G_1/G_{total} ratios as a function of Δ/h for this case. This plot is similar to one provided in Ref 8 and the results in Fig. 7 for application of the traditional VCCT agree with those in Ref 8. The result for application of the modified VCCT

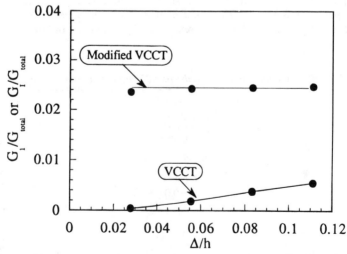

FIG. 7—*Ratio of Mode I energy release rate to total energy release rate versus normalized virtual crack extension length,* Δ, *for the [0/±35/90]ₛ case.*

using only the first element from the crack tip differs slightly from results obtained using multiple elements. This can be attributed to the fact that rapid changes in the displacements as the crack tip is approached can cause the first element from the tip to give less accurate nodal force and displacement results than subsequent elements away from it. As these plots and values indicate, application of the modified VCCT results in a G_1/G_{total} ratio of 0.0247. The $\beta = 0$ method [8] yields a G_I/G_{total} value of approximately 0.0442. Reference 8 also indicates that the resin interlayer method gives a similar prediction. The mode mix prediction from application of the modified VCCT is comparable to that from the $\beta = 0$ and resin interlayer methods. All three methods are predicting Mode II-dominated loading, with the magnitude of the Mode I energy release rate small relative to G_{total}. In this case, however, both the $\beta = 0$ and resin interlayer methods are providing predictions that would correspond to use of a traditional VCCT with a value of Δ greater than h.

In the Theory section of this paper, it is stated that the energy release rate components, G_1 and G_2, obtained by use of the modified VCCT correspond to the energy release rates, G_I and G_{II}, one would obtain using the traditional VCCT with $\Delta = h$, if the near-crack-tip fields extended to a distance, h, from the crack tip. This is due to the normalization by h used in the modified VCCT (see Eqs 13, 14, and 18). It is also stated that near-tip fields will generally not extend to a distance, h, from the crack tip, and G_1 and G_2 will not agree exactly with G_I and G_{II} evaluated using $\Delta = h$. This issue is explored further in Fig. 8. In the figure, the G_{II}/G_I ratio is plotted versus Δ/h, over the range $\Delta = h/18$ to $\Delta = h$ for the $[0_8/90_4]_s$ case. To obtain data over this range, a finite element model is used with 18 rings of elements meshed over the layer thickness, $h = 1120$ μm. Additionally, quantities designated as G_2^ℓ/G_1^ℓ versus ℓ/h are plotted, where G_2^ℓ/G_1^ℓ designates an energy release rate ratio obtained using Eq 18 but with a normalization by $\ell \leq h$ used to calculate Φ_2'. In Fig. 8, values of G_2^ℓ/G_1^ℓ are calculated using a value of Δ equal to $h/9$, using the same mesh as is used to obtain the G_{II}/G_I values. The quantity, G_2^ℓ/G_1^ℓ, for a particular value of ℓ/h corresponds to the quantity, G_{II}/G_I, for $\Delta/h = \ell/h$ and, by definition, G_2^ℓ/G_1^ℓ with $\ell/h = 1$ is equal to G_2/G_1.

It can be seen that G_2^ℓ/G_1^ℓ with $\ell/h = 1$ (that is G_2/G_1) corresponds to, but does not equal the value of G_{II}/G_I from a traditional VCCT applied using $\Delta = h$. The difference (which is

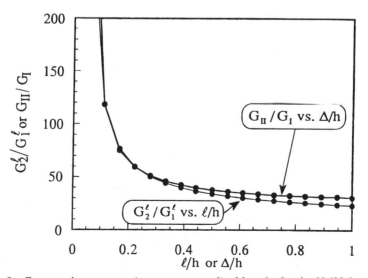

FIG. 8—*Energy release rate ratios versus normalized lengths for the $[0_8/90_4]_s$ case.*

relatively small) between G_2/G_1 evaluated using a small value of Δ/h and G_{II}/G_I evaluated using $\Delta/h = 1$ is due to the fact that nodal forces and displacements outside of the region dominated by the near-tip fields are used in calculating the G_{II}/G_I values. This is further illustrated by the trends in the plots as Δ/h and ℓ/h are reduced. As Fig. 8 shows, the G_{II}/G_I versus Δ/h data smoothly approaches the G_2^ℓ/G_1^ℓ versus ℓ/h data as Δ/h and ℓ/h are reduced. This is due to the increased dominance of the near-tip stresses and displacements as the crack tip is approached.

Comparing a plot of the quantity, G_2^ℓ/G_1^ℓ versus ℓ/h, with that of G_{II}/G_I versus Δ/h is illustrative in pointing out the role of normalization by h in defining Φ_2'. It is also useful in illustrating the region of dominance of the near-crack-tip fields. The merging of the G_2^ℓ/G_1^ℓ versus ℓ/h and G_{II}/G_I versus Δ/h data as ℓ/h and Δ/h are reduced illustrates the increasing role of the singular near-tip stress fields as the crack tip is approached. Finally, the plot of Fig. 8 demonstrates that normalizations other than by the critical length, h, can be used within the methods outlined in this paper. As mentioned in the Theory section, however, alternative normalizations should only be pursued in response to clear experimental evidence that such efforts are needed. Unless or until such evidence exists, consistent normalization by h will offer consistent comparisons between different delamination problems, which is the principal task in delamination analyses.

For the cases considered in this study, some (but not all) of the predictions of the $\beta = 0$ and resin interlayer methods corresponded to a value of Δ greater than h. This would suggest that these methods are not correctly modeling the physics of the problem. As mentioned in the introduction, the resin interlayer method has drawbacks, including the sensitivity of the method to changes in layer thickness and interlayer properties and its relationship to actual crack extension on an interlayer scale. Concern with application of the $\beta = 0$ method rests with whether or not changes in layer properties significantly change the physics of the analysis. For the $[0_8/90_4]_s$ case, the result for the $\beta = 0$ method was obtained in Ref 6 by changing the property ν_{zy} for the $0°$ plies (see Table 1) from 0.25 to 2.474. For the $[0/\pm35/90]_s$ case, ν_{zy} for the ±35 ply group (see Table 2) was changed in Ref 8 from -0.124 to 0.5632. This property change does not affect G_{total}. The interfacial fracture problems modeled in this study are steady-state in nature (independent of crack length). The total energy release rate can be calculated as the difference in potential energy in regions far ahead and far behind the crack tip (because the applied loads are self-equilibrating for this problem, the potential energy ahead of the tip equals zero). This potential energy difference can be calculated using laminate analysis, and laminate stresses are independent of ν_{zy}. In contrast, the mode mix should depend at least somewhat on ν_{zy} that couples normal strains (such as those near the crack tip) with the strain in the z-direction that is constrained to equal zero.

The results in Figs. 5, 6, and 7 point out another use for the Δ-independent energy release rate quantities presented in this work, related to the task of verifying the near-tip convergence of a finite element model. By looking at the modified VCCT results in Figs. 5 through 7, one can be reasonably well-assured that the near-tip mesh is sufficiently refined. Because the ratio of G_2/G_1 is constant for the modified VCCT applied over a number of elements, it is clear that multiple elements are within the region dominated by the near-crack-tip fields. Also, the near-tip elements are correctly picking up the variations in energy release rates due to the oscillatory nature of the solution (because when the modified VCCT factors out oscillatory quantities, a constant value results). Thus, by applying the modified VCCT over multiple elements, numerical modelers can determine if they have a sufficiently refined near-tip mesh, substantially reducing the need to compare results from meshes with many levels of refinement to explicitly verify convergence.

Conclusions

The problem of consistently separating crack extension modes in oscillatory composite delamination models has been addressed. A recently-developed modified virtual crack closure technique (VCCT) from the literature has been described and used to analyze results from two numerical models. The methods described in this study can be applied to planar composite delamination problems where the plies or ply groups are modeled as in-plane orthotropic layers (with one principal axis aligned with the crack front).

The modified VCCT has been shown to extract Δ-independent energy release rate quantities from oscillatory finite element models. Mode mix values from the modified technique are obtained by defining energy release rate quantities that are separated from oscillatory effects of the solution. These energy release rate components correspond to those that would be obtained from a traditional VCCT using $\Delta = h$, if the near-crack-tip fields extended to that distance. Predicted mode mix values from the modified VCCT have been compared to those from the existing $\beta = 0$ and resin interlayer methods. Although results from all of the methods are comparable, some (but not all) of the results from the $\beta = 0$ and resin interlayer methods correspond to a mode mix that would be obtained from a traditional VCCT using a value of Δ greater than h. Such predictions appear to be outside the range of reasonable values for these problems. The ability to extract Δ-independent energy release rate quantities using the modified VCCT can aid numerical modelers in evaluating near-tip solution convergence. The definition of mode mix used in this study also has potential as a consistent definition to be used in comparing composite delamination problems.

Acknowledgment

The authors gratefully acknowledge the support of the Army Research Office, Grant Number DAAH-04-95-1-0163. The authors would also like to thank Kevin O'Brien for his thoughts and insight related to this research.

References

[1]   Wang, A. S. D. and Crossman, F. W., "Initiation and Growth of Transverse Cracks and Edge Delamination in Composite Laminates Parts 1 and 2," *Journal of Composite Materials*, Supplemental Volume, 1980, pp. 71–106.
[2]   O'Brien, T. K., "Characterization of Delamination Onset and Growth in a Composite Laminate," *Damage in Composite Materials, ASTM STP 775*, K. Reifsnider, Ed., American Society for Testing and Materials, Philadelphia, 1982, pp. 140–147.
[3]   Wang, S. S., "Fracture Mechanics for Delamination Problems in Composite Materials," *Progress in Science and Engineering of Composites*, T. Hayashi, K. Kawata, and S. Umekawa, Eds., *Proceedings*, ICCM-IV, Tokyo, 1982, pp. 287–296.
[4]   Atkinson, C., "On Stress Singularities and Interfaces in Linear Elastic Fracture Mechanics," *International Journal of Fracture*, Vol. 13, 1977, pp. 807–820.
[5]   Raju, I. S., Crews, J. H., Jr., and Aminpour, M. A., "Convergence of Strain Energy Release Rate Components for Edge-Delaminated Composite Laminates," *Engineering Fracture Mechanics*, Vol. 30, 1988, pp. 383–396.
[6]   Davidson, B. D., "Prediction of Edge Delamination for Combined In-Plane, Bending and Hygrothermal Loading, Part I Delamination at a Single Interface and Part II Two Symmetrically Located Delaminations," *Journal of Composite Materials*, Vol. 28, 1994, pp. 1009–1031 and pp. 1371–1392.
[7]   Chai, H., "Experimental Evaluation of Mixed-Mode Fracture in Adhesive Bonds," *Experimental Mechanics*, Vol. 32, 1992, pp. 296–303.
[8]   Davidson, B. D., "Prediction of Energy Release Rate for Edge Delamination Using a Crack Tip Element Approach," *Composite Materials: Fatigue and Fracture: Fifth Volume, ASTM STP 1230*, R. H. Martin, Ed., American Society for Testing and Materials, Philadelphia, 4–6 May 1993, pp. 155–175.

[9] He, M. Y. and Hutchinson, J. W., "Kinking of a Crack Out of an Interface," *Journal of Applied Mechanics*, Vol. 56, 1989, pp. 270–278.

[10] Suo, Z. and Hutchinson, J. W., "Sandwich Test Specimens for Measuring Interface Crack Toughness," *Materials Science and Engineering*, Vol. A107, 1989, pp. 135–143.

[11] Beuth, J. L., "Separation of Crack Extension Modes in Orthotropic Delamination Models," *International Journal of Fracture*, Vol. 77, No. 4, 1996, pp. 305–321.

[12] Suo, Z., "Singularities, Interfaces and Cracks in Dissimilar Anisotropic Media," *Proceedings*, Royal Society of London, Vol. A 427, 1990, pp. 331–358.

[13] Rice, J. R., "Elastic Fracture Mechanics Concepts for Interfacial Cracks," *Journal of Applied Mechanics*, Vol. 55, 1988, pp. 98–103.

[14] Suo, Z. and Hutchinson, J. W., "Interface Crack Between Two Elastic Layers," *International Journal of Fracture*, Vol. 43, 1990, pp. 1–18.

[15] Irwin, G. R., "Analysis of Stresses and Strains Near the End of a Crack Traversing a Plate," *Journal of Applied Mechanics*, Vol. 24, 1957, pp. 361–364.

[16] Sun, C. T. and Jih, C. J., "On Strain Energy Release Rates for Interfacial Cracks in Bi-Material Media," *Engineering Fracture Mechanics*, Vol. 28, No. 1, 1987, pp. 13–20.

[17] Sun, C. T. and Manoharan, M. G., "Strain Energy Release Rates of an Interfacial Crack Between Two Orthotropic Solids," *Journal of Composite Materials*, Vol. 23, 1989, pp. 460–478.

[18] Hwu, C. and Hu, J. S., "Stress Intensity Factors and Energy Release Rates of Delaminations in Composite Laminates," *Engineering Fracture Mechanics*, Vol. 42, No. 6, 1992, pp. 977–988.

[19] Rybicki, E. F. and Kanninen, M. F., "A Finite Element Calculation of Stress Intensity Factors by Modified Crack Closure Integral," *Engineering Fracture Mechanics*, Vol. 9, 1977, pp. 931–938.

David W. Palmer,[1] Erian A. Armanios,[1] and David A. Hooke[1]

Fracture Analysis of Internally Delaminated Unsymmetric Laminated Composite Plates

REFERENCE: Palmer, D. W., Armanios, E. A., and Hooke, D. A., **"Fracture Analysis of Internally Delaminated Unsymmetric Laminated Composite Plates,"** *Composite Materials: Fatigue and Fracture (Sixth Volume), ASTM STP 1285*, E. A. Armanios, Ed., American Society for Testing and Materials, 1997, pp. 343–363.

ABSTRACT: A sublaminate analysis is used to develop an analytical solution that determines the effect of midplane internal delamination on unsymmetric elastically tailored laminated plates. A displacement field that recognizes transverse shear deformation is proposed, and an elasticity boundary value problem is developed. A force-deformation relationship for the delaminated plate is derived. A change in total strain energy release rate and extension-twist coupling with crack length is calculated. The distribution of interlaminar shear and peel stresses on the interface containing the delamination is determined. The analysis is applied to a class of hygrothermally stable unsymmetric laminates under uniform extension. Results show that extension-twist coupling decays monotonically and total strain energy release rate increases monotonically, respectively, with delamination width. Interlaminar shear and peel stress are shown to be of larger magnitude at the free edge than at the delamination tip. The variation of extension-twist coupling with delamination width was verified experimentally. Test results indicate that for the cases under consideration, Teflon FEP film is inadequate for the simulation of internal delamination.

KEYWORDS: composite (materials), delamination, unsymmetric plates, interlaminar stress, strain energy release rate, extension-twist coupling, fatigue (materials), fracture (materials)

With the increasing use of composite materials in engineering applications, more attention is being focused on unsymmetric laminates, for two reasons. First, an initially symmetric laminate may have its symmetry altered by the introduction of damage that is not symmetrically distributed about the laminate midplane. A model capable of capturing unsymmetric effects would be required in such a situation. Second, unsymmetric constructions couple in-plane to out-of-plane deformations and therefore provide unique design opportunities. Through the proper selection of a material and stacking sequence, structures can be tailored elastically to capitalize on a given loading environment in a way not possible with conventional materials.

During the service life of any type of composite structure, there is the possibility that the structure will sustain some type of damage. To be able to ensure the ability of the structure to meet design requirements while damaged, methodologies to characterize the effect of structural damage must be developed and incorporated into the design process. To this end, the objective of this work is to determine the response of unsymmetric laminated composites in the presence of internal delamination.

[1]Graduate research assistant, associate professor, and graduate research assistant, respectively, Georgia Institute of Technology, School of Aerospace Engineering, Atlanta, GA 30332-0150.

Separation of plies, or delamination, is a predominant mode of damage in laminated composite structures. The phenomena of interlaminar stress-driven free-edge delamination has been well studied and documented for symmetric laminates. Less attention, however, has been paid to internal delamination in unsymmetric laminates. Interlaminar stresses and strain energy release rates have been calculated for unsymmetric laminates using numerical techniques [1–4]. Analytical solutions for interlaminar stresses and fracture behavior have also been developed. Bottega [5] has generated a growth law for a delamination of arbitrary shape embedded in a layered plate of arbitrary shape. Nilsson and Storakers [6] considered arbitrarily shaped layered composite plates with arbitrarily shaped internal delaminations.

By taking multiple Fourier transforms of the equilibrium equations of three-dimensional anisotropic elasticity, Chatterjee [7] has reduced the three-dimensional elasticity problem of a group of elliptic delaminations in an infinite laminated plate to the solution of a set of integral equations. Two-dimensional problems with disbonds infinitely long in one direction were also considered. Final solutions allow for the evaluation of stress intensity factors and strain energy release rates at delamination tips.

A sublaminate approach was used by Chatterjee and Ramnath [8] for laminates with internal delaminations of circular or elliptical shapes. An exact solution is developed for an infinitely large symmetric laminate with a midplane disbond of finite width. The disbond surfaces were loaded with a normal pressure distribution. Displacement gradients and strain energy release rates were calculated and compared with elasticity and finite element solutions. For finite delaminations or unsymmetric sequences, a finite element solution is required.

The free-edge delamination behavior of unsymmetric laminates has been considered by Armanios and Li [9]. Using a sublaminate modeling scheme, interlaminar stress fields near the crack tip were calculated as well as strain energy release rates and trends of extension-twist coupling decay. The analysis is simplified by neglecting thickness strain. Using a quasi-three-dimensional finite element scheme, Leger and Chan [1] report good correlation with Li [10] in free-edge peel stress predictions.

The approach of Ref 9 allows for a simple and efficient closed-form solution where cause-effect relationships can be established as a result of isolating parameters that control behavior. It will be adopted and expanded in this paper to evaluate the behavior of internally delaminated unsymmetric laminates.

Analytical Model

The laminate under consideration will be subjected to extension force, F, bending moment, M, and twisting moment, T, applied at the ends as shown in Fig. 1. A midplane internal delamination extends along the length of the laminate. A sublaminate modeling scheme is used in which the damaged laminate is considered to be made up of a group of separate laminated units connected with the appropriate continuity conditions on stresses and displacements.

Kinematic Relationships

In this analysis, the laminate length, taken along the x-axis is much larger than its cross-sectional dimensions. Subsequently, the variation of the stress tensor along the x-direction is negligible compared to its variation along the cross-sectional coordinates, y and z.

The laminate is divided into sublaminates representing the regions above and below the delaminated interface. Each sublaminate will be considered as a cylindrical, homogeneous anisotropic elastic body. As the ratio of extensional to shear modulus can be high for a composite material, transverse shear deformation is taken into account. The assumed sublaminate displacement field is written as

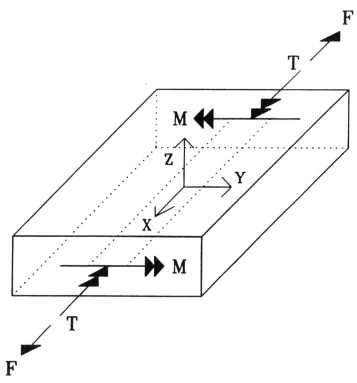

FIG. 1—*Generic damaged laminate and loads.*

$$u(x, y, z) = \epsilon_0 x + \kappa x(z + \delta) + U(y) + z\beta_x(y)$$

$$v(x, y, z) = V(y) + z\beta_y(y) + Cx(z + \delta) \qquad (1)$$

$$w(x, y, z) = -\frac{1}{2}\kappa x^2 - Cx(y + \rho) + W(y)$$

where u, v, and w designate displacements along the x, y, and z-axes, respectively. Equation 1 represents a first-order shear deformation field, using midplane functions and a linear variation of rotation through the sublaminate thickness to describe displacements. The extensional strain, bending curvature in the x-z-plane and relative angle of twist about the x-axis are denoted by ϵ_0, κ, and C, respectively. The constants, δ and ρ, fix the location of the local sublaminate coordinate system relative to the global coordinate system. The strain associated with the displacement field of Eq 1 may be written as

$$\epsilon_{xx} = \epsilon_{xx}^0 + z\kappa_x \qquad \epsilon_{yy} = \epsilon_{yy}^0 + z\kappa_y \qquad \epsilon_{zz} = 0$$

$$\gamma_{xy} = \gamma_{xy}^0 + z\kappa_{xy} \qquad \gamma_{yz} = \gamma_{yz}^0 \qquad \gamma_{xz} = \gamma_{xz}^0 \qquad (2)$$

where

$$\epsilon_{xx}^0 = \epsilon_0 + \kappa\delta \qquad \epsilon_{yy}^0 = V_{,y} \qquad \gamma_{xy}^0 = U_{,y} + C\delta$$

$$\gamma_{yz}^0 = \beta_y + W_{,y} \qquad \gamma_{xz}^0 = \beta_x - C(y + \rho) \qquad (3)$$

$$\kappa_x = \kappa \qquad \kappa_y = \beta_{y,y} \qquad \kappa_{xy} = \beta_{x,y} - C$$

and those strain components associated with the sublaminate midplane are denoted with the superscript, 0. Differentiation is denoted with a comma.

Constitutive Relationships

The constitutive relationships for the generic sublaminate subjected to internal resultant forces and moments as shown in Fig. 2, may be written as

$$
\begin{Bmatrix} N_x \\ N_y \\ N_{xy} \\ M_x \\ M_y \\ M_{xy} \end{Bmatrix} =
\begin{vmatrix}
A_{11} A_{12} A_{16} B_{11} B_{12} B_{16} \\
A_{12} A_{22} A_{26} B_{12} B_{22} B_{26} \\
A_{16} A_{26} A_{66} B_{16} B_{26} B_{66} \\
B_{11} B_{12} B_{16} D_{11} D_{12} D_{16} \\
B_{12} B_{22} B_{26} D_{12} D_{22} D_{26} \\
B_{16} B_{26} B_{66} D_{16} D_{26} D_{66}
\end{vmatrix}
\begin{Bmatrix} \epsilon_{xx}^0 \\ \epsilon_{yy}^0 \\ \gamma_{xy}^0 \\ \kappa_x \\ \kappa_y \\ \kappa_{xy} \end{Bmatrix} \qquad (4)
$$

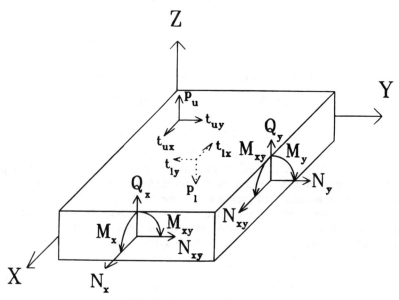

FIG. 2—*Generic sublaminate.*

and

$$\begin{Bmatrix} Q_y \\ Q_x \end{Bmatrix} = \begin{vmatrix} A_{44} A_{45} \\ A_{45} A_{55} \end{vmatrix} \begin{Bmatrix} \gamma^0_{yz} \\ \gamma^0_{xz} \end{Bmatrix} \tag{5}$$

The in-plane stress resultants are denoted by N_x, N_y, and N_{xy}; the moments by M_x, M_y, and M_{xy}; and the transverse shear stress resultants by Q_x and Q_y, as shown in Fig. 2.

The in-plane, coupling, and bending stiffness coefficients, A_{ij}, B_{ij}, and D_{ij}, in Eqs 4 and 5 are defined in terms of the in-plane reduced stiffness coefficients, $\overline{Q}_{ij}$, as [11]

$$(A_{ij}, B_{ij}, D_{ij}) = \int_{-h/2}^{h/2} \overline{Q}_{ij}(1, z, z^2) \, dz \tag{6}$$

Equilibrium Relationships and Boundary Conditions

Using the principle of virtual work, the equilibrium equations for the generic sublaminate are derived as

$$N_{y,y} + t_{uy} - t_{1y} = 0 \tag{7}$$

$$N_{xy,y} + t_{ux} - t_{1x} = 0 \tag{8}$$

$$Q_{y,y} + p_u - p_1 = 0 \tag{9}$$

$$M_{y,y} - Q_y + \frac{h}{2}(t_{uy} + t_{1y}) = 0 \tag{10}$$

$$M_{xy,y} - Q_x + \frac{h}{2}(t_{ux} + t_{1x}) = 0 \tag{11}$$

As shown in Fig. 2, the interlaminar shear stresses t_{ux}, t_{uy}, and the peel stress p_u act on the upper surface of the sublaminate while t_{1x}, t_{1y}, p_1 act on the lower surface. The boundary conditions resulting from the virtual work analysis, to be applied at constant values of y, are the specification of one member of the following pairs

$$N_y \quad \text{or} \quad V, \quad N_{xy} \quad \text{or} \quad U, \quad Q_y \quad \text{or} \quad W, \quad M_{xy} \quad \text{or} \quad \beta_x, \quad M_y \quad \text{or} \quad \beta_y \tag{12}$$

Analysis of Delaminated Plate

The generic sublaminate model will now be used to determine the effect of the delamination on the laminate response. The damage under consideration will be on the midplane of the laminate, symmetric with respect to the X-Z plane and spans the entire length. By applying the generic model to each sublaminate and enforcing appropriate conditions on stress and displacement, a force-deformation relationship describing the behavior of the delaminated plate can be derived. From that relationship, the effect of the delamination on the elastic coupling and an estimate of the total strain energy release rate can be determined. Finally, the shear and peel stresses on the interface containing the delamination will be calculated.

In modeling the delaminated plate, the conditions that exist at the midsection of the laminate (on the X-Z plane) will be assumed to be those that exist on the X-Z plane of a completely delaminated strip. This will allow the analysis to be completed with consideration of only half the cross section [12].

Kinematic Relationships

The damaged laminate will be divided into sublaminates as shown in Fig. 3. A local coordinate system is established at the center of each sublaminate by specifying the constants, δ and ρ, as

$$\delta_1 = \delta_2 = \frac{h_3}{2}, \qquad \delta_3 = \delta_4 = -\frac{h_2}{2}$$

$$\rho_1 = \rho_4 = -\frac{a}{2} \tag{13}$$

$$\rho_2 = \rho_3 = -\left(a + \frac{b-a}{2}\right)$$

The subscripts associated with the variables in Eq 13 refer to the respective sublaminate.

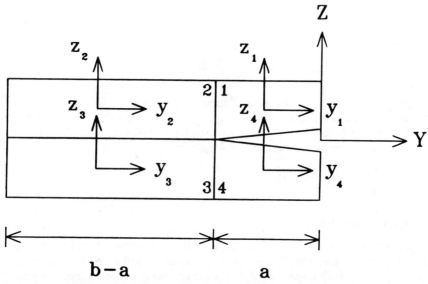

FIG. 3—*Sublaminate numbering and coordinate systems.*

Sublaminates 1 and 4 represent the group of plies above and below the delamination, respectively, while Sublaminates 2 and 3 represent the same group of plies in the uncracked region, as shown in Fig. 3. As Sublaminates 2 and 3 comprise the uncracked region, continuity in displacement at the 2–3 interface must be enforced.

Sublaminate Equilibrium Equations

Sublaminates 1 and 4, above and below the delamination, are separated from one another and will be considered to have stress-free upper and lower surfaces. As a result, the responses of the cracked region sublaminates are independent of each other, and Eqs 7 through 11 may be applied directly. At the laminate X-Z plane, the following conditions hold

$$N_y = N_{xy} = Q_y = M_y = \beta_x = 0 \tag{14}$$

Equations 7 through 11 may be reduced by integrating and enforcing the conditions of Eq 14. They then become

$$N_{xy} = Q_y = 0$$
$$N_y = M_y = 0 \tag{15}$$
$$M_{xy,y} - Q_x = 0$$

In the uncracked region, however, the sublaminate responses are coupled to one another. Enforcing reciprocity of interlaminar stresses at the interface between Sublaminates 2 and 3 allows the equilibrium equations in the uncracked region to be combined in the following form

$$N_{xy2} + N_{xy3} = 0 \tag{16}$$

$$N_{y2} + N_{y3} = 0 \tag{17}$$

$$Q_{y2} + Q_{y3} = 0 \tag{18}$$

$$M_{xy2,y} + \frac{h_2}{2} N_{xy2,y} - Q_{x2} = 0 \tag{19}$$

$$M_{y2,y} + \frac{h_2}{2} N_{y2,y} - Q_{y2} = 0 \tag{20}$$

$$M_{xy3,y} - \frac{h_3}{2} N_{xy3,y} - Q_{x3} = 0 \tag{21}$$

$$M_{y3,y} - \frac{h_3}{2} N_{y3,y} - Q_{y3} = 0 \tag{22}$$

The subscripts associated with the variables in Eqs 16 through 22 denote the respective sublaminate.

Solution for the Rotation Functions

By combining the equilibrium equations with the constitutive and strain-displacement relationships, it will be shown that the sublaminate rotation functions are governed by ordinary differential equations with constant coefficients. Exact solutions can be written for these equations, allowing for a simple and efficient analysis.

Uncracked Region—Since the responses of Sublaminates 2 and 3 are coupled, determination of the rotation functions in the uncracked region involves the solution of a system of seven equations, Eqs 16 through 22, in seven unknowns: U_2, V_2, W_2, β_{2x}, β_{2y}, β_{3x}, and β_{3y}. Using Eqs 16, 17, and 18 to eliminate $U_{2,y}$, $V_{2,y}$, and $W_{2,y}$ in terms of the rotation functions, Eqs 19 through 22 yield the system of equations governing the rotation functions. The system may be written as

$$[P]\{\beta\}_{,yy} - [Q]\{\beta\} = \{R\}C(y + \rho) \tag{23}$$

The solution of Eq 23 may be written as

$$\begin{Bmatrix} \beta_{2x} \\ \beta_{2y} \\ \beta_{3x} \\ \beta_{3y} \end{Bmatrix} = [\Phi] \left\{ [\Omega_1] \begin{Bmatrix} c_1 \\ c_2 \\ c_3 \\ c_4 \end{Bmatrix} + [\Omega_2] \begin{Bmatrix} c_5 \\ c_6 \\ c_7 \\ c_8 \end{Bmatrix} - C \begin{Bmatrix} t_{11} \dfrac{(y+\rho)}{\lambda_1} \\ t_{21} \dfrac{(y+\rho)}{\lambda_2} \\ t_{31} \dfrac{(y+\rho)}{\lambda_3} \\ -\dfrac{t_{41}}{6}(y^3 + 3y^2\rho) \end{Bmatrix} \right\} \tag{24}$$

Matrices Ω_1 and Ω_2 are 4×4 diagonal with the following form

$$[\Omega_1] = \begin{bmatrix} e^{y\sqrt{\lambda_i}} & 0 \\ 0 & y \end{bmatrix}, \quad [\Omega_2] = \begin{bmatrix} e^{-y\sqrt{\lambda_i}} & 0 \\ 0 & 1 \end{bmatrix} \tag{25}$$

Subscript i ranges from 1 to 3 and takes the value of the corresponding row number. The t_{i1} ($i = 1$–4) are the elements of the $[\Phi]^{-1}[P]^{-1}\{R\}$ vector. Parameters λ_i and matrix Φ comprise the eigensystem of the matrix $P^{-1}Q$. The columns of Φ are the eigenvectors, and $\lambda_1 > \lambda_2 > \lambda_3$. Adding Eqs 20 and 22, integrating and enforcing free-edge boundary conditions, reveals that there will be only three nonzero eigenvalues.

Cracked Region—Since the responses of Sublaminates 1 and 4 are uncoupled, determination of the rotation functions in the cracked region involves the solution of five equations, Eqs 15, in five unknowns, U, V, W, β_x, and β_y. The first four of Eqs 15 may be used to solve for the kinematic variables $U_{1,y}$, $V_{1,y}$, $W_{1,y}$, and $\beta_{1y,y}$ associated with Sublaminate 1 in terms of the rotation, β_{1x}. Elimination of the midplane displacement functions from the last of Eqs 15 will yield the following differential equation governing β_{1x}

$$\gamma_1\beta_{1x,yy} - \gamma_2\beta_{1x} = -\gamma_2 C(y + p) \tag{26}$$

The solution of Eq 26 may be written as

$$\beta_{1x} = c_9 e^{y\sqrt{\gamma_2/\gamma_1}} + c_{10}e^{-y\sqrt{\gamma_2/\gamma_1}} + C(y + p) \tag{27}$$

Using a similar procedure for Sublaminate 4

$$\beta_{4x} = c_{11}e^{y\sqrt{\gamma_4/\gamma_3}} + c_{12}e^{-y\sqrt{\gamma_4/\gamma_3}} + C(y + p) \tag{28}$$

Solution for the Integration Constants

The sublaminate constitutive equations may now be expressed in terms of the constants of integration by using Eqs 24, 27, 28, 4, and 5. It should be noted that the force and moment resultants in the uncracked region do not depend on c_8. Further, since the calculation of subsequent results depends only on the sublaminate force and moment resultants, only eleven constants need be determined. The solution for the remaining eleven integration constants in Eqs 24, 27, and 28 may be determined by simultaneously solving the boundary and continuity conditions [12] that result from the derivation of Eqs 7 through 11 and 16 through 22. At the free edge

$$M_{xy}^2 + \frac{h_2}{2} N_{xy}^2 = 0$$

$$M_y^2 + \frac{h_2}{2} N_y^2 = 0$$

$$M_{xy}^3 - \frac{h_3}{2} N_{xy}^3 = 0 \tag{29}$$

$$M_y^3 - \frac{h_3}{2} N_y^3 = 0$$

At the X–Z plane

$$\beta_{1x} = 0$$

$$\beta_{4x} = 0 \tag{30}$$

At the 1–2 interface

$$M_{xy2} + \frac{h_2}{2} N_{xy2} = M_{xy1}$$

$$\beta_{1x} = \beta_{2x} \tag{31}$$

$$Q_{y2} = 0$$

and at the 3–4 interface

$$M_{xy3} - \frac{h_3}{2} N_{xy3} = M_{xy4}$$

$$\beta_{3x} = \beta_{4x} \tag{32}$$

Equation 29 represents the effective free-edge conditions. The vanishing rotations at $y_1 = a/2$ (Eq 30) are the remaining unused conditions of Eq 14. Equations 31 and 32 equate the effective twisting moments and associated rotation functions at the 1–2 and 3–4 interfaces. Equations 29 through 32 represent an 11×11 system of linear algebraic equations in the constants of integration. A numerical solution is obtained via the IMSL DLSARG subroutine [13].

Loading Conditions

The laminate loading conditions may be expressed as

$$F = 2 \int_{(a-b)/2}^{(b-a)/2} (N_{x2} + N_{x3}) dy + 2 \int_{-a/2}^{a/2} (N_{x1} + N_{x4}) dy \tag{33}$$

$$M = 2 \int_{(a-b)/2}^{(b-a)/2} (M_{x2} + M_{x3} + \delta_2 N_{x2} + \delta_3 N_{x3}) dy +$$

$$2 \int_{-a/2}^{a/2} (M_{x1} + M_{x4} + \delta_1 N_{x1} + \delta_4 N_{x4}) dy \tag{34}$$

$$T = 2 \int_{(a-b)/2}^{(b-a)/2} (M_{xy2} + M_{xy3} + \delta_2 N_{xy2} + \delta_3 N_{xy3} - (Q_{x2} + Q_{x3})(y - a)) dy +$$

$$2 \int_{-a/2}^{a/2} (M_{xy1} + M_{xy4} + \delta_1 N_{xy1} + \delta_4 N_{xy4} - (Q_{x1} + Q_{x4})(y - a)) dy \tag{35}$$

Using Eqs 15, 19, 21, and 29, the twisting moment may be reduced to

$$T = 4 \int_{(a-b)/2}^{(b-a)/2} (M_{xy2} + M_{xy3} + \delta_2 N_{xy2} + \delta_3 N_{xy3}) dy + 4 \int_{-a/2}^{a/2} (M_{xy1} + M_{xy4}) dy \tag{36}$$

Laminate Force-Deformation Relationship

After obtaining the constants of integration, the sublaminate force and moment resultants are expressed in terms of the strain and curvatures ϵ_0, κ, and C. Substitution into the loading conditions yields the force-deformation relationship for the damaged laminate as

$$\begin{Bmatrix} F \\ M \\ T \end{Bmatrix} = \begin{bmatrix} \psi_{11} & \psi_{12} & \psi_{13} \\ \psi_{21} & \psi_{22} & \psi_{23} \\ \psi_{31} & \psi_{32} & \psi_{33} \end{bmatrix} \begin{Bmatrix} \epsilon_0 \\ \kappa \\ C \end{Bmatrix} \tag{37}$$

From Eq 37, the effect of the delamination on the laminate stiffnesses or coupling may be determined.

This study will pay particular attention to the effect of the damage on extension-twist coupling. In order to isolate that effect, set M and T to zero in Eq 37 and solve for κ and C as a function of ϵ_0

$$\begin{Bmatrix} \kappa \\ C \end{Bmatrix} = - \begin{bmatrix} \psi_{22} & \psi_{23} \\ \psi_{32} & \psi_{33} \end{bmatrix}^{-1} \begin{Bmatrix} \psi_{21} \\ \psi_{31} \end{Bmatrix} \epsilon_0 \tag{38}$$

Total Strain Energy Release Rate

The total strain energy release rate of the delaminated plate for extension loading may be calculated as

$$G_T = - \frac{\epsilon_0}{2} \frac{\partial F}{\partial a} \tag{39}$$

Substitute from Eq 37 to get

$$G_T = - \frac{\epsilon_0}{2} (\psi_{11,a} \epsilon_0 + \psi_{12,a} \kappa + \psi_{13,a} C) - \frac{\epsilon_0}{2} (\psi_{12} \kappa_{,a} + \psi_{13} C_{,a}) \tag{40}$$

Interlaminar Stresses

The shear and peel stresses acting on the interface between Sublaminates 2 and 3 may be evaluated directly from Eqs 7 through 9. As the upper surface of Sublaminate 2 is stress free, we have

$$\begin{Bmatrix} t_{1y} \\ t_{1x} \\ p_1 \end{Bmatrix} = \begin{Bmatrix} N_{2y,y} \\ N_{2xy,y} \\ Q_{2y,y} \end{Bmatrix} \tag{41}$$

The displacement field of Eq 2 does not allow for ϵ_{zz}. As a result, the peel stress distribution of Eq 41 is not expected to be accurate. With the addition of a boundary layer function to the shear stress resultant, the peel stress distribution can be corrected. The modified resultant will be written as

$$Q_y^* = Q_y + a_1 e^{\lambda_3 y} + a_2 e^{-\lambda_3 y} \tag{42}$$

The coefficients, a_1 and a_2, are determined by enforcing the free-edge condition and satisfying moment equilibrium for the sublaminate

$$\int_{(a-b)/2}^{(b-a)/2} p y \, dy + \frac{b-a}{2}\left(Q^*\left(\frac{b-a}{2}\right) + Q^*\left(\frac{a-b}{2}\right)\right)$$

$$-\frac{h}{4} N_{2y}\left(\frac{b-a}{2}\right) - M_{2y}\left(\frac{b-a}{2}\right) = 0 \tag{43}$$

In a similar fashion, the inability of the current model to satisfy individual boundary and continuity conditions on N_{xy} can be addressed by modifying the resultant as

$$N_{xy}^* = N_{xy} + b_1 e^{\lambda_3 y} + b_2 e^{-\lambda_3 y} \tag{44}$$

and obtaining b_1 and b_2 by setting N_{xy} to zero at the free edge and delamination front.

Application and Results

The model is applied to a class of hygrothermally stable unsymmetric laminates exhibiting extension twist coupling. The stacking sequence was developed by Winckler [14] and is given as

$$[\theta/(90 + \theta)_2/\theta/-\theta/(90 - \theta)_2/-\theta]_T \tag{45}$$

The sequence consists of two $[0/90]_s$ laminates atop one another that have undergone an opposing θ-degree rotation. Once rotated, the upper and lower halves produce opposing in-plane extension-shear coupling. Under extension, the opposing in-plane shear forces then produce a twisting couple, and the laminate exhibits extension-twist coupling. As each half of the laminate is symmetric, the introduction of midplane delamination will not cause a loss of hygrothermal stability. Hence, the model just developed remains valid. Results will be calculated for a laminate with $\theta = 30°$ undergoing a uniform extension, ϵ_0.

The material system for the laminates studied is AS4/3502 graphite/epoxy with properties provided in Table 1.

Figure 4 plots normalized extension twist coupling, Ch/ϵ_0, where h is the laminate thickness, versus delamination width. From Fig. 4, it can be seen that the extension-twist coupling decays monotonically with delamination width. The coupling value corresponding to no delamination is identical to the value obtained by classical lamination theory (CLT). This indicates that the influence of shear deformation on the extension-twist coupling in the undamaged state is negligible. It may be noted that the influence of shear deformation on extension-twist coupling vanishes for an undamaged laminate with a general stacking sequence, under axial loading only [10]. A quantitative measure for the coupling loss is provided in Table 2 where the

TABLE 1—*Properties of AS4/3502 graphite/epoxy material system.*

$E_{11} = 113.31$ GPa (16.44 Msi)
$E_{22} = 9.73$ GPa (1.41 Msi)
$G_{12} = 5.66$ GPa (0.82 Msi)
$G_{23} = 3.4$ GPa (0.50 Msi)
$\nu_{12} = 0.25$
$\nu_{23} = 0.42$
Ply thickness, $t = 0.14$ mm (0.0055 in.)

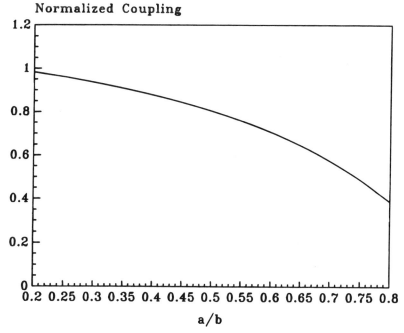

FIG. 4—*Variation of extension-twist coupling with delamination.*

TABLE 2—*Percentage coupling loss.*

a/b	Loss, %
0.2	7.4
0.3	11.8
0.4	17.3
0.5	24.2
0.6	33.4
0.7	45.9
0.8	63.6

coupling loss percent relative to CLT is calculated for delamination lengths ranging from 20 to 80% of the laminate width, b.

The interlaminar stresses, p and t_x, normalized by E_{11} are plotted in Figs. 5 and 6, respectively, for a plate with $a/b = 20\%$. The ply thickness is denoted by t in the figures, and the laminate free edge is located at $(y/t) = -36.4$ while the internal delamination tip at $(y/t) = 36.4$. The predictions of the modified interlaminar stresses based on Eqs 42 and 44 are labeled "corrected" in Figs. 5 and 6. It should be noted that for both stresses, the effect of the boundary layer modifications is to increase the criticality of the free edge relative to the delamination tip. For the stacking sequence under consideration, the interlaminar shear stress, t_y, depends on the induced bending curvature, κ, only. As the sequence of Eq 45 produces no extension-bending coupling, t_y is zero.

The total strain energy release rate (SERR), normalized by $E_{11}h\epsilon_0^2$, versus delamination width is plotted in Fig. 7. It is seen to increase monotonically with delamination width.

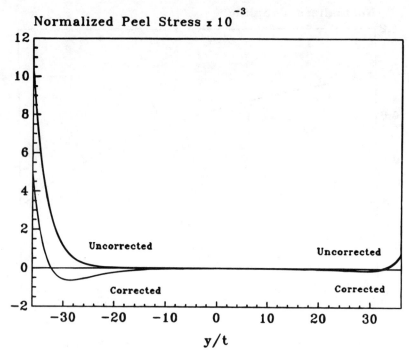

FIG. 5—*Interlaminar peel stress distribution along the delaminated interface.*

Experimental Verification

An experimental program was developed in order to verify the analytical prediction of variation of extension-twist coupling with delamination width. Using the stacking sequence of Eq 45, four damaged and four undamaged laminates were constructed for $\theta = 30°$. The test specimens measured 25.4 mm (1 in.) wide by 279.4 mm (11 in.) long, and the damaged specimens contained 12.7-mm (0.5-in.) wide internal midplane delamination. Teflon FEP film was used to simulate the delamination and was placed along the full length of the specimen. The specimens were constructed of ICI Fiberite T300/954-3 graphite/cyanate prepreg material with properties listed in Table 3.

Three experimental methods for measuring the twist induced in a laminate with extension-twist coupling have been developed by Hooke and Armanios [15]. Two of the methods involve the use of transducers that allow one end of the test specimen to twist as an extensional load is applied. In the third method, the specimen is loaded in tension by attaching a mass to one end and spinning the laminate about the other end, in the fashion of a helicopter blade. For the current set of tests, a rotary transducer-based test platform was used to collect twist data.

The transducer under consideration is referred to as the Improved Thrust Bearing Transducer in Ref *15* and is shown to provide freedom of end twist with minimal friction. The transducer is loaded into a universal test machine, and the upper end of the specimen is clamped to the transducer shaft, as shown in Fig. 8. In taking the experimental data, the twist at the transducer end of the specimen was recorded at load steps of 222.4 N (50 lb).

Tables 4 and 5 present a comparison of the analytical predictions and average experimental measurements of end twist for the undamaged and damaged specimens, respectively. The percent difference is relative to the average end-twist test data for the corresponding four-

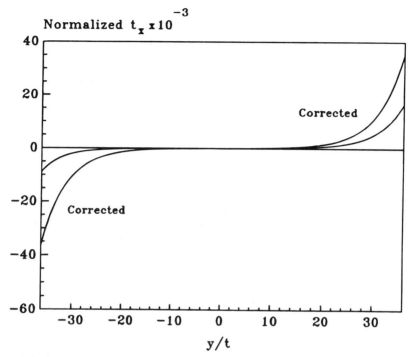

FIG. 6—*Interlaminar shear stress distribution along the delaminated interface.*

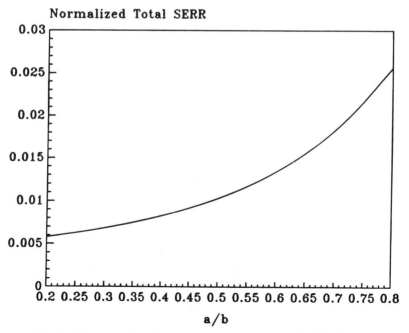

FIG. 7—*Variation of total strain energy release rate with delamination.*

TABLE 3—*Properties of T300/954-3 graphite/cyanate material system.*

$$E_{11} = 135.6 \text{ GPa } (19.7 \text{ Msi})$$
$$E_{22} = 9.9 \text{ GPa } (1.4 \text{ Msi})$$
$$G_{12} = 4.2 \text{ GPa } (0.6 \text{ Msi})$$
$$G_{23} = 2.5 \text{ GPa } (0.36 \text{ Msi})$$
$$\nu_{12} = 0.3$$
$$\nu_{23} = 0.5$$
$$\text{Ply thickness} = 0.15 \text{ mm } (0.006 \text{ in.})$$

specimen group. Figure 9 presents the results of the test data from the four delaminated specimens, labeled 1–4 in the figure, and the predictions of end twist provided by the model for the undamaged and damaged states. As the model does not account for large deflections, a linear relationship between extension force and end twist is predicted. It is clear from the figure that the end twist data shows a nonlinear trend and that the analytical model underpredicts end twist when damage is present. The nonlinearity in the data is a result of the stiffening effect produced by the large values of cross-sectional rotation. The test data falls between the analytical predictions for the undamaged and damaged laminates.

Influence of the Teflon Film

The data in Tables 4 and 5 indicate that the undamaged specimens exhibit smaller amounts of coupling experimentally than analytically. The trend is reversed for the damaged specimens. It is also seen from Table 5 that the analytical model underestimates the amount of end twist, although the absolute difference decreases as load increases. In order to address this issue, the use of the Teflon film to simulate internal delamination was reevaluated.

In the development of the analytical model, it is assumed that the surfaces above and below the delamination are stress free. The difference between the analytical predictions and experimental values of end twist is a result of the breakdown of this stress-free assumption. After sectioning a specimen that contained simulated delamination, it was found that the Teflon film is deflected in the thickness direction during the curing process, where the pressure applied to the material reaches a peak of 689.5 KPa (100 psi), so that it no longer presents a smooth, planar surface.

Figures 10 and 11 display photographs of the Teflon film at magnifications of 20 and 50 times, respectively. The figures reveal that after having gone through the cure cycle the opposing sides of the film are left with an impression of the adjacent fibers. This indicates that the fibers are partially embedded in the Teflon film after curing, and the film serves to mechanically couple the material on the opposing sides of the film. Under such conditions, interfacial stresses are transferred across the area of the Teflon, and the assumption of stress-free surfaces taken in the model breaks down.

In order to further verify the transfer of stress across the Teflon film, and hence the generation of extension-twist coupling, additional experimental steps were taken. Two of the damaged test specimens were machined to widths of 12.7 mm (1/2 in.) and 14.29 mm (9/16 in.), corresponding to a/b ratios of 1.0 and 0.9, respectively. In addition, two four-ply laminates corresponding to the upper and lower halves of the stacking sequence in Eq 45 with $\theta = 30°$ were fabricated and cured separately. One of the undamaged specimens was also machined to a width of 12.7 mm ($^1/_2$ in.).

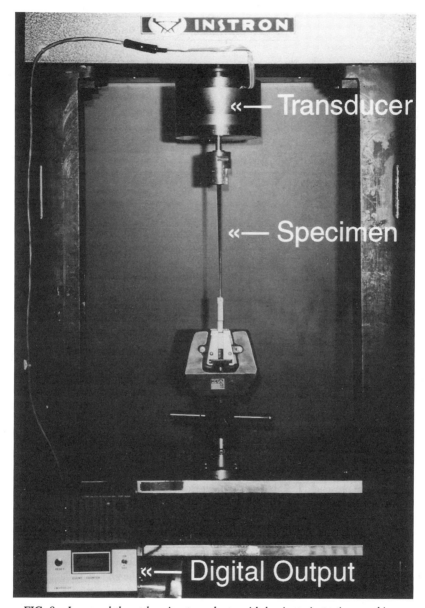

FIG. 8—*Improved thrust bearing transducer with laminate in testing machine.*

The first trial consisted of mounting the two four-ply laminates into the test platform at once, so that their surfaces were in contact. Extensional loading was applied up to 1.112 KN (250 lb), and no end twist was produced. This verified the analytical result indicating that a specimen fully delaminated at the midplane with the stacking sequence of Eq 45 produces no extension-twist coupling.

TABLE 4—*Analytical and experimental end twist, undamaged.*

Load, N (lb)	End Twist, °		
	Theory	Experiment[a]	Difference, %
222.4 (50)	4.9	4.7	−4.3
444.8 (100)	9.7	8.7	−11.5
667.2 (150)	14.6	12.8	−14.1
889.6 (200)	19.4	16.4	−18.3
1112.1 (250)	24.3	19.7	−23.4
1334.5 (300)	29.2	22.5	−29.7
1556.9 (350)	34.0	25.6	−32.8

[a]Average from four specimens.

TABLE 5—*Analytical and experimental end twist, damaged.*

Load, N (lb)	End Twist, °		
	Theory	Experiment[a]	Difference, %
222.4 (50)	3.4	4.0	15.0
444.8 (100)	6.7	8.2	18.3
667.2 (150)	10.1	12.0	15.8
889.6 (200)	13.5	15.4	12.3
1112.1 (250)	16.9	18.8	10.1
1334.4 (300)	20.2	21.8	7.3
1556.9 (350)	23.6	24.7	4.5

[a]Average from four specimens.

The machined specimens were then tested in the same manner as the original specimens, and the results are presented in Table 6. It is clear from those results that even at a delamination level of 90%, the specimen produced at least 66.7% of the undamaged end twist at each load. It should be noted that the 14.29-mm (9/16-in.) wide specimen experienced catastrophic failure at a load of 1 KN (225 lb), as it became fully delaminated and lost all extension-twist coupling.

The analysis indicates that the end twist is inversely proportional to the width of the specimen. That is, an undamaged specimen trimmed to half its width should exhibit twice as much twist as the original specimen for the same load. Table 7 provides the end twist data of the original undamaged specimen with a width of 25.4 mm (1 in.) and the trimmed 12.7-mm (1/2-in.) wide specimen for the same load. The twist associated with the original and trimmed specimens are denoted by θ_o and θ_h, respectively. The relative percent difference between twice the original and trimmed specimen twist data at each load is given in Table 7. The maximum absolute difference is 5.4%. This provides further verification of the accuracy of the test procedure and designed transducer.

The end-twist data produced by the machined fully delaminated specimen verifies that the Teflon film is indeed permitting the transfer of stresses, and that for laminates of the material in Table 3 and the stacking sequence of Eq 45 under extension loading, Teflon FEP film is inadequate for simulating delamination.

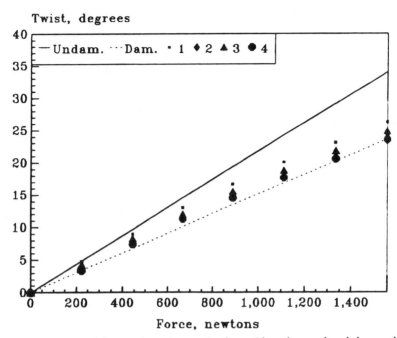

FIG. 9—*Comparison of damaged specimen twist data with undamaged and damaged analytical predictions.*

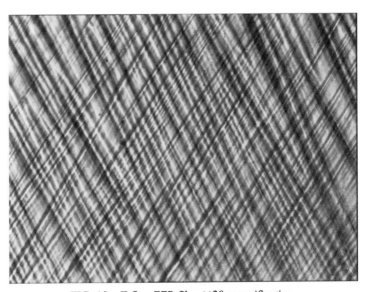

FIG. 10—*Teflon FEP film, ×20 magnification.*

FIG. 11—*Teflon FEP film, ×50 magnification.*

TABLE 6—*Experimental end twist for partially and fully delaminated specimens.*

	End Twist, °		
Load, N (lb)	$a/b = 0.0$	$a/b = 0.9$	$a/b = 1.0$
222.4 (50)	8.7	6.8	1.4
444.8 (100)	17.3	13.7	2.5
667.2 (150)	25.4	20.5	3.6
889.6 (200)	33.9	22.6	4.2
1112.1 (250)	42.6	⋯	4.9

TABLE 7—*Experimental end twist for original and trimmed specimens.*

	End Twist, °		
Load, N (lb)	12.7 mm (0.5 in.), θ_0	25.4 mm (1 in.), θ_h	Difference, %, ×100 $(2\theta_0 - \theta_h)/2\theta_0$
222.4 (50)	8.7	4.5	3.3
444.8 (100)	17.3	8.9	2.8
667.2 (150)	25.4	13.1	3.1
889.6 (200)	33.9	16.8	−1.0
1112.1 (250)	42.6	20.2	−5.4

Conclusion

An analytical model has been developed to predict the effect of internal midplane delamination in unsymmetric laminated plates. The model is based on a sublaminate approach that accounts for transverse shear deformation. The model was applied to a class of hygrothermally stable unsymmetric laminates made of AS4/3502 graphite/epoxy. It was found that extension-twist coupling decreases monotonically with increasing delamination width. Shear and peel stresses were found to be of larger magnitude at the free edge than at the delamination tip, and total strain energy release rate increased monotonically with delamination width.

Experimental verification of extension-twist coupling variation with delamination was performed for specimens made of T300/954-3 graphite/ cyanate material system. Extension-twist coupling data exhibited a nonlinear trend. Test data from specimens with internal delamination exhibited more coupling than predicted analytically. This is attributed to the interfacial stresses generated as a result of the nonplanar deformation during the cure cycle of the Teflon FEP film used to simulate delamination.

It is recommended that a geometrically nonlinear model be developed in order to assess the extension-twist coupling behavior of specimen undergoing finite twisting rotation.

References

[1] Leger, C. A. and Chan, W. S., "Analysis of Interlaminar Stresses in Symmetric and Unsymmetric Laminates Under Various Loadings," *Proceedings*, Thirty-fourth AIAA/ASME/ASCE/AHS/ASC Structures, Structural Dynamics and Materials Conference, La Jolla, CA, 1993, pp. 1770–1776.

[2] Norwood, D. S., Shuart, M. J., and Herakovich, C. T., "Geometrically Nonlinear Analysis of Interlaminar Stresses in Unsymmetrically Laminated Plates Subjected to Uniform Thermal Loading," *Proceedings*, Winter Annual Meeting, American Society of Mechanical Engineers, Atlanta, GA, Dec. 1991, pp. 91–109.

[3] Chaudhuri, R. A. and Seide, P., "Approximate Semi-Analytical Method for Prediction of Interlaminar Shear Stresses in an Arbitrarily Laminated Thick Plate," *Computers and Structures*, Vol. 25, No. 4, 1987, pp. 627–636.

[4] Ochoa, O. O. and Ross, G. R., "Unsymmetric Laminates," *Proceedings*, Fourteenth Annual Energy-Sources Technology Conference and Exhibition, Houston, TX, Oct. 1991, pp. 229–235.

[5] Bottega, W. J., "A Growth Law for Propagation of Arbitrary Shaped Delaminations in Layered Plates," *International Journal of Solids and Structures*, Vol. 19, No. 11, 1983, pp. 1009–1017.

[6] Nilsson, K.-F. and Storakers, B., "On Interface Crack Growth in Composite Plates," *ASME Journal of Applied Mechanics*, Vol. 59, Sept. 1992, pp. 530–538.

[7] Chatterjee, S. N., "Three and Two Dimensional Stress Fields Near Delaminations in Laminated Composite Plates," *International Journal of Solids and Structures*, Vol. 23, No. 11, 1987, pp. 1535–1549.

[8] Chatterjee, S. N. and Ramnath, V., "Modeling Laminated Composite Structures as Assemblage of Sublaminates," *International Journal of Solids and Structures*, Vol. 24, No. 5, 1988, pp. 439–458.

[9] Armanios, E. A. and Li, J., "Interlaminar Fracture Analysis of Unsymmetrical Laminates," *Composite Materials: Fatigue and Fracture, Fourth Volume, ASTM STP 1156*, W. W. Stinchcomb and N. E. Ashbaugh, Eds., American Society for Testing and Materials, Philadelphia, 1993, pp. 241–360.

[10] Li, J., "Interlaminar Fracture Analysis of Laminated Composites Under Combined Loading," Ph.D. thesis, School of Aerospace Engineering, Georgia Institute of Technology, Atlanta, GA, 1992.

[11] Vinson, J. R. and Sierakowski, R. L., *The Behavior of Structures Composed of Composite Materials*, Martinus Nijhoff Publishers, Boston, 1987, p. 46.

[12] Palmer, D. W., "The Effect of Internal Delamination on Unsymmetric Laminated Composite Plates," Ph.D. thesis, School of Aerospace Engineering, Georgia Institute of Technology, Atlanta, GA, 1995.

[13] *Math/Library FORTRAN Subroutines for Mathematical Applications*, Version 1.0, International Mathematical and Statistical Libraries, Inc., Houston, TX, 1987, pp. 278–281.

[14] Winckler, S. J., "Hygrothermally Curvature Stable Laminates with Tension-Torsion Coupling," *Journal*, American Helicopter Society, Vol. 30, No. 3, July 1985, pp. 56–58.

[15] Hooke, D. A. and Armanios, E. A., "Design and Evaluation of Three Methods for Testing Extension-Twist Coupled Laminates," *Composite Materials: Testing and Design: Twelfth Volume, ASTM STP 1274*, C. R. Saff and R. B. Deo, Eds., American Society for Testing and Materials, 1996, pp. 340–357.

Elizabeth A. Friis,[1] Dustan L. Hahn,[2] Francis W. Cooke,[1] and Steven J. Hooper[3]

Modeling Crack Extension in Chopped-Fiber Composites

REFERENCE: Friis, E. A., Hahn, D. L., Cooke, F. W., and Hooper, S. J., **"Modeling Crack Extension in Chopped-Fiber Composites,"** *Composite Materials: Fatigue and Fracture (Sixth Volume), ASTM STP 1285*, E. A. Armanios, Ed., American Society for Testing and Materials, 1997, pp. 364–378.

ABSTRACT: A finite element model for predicting the effect of fiber bridging, fiber properties, and fiber-matrix interface strength on the crack-tip stresses and crack propagation potential of a chopped-fiber composite is proposed. The method of virtual crack extension was used to model crack growth in a micromechanics composite material model. The purpose of this study was to investigate the effects of fiber bridging of a crack and fiber-matrix interface strength on the stress intensity and strain energy release rate in virtual crack extension.

A model of an aligned, 1% by volume fiber loading, chopped-fiber composite with a pre-existing crack was developed to represent a portion of a fracture toughness or fatigue crack propagation specimen. Nonlinear contact elements were used to model fiber-matrix interface strengths. The von Mises stress at the crack tip was calculated for each configuration before crack extension, and the strain energy release rate was calculated for each crack step.

The presence of fibers without bridging of the crack did not greatly affect the stress at the crack tip. However, fiber bridging of the crack reduced the crack tip stress by a factor of seven. The magnitude of strain energy release rate was greatly reduced and the sign of the slope of the strain energy release rate versus crack length curve was changed from positive to negative by fiber bridging. In accordance with the theory of tough fiber reinforcement of brittle matrices, the results of applying nonlinear contact elements with varying coefficients of friction predict that an intermediate fiber-matrix interface strength will be most effective in toughening a brittle composite.

KEYWORDS: finite element analysis, virtual crack extension, brittle matrix composites, crack propagation, polymethyl methacrylate, tough fibers, fiber bridging, fiber-matrix interface strength, fatigue (materials), fracture (materials), composite materials

The use of a long fiber composite is not practical in applications where the composite must be formed in situ, especially where irregular shapes are involved. An example of this is reinforced bone cement. Bone cement is a polymethyl methacrylate (PMMA) based material used as a grouting agent to fix total joint replacement implants to bone. It would be impossible for an orthopaedic surgeon to mix and align long fiber impregnated bone cement in an operating room setting. Even chopped- or short-fiber reinforcement of bone cement is severely limited; addition of even a small percentage of chopped fibers increases the viscosity of the cement

[1]Research scientist, and research director, respectively, Orthopaedic Research Institute, Inc., Wichita, KS 67214.

[2]Director of Engineering, Voranado, 550 N. 159th St. East, Wichita, KS 67230.

[3]Associate professor, Wichita State University, Department of Aerospace Engineering, Wichita, KS 67260.

drastically. While this may not be of concern in commercial manufacturing where high pressure injection molding techniques can be implemented, it is of great consequence in the operating room.

Overview

The classical mechanism by which chopped-fiber additions increase the fracture and fatigue properties of a tough matrix material depends on using fibers that are much stiffer than the matrix material and are strongly bonded to the matrix, thus reducing the strain in the matrix. Implementing this approach for bone cement using various stiff fibers, such as carbon, glass, Kevlar, metal, etc., was confounded because even very small fiber additions made intraoperative mixing and injection of the cement nearly impossible. Furthermore, the fibers clumped and prevented cement intrusion into the interstices of the bone. To circumvent these obstacles, relatively low volume percentages of very flexible but tough fibers are proposed for toughening and fatigue life elevation of bone cement. Use of flexible fibers greatly improves intraoperative mixing and cement intrusion into the bone. Since flexible fibers are not stiff and the PMMA matrix is itself brittle, the classical composite strengthening (that is, stiff brittle fibers in a tough weak matrix) does not apply. In the case of tough fiber reinforcement of bone cement, the mechanism of crack energy dissipation is the disruption of fiber-matrix interfaces with controlled interface strength. Because the tough flexible fiber reinforcement of brittle matrices theory has not been greatly explored, a finite element model was developed in an attempt to predict how variation of the fiber-matrix interface strength and fiber properties could affect the fracture toughness and fatigue crack propagation rate of the composite.

Short Fiber Composite Reinforcement of Brittle Matrices

In brittle matrix composites, energy can be absorbed from the crack by four basic mechanisms: (1) deviation of the crack from the self-similar direction (which includes the crack following the fiber-matrix interfaces), (2) strain reduction in the matrix material in front of the crack by the fiber reinforcement, (3) pullout and fracture of the fibers behind a growing crack, and (4) stretching of bridging fibers behind a growing crack. In such a composite, if the fibers are weakly bonded to the brittle matrix, the crack may deviate from the self-similar direction to propagate through the weak fiber-matrix interface. This deviation results in an increase in energy absorption in comparison to cracking of the neat material. However, with a weak fiber-matrix interface strength, the fibers will tend to pull out easily, not absorbing much energy. Failed fiber-matrix interfaces will also limit the ability of bridged fibers to absorb energy or shield the crack tip from stress or both [1].

If the fibers are strongly bonded to the matrix, a crack will propagate right through the fibers without deviation. Such an arrangement might seem to be favorable, but in fact, the potential for energy absorption by deviation of the crack may be lost. Some energy could be absorbed from deformation and fracture of the fiber during bridging, but little or no energy will be expended in pulling the fiber out of the matrix.

Therefore, to maximize toughness and decrease the rate of crack propagation, it may be desirable to form a bond of intermediate strength between the fiber and matrix so that the energy can be absorbed in interface fracture as well as by fiber slip or pullout. In this mechanism, energy is absorbed by the fiber pullout and deformation of the fiber bridging behind the crack, deformation of the fibers in front of the crack, and deflection of the crack along the length of the fibers. An intermediate fiber-matrix bond strength might be able to take advantage of all modes of energy dissipation, and it is even possible that there is an optimum interface strength.

Finite Element Modeling of a Crack

The finite element technique of modeling cracks in homogeneous materials is now common-place and has been used by many investigators to study the effect of various geometries on parameters such as the stress intensity factors in Modes I, II, and III, the *J*-integral, and the strain energy release rate in the region of the crack tip [2]. Many finite element analysis (FEA) codes (including the one used in this study) include an automated routine for determining parameters such as the stress intensity of a crack; however, this routine can be used only in static linear analysis [2]. Investigators have used this common technique to analyze the crack-tip stress intensity effect of such geometries as changing thickness sections, inclusions, voids, and fiber bridging of the crack [3–6].

The technique of virtual crack extension to determine the strain energy release rate as a function of crack length of a structure has become common [1,7]. Generally, in virtual crack extension, the region in front of the crack tip is modeled by a series of elements of equal spacing. Virtual crack step sizes are based on the initial crack length. Displacement constraints along an initial crack face are removed incrementally to simulate crack growth and the total strain energy in the model is measured in each crack extension. The strain energy release rate is defined as the rate of change in strain energy in the total model per step as the crack grows [2]. This modeling technique analytically determines the strain energy release rate and is used for predicting a material's potential for unstable or stable crack propagation or crack arrest. A positive slope in strain energy release rate (*G*) versus crack length (*a*) implies the likelihood of unstable crack propagation. A zero or negative slope of *G* versus *a* implies the possibility of stable crack growth or crack arrest in a structure [8,9].

Finite Element Modeling Fiber-Matrix Interface in a Composite Material

Several investigators have performed finite element and mathematical modeling of a crack in a composite material [6,10–16]. All of these studies cite the common problem of not being able to model variable (and changing) interface shear stresses along the fiber length.

Finite element techniques exist that model the contacting interface between two surfaces. In contact modeling, a coefficient of friction (to control shear) and normal displacements can be specified. Contact elements are often used to model bimaterial interfaces [2,17,18]. In these elements, a very high coefficient of friction would represent a strongly bonded interface while a zero friction coefficient would model no bonding. The model proposed in this study employed the concept that given identical model geometries and initial precompression across the inter-face, the normal forces across the interface would be initially identical for each model. The coefficient of friction, therefore, governs the maximum frictional force that must be overcome in order for the interface to shear. This concept is the basis of the inferred relationship between fiber-matrix interface strength (that is, the maximum stress before the fiber starts to slide) and the coefficient of friction. In this study then, the contact element technique was combined with the crack modeling technique to determine the effect of fiber interface strength and fiber bridging on stresses at the crack-tip strain energy release rates.

Statement of Problem Addressed in this Work

The fundamental purpose of this study was to analytically examine the effect of fiber reinforcement, fiber bridging, and especially fiber-matrix interface strength on crack propaga-tion potential and parameters that can infer fracture toughness. Optimum bonding of these interfaces can maximize the contribution of the various energy absorbing mechanisms including dissipation of energy through fiber pullout and fiber bridging of the crack, thereby increasing the fracture toughness of the material. An attempt was also made to define an optimum fiber-matrix interface strength by performing a finite element analysis of a simple model with varying fiber-matrix interface friction and fiber bridging.

Methods

Two-dimensional plane-stress finite element models were formed to represent the following conditions:

1. neat bone cement (no fiber reinforcement).
2. fiber-reinforced bone cement (fiber-matrix interface strength assumed to be the strength of the matrix material) with no fibers bridging the open crack.
3. fiber-reinforced bone cement (as in Condition 2) but with fiber bridging of the open crack.
4. Fiber-reinforced bone cement with fiber bridging (as in Condition 3) but with the fiber-matrix interface modeled with complex interface contact elements assigned a coefficient of friction to model a finite shear strength

All models were analyzed using ANSYS 5.0 on a 486DX2-66V personal computer with a 1 gigabyte harddrive and 32 megabytes of RAM.

The basic model geometry consisted of a composite with 1% by volume fiber loading, aligned, chopped (5 mm) polyethylene terephthalate (PET) fibers, and a pre-existing crack as shown schematically in Fig. 1. A more detailed model schematic is shown in Fig. 2. The model represents a portion of a fracture toughness or fatigue crack propagation specimen. Constraints were applied in the direction of the applied force with freedom of movement in the perpendicular direction [2]. Two-dimensional, polynomial plane-stress triangular elements were used to model the matrix material. The plane-stress criterion on the elements was used so that the area of the springs used to represent fibers in conjunction with a specific model thickness could be chosen to produce a 1% by volume fiber loading. The model dimensions were width = 3.75 mm, thickness = 2.00 mm, and height = 2.825 mm; the 2.00 mm thickness;

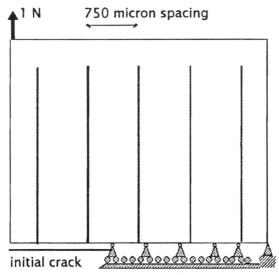

FIG. 1—*Schematic of the basic finite element model of an aligned chopped-fiber composite with pre-existing crack.*

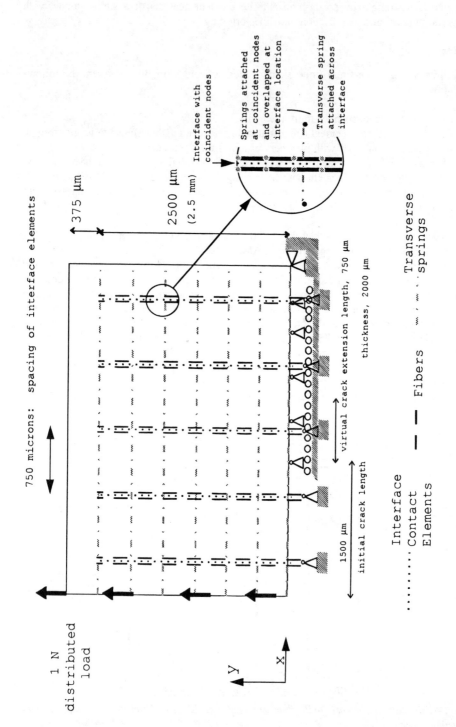

FIG. 2—*Detailed schematic of the finite element model with transverse springs.*

the ratio of these dimensions is equal to the ratio of dimension in a standard fracture toughness specimen. Fixed vertical translation on the lower edge was used to model the lower half of the symmetric specimen. The vertical lines represent the location of fibers. Spring elements (tension only) were superimposed on the matrix to model the fibers; the 0.75-mm fiber spacing and spring area of 0.08 mm^2 were chosen based on the model thickness and the 1% by volume fiber loading. A value of 6 GPa was used for the spring stiffness to represent a polyethylene terephthalate fiber. The matrix stiffness was set equal to the average elastic modulus of neat bone cement (2.2 GPa) with a Poisson's ratio of 0.3. A total unit force of 1 N was applied to the model by distributing it along the left edge. Virtual crack extension was performed for 0.75 mm, that is, one fiber spacing. Forty virtual crack steps of 0.01875 mm were performed and analyzed for each model configuration. The model geometry, boundary conditions, and mesh with a close-up of the virtual crack extension region are shown in Fig. 3. The basic model had over 10 000 degrees of freedom.

Modeling the Effect of Fiber Bridging

Three model types were analyzed to determine the effect of fiber bridging: (1) neat bone cement, (2) composite (fiber-reinforced) bone cement with no fiber bridging, and (3) composite bone cement with fiber bridging. Neat bone cement was modeled by removing the spring elements. In the composite cement, a fiber-matrix interface location was modeled by placing contact elements between two coincident edges of adjacent matrix elements; unique yet coincident nodes were used between the matrix elements to make them completely separate in designation (note that the contact elements were not made active for the case of infinite fiber-matrix interface strength). Composite cement with no fiber bridging was formed by superimposing springs on the matrix at these interface locations. Fiber bridging was modeled by inserting additional spring elements at the fiber interface location below the initial crack surface; the end of these spring elements remained attached to the displacement constraint throughout the virtual crack extension (Fig. 2). The von Mises stress at the initial crack tip was calculated for each configuration before crack extension. The virtual crack extension method was then applied to each model. In this method, crack extension is modeled by incrementally varying the crack length in the model by release of the equally spaced nodes one at a time. Strain energy release rates (G) were calculated for each crack step by dividing the difference in total model strain energies, U, in two consecutive crack steps by the model thickness and the crack step length (Δa) as shown in Eq 1

$$G = -(U_{a+\Delta a} - U_a)/(B\Delta a) \qquad (1)$$

where

$U_{a+\Delta a}$ = elastic strain energy at crack length ($a + \Delta a$),
  U_a = elastic strain energy at crack length (a),
  B = thickness of fracture model, and
  Δa = change in crack length from Step n to $n + 1$.

The strain energy release rate was plotted as a function of crack length, a. The onset of the potential for crack arrest is defined by a negative slope of this curve. The maximum value of G, the slope of the G versus a curve, and the von Mises stress at the initial crack tip were compared between the three models.

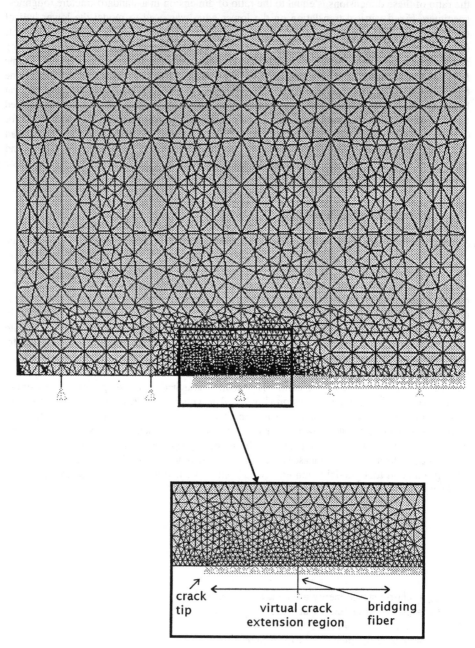

FIG. 3—*Finite element mesh of the composite model with a detailed examination of the virtual crack extension region.*

Modeling Fiber-Matrix Interface Strength

The effect of fiber-matrix interface strength was determined using nonlinear point-to-surface contact elements at the interfaces of the fibers. The coefficient of friction in these elements was intended to represent the shear strength of the fiber-matrix interface. The friction and shear strength are obviously not identical. The basic assumption for this model is that when only the coefficient of friction is varied (that is, identical material properties, loading conditions, and normal force across the interface), the amount of force required to overcome friction and allow the interface to slide is related directly to the coefficient of friction. Sliding of the interface is meant to model pullout of the fiber from the matrix, hence, the larger the coefficient of friction of the interface, the higher the fiber-matrix interface strength.

Patch tests (that is, analysis of a very small simple model with element geometry and type similar to that used in the large model) were performed using contact element interfaces in simple geometries under various loading conditions (shear, tension, compression, and bending). In normal tensile loading, models of contact geometry were not stable unless transverse spring elements were superimposed across the contact geometry to keep it from separating. In addition, it was necessary to use short transverse springs with a slight tensile prestrain therefore transmitting a slight compressive load to the interface.

In all transverse spring interface models, several equally spaced transverse springs with 0.001 prestrain were placed across each interface to provide numerical stability (Fig. 2). The transverse springs represented a fiber-matrix interface resistance to normal tensile loads. An elastic modulus equal to the matrix material stiffness was used for the transverse springs. Only one spring was applied at the bottom of the model along the line of crack growth across the entire model width. This bottom spring placement eliminated the effect of transverse spring attachment node release during virtual crack extension and eliminated the aberration of the true behavior of the crack extension.

A second, numerically stable interface model was formed without using transverse springs. In this model, equal displacement constraints normal to the fiber direction were applied to the originally coincident nodes on each side of an interface location; the interfaces were allowed to shear in the fiber direction, but not allowed to have normal displacement to the fiber direction between the interface nodes.

The coefficient of friction of the interface in both interface models was varied parametrically. As with the noninterface models, von Mises stress at the crack tip was calculated for each configuration before crack extension. The strain energy in the model was determined for each crack step and the strain energy release rate calculated. The relationship between stress at the initial crack tip and fiber-matrix interface shear strength (coefficient of friction of contact element) was determined.

Results and Discussion

The Effect of Fiber Bridging

Figure 4 shows an exaggerated deformed mesh plot of the first principal stress in the neat bone cement model before crack extension. A detailed examination of the stresses at the crack tip region in this figure shows that the shape of the enhanced stress field is that expected from fracture mechanics theory [8,9,19]. In the fiber bridging model, a relatively low 1-N unit load was applied so that the stresses and strains in the bridging fibers were well within the limits of elastic behavior for PET fibers. No attempt was made to model breakage of the bridging fibers in higher loading conditions.

A plot of G versus Δa (strain energy release rate versus change in crack length) for the three noninterface models is shown in Fig. 5. The magnitude of the strain energy release rate

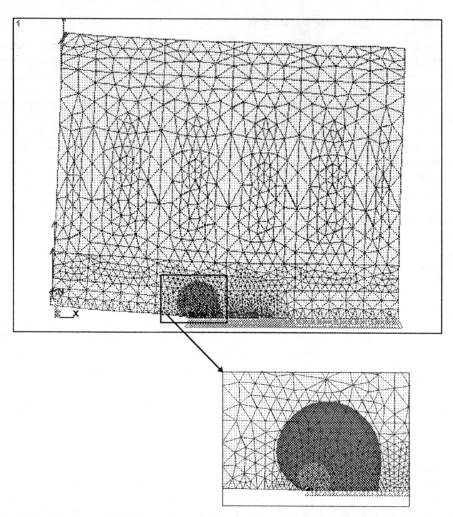

FIG. 4—*Exaggerated displacement, first principal stress plot of the neat bone cement model before crack extension with a detailed view of the stress region at crack tip.*

(G) was substantially reduced by fiber bridging; it is hard to distinguish from the abscissa on this graph. In the fiber model with no bridging, when the crack extends close to the fiber location, the level of G does decrease. During release of the fiber, G varies rapidly; no interpretation of the meaning of this region is made because the crack step size was probably too large to determine the true nature of the effect of the fiber release. Once the fiber is released, that is, the propagating crack releases it from its constraint, G returns to its higher level, almost equal to that of the neat model. From this model, it is clear that there is little effect from fiber reinforcement without fiber bridging.

A semilog scale plot of this same G versus Δa data in given in Fig. 6. From this figure, it is clear that not only was G reduced, but the slope of the G versus Δa curve was also changed from positive to negative by fiber bridging. The negative slope of the fiber-bridged curve indicates a potential for stable crack propagation or crack arrest that was not present in the

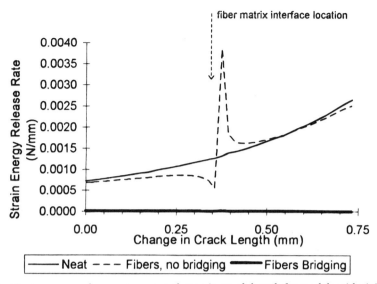

FIG. 5—*Strain energy release rate versus change in crack length for models with rigid fixation at the fiber interface (noninterface models).*

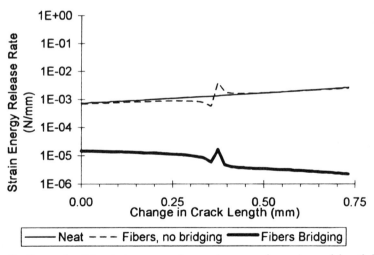

FIG. 6—*Semilog scale of the strain energy release rate versus change in crack length for models with rigid fixation at the fiber interface (noninterface models).*

neat cement or the non-fiber-bridged model. This finding supports the theory that bridging and deformation of well-bonded fibers together is an important mechanism of energy dissipation in short fiber composites [1].

The von Mises stress values at the crack tip prior to crack extension for the noninterface models is given in Table 1. The stress at the initial crack tip with nonbridging fibers (9.40 MPa) was not much lower than the crack tip stress in fiber-free (neat) material (9.72 MPa). This is not surprising considering the 1% by volume fiber loading in the model and the small

TABLE 1—*Initial von Mises stress at crack tip for noninterface models.*

Model Configuration	von Mises Stress, MPa
Neat bone cement	9.72
PET fiber, no fiber bridging	9.40
PET fiber, fiber bridging	1.40

difference in stiffness of the fiber (6.0 GPa) and the matrix (2.2 GPa). In addition, the initial crack tip was located as far away from the fibers as possible, that is, halfway between two fibers, further minimizing the effect of the fiber. However, when fiber bridging was modeled, the crack-tip stress was reduced by nearly a factor of seven (1.40 MPa). The ability of bridging fibers to shield the stress at the crack tip is a paramount feature of short fiber reinforcement theory and has been shown analytically by several investigators [6,14].

Figure 7 is a plot of von Mises stresses directly in front of the crack versus inverse crack length ($\sigma_{\text{von Mises}}$ versus $1/a$) for the noninterface models. This figure illustrates that the von Mises stress from the crack tip is a linear function of (1/crack length). The slope of the lines

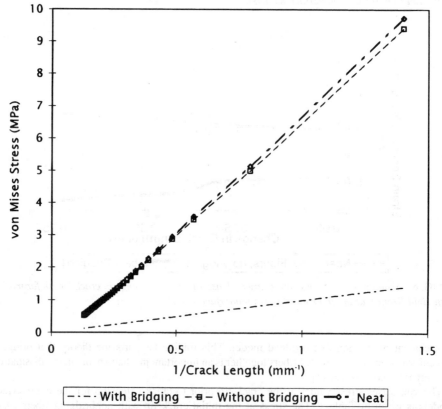

FIG. 7—*The von Mises stress directly in front of the crack tip versus inverse crack length for the noninterface models.*

in Fig. 7 are related to the stress intensity factor [9]. The initial crack-tip stress and higher slopes of these lines correspond directly, showing that initial crack-tip von Mises stress is a relative indication of stress intensity.

The model predicted the effect of fiber bridging on the initial crack-tip stress and the potential for crack growth. This model could be used to further predict the effect of fiber bridging with varying parameters such as fiber loading and stiffness. In the future, other features may be added to this basic model to simulate other short fiber reinforcement phenomena. One such example is "death options" applied to the springs to simulate the effect of breakage of brittle fibers rather than extensive elongation of tough fibers.

The Effect of the Fiber-Matrix Interface

In the second phase of the finite element work, the effect of fiber-matrix interface strength was modeled by nonlinear point-to-surface contact elements generated at the interface locations of the fibers. Deformation plots of the loaded interface models show that the contact surfaces at the interface have displaced or slid past each other slightly. The sliding of contact surface and the adjustment of the fibers due to the sliding is the model's representation or allowance for fiber pullout.

The relationships between stress at the initial crack tip and fiber-matrix interface shear strength (coefficient of friction of contact element) as determined by the two interface models are shown in Figs. 8a and b. From the transverse spring model (Fig. 8a), one may conclude that the stress at the crack tip is minimized at an intermediate fiber-matrix interface strength. Little change in the initial stress was seen for values of friction above 0.15. The highest coefficient of friction (1.0) did not behave as a rigid interface. This behavior is a manifestation of a limitation of the contact elements to model actual situations at extremes. This limitation was observed in the preliminary patch tests as well. The alternative constrained displacement interface model yielded a parabolic relationship between initial crack tip stress and interface friction (Fig. 8b). No minimum initial crack tip stress was predicted at an intermediate fiber-matrix interface strength in the constrained displacement interface model.

The results from the transverse spring interface model support the hypothesis of optimum reinforcement of short, tough fiber composites by an intermediate fiber-matrix interface strength. A reduction of the stress at the crack tip with an intermediate fiber-matrix interface strength directly relates to the concept of maximizing fracture toughness at an intermediate interface strength. The magnitude of the coefficient of friction applied in this model had no direct correlation with the actual optimum value of interface strength, but it does support the contention that an optimum does exist that is not at zero nor at high interface strengths.

Figure 9 shows a plot of the G versus Δa curves for models with varying coefficients of friction (CF) for the constrained displacement model. The transverse spring model displayed a similar relationship as the constrained displacement model. Strain energy release rates were nearly independent of the coefficient of friction. In both interface model techniques, the magnitude of strain energy release rate was greatly lowered by the presence of the interface. Reduction in the magnitude of the strain energy release rate may indicate an increased crack propagation resistance with finite shear strength interfaces.

Conclusions

A new finite element model technique was formulated to test the effect of fiber bridging and fiber matrix interface strength on fracture toughness and crack propagation behavior. A novel technique for modeling variable fiber-matrix interface strength with contact elements of variable coefficient of friction is presented. Results from the models showed that neat and

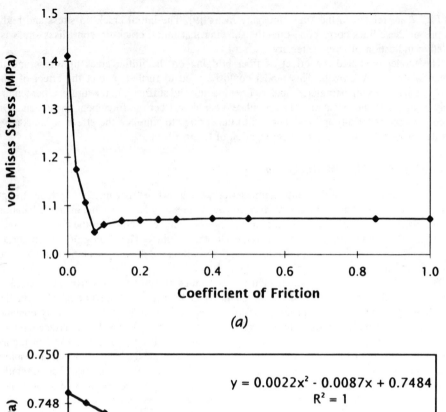

(a)

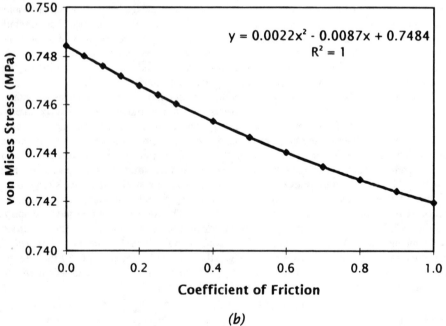

(b)

FIG. 8—*The von Mises stress at the crack tip before crack extension for varying coefficients of friction at the interface using the* (a) *transverse spring model and* (b) *constrained displacement model.*

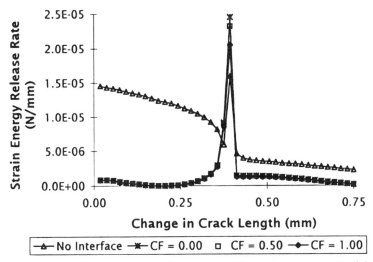

FIG. 9—*Strain energy release rate versus change in crack length for the constrained displacement interface mode for varying coefficients of friction (CF) at the fiber-matrix interface location.*

fiber-reinforced material, with no fibers bridging the crack and fiber-matrix interface strengths equal to the matrix strength, displayed nearly the same initial stress at the crack tip and the same strain energy release rate behavior. Fiber bridging reduced the crack-tip stress in these models by nearly a factor of seven. In addition, the fiber bridged model displayed a potential for stable crack propagation or crack arrest. These simple models worked well to qualitatively show the potential effects of fiber bridging on crack-tip stress shielding and crack propagation potential.

Two nonlinear finite element techniques with nonlinear contact elements at the fiber-matrix interface were developed to represent variable fiber-matrix interface strengths and the contribution of fiber pullout. The results from the transverse spring interface model predicted a maximum stress shielding at the crack tip (or maximum fracture toughness) at an intermediate value of friction (or intermediate fiber-matrix interface shear strength). These results correlate with short fiber composite theory predictions made by Piggott [1]. Both interface models predict a potential for reduction in crack propagation potential with finite strength interfaces. Physical fracture toughness and fatigue crack propagation experiments on the chopped fiber composites with varying fiber-matrix interface strength (the system these models were meant to simulate) are currently in progress. It is hoped that the experiments can be guided with greater efficiency and that the results can be more generally understood because of the insights provided by these analyses.

With further investigation, it is possible that this micromechanics model of crack extension in chopped-fiber composites may become a practical tool for predicting the effect of fiber properties on the potential for increased fracture resistance. It may also be possible to perform useful analyses of more complex composites by modification of the model to include nonlinear fiber material properties, varying fiber orientations, and techniques to simulate fiber breakage.

Acknowledgments

This work was supported in part by grants from the University of Kansas School of Medicine-Wichita and the St. Francis Research Institute.

References

[1] Piggott, M. R., "Expressions Governing Stress-Strain Curves in Short Fiber Reinforced Polymers," *Journal of Materials Science*, Vol. 13, 1978, pp. 1709–1716.
[2] *ANSYS User's Manual*, Revision 5.0, Vol. 1: Procedures Section 3.9: Fracture Mechanics, Swanson Analysis Systems, Inc., Houston, PA, 1992.
[3] Sides, A., Perl, M., and Uzan, J., "The Effect of a Disturbance and Its Location on the Stress Intensity Factor in the Three-Point Bend Specimen," *Engineering Fracture Mechanics*, Vol. 36, No. 3, 1990, pp. 365–371.
[4] Gharpuray, V. M., Keer, L. M., and Lewis, J. L., "Cracks Emanating From Circular Voids or Elastic Inclusions in PMMA Near a Bone-Implant Interface," *Journal of Biomechanical Engineering*, Vol. 112, Feb. 1990, pp. 22–28.
[5] Gharpuray, V. M., Keert, L. M., and Lewis, J. L., "Cracks Emanating from Defects in PMMA Near a Bone-Implant Interface," *Transactions*, Seventeenth Annual Meeting of the Society for Biomaterials, Scottsdale, AZ, May 1991, p. 203.
[6] Yin, S. W., "A Fiber Bridging Model for the Fracture of Brittle Matrix Composites," *Engineering Fracture Mechanics*, Vol. 46, No. 5, 1993, pp. 887–894.
[7] Rybicki, E. F. and Kanninen, M. F., "A Finite Element Calculation of Stress Intensity Factors by a Modified Crack Closure Integral," *Engineering Fracture Mechanics*, Vol. 9, 1977, pp. 931–938.
[8] Broek, D., *Elementary Engineering Fracture Mechanics*, Kluwer Academic Publishers, Dordrecht, 1986.
[9] Kanninen, M. F. and Popelar, C. H. *Advanced Fracture Mechanics*, Oxford University Press, New York, 1985, pp. 234–238.
[10] Ballarini, R. and Muju, S., "Stability Analysis of Bridged Cracks in Brittle Matrix Composites," *Journal of Engineering Gas Turbines and Power*, Vol. 115, 1993, pp. 127–138.
[11] Mohammadi, J. and Kuraydlo, A. S., "Stochastic Modeling of Short Fiber Reinforced Composites—A Review," *Probabilistic Mechanics and Structural Geotechnical Reliability*, American Society of Civil Engineering, New York, 1992, pp. 479–482.
[12] Courage, W. M. G. and Schreurs, P. J. G., "Effective Material Parameters for Composites with Randomly Oriented Short Fibers," *Computers and Structures*, Vol. 44, No. 6, 1992, pp. 1179–1185.
[13] Murat, M., Anhold, M., and Wagner, H. D., "Fracture Behavior of Short-Fiber Reinforced Materials," *Journal of Materials Research*, Vol. 7, No. 11, 1992, pp. 3120–3131.
[14] Bakuckas, J. G. and Johnson, W. S., "Application of Fiber Bridging Models to Fatigue Crack Growth in Unidirectional Titanium Matrix Composites," *Journal of Composite Technology and Research*, Vol. 15, No. 3, 1993, pp. 242–255.
[15] Cox, B. N. and Lo, C. S., "Simple Approximations for Bridged Cracks in Fibrous Composites," *Acta Metallurgica Material*, Vol. 40, No. 7, 1992, pp. 1487–1496.
[16] Telesman, J., Ghosn, L. J., and Kantzos, P., "Methodology for Prediction of Fiber Bridging Effects in Composites," *Journal Composite Technology and Research*, Vol. 15, No. 3, 1993, pp. 234–241.
[17] Harrigan, T. P. and Harris, W. H., "A Three-Dimensional Non-Linear Finite Element Study of the Effect of Cement-Prosthesis Debonding in Cemented Femoral Total Hip Components," *Journal of Biomechanics*, Vol. 24, No. 11, 1991, pp. 1047–1058.
[18] Harrigan, T. P. and Harris, W. H., "A Finite Element Study of the Effect of Diametral Interface Gaps on the Contact Areas and Pressures in Uncemented Cylindrical Femoral Total Hip Components," *Journal of Biomechanics*, Vol. 24, No. 1, 1991, pp. 87–91.
[19] Hertzberg, R. W., *Deformation and Fracture Mechanics of Engineering Materials*, 3rd ed., Wiley, New York, 1989.

Environmental Effects

Allan S. Crasto[1] and Ran Y. Kim[1]

Hygrothermal Influence on the Free-Edge Delamination of Composites Under Compressive Loading

REFERENCE: Crasto, A. S. and Kim, R. Y., **"Hygrothermal Influence on the Free-Edge Delamination of Composites Under Compressive Loading,"** *Composite Materials: Fatigue and Fracture (Sixth Volume), ASTM STP 1285*, E. A. Armanios, Ed., American Society for Testing and Materials, 1997, pp. 381–393.

ABSTRACT: This paper reports on determination of the individual and combined effects of temperature and moisture on the initiation of free-edge delamination in a $[90_3/30_3/-30_3]_s$ graphite/epoxy laminate. Dry and wet specimens were tested at various temperatures under uniaxial compression. The onset of delamination was determined by monitoring axial and transverse strains and confirmed by microscopic examination of the polished specimen edges. Interlaminar free-edge stresses were analyzed using a global-local model, and the elastic constants for these calculations were determined experimentally for the various hygrothermal conditions. The results of this stress analysis were applied to predict the onset of delamination in conjunction with the maximum effective stress criterion. Delamination in all cases occurred at the interfaces predicted, and the predicted stress for the onset of delamination compared reasonably well with experiments. While the curing residual stresses decrease with increasing moisture content and test temperature, so does the interlaminar tensile strength, and these two effects compete in their influence on the stress for the initiation of delamination.

KEYWORDS: graphite/epoxy laminates, hygrothermal conditions, free-edge stresses, delamination, analytical global-local model, failure prediction, fatigue (materials), fracture (materials), composite materials

Delamination of rectangular laminated composite specimens at the free edges under in-plane uniaxial loading has been studied at length [1–3]. This phenomenon results primarily from interlaminar stresses that are associated with the interactions of the various laminating plies and is intensified at locations where there is an abrupt change in material or in geometry, such as a free edge. The magnitude and sign of the interlaminar stress components vary widely with laminate layup, stacking sequence, constituent properties, and the nature of the uniaxial loading (that is, tension or compression). The occurrence of free-edge delamination governs the ultimate uniaxial strength of a laminate coupon. For example, the tensile strength of a $[\pm45/0/90]_s$ graphite/epoxy coupon was found to be much smaller than a $[0/90/\pm45]_s$ coupon of the same material; a stress analysis revealed large tensile interlaminar stresses in the former that precipitated delamination at the free edges prior to ultimate failure [4]. These interlaminar stresses can be calculated accurately with an analytical global-local model developed by Pagano and Soni [5]. By assuming an appropriate failure criterion for the laminate under the loading conditions employed, the stress level at the onset of delamination can be predicted.

[1]Materials engineer and senior research engineer, respectively, University of Dayton Research Institute, Dayton, OH 45469-0168.

When a composite is subjected to a hygrothermal environment, the material elastic properties as well as the matrix-dominated strengths and residual stresses change significantly, and these changes influence the stress for the onset and growth of free-edge delamination in the laminate. While the residual stresses decrease with increasing moisture content and test temperature, so does the transverse ply strength, and these two effects compete in their influence on the stress for the initiation of delamination. However, the effect of a hygrothermal environment on the initiation of free-edge delamination has received little attention in the literature [6–8]. In an earlier study, we investigated the free-edge delamination of composites subjected to uniaxial tensile loading under hygrothermal conditions [9]. In the present study, we have extended that investigation to determine the individual and combined effects of temperature and moisture on the initiation of free-edge delamination in a composite laminate under uniaxial compressive loading. The interlaminar stresses at the free edge were analyzed using a global-local model [5]. The results of this stress analysis were applied to predict the onset of delamination in conjunction with a maximum stress criterion, and these predictions are compared with experiments.

Experimental Procedure

A graphite/epoxy composite material, AS4/3501-6, from Hercules, Inc., was utilized in this study. Multidirectional laminates with a $[90_3/30_3/-30_3]_s$ stacking sequence were fabricated from unitape prepreg using the manufacturer's recommended cure cycle. When this laminate is subjected to a uniaxial compressive load, significant interlaminar shear and normal tensile stresses are generated at the free edges that precipitate free-edge delamination. Specimens, 76.2 mm long, 25.4 mm wide, and 2.3 mm thick, were sectioned from the panels and their edges polished to enhance the microscopic detection of delamination cracks. Additional composites were also fabricated to determine the elastic constants necessary for calculation of the interlaminar free-edge stresses and to measure the material's 90° tensile strength under the same hygrothermal conditions utilized to investigate free-edge delamination. The average fiber content was determined to be 67% by volume. The Poisson's ratio, ν_{LT}, was determined from the tensile loading of a $[90]_{16T}$ specimen with strains measured in both the y- and z-directions and utilized to calculate G_{TT}. The strain in the z-direction was measured with a miniature strain gage (of gage length, 0.81 mm) mounted on the free edge of the specimen.

Some specimens were conditioned in distilled water at 66°C for an extended period of time and the moisture absorption determined from intermittent measurements of weight gain. A second batch of specimens was desiccated prior to testing. Both dry and wet specimens were tested in uniaxial compression in an environmental test chamber on an MTS test machine at temperatures of −25, 24, 75, and 125°C. After several trials, a gage length of 1.9 cm was employed (to precipitate free-edge delamination prior to specimen buckling), and specimens were directly gripped in hydraulic grips with Surfalloy faces. The specimens tested at elevated temperatures (75 and 125°C) were heated rapidly to the desired temperature and allowed to equilibrate before testing. During this time period, the decrease in moisture content near the edge of a wet specimen was determined experimentally using a wet dummy specimen. After exposure to the same heat excursion as the test specimen, a 2-mm-wide strip was cut from the edge of the dummy specimen and dried at 75°C to determine the residual moisture content. These results compared well with predictions of moisture content in the free-edge region based on Fick's law and published data on moisture diffusivity at various temperatures [10]. Figure 1 shows typical stress-strain curves in the axial and transverse directions at 24°C. The responses of strain gages mounted in both the axial and transverse directions (on the flat surface) and the thickness direction (on the specimen edge) were used to detect the onset of delamination. A sudden increase in both axial strain and transverse strain, as seen in this figure, indicates

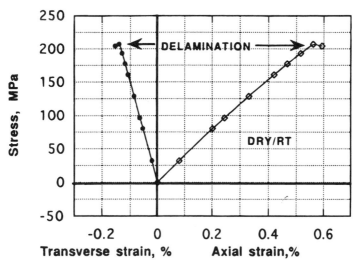

FIG. 1—*Typical axial and transverse stress-strain curves for a dry specimen loaded in axial compression at 24°C.*

the onset of delamination and the corresponding stress level. At the first indication of delamination, the specimen was unloaded and the delamination confirmed by microscopic examination of the specimen edge.

Results and Discussion

Typical moisture absorption profiles for the $[90]_{8T}$ and $[90_3/30_3/-30_3]_s$ laminates are plotted as functions of the square root of time in Fig. 2. After about 60 days of immersion in water

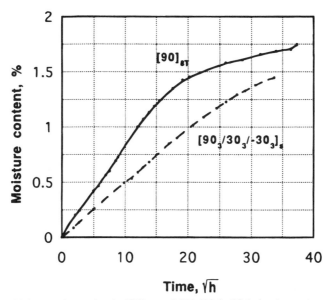

FIG. 2—*Moisture absorption in $[90]_{8T}$ and $[90_3/30_3/-30_3]_s$ laminates versus time.*

at 66°C, the [90]$_{8T}$ specimens almost reached saturation, whereas the [90$_3$/30$_3$/−30$_3$]$_s$ specimens continued to absorb moisture. The average moisture content at the time of testing was 1.7% for the [90]$_{8T}$ specimens and 1.5% for the [90$_3$/30$_3$/−30$_3$]$_s$ specimens. It should be noted that the moisture content in the free-edge region, where the interlaminar stress components are critical, is greater than the measured value that is the average over the entire specimen volume. For tests conducted above ambient temperature, the moisture content in the free-edge region reduced to approximately 1.5 and 1.4% for the [90]$_{8T}$ laminates tested at 75 and 125°C, respectively, and 1.3 and 1.2% for the [90$_3$/30$_3$/−30$_3$]$_s$ laminates tested at 75 and 125°C, respectively. A moisture content of 1.2% was used in the calculation of the free-edge stresses for both these test temperatures.

The elastic constants and transverse strengths that were determined experimentally at various hygrothermal conditions are shown in Table 1. The transverse Young's modulus, E_T, and transverse shear moduli, G_{LT} and G_{TT}, are dependent on the hygrothermal conditions, while the Poisson's ratio, ν_{TT}, is independent of the hygrothermal conditions. The values of the shear moduli in this table are averages from three or more specimens for each hygrothermal condition. The other properties necessary for calculation of the free-edge stresses were assumed to be independent of the hygrothermal conditions and are given in Table 2.

The distributions of the interlaminar stresses, σ_z and τ_{xz}, at selected interfaces of dry specimens at 24°C, due to the independent influences of mechanical loading (application of

TABLE 1—*Hygrothermal effect on transverse and shear properties of AS4/3501-6.*

Condition	Test Temperature, °C	Transverse Modulus, E_T (GPa)	Poisson's Ratio, ν_{TT}	Shear Modulus, G_{LT} (GPa)	Calculated Shear Modulus, G_{TT} (GPa)	Transverse Strength, σ_T (MPa)
Dry	−25	9.4	0.52	7.1	3.1	68.1
	24	9.4	0.52	6.9	3.1	65.3
	75	7.2	0.52	6.6	2.4	57.6
	125	6.3	0.52	6.1	2.1	48.7
Wet	−25	9.0	0.52	7.0	3.0	43.8
	24	9.2	0.52	5.7	3.0	46.2
	75	7.5	0.52	5.2	2.5	23.0
	125	4.1	0.52	3.7	1.3	14.0

TABLE 2—*AS4/3501-6 material properties independent of hygrothermal conditions.*

Property	Value
Longitudinal modulus, E_L, (GPa)	138
Longitudinal Poisson's ratio, ν_{LT}	0.3
Coefficient of thermal expansion, 1×10^{-6}/°C	
longitudinal	−0.7
transverse	15
Stress-free temperature, °C	178
Coefficient of moisture expansion, 1×10^{-6}/%	
longitudinal	0
transverse	4000
Ply thickness, mm	0.127

a compressive strain of 0.001) and thermal residual stresses, are shown in Figs. 3 and 4, respectively. (The interlaminar shear stress component, τ_{yz}, was found to be too small to influence free-edge delamination and is not considered further.) There is a large gradient in the interlaminar stress components in the vicinity of the free-edge region. The interlaminar

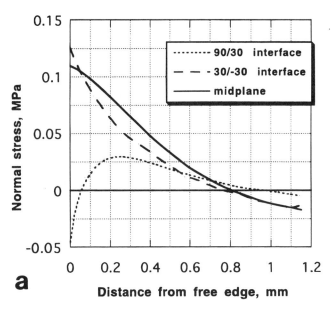

a

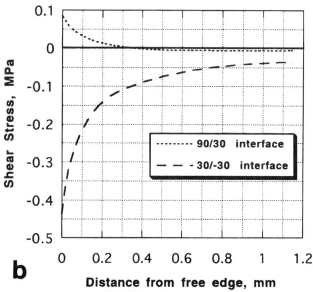

b

FIG. 3—*(a) Interlaminar normal tensile stresses, σ_2 and (b) interlaminar shear stresses, τ_{xz}, near the free edges of selected interfaces of dry specimens at 24°C, with only an axial compressive strain of 0.001.*

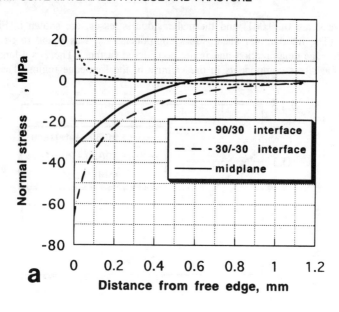

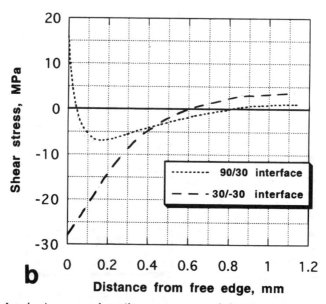

FIG. 4—(a) Interlaminar normal tensile stresses, σ_z, and (b) interlaminar shear stresses, τ_{xz}, near the free edges of selected interfaces of dry specimens at 24°C, resulting from thermal residual stresses alone.

normal stresses at both the midplane and the 30/−30 interface are tensile when considering mechanical loading alone (Fig. 3a) but compressive under the influence of thermal loading (Fig. 4a). The only interlaminar shear stress of significance near the free edge is τ_{xz} at the 30/−30 interface (Figs. 3b and b). The influence of moisture alone (thermal residual stresses are not considered) on the interlaminar normal and shear stresses at selected interfaces at 24°C is shown in Fig. 5. The interlaminar normal stresses at both the midplane and the 30/−30 interface are tensile, in contrast to the compressive stresses generated by the thermal residual stresses. From a comparison of Figs. 4 and 5, it is clear that the absorbed moisture reduces the shear stress, τ_{xz}, and the normal compressive stress at this interface; consequently, the net shear stress (due to thermal and mechanical loading) is reduced, while the net normal tensile stress is increased with moisture absorption.

In addition to influencing the total residual stress state in the composite, the test temperature and absorbed moisture also influence the material elastic and strength properties (as seen in Table 1) that are reflected in a change in interlaminar stresses. The influence of a change in material properties on interlaminar stresses is exemplified in Figs. 6a and b that depict the interlaminar normal stress distributions at the 30/−30 interface of dry and wet specimens, respectively, at various test temperatures due to an applied compressive strain of 0.001. The stresses are shown up to a distance from the free edge equivalent to the thickness of nine plies. With increasing test temperature, there is a significant decrease in interlaminar stresses that partially offsets the reduction in transverse strength.

The results of the interlaminar stress analyses are applied to predict the onset of delamination in conjunction with failure theory [11,12]. The average value of each stress component at the free edge, instead of the maximum value, is assumed to be the effective stress level, that is, that stress which is responsible for failure. The effective stress is obtained by averaging the interlaminar stress components over a fixed distance, h_0, from the free edge along the width of the laminate at the interface in question, as depicted in Fig. 7 and is given by

$$\bar{\sigma}_i(z) = \frac{1}{h_0} \int_0^{h_0} \sigma_i(y, z)\, dy \tag{1}$$

The fixed distance, h_0, was taken as the thickness of one ply (0.127 mm) in all cases. The effective free-edge interlaminar stresses at selected interfaces corresponding to the strains for

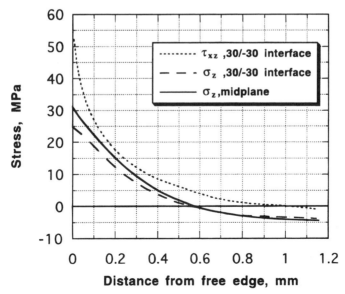

FIG. 5—*Interlaminar normal and shear stresses near the free edges of selected interfaces at 24°C, resulting from absorbed moisture alone.*

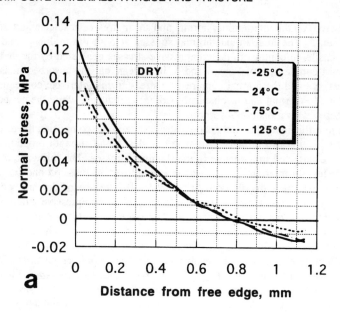

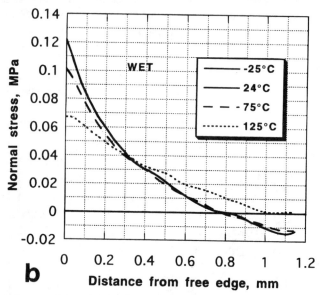

FIG. 6—*The influence of changing mechanical properties on the interlaminar normal stress near the free edge of the 30/−30 interface in (a) dry and (b) wet specimens.*

the observed onset of delamination are shown in Table 2. Each measured strain to delamination is the average from at least three tests.

For prediction of the stress level and location (interface) for the onset of delamination, the maximum stress criterion was applied. Failure via delamination was assumed to occur when any of the effective stresses, $\bar{\sigma}_i(z)$, at each interface reached the corresponding uniaxial strength.

AVERAGING INTERLAMINAR STRESS

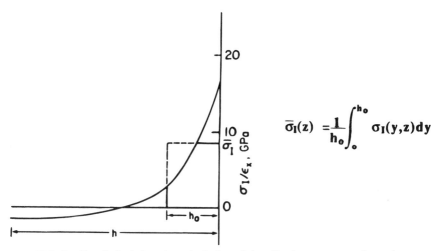

$$\bar{\sigma}_I(z) = \frac{1}{h_o} \int_0^{h_o} \sigma_I(y,z)dy$$

FIG. 7—*Sketch depicting the calculation of the effective stress at a free edge.*

For the laminate used in this study, the critical effective stress components were the interlaminar shear stress, τ_{xz}, at the $30/-30$ interface for the dry specimens at all test temperatures and the interlaminar normal tensile stresses, σ_z, at either the $30/-30$ interface or the laminate midplane for the wet specimens. The onset of delamination was predicted when these effective stresses exceeded the corresponding interlaminar (tensile or shear) strength. The interlaminar tensile strength of the laminate was assumed to be equal to the transverse tensile strength of the material system. Results from Ref *13* demonstrate that the difference between the interlaminar tensile strength and transverse tensile strength is less than 6% for the graphite/epoxy AS4/3501-6 composite system.

Table 3 shows a comparison of the experimental results with predictions. For the dry specimens, failure is predicted via interlaminar shear at the free edge of the $30/-30$ interface. This failure location was verified from optical micrographs of the polished edges of delaminated specimens (Figs. 8 and 9). The predicted failure stress is significantly larger than the laminate shear strength, although the difference decreases with increasing test temperature. The reduction in thermal residual stresses with increasing test temperature reduces the critical interlaminar shear stress component but has no apparent influence on the failure strain. For the wet specimens, failure is predicted via interlaminar tension at either the laminate midplane or the $30/-30$ interface. Once again, this locus of failure is confirmed from experimental observations of polished specimen edges following delamination. In this case, there is reasonable agreement between the predicted failure stress and the transverse tensile strength. Although the hygrothermal environment reduces the critical interlaminar stress component considerably through the reduction of residual stresses and elastic properties, the significant decrease in transverse tensile strength causes the onset of delamination at lower applied strains. These two effects, therefore, compete in their influence on the stress for the initiation of delamination. No transverse ply cracks were observed in the (delaminated) wet specimens tested at 75 and 125°C. This is in line with analytical predictions of the stress to first-ply failure being greater than the stress for the onset of delamination under these conditions.

TABLE 3—*Comparison of calculated critical interlaminar stresses at the onset of delamination and measured interlaminar strengths.*

Test Condition	Temperature, °C	Strain at Delamination Onset, %	Stress Component	Interlaminar Stresses, MPa						Interlaminar Strength, MPa	
				30/−30 Interface			Midplane				
				Thermal	Mechanical	Total	Thermal	Mechanical	Total	Y	S
Dry	−25	0.46	τ_{xx}	−39.9	−117.0	−156.9					
			σ_z	−24.5	46.6	22.1	−28.6	47.0	18.4	68.1	98.2
	24	0.47	τ_{xx}	−30.2	−118.8	−149.0					
			σ_z	−18.5	47.3	28.8	−21.6	47.7	26.1	65.3	88.7
	75	0.44	τ_{xx}	−18.6	−106.9	−125.5					
			σ_z	−8.2	37.1	28.9	−10.8	39.0	28.2	57.6	102.0
	150	0.45	τ_{xx}	−8.3	−104.9	−113.2					
			σ_z	−3.8	35.0	31.2	−4.2	36.9	32.7	48.7	94.5
Wet	−25	0.38	τ_{xx}	−10.8	−94.9	−72.2					
			σ_z	−4.8	37.8	53.2	−9.3	38.0	53.5	43.8	112.7
	24	0.41	τ_{xx}	−10.9	−102.9	−93.7					
			σ_z	−4.8	41.0	56.3	−7.0	41.2	41.3	46.2	116.6
	75	0.30	τ_{xx}	−17.7	−76.5	−65.0					
			σ_z	−9.4	27.0	34.4	−11.3	−27.8	36.5	23.0	97.0
	150	0.21	τ_{xx}	−6.2	−41.8	−29.9					
			σ_z	−2.3	13.5	17.6	−2.4	13.3	20.6	14.0	71.6

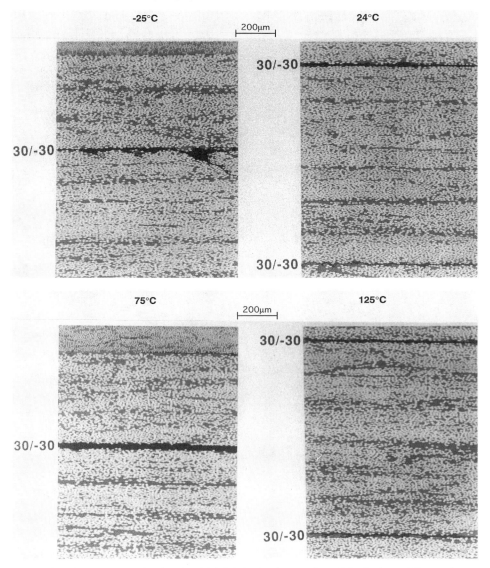

FIG. 8—*Photomicrographs showing initial delamination on the polished edges of dry specimens at the interfaces predicted.*

Summary and Conclusions

An investigation was conducted to determine the individual and combined effects of temperature and moisture on the initiation of free-edge delamination in an AS4/3501-6 graphite/epoxy laminate. The onset of delamination in this laminate can be accurately determined experimentally by monitoring axial and transverse strains. The change in material properties due to the hygrothermal environment considerably influences the magnitude and distribution of interlaminar stresses in the free-edge region. Absorbed moisture as well as an elevated test temperature combine to reduce interlaminar stress components through a reduction in the

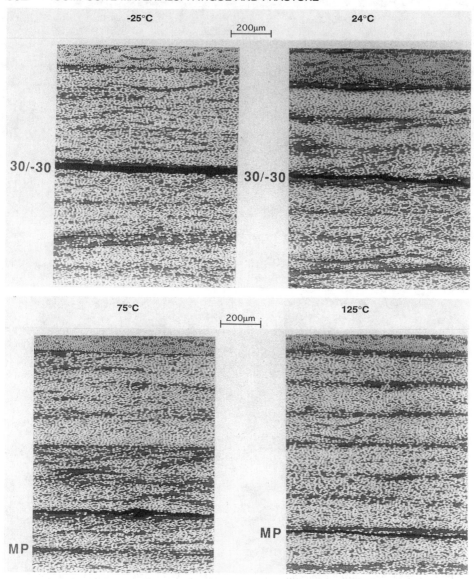

FIG. 9—*Photomicrographs showing initial delamination on the polished edges of wet specimens at the interfaces predicted.*

residual stresses. In spite of this reduction in the interlaminar stress components, delamination initiates at significantly lower stresses with increasing test temperature, particularly for wet specimens, due to the significant decrease in interlaminar transverse tensile strength of the laminate under these conditions. Predictions regarding the stress level and location (interface) for the onset of delamination agree fairly well with the experiment for wet specimens; for dry specimens, the failure mode and location are accurately predicted, although there are significant differences between predicted stresses and corresponding laminate strengths. These predictions would improve with a more accurate and reliable determination of the hygrothermomechanical properties, transverse strengths at various test conditions, and moisture content at the time of failure.

Acknowledgments

This work was performed under U.S. Air Force Contract F33615-91-C-5618. The authors wish to express their appreciation to Mr. Ron Esterline of the University of Dayton Research Institute for preparing and testing the specimens.

References

[1] Pipes, R. B. and Pagano, N. J., "Interlaminar Stresses in Composite Laminates under Uniform Axial Extension," *Journal of Composite Materials*, Vol. 4, 1960, pp. 537–548.

[2] Pagano, N. J. and Pipes, R. B., "The Influence of Stacking Sequence on Laminate Strength," *Journal of Composite Materials*, Vol. 5, 1971, pp. 50–57.

[3] Whitney, J. M. and Browning, C. E., "Free-Edge Delamination of Tensile Coupons," *Journal of Composite Materials*, Vol. 5, 1973, pp. 300–303.

[4] Pagano, N. J., "Stress Field in Composite Laminates," *International Journal of Solids and Structures*, Vol. 14, 1978, pp. 385–400.

[5] Pagano, N. J. and Soni, S. R., "Global-Local Laminate Variation Model," *International Journal of Solids & Structures*, Vol. 19, 1983, p. 207.

[6] O'Brien, T. K., Raju, I. S., and Garber, D. P., "Residual Thermal and Moisture Influence on the Strain Energy Release Rate Analysis of Edge Delamination," *Journal of Composite Technology & Research*, Vol. 8, No. 2, 1986. pp. 37–47.

[7] Armanios, E. A. and Badir, A. M., "Hygrothermal Influence on Mode I Edge Delamination in Composites," *Composite Structure*, No. 15, 1990. pp. 323–342.

[8] Armanios, E. A., Sriram, P., and Badir, A. M., "Fracture Analysis of Transverse Crack-Tip and Free-edge Delamination in Laminated Composites," *Composite Materials: Fatigue and Fracture, Third Volume, ASTM STP 1110*, T. K. O'Brien, Ed., American Society for Testing and Materials, Philadelphia, 1990, pp. 269–286.

[9] Kim, R. Y. and Crasto, A. S., "Hygrothermal Effects on the Onset of Free-Edge Delamination in Composites," *Proceedings*, International SAMPE Symposium, Vol. 39, 1994, pp. 2935.

[10] Chen, C. H. and Springer, G. S., "Moisture Absorption of Graphite-Epoxy Composites Immersed in Liquids and in Humid Air," *Journal of Composite Materials*, Vol. 13, 1979, pp. 131–147.

[11] Kim, R. Y. and Soni, S. R., "Experimental and Analytical Studies on the Onset of Delamination in Laminated Composites," *Journal of Composite Materials*, Vol. 18, 1984, pp. 70–80.

[12] Kim, R. Y., "Experimental Observation of Free-Edge Delamination," *Interlaminar Response of Composite Materials*, N. J. Pagano, Ed., Elsevier, New York, 1989, pp. 111–160.

[13] Kim, R. Y., Abrams, F., and Knight, M., "Mechanical Characterization of a Thick Composite Laminate," *Proceedings*, American Society for Composites, Technomic Publishing, Lancaster, PA, 1988, pp. 711–718.

Andrew H. Rosenberger[1] *and Theodore Nicholas*[1]

Environmental Effects on the Isothermal and Thermomechanical Fatigue of SCS-6/ TIMETAL 21S Unidirectional Composites

REFERENCE: Rosenberger, A. H. and Nicholas, T., **"Environmental Effects on the Isothermal and Thermomechanical Fatigue of SCS-6/TIMETAL 21S Unidirectional Composites,"** *Composite Materials: Fatigue and Fracture (Sixth Volume), ASTM STP 1285*, E. A. Armanios, Ed., American Society for Testing and Materials, 1997, pp. 394–408.

ABSTRACT: The effect of environment on the fatigue behavior of SCS-6/TIMETAL 21S $[0]_4$ composites is examined through a comparison of fatigue lives and damage progression for tests performed in air and high-purity helium. Isothermal and thermomechanical tests were conducted at 650°C and 150 to 650°C, respectively. Out-of-phase thermomechanical fatigue (TMF) lives of specimens tested in the inert environment show a ×2 increase in life. In-phase TMF lives in inert and air environments, which are governed primarily by fiber bundle strength and matrix stress relaxation, are comparable.

Isothermal fatigue tests in the inert environment performed at 1.0 Hz show that at high stresses the life is not affected by the environment but at lower stresses a ×3.5 increase in life is observed. Reducing the fatigue frequency to 0.01 Hz causes no change in life at low stresses in the inert condition as compared to the low-frequency air condition. At high stresses, the behavior is governed primarily by the statistics of fiber bundle strength. Life fraction models that consider time-dependent and cycle-dependent behavior as well as fiber- and matrix-dominated failure modes are used to correlate observed behavior in these experiments.

KEYWORDS: metal-matrix composites, silicon-carbide fibers, thermomechanical fatigue, environmental effects, inert environment, life fraction models, fatigue (materials), fracture (materials), composite materials

High temperature fatigue of titanium-matrix composites (TMCs) is affected by environment primarily in terms of the matrix embrittlement and subsequent crack growth. This has been observed in a number of studies on various TMC matrix/fiber combinations. Thermal fatigue of SCS-6/TIMETAL 21S[2] composites between the temperature range of 150 and 815°C resulted in a significant loss of tensile strength that has been linked to environmental embrittlement of the matrix and matrix crack formation [1]. Unidirectional, $[0]_4$, composites retained more of their strength than cross-ply composites, $[0/90]_s$, for a given test condition that indicates a more rapid environmental embrittlement due to short-circuit oxygen diffusion paths down debonded, exposed 90° fibers. In fact, specimens thermally cycled after encapsulation in a low-pressure inert environment maintained a substantial fraction of their strength as compared to air-tested specimens. Similar environmental synergism has been demonstrated by Revelos and Smith [2] on a SCS-6/Ti-24Al -11Nb (a/o) composite.

[1]Visiting scientist from Systran Corporation and senior scientist, respectively, Wright Laboratory, WL/MLLN, Wright-Patterson AFB, OH 45433-7817.
[2]TIMETAL 21S is a trademark of Timet Corporation, Henderson, NV.

In-phase (IP) and out-of-phase thermomechanical fatigue (OP-TMF), where the maximum stress occurs at maximum and minimum temperature, respectively, of SCS-6/TIMETAL 21S composites in air has been examined by Neu and Nicholas [3] for [0]$_4$, [0/90]$_s$, and [0/±/45/90]$_s$ orientations. Both minimum and maximum strain increase under in-phase cycling resulting in a constant loading modulus but only the maximum strain increases under out-of-phase cycling resulting in a decrease in the loading modulus. This behavior is a result of the observed damage mechanisms; OP-TMF is dominated by environmentally assisted matrix crack growth while IP-TMF is controlled by fiber failure induced by matrix stress relaxation. Similar strain responses and damage mechanisms have been found for all lay-ups examined. A mechanistic-based thermomechanical fatigue life prediction model for TMCs has been developed by Neu [4] that accounts for three dominant damage mechanisms; (1) matrix fatigue damage, (2) surface-initiated environmental damage, and (3) fiber-dominated damage. This model, and an associated mechanism map [5], shows fatigue life of OP-TMF and lower stress isothermal fatigue to be controlled by surface-initiated environmental damage. A life fraction model, which considers time-dependent and cycle-dependent mechanisms as well as fiber and matrix dominated modes, has also been used to explain observed damage mechanisms and to predict fatigue lives [6,7]. In a recent study, from modeling and experimental data of TMCs, it was deduced that high-stress, low-frequency fatigue and IP-TMF are controlled by the same mechanisms, namely, matrix stress relaxation and subsequent failure of the fiber bundle [8].

Fatigue-environment interactions have been examined by Gayda et al. [9] in SCS-6/Ti-15V-3Cr-3Sn-3Al (percent by weight) composites under isothermal fatigue conditions at 300°C and nonisothermal (essentially OP-TMF) at a temperature range of 300 to 550°C. In both test conditions, there was a strong environmental effect on the fatigue lives. There were substantially fewer cracks initiated at fiber-matrix interfaces in the vacuum experiments. The relatively poor environmental resistance of the Ti-15-3 matrix material makes it environmentally sensitive even at these low temperatures. Environmental effects on the thermomechanical fatigue of SCS-6/Ti-24Al-11Nb (atomic percent) was examined by Bartolotta and Verrilli [10] under in-phase and out-of-phase conditions. They found, consistent with the damage mechanisms described earlier, that there is no environmental effect on IP-TMF but under OP-TMF, the environment plays an important role in controlling damage mechanisms and life with an increasing effect for decreasing stress levels. At high stress levels, there was essentially no difference in OP-TMF life while at low stress levels, tests performed in flowing argon had an order of magnitude increase in life compared to tests in air. Using nondestructive examination techniques, surface connected cracks were found to be longer in specimens tested in air compared to those tested in argon [11]. The longer cracks present in the air specimens also tended to promote fiber/matrix interface damage in the form of debonds or reduced interfacial debond strength.

The objective of this study was (1) to determine the environmental effect on the isothermal and thermomechanical fatigue of SCS-6/TIMETAL 21S, and (2) to determine the mechanisms of the fatigue damage accumulation in these composites. The next section discusses the materials and experimental techniques used in this study. This is followed by a description and discussion of the results of the different aspects of this study.

Materials and Experimental System

The material used in this study consisted of continuous SiC (SCS-6) fibers in a beta titanium matrix, TIMETAL 21S Ti-15Mo-3Nb-3Al-0.2Si (percent by weight) and sheet TIMETAL 21S (0.5 mm thick) from Heat G-1664. Composites were fabricated by the foil-fiber-foil method using rolled foils of TIMETAL 21S and fiber mats of SCS-6 fibers held together by titanium-niobium (Ti-Nb) ribbon and consolidated by hot isostatic pressing. Samples were wrapped in tantalum foil, to prevent any oxidation damage, and heat treated at 621°C for 8 h in a vacuum furnace. The resulting transverse matrix microstructure and composite structure is shown in Fig. 1. Unidirectional [0]₄ straight-sided specimens were used with a width of 10 mm, length of 110 mm, and nominal thickness of 0.9 mm. The axial strain was monitored during the tests by a 12-mm nominal gage length quartz rod extensometer. Specimens were tested in an

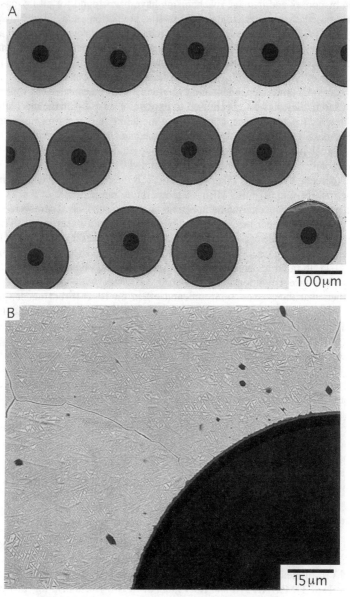

FIG. 1—*Matrix microstructure and composite structure.*

environmental chamber mounted on a computer controlled, servohydraulic load frame. They were heated using a four-zone quartz lamp system with independent temperature zone control via thermocouples welded to the specimen [12]. The specimens were loaded using water-cooled, precisely-aligned, hydraulic friction grips. Tests were conducted using MATE [13] software that controlled the symmetric triangular wave forms used for both load and temperature (TMF). The software automatically adjusted the four temperature setpoints to maintain proper temperature ramp rates and load/temperature phasing.

An inert testing system constructed for this investigation is shown schematically in Fig. 2. The environmental chamber is constructed of double-walled stainless steel and has a volume of approximately 291 L and a leak rate of less than 2×10^{-4} torr L/s. The chamber is pumped by a 6-in. diffusion pump with liquid nitrogen (LN_2) cold trap. This system is capable of maintaining a vacuum of 10^{-6} torr, at temperature, which results in an oxygen partial pressure of approximately 10^{-8} torr as monitored using a quadrapole mass spectrometer. In inert testing mode, the chamber is first evacuated to 10^{-5} torr and back filled with Grade 5 helium (99.999% pure). The helium is then circulated through the system at 15 to 35 kPa (gage) using a metal diaphragm recirculation pump with a nominal capacity greater than 3 L/s. Within this circuit is a titanium-based gettering system, where oxygen, nitrogen, and carbon are actively sorbed by the titanium sponge held at 825°C. This improves the purity of the helium and reduces the influence of contaminants in the system and off-gassing of the specimen during the cyclic heating. The helium is then cooled using a chilled serpentine tube that is necessary since the gas is used to cool the specimen during TMF cycling. The high mass flow of helium and its high specific heat make this an excellent inert medium for TMF testing.

Specimen heating and cooling profiles during the thermal cycling in helium were held within ±5°C of the command signal during thermal cycles. Three helium jets cool the specimen during TMF that allowed fine tuning of the specimen temperature profile during the test using valving external to the chamber. At the end of the tests, no visible indication of oxidation of

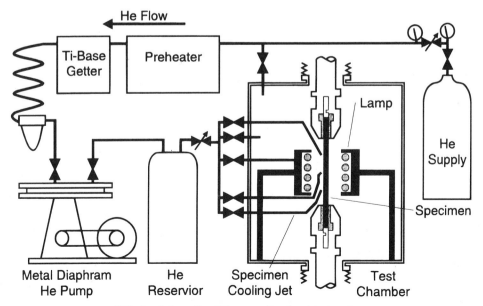

FIG. 2—*Schematic of the recycling inert test system.*

the specimens was apparent, which indicates the absence of corrosive species in the environment, that is, oxygen, nitrogen, or water vapor.

Tests were conducted in load control at a stress ratio of 0.1. Isothermal testing was conducted at 650°C at 1 and 0.01 Hz. TMF cycling was performed over the temperature range 150 to 650°C at a cyclic frequency of 0.00556 Hz (180-s cycle). Failure was defined as complete separation of the specimen into two pieces.

Experimental Results

The results are presented in three sections. The first briefly describes an examination of the effect of the test system environment on the embrittlement of the matrix material, TIMETAL 21S. The second discusses the baseline TMF behavior of the composite system in inert conditions and a comparison to TMF in air. The final section discusses high- and low-frequency isothermal fatigue tests performed at 650°C. In all cases, comparison data for fatigue in air have been obtained from previous research [3,14,15].

Effect of Test Environment on the Embrittlement of TIMETAL 21S

Sheet tensile specimens were exposed in the test system described earlier in a stress-free state for 50 h at 760°C in various vacuum and inert environments. This exposure was followed by a room-temperature tension test using a displacement rate of 0.0021 mm/s. Table 1 shows the exposure environment, tensile strength, and elongation for each specimen; only single tests were performed. These exposures represent the spectrum of the testing capabilities possible in the inert testing machine from high-vacuum (10^{-6} torr) to gettered, recirculated helium.

The air exposure at 760°C produces a significant reduction in ductility and results in completely brittle behavior whereas similar exposure in high-vacuum results in a normal level of ductility and usual strength levels for TIMETAL 21S. Exposure in a low vacuum environment, 8×10^{-4} torr, results in a ductility reduction similar to that seen in the air exposure. This illustrates that at 760°C, the transition from environmental sensitivity to non-sensitivity is within the range from 10^{-6} to 10^{-4} torr which is comparable to the operation region of most material processing equipment. Exposures in gettered, recirculated helium indicate similar strength and ductility levels as seen after the high-vacuum exposure while the exposure in the ungettered helium exhibits embrittlement similar to air. These observations demonstrate that the gettered helium system produces no significant environmental degradation of the matrix material and that the tests were conducted under "inert" conditions.

TABLE 1—*Tensile results afer 50-h/760°C exposure in various environments.*

Environment	Yield Strength, MPa	Ultimate Tensile Strength, MPa	Elongation, %
As-heat treated (no exposure)	1162	1211	9.0
Air	···	225	[a]
Low vacuum (8×10^{-4} torr)	···	363	0.4
High vacuum (2×10^{-6} torr)	867	897	16.4
Helium, un-gettered	···	344	0.4
Helium, gettered	811	832	9.0

[a]Warped specimen prevented accurate strain measurement.

Baseline In-Phase and Out-of-Phase Thermomechanical Fatigue

Four baseline inert TMF tests were performed and compared to results obtained in air on similar material [*3,14*]. All specimens tested in this investigation had a nominal fiber volume fraction of 0.32. However, the air specimens tested previously were from different composite panels reported to have fiber volume fractions ranging from 0.32 to 0.38 [*15*]. The results from two in-phase and two out-of-phase tests are shown in Fig. 3 in the form of number of cycles to failure as a function of the maximum cyclic stress. As can be seen, the high-stress, 900-MPa, in-phase inert test shows an increase in life compared to air testing, however, the lower stress test, 820 MPa, shows a life slightly lower than the air test (in fact, shorter than the higher stress inert test). This demonstrates the sensitivity of IP-TMF lives to a statistical variation of the SCS-6 fiber strength or volume fraction differences. The inert, 820 MPa, specimen was polished down to the first fiber layer and numerous broken fibers were found, Fig. 4*a*. It has not been proven, however, that many of these hairline cracks are not a result of mechanical polishing. A close examination of one fiber fracture on this specimen showed no fretting damage of the fiber coatings but crack initiation into the surrounding matrix, Fig. 4*b*. This form of internal fiber damage and crack growth is similar to that seen in specimens tested in air [*3*], and would be protected from environmental damage, leading to similar fatigue lives in the two environments. The fracture surfaces of the air and inert IP-TMF specimens, (typical example shown in Fig. 5) show ductile overload of the specimen with no indications of surface-connected crack growth that could be affected by the environment.

The inert OP-TMF tests in Fig. 3 show a constant ×2 increase in life over the stress range from 1000 to 700 MPa. These two inert specimens, however, failed near the grip, indicating that damage had accumulated in this highly stressed grip region of these straight-sided specimens at a rate greater than in the gage section. Mechanical strain accumulation trends observed in the inert specimens indicate that considerable damage had accumulated within the gage length, indicating that failure within the gage was imminent [*16*]. Figure 6 shows a typical surface matrix crack from the gage section of the 700 MPa maximum stress test. The typical crack length in the inert tests was 150 μm in length and typically initiated at grain boundaries. Specimens tested in air had a considerably higher density of surface cracking with crack lengths up to 500 μm in length—also grain boundary initiated. The fracture behavior of specimens tested in air indicates that numerous cracks had initiated at the surface and grown

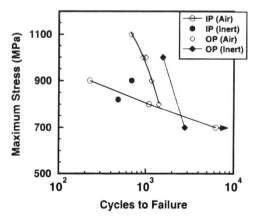

FIG. 3—*Comparison of inert IP- and OP-TMF results with those obtained in air [3,14] in terms of maximum applied stress versus cycles to failure.*

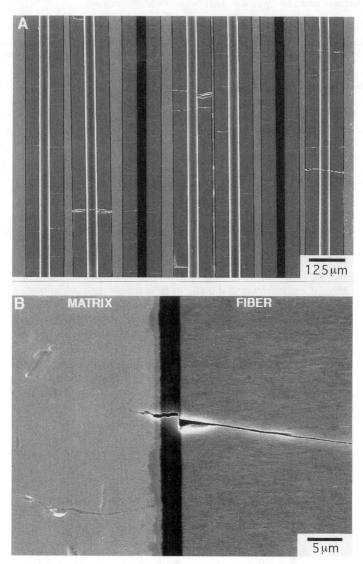

FIG. 4—*Longitudinal cross section of IP-TMF specimen tested at 820 MPa; (a) numerous broken fibers are present and (b) crack growth into the matrix from a broken fiber.*

to a depth greater than the first layer of fibers. This enhanced matrix crack growth is consistent with the embrittlement phenomena seen in TIMETAL 21S. The surface cracking seen in the inert tested specimens is similar, but the cracks are smaller and fewer.

Environmental Effects in Isothermal Fatigue

Four baseline inert isothermal fatigue tests were performed and compared with results obtained in air on similar materials [15]. The results from 1- and 0.01-Hz tests are shown in

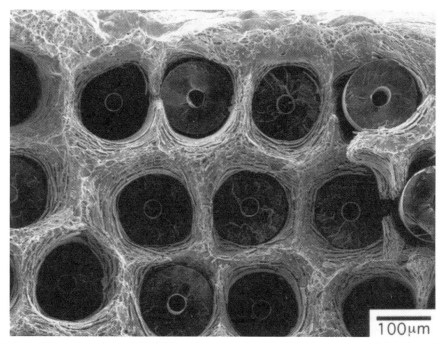

FIG. 5—*Fracture surface of IP-TMF specimen tested at 820 MPa showing typical ductile overload fractures and no apparent crack growth region.*

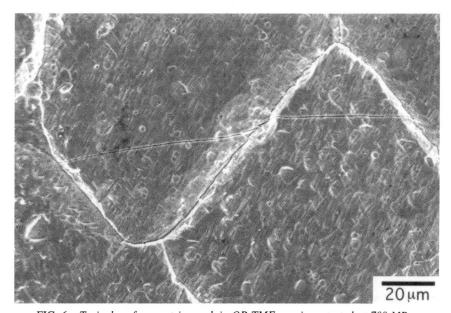

FIG. 6—*Typical surface matrix crack in OP-TMF specimen tested at 700 MPa.*

Fig. 7 in the form of the number of cycles to failure as a function of the maximum cyclic stress. The 1-Hz data indicate that at a high stress level (900 MPa) there is no difference in fatigue life in air or inert environments while the lower stress test, 700 MPa, shows more than a ×3 increase in life under the inert conditions. The fracture behavior of the specimen tested at 900 MPa in air indicates localized, surface crack growth was present as well as internally initiated crack growth. The specimen tested in an inert environment shows no indications of crack growth but has tensile overload indications similar to that seen in IP-TMF (Fig. 5). At the lower stress, localized crack growth covered a greater fraction of the fracture surface for specimens tested in air, however, failure occurred outside the hot zone in the specimen tested in an inert environment with no apparent damage in the gage length of the specimen.

The low frequency (0.01-Hz) inert isothermal fatigue tests, at both 900 and 700 MPa, show a decrease in fatigue life compared to the baseline air data, Fig. 7. Since the composite panels for the air and inert tests had different fiber volume fractions, 0.37 and 0.32, respectively, a debit in fatigue life for the inert tested specimens may be expected due to faster relaxation of the matrix stress-carrying capacity. Fracture behavior was similar to that seen in the higher frequency fatigue tests. Localized crack growth was present in specimens tested in air while the high-stress inert-tested specimen exhibited no crack growth region but only tensile overload.

Discussion

The high temperature air exposure of the TIMETAL 21S matrix material produces a severe embrittlement that occurs at 760°C after as little as 50 h in a low-vacuum environment. The observed failure strain of 0.4% and completely brittle behavior indicate that the matrix cracking and crack growth can be a significant form of damage in TMCs. The larger matrix strain ranges seen in OP-TMF as compared to IP-TMF [17] indicate matrix cracking will be more prevalent in OP-TMF as observed. Kroupa and Neu [17] modeled the fiber and matrix stress and strain ranges in SCS-6/TIMETAL 21S unidirectional composites and found the matrix strain range in OP-TMF at a maximum stress of 1000 MPa to be greater than 0.6%. Considering that the maximum matrix stress occurs at the minimum temperature, where the matrix is less ductile, supports the observation that relatively easy environmentally-induced crack initiation and growth can occur in an OP-TMF cycle. Lower matrix strain ranges in IP-TMF would lessen the drive for crack initiation. The maximum stress in this cycle occurs at maximum

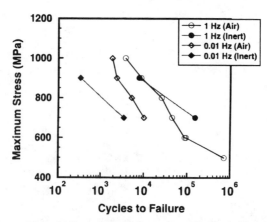

FIG. 7—*Comparison of inert isothermal fatigue results with those obtained in air [15] in terms of maximum applied stress versus cycles to failure for 1.0 and 0.01 Hz.*

temperature where the matrix ductility is higher—even in an embrittled state. Also, Kroupa and Neu have shown that the matrix stresses relax quickly in this cycle. The consistency of damage mechanisms observed in both IP- and OP-TMF in air and inert conditions indicates that the environment alone does not promote damage but affects the rate of damage accumulation. Typical strain accumulation during TMF in air and inert environments is shown in Figs. 8a and b. Under IP-TMF, similar increases in maximum and minimum mechanical strain are observed for the air and inert conditions. This progressive strain ratcheting seen in IP-TMF, Fig. 8a, has been attributed to periodic fiber fracture during the entire fatigue life based on

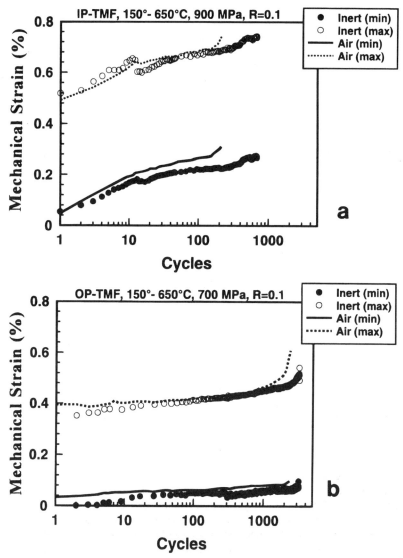

FIG. 8—*Comparison of strain accumulation in air and inert OP/IP TMF tests; (a) IP-TMF 150 to 650°C, 900 MPa, R = 0.1 and (b) OP-TMF 150 to 650°C, 700 MPa, R = 0.1.*

acoustic emission monitoring of TMF [18]. As the IP-TMF behavior is predominately controlled by the fiber strain range, a negligible environmental influence is expected and has been observed. Internal matrix crack initiation is possible at fiber fracture locations, however, this is also an environmentally protected location.

Under OP-TMF, the dominant damage mechanism is matrix crack initiation and growth due to the high matrix strain range [4,14,17]. The mechanical strain response for this TMF cycle is shown in Fig. 8b for air and inert conditions. Here, again, the response in the two environments is similar. The maximum strain increases during cycling while the minimum strain remains relatively constant. This gradual increase in compliance is consistent with the matrix cracking observed. The specimens tested in air had considerable matrix cracking such that upon failure, the fracture surface showed complete matrix crack penetration beyond the first layer of fibers due to the joining of many approximately co-planar matrix cracks. Crack initiation at grain boundaries in this case is consistent with environmentally assisted damage. Specimens tested in helium showed a lesser degree of cracking due to the slower crack initiation and growth in the inert environment. Failure at the grip section of the inert specimens indicates that the constant $\times 2$ increase in life may be conservative, however, the strain accumulation seen in Fig. 8b indicates that damage accumulated in the gage section of the specimens is due to fatigue of the matrix alone.

Under isothermal fatigue conditions, damage accumulation is a combination of fiber-dominated and matrix-dominated mechanisms. Neu's mechanistic-based model for life prediction in MMCs indicates that a transition from surface-initiated environmental damage to fiber-dominated damage occurs with an increase in maximum applied stress, increase in R-ratio, and a decrease in cyclic frequency [4]. In a SCS-6/Ti-24Al-11Nb (atom %) unidirectional composite, the transition occurs above 1000 MPa for 3-Hz cycling and at approximately 760 MPa for 0.003-Hz cycling at a test temperature of 650°C and $R = 0.1$. In a TIMETAL 21S composite, with lower creep resistance, the transition stresses would be lower—such that at 1 Hz it could occur between 900 and 700 MPa and at 0.01 Hz it could fall below 700 MPa. Experimental observations by Nicholas et al. [6] show that a transition occurs from mostly cycle-dependent to mostly time-dependent behavior in going from 1 Hz to 0.01 Hz in the [0/90]$_s$ lay-up of the same material used in this investigation at a test temperature of 650°C and a stress level of 470 MPa.

Fractography of the specimens tested under inert conditions at 1 Hz in the current study indicates that fiber-dominated damage occurs at a maximum applied stress of 900 MPa as is evident by the lack of crack growth on the fracture surface. The specimen tested in air had a limited region of crack growth and an area of internal crack growth that is probably the result of a fiber and reaction zone fracture. This indicates that both surface-initiated environmental and fiber-dominated damage were active. The similarity in fatigue lives also lends support to fiber-dominated damage mechanisms. At the lower applied maximum stress of 700 MPa, the greater than $\times 3$ increase in life as compared to air testing indicates that an environmental-assisted crack growth mechanism was active. Failure at the grip section of the inert specimen also indicates that the $\times 3$ increase in life may be conservative. Strain monitoring of this inert specimen failed to show an increase in compliance that would indicate damage accumulation within the gage section.

Under low-frequency fatigue conditions, the influence of fiber-dominated damage should be greater than at the higher frequency. This will lead to a strong influence of fiber volume fraction and matrix stress relaxation, as seen in IP-TMF of this composite system [17]. Based on FIDEP [19] concentric cylinder model simulations of the micromechanical matrix and fiber stress ranges, a reduction in life for the inert specimens of as much as $\times 2$ could be attributed to the fiber volume fraction effects based on a current life prediction model [14]. However, this would not fully account for the life reduction seen in the inert specimens. In fact, a fatigue test (0.01 Hz, 700 MPa, σ_{max}) in air was performed on a specimen from the same composite panel as the inert specimens that resulted in a fatigue life within 5% of the higher fiber volume fraction composite.

To explore the effect of volume fraction on fatigue life under isothermal conditions, predictions were made for the number of cycles to failure using a model developed and calibrated for this material and layup [7]. The model constants were used to predict the life under inert conditions, assuming no environmental effects were contributing to the tests in air. All calculations are made using the actual volume fraction of each test specimen, in order to eliminate any bias due to effects of volume fraction. A comparison of predicted lives for tests in air and inert environment for two frequencies, 0.01 and 1.0 Hz, is shown in Fig. 9, where the straight line represents perfect correlation. It should be noted first that the model was calibrated to fit a much larger data set and that the 0.01-Hz data in air are generally underpredicted by a factor from two to three. The comparison of data from air and inert environment tests shows that two of the four inert tests correlate well with the model and air data, indicating that environmental effects were absent. The short-life test at 0.01 Hz (900 MPa) and the long-life test at 1.0 Hz (700 MPa) both deviate somewhat from the air data. The 900 MPa test, however, represents a test condition near the maximum strength of the material or Region I of a Talreja diagram where large scatter in observed lives frequently occurs [20]. On the basis of this analysis, it is concluded that environmental effects do not significantly influence the fatigue behavior of the material under isothermal conditions except for the long-life test at 1.0 Hz. In fact, the low frequency behavior may be attributed primarily to creep or stress relaxation in the matrix material and the subsequent breakage of a bundle of fibers, irrespective of environment [8].

A comparison of the maximum and minimum strains from the two isothermal tests at 700 MPa, 0.01 Hz (air and inert), shows that there is a substantial increase in the maximum strain in the inert condition during the first 1000 cycles (27 h), Fig. 10. A possible explanation for this is that air exposure during the test startup and soak (1/2 h) resulted an interstitial oxygen pickup that improved the matrix creep behavior under the low-frequency cyclic loading. No difference in relative amounts of alpha and beta phases was observed in specimens tested in

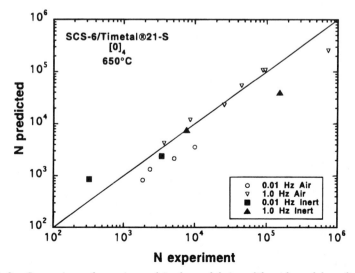

FIG. 9—*Comparison of experimental isothermal fatigue life with model prediction.*

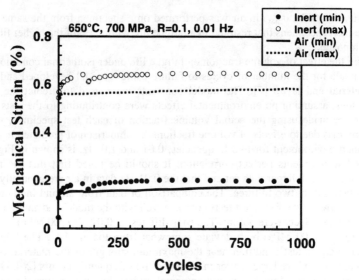

FIG. 10—*Comparison of strain accumulation in air and inert isothermal fatigue tests (650°C, σ_{max} = 700 MPa, R = 0.1, 0.01 Hz).*

the different environments, however, microhardness measurements indicated a greater hardened depth and hardness level in the air-tested specimens as compared to the inert-tested specimens that lends some support to this hypothesis. A detailed examination of this phenomena, however, is beyond the scope of this investigation.

Summary

The influence of an oxidizing environment on the isothermal and thermomechanical fatigue of unidirectional SCS-6/TIMETAL 21S composites was examined through a comparison of tests performed in air and inert environments. In general, an environmental influence on fatigue life was found under conditions in which matrix crack initiation and crack growth are dominant. Fatigue conditions in which life is dominated by fiber strength were not affected by the test environment.

Under IP-TMF conditions, no influence of the environment was observed. This is based on the fiber-dominated fatigue behavior and possible internal crack initiation at fiber fracture locations. Under OP-TMF conditions, a ×2 increase in fatigue life was found under inert conditions at both high and low stresses. The composite was still life limited by matrix crack initiation and growth—indicating that matrix material substitutions could lead to improvements in life. Isothermal fatigue tests demonstrated that the environment affects life at low stresses and high frequencies where significant stress relaxation of the matrix did not occur. However, at high stresses or low frequencies, or both composite life is fiber dominated and not influenced by the environment. Analysis of the role of fiber volume fraction on fatigue life helped explain some of the experimental observations. A possible mechanism of environmental matrix creep modification was discussed but remains to be explored further.

Acknowledgments

The support of the U.S. Air Force Wright Laboratory, Materials Directorate, Wright-Patterson Air Force Base, Ohio, where this work was performed, is gratefully acknowledged. Dr. Rosenberger would like to acknowledge the visiting scientist support of U.S. Air Force Contract F33615-90-C-5944. The authors would also like to thank Dr. Rick Neu and Mr. Steve Russ for helpful discussions and fatigue data in air.

References

[1]   Revelos, W. C., Jones, J. W., and Dolley, E. J., "Thermal Fatigue of a SiC/Ti-15Mo-2.7Nb-3Al-0.2Si (wt. %) Composite," *Metallurgical Transactions A*, in press.

[2]   Revelos, W. C. and Smith, P. R., "Effect of Environment on the Thermal Fatigue Response of an SCS-6/Ti-24Al-11Nb Composite," *Metallurgical Transactions A*, Vol. 23A, 1992, pp. 587–595.

[3]   Neu, R. W. and Nicholas, T., "Effect of Laminate Orientation on the Thermomechanical Fatigue Behavior of a Titanium Matrix Composite," *Journal of Composites Technology & Research*, Vol. 16, No. 3, July 1994, pp. 214–224.

[4]   Neu, R. W., "A Mechanistic-Based Thermomechanical Fatigue Life Prediction Model for Metal Matrix Composites," *Fatigue and Fracture of Engineering Materials and Structures*, Vol. 16, No. 8, 1993, pp. 811–828.

[5]   Neu, R. W., "Thermomechanical Fatigue Damage Mechanism Maps for Metal Matrix Composites," *Thermomechanical Fatigue Behavior of Materials: 2nd Volume, ASTM STP 1263*, M. J. Verrilli and M. G. Castelli, Eds., American Society for Testing and Materials, West Conshohochen, PA, 1996, pp. 3–26.

[6]   Nicholas, T., Russ, S. M., Neu, R. W., and Schehl, N., "Life Prediction of a [0/90] Metal Matrix Composite Under Isothermal and Thermomechanical Fatigue," *Life Prediction Methodology for Titanium Matrix Composites, ASTM STP 1253*, W. S. Johnson, J. M. Larsen, and B. N. Cox, Eds., American Society for Testing and Materials, West Conshohochen, PA, 1996, pp. 595–617.

[7]   Nicholas, T., "An Approach to Fatigue Life Modeling in Titanium Matrix Composites," *Materials Science and Engineering*, Vol. 200, No. 1, 1995, pp. 29–37.

[8]   Nicholas, T. and Johnson, D. A., "Time- and Cycle-Dependent Aspects of Thermal and Mechanical Fatigue in a Titanium Matrix Composite," *Second Symposium on Thermomechanical Fatigue Behavior of Materials*, Phoenix, AZ, 14–15 Nov. 1994.

[9]   Gayda, J., Gabb, T. B., and Lerch, B. A., "Fatigue-Environment Interactions in SiC/Ti-15-3 Composite," *International Journal of Fatigue*, Vol. 15, No. 1, 1993, pp. 41–45.

[10]   Bartolotta, P. A. and Verrilli, M. J., "Thermomechanical Fatigue Behavior of SiC/Ti-24Al-11Nb [0]$_8$ in Air and Argon Environments," *Composite Materials: Testing and Design (Eleventh Volume), ASTM STP 1206*, E. T. Camponeschi, Jr., Ed., American Society for Testing and Materials, Philadelphia, 1993, pp. 190–201.

[11]   Bartolotta, P., Kantzos, P., Verrilli, M., and Dickerson, R., "TMF Damage Mechanisms of SCS-6/Ti-24Al-11Nb in Air and Argon Environments," *Sixth Annual HITEMP Review NASA CP 19117*, Vol. 2, 1993, pp. 39.1–39.10.

[12]   Hartman, G. A. and Russ, S. M., "Techniques for Mechanical and Thermal Testing of Ti$_3$Al/SCS-6 Metal Matrix Composites," *Metal Matrix Composites: Testing, Analysis, and Failure Modes, ASTM STP 1032*, W. S. Johnson, Ed., American Society for Testing and Materials, Philadelphia, 1989, pp. 43–53.

[13]   Hartman, G. A. and Ashbaugh, N. E., "A Fracture Mechanics Test Automation System for a Basic Research Laboratory," *Applications of Automation Technology to Fatigue and Fracture Testing, ASTM STP 1092*, A. A. Braun, N. E. Ashbaugh, and F. M. Smith, Eds., American Society for Testing and Materials, Philadelphia, 1990, pp. 95–110.

[14]   Neu, R. W. and Nicholas, T., "Thermomechanical Fatigue of SCS-6/TIMETAL®21S Under Out-of-Phase Loading," *Thermomechanical Behavior of Advanced Structural Materials*, AD-Vol. 34, AMD-Vol. 173, W. F. Jones, Ed., American Society of Mechanical Engineers, New York, 1993, pp. 97–111.

[15]   Neu, R. W. and Nicholas, T., "Methodologies for Predicting the Thermomechanical Fatigue Life of Unidirectional Metal Matrix Composites," *Advances in Fatigue Lifetime Predictive Techniques Third Volume ASTM STP 1292*, M. R. Mitchell and R. W. Landgraf, Eds., American Society for Testing and Materials, West Conshohocken, PA, 1996, pp. 1–23.

[16]   Castelli, M. G., "Characterization of Damage Progression in SCS-6/TIMETAL 21S [0]$_4$ Under Thermomechanical Fatigue Loading," *Life Prediction Methodology for Titanium Matrix Composites, ASTM 1253*, W. S. Johnson, J. M. Larsen, and B. N. Cox, Eds., American Society for Testing and Materials, West Conshohocken, PA, 1996, pp. 412–431.

[17]   Kroupa, J. L. and Neu, R. W., "The Nonisothermal Viscoplastic Behavior of a Titanium Matrix Composite," *Composites Engineering*, Vol 4, No. 9, 1994, pp. 965–977.

[*18*]   Neu, R. W. and Roman, I., "Acoustic Emission Monitoring of Damage in Metal Matrix Composites Subjected to Thermomechanical Fatigue," *Composite Science and Technology*, Vol. 52, No. 1, 1994, pp. 1–8.

[*19*]   Coker, D., Ashbaugh, N. E., and Nicholas, T., "Analysis of the Thermomechanical Cyclic Behavior of Unidirectional Metal Matrix Composites," *Thermomechanical Fatigue Behavior of Materials, ASTM STP 1186*, H. Sehitoglu, Ed., American Society for Testing and Materials, Philadelphia, 1993, pp. 50–69.

[*20*]   Talreja, R., *Fatigue of Composite Materials*, Technomic Publishing Co., Lancaster, PA, 1987.

Jian Li[1] and T. Kevin O'Brien[2]

Analysis of the Hygrothermal Effects and Parametric Study of the Edge Crack Torsion (ECT) Mode III Test Layups

REFERENCE: Li, J. and O'Brien, T. K., **"Analysis of the Hygrothermal Effects and Parametric Study of the Edge Crack Torsion (ECT) Mode III Test Layups,"** *Composite Materials: Fatigue and Fracture (Sixth Volume), ASTM STP 1285*, E. A. Armanios, Ed., American Society for Testing and Materials, 1997, pp. 409–431.

ABSTRACT: A shear deformation theory including residual thermal and moisture effects is developed for the analysis of either symmetric or asymmetric laminates with midplane edge delamination under torsional loading. The theory is based on an assumed displacement field that includes shear deformation. The governing equations and boundary conditions are obtained from the principle of virtual work. The analysis of the $[90/(\pm45)_n/(\mp45)_n/90]_s$ edge crack torsion (ECT) Mode III test layup indicates that there are no hygrothermal effects on the Mode III strain energy release rate because the laminate, and both sublaminates above and below the delamination, are symmetric layups. A further parametric study reveals that some other layups can have negligible hygrothermal effects even when the sublaminates above and below the delamination are not symmetric about their own midplanes. However, these layups may suffer from distortion after the curing process. Another interesting set of layups investigated is a class of antisymmetric laminates with $[\pm(\theta/(\theta - 90)_2/\theta)]_n$ layups. It is observed that when n takes on even numbers (2 and 4), both hygrothermal and Mode I effects can be neglected. From this point of view, these layups provide a way to determine the Mode III toughness between two dissimilar layers. However, when n takes on odd numbers (1 and 3), both hygrothermal and Mode I effects may be strong in these layups. In particular, when θ equals 45°, the layups are free from both hygrothermal and Mode I effects irrespective of n.

KEYWORDS: fatigue (materials), fracture (materials), composite materials, laminated composites, fracture toughness, delamination, strain energy release rate, Mode III fracture toughness tests, torsion

Nomenclature

a Delamination length, m
A_{ij} Extensional stiffness coefficients, $N \cdot m^{-1}$
b Specimen width, m
B_{ij} Coupling stiffness coefficients, N
D_{ij} Bending and twisting stiffness coefficients, $N \cdot m$
E_{ii} Young's moduli, GPa

[1] National Research Council resident research associate, NASA Langley Research Center, Hampton, VA 23681–0001.
[2] Senior research scientist, U.S. Army Research Laboratory, Vehicle Structures Directorate, NASA Langley Research Center, Hampton, VA 23681–0001.

F_X Applied tension force, N

g_2 Strain energy release rate parameter due to mechanical effect, N·m

g_1 Strain energy release rate parameter due to hygrothermal and mechanical coupling effect, N

g_0 Strain energy release rate parameter due to hygrothermal effect, N·m^{-1}

G_T Total strain energy release rate, N·m^{-1}

h Half the laminate thickness, m

ΔH Percentage moisture weight gain, % by weight

I Integration constant, m

M_B Applied bending moment, N·m

M_T Applied twisting moment, N·m

$\{M\}$ Vector of stress couples, N

n Number of repeats of a group of layers

$\{N\}$ Vector of stress resultants, N·m^{-1}

$\overline{Q}_{ij}$ Three-dimensional transformed reduced stiffness, N·m^{-2}

s Characteristic root, m^{-1}

ΔT Temperature change from cure temperature to test temperature, °C

x, y, z Cartesian coordinates

Greek Letters (Symbols)

$\{\alpha\}, \{\overline{\alpha}\}$ Lamina coefficients of thermal expansion in lamina and laminate coordinate systems, respectively, per °C

$\{\beta\}, \{\overline{\beta}\}$ Lamina coefficients of moisture expansion in lamina and laminate coordinate systems, respectively, per % by weight

ϵ_0 Laminate midplane extension strain

ϵ_{ij} Strain tensor

$\{\epsilon\}$ Vector of in-plane strain components

ϕ_1 Extension-twist coupling, m

ϕ_2 Bending-twist coupling

$\{\gamma\}$ Vector of out-of-plane shear strain components

η Normalized constant term in M_{xy}, m^{-1}

$\varphi^1 = \varphi^2$ Delaminated portion of laminate width (or a), m (see Eq 82)

φ^3 Undelaminated laminate width (or $b-a$), m (see Eq 82)

λ Laminate torsional stiffness, N·m^2

Λ_1 Strain energy parameter due to hygrothermal effect, N·m

Λ_0 Strain energy parameter due to hygrothermal effect, N

κ Bending curvature, m^{-1}

μ_{ij} Shear moduli, GPa

Π Total strain energy per unit laminate length, N

Θ Twisting angle per unit length (or twist), m^{-1}

Θ^{HT} Twist due to hygrothermal effects, m^{-1}

θ Fiber angle with respect to x-axis for a ply in the laminate

σ_{ij} Stress tensor

$\{\sigma\}$ Vector of in-plane stress components

$\{\tau\}$ Vector of out-of-plane shear stress components

$[\Psi_{ij}]$ Coefficients that relate the end tension, bending, and twisting moments to extension strain, bending curvature, and twist

$\{\Psi_j\}^{HT}$ Force and moments due to hygrothermal effects

Ξ_1, Ξ_2 Extension-twist and bending-twist coupling indicators, respectively
Ψ_x, Ψ_y Rotation functions in warping and transverse directions, respectively

Delamination is a common failure mode in laminated composites, and often the most serious. Characterization of a laminated composite material for its resistance against such failure has become an important task for damage tolerance design purposes. Delamination in composites is characterized in the fracture mechanics sense by the three modes: Mode I (opening), Mode II (in-plane shear), and Mode III (out-of-plane shear). In general, the stress field at the delamination front could be a mixture of all three loading modes. Characterization of fracture resistance, or fracture toughness, in each individual mode for laminated composites can help a designer in materials selection and as a foundation for fracture mechanics criterion used to design against delamination failures.

Analyses and test techniques for Mode I delamination fracture toughness have been established [1], and resulted in the publication of ASTM Test Method for Mode I Interlaminar Fracture Toughness of Unidirectional Fiber-Reinforced Polymer Matrix Composites (D 5528-94a). Mode II fracture toughness characterization [3–6] has also received extensive attention, and standardization development is under way. In the past, the study of Mode III fracture toughness had been largely ignored due to the fact that the Mode I fracture toughness is much lower than Mode III in the traditional laminated composites with epoxy matrices. The lack of a suitable Mode III test method at the time was also a factor [7]. As the Mode I fracture toughness improves with the development of enhanced resin systems, the two shear failure modes may become more important. Furthermore, all three failure modes are important in the establishment of a mixed-mode fracture criterion for structural design purposes.

Recently, an edge crack torsion (ECT) test was proposed as a Mode III toughness test method [8]. This test is based on a class of $[90/(\pm45)_n/(\mp45)_n/90]_s$ layup specimens with a midplane free-edge delamination subjected to torsion. The ECT test was analyzed based on a shear deformation theory, for its validity as a Mode III test [9]. In addition, a simplified solution for the $[90/(\pm45)_n/(\mp45)_n/90]_s$ ECT layups was obtained, and a parametric study on the shear modulus effect was also performed [10]. The analytical solution developed in Ref 9 was restricted to symmetric laminates. Asymmetry in sublaminates (layups above and below the delamination) was considered, but the residual thermal and moisture effects were neglected. The present work is based on the shear deformation theory developed in Ref 9, but includes the analysis of asymmetrical laminates. The hygrothermal effects are also included. The present solution should be identical to the solution of Ref 9 on laminate and sublaminate symmetric layups. The present solution will be able to identify the hygrothermal effects neglected in Ref 9 when the sublaminates are not symmetric. In addition, the present analysis can deal with asymmetric layups as well. In the following, a shear deformation theory including thermal and moisture effects is developed for the analysis of arbitrary composite layup with midplane edge delamination under torsion loading. The theory is applied to the ECT layups first. Next, a study of some altered ECT layups is presented to assess the thermal and moisture effects on these layups. Finally, the theory is applied to a class of antisymmetric layups.

Mathematical Model

A symmetric or asymmetric laminate with midplane free-edge delamination is subjected to torsional loading as shown in Fig. 1. The delamination is assumed to run through the entire length of the laminate. The laminate is modeled as three sublaminates as shown in Fig. 2. Sublaminates 1 and 2 represent the portions of the laminate above and below the delamination, while Sublaminate 3 represents the remaining portion.

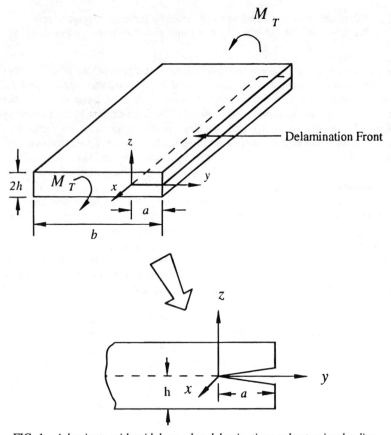

FIG. 1—*A laminate with midplane edge delamination under torsion loading.*

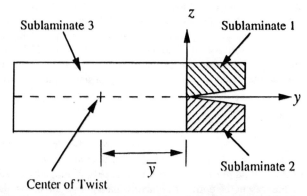

FIG. 2—*A typical* x = *constant plane of the delaminated laminate.*

Displacement Field

The laminate is assumed to be loaded at its ends in such a way that the stresses do not vary along the *x*-axis. The result of the stresses at a typical cross section reduce to a twisting moment as shown in Fig. 1. If the laminate is loaded by a static equivalent twisting moment at the ends, the stresses do not depend on the *x*-axis at locations away from both ends by virtue of the St. Venant principle [*11*]. Such a state of deformation fits the definition of the so-called generalized plane deformation [*12*]. By following the derivation of Ref *12*, the displacement field may be written as

$$
\left.
\begin{aligned}
u(x, y, z) &= \epsilon_0 \cdot x + \kappa \cdot x \cdot (z + \bar{z}) + \kappa_1 \cdot x \cdot (y + \bar{y}) + U_0(y, z) \\
v(x, y, z) &= -\frac{1}{2} \kappa_1 \cdot x^2 + \Theta \cdot x \cdot (z + \bar{z}) + V_0(y, z) \\
w(x, y, z) &= -\frac{1}{2} \kappa \cdot x^2 - \Theta \cdot x \cdot (y + \bar{y}) + W_0(y, z)
\end{aligned}
\right\}
\tag{1}
$$

where *u*, *v*, and *w* denote displacements along the *x*-, *y*-, and *z*-axes, respectively. The axial extension is denoted by ϵ_0, while the bending curvatures in the *x-z* plane and *x-y* plane are denoted by κ and κ_1, respectively. The angle of rotation per unit length about the *x*-axis, or twist, is denoted by Θ.

The constant, $\bar{z}$, represents the *z*-coordinate distance of sublaminate midplane with respect to the delamination plane; $\bar{z} = 0$ for Sublaminate 3 because the sublaminate midplane coincide with the delamination plane, $\bar{z} = h/2$ for Sublaminate 1, and $\bar{z} = -h/2$ for Sublaminate 2. The constant, $\bar{y}$, is the distance between the twisting center and the delamination tip and, hence, is identical for all three sublaminates. The twisting center is initially unknown, and $\bar{y}$ does not show up in the expressions of torsional stiffness and total strain energy release rate of the laminate as will be seen in the final analysis.

The bending curvature, κ_1, is neglected because the thickness-to-width ratio of the laminate is much smaller than unity. Further simplification is achieved by considering first-order terms only in the Taylor expansion of the displacement functions, U_0 and V_0, with respect to the reference surface, and neglecting thickness (*z*) dependence in the function, W_0, as the following

$$
\left.
\begin{aligned}
U_0(y, z) &= U_0(y, 0) + \frac{\partial U_0(y, 0)}{\partial z} \cdot z + O(z^2) + \cdots \\
V_0(y, z) &= V_0(y, 0) + \frac{\partial V_0(y, 0)}{\partial z} \cdot z + O(z^2) + \cdots \\
W_0(y, z) &= W_0(y, 0) + O(z) + \cdots
\end{aligned}
\right\}
\tag{2}
$$

Neglecting the terms of $O(z^2)$ and higher for U_0, V_0, and $O(z)$ and higher for W_0 in Eq 2 and substituting the rest into Eq 1 yields

$$
\left.
\begin{aligned}
u(x, y, z) &= \epsilon_0 \cdot x + \kappa \cdot x \cdot (z + \bar{z}) + U(y) + z \cdot \psi_x(y) \\
v(x, y, z) &= \Theta \cdot x \cdot (z + \bar{z}) + V(y) + z \cdot \psi_y(y) \\
w(x, y, z) &= -\frac{1}{2} \kappa \cdot x^2 - \Theta \cdot x \cdot (y + \bar{y}) + W(y)
\end{aligned}
\right\}
\tag{3}
$$

where the following simplified notations, $U(y) = U_0(y, 0)$, $V(y) = V_0(y, 0)$, $W(y) = W_0(y, 0)$, $\psi_x = \partial U_0(y, 0)/\partial z$, and $\psi_y = \partial V_0(y, 0)/\partial z$ are used.

Shear deformation is recognized through the rotations, ψ_x and ψ_y. Displacement Functions U, V, W, ψ_x, and ψ_y are functions of y only. These displacement functions will be determined from the governing equations and boundary conditions after the section on constitutive relationships. The strains corresponding to this displacement field can be written as

$$\{\epsilon\} = \begin{Bmatrix} \epsilon_{xx} \\ \epsilon_{yy} \\ \gamma_{xy} \end{Bmatrix} = \begin{Bmatrix} \epsilon_{xx}^o \\ \epsilon_{yy}^o \\ \gamma_{xy}^o \end{Bmatrix} + z \cdot \begin{Bmatrix} \kappa_{xx} \\ \kappa_{yy} \\ \kappa_{xy} \end{Bmatrix} \tag{4}$$

$$\{\gamma\} = \begin{Bmatrix} \gamma_{yz} \\ \gamma_{xz} \end{Bmatrix} = \begin{Bmatrix} \gamma_{yz}^o \\ \gamma_{xz}^o \end{Bmatrix} \tag{5}$$

where

$$\left. \begin{aligned} \epsilon_{xx}^o &= \epsilon_0 + \kappa \cdot \bar{z} & \kappa_{xx} &= \kappa & \gamma_{yz}^o &= \psi_y + W_{,y} \\ \epsilon_{yy}^o &= V_{,y} & \kappa_{yy} &= \psi_{y,y} & \gamma_{xz}^o &= \psi_x - \Theta \cdot (y + \bar{y}) \\ \gamma_{xy}^o &= U_{,y} + \Theta \cdot \bar{z} & \kappa_{xy} &= \psi_{x,y} + \Theta \end{aligned} \right\} \tag{6}$$

where partial differentiation by a variable is denoted by the subscript comma followed by that variable.

Constitutive Relationships

A generic sublaminate along with stress resultants and moment couples is shown in Fig. 3. The stress resultants and moment couples are denoted by N_x, N_y, N_{xy}, Q_x, Q_y, M_x, M_y, and M_{xy}

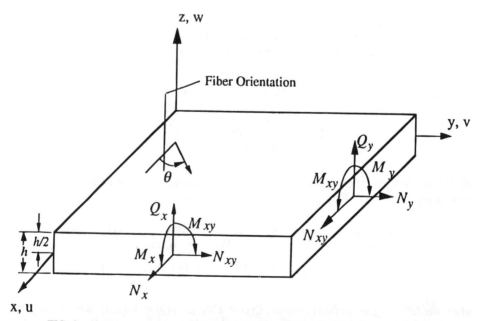

FIG. 3—*Notation and sign convention for stress resultants and moment couples.*

as shown in Fig. 3. The constitutive relationships can be written in terms of stress resultants and moment couples, and associated strains and curvatures as follows

$$
\begin{Bmatrix} N_x \\ N_y \\ N_{xy} \\ \hline M_x \\ M_y \\ M_{xy} \end{Bmatrix} = \left[\begin{array}{ccc|ccc} A_{11} & A_{12} & A_{16} & B_{11} & B_{12} & B_{16} \\ A_{12} & A_{22} & A_{26} & B_{12} & B_{22} & B_{26} \\ A_{16} & A_{26} & A_{66} & B_{16} & B_{26} & B_{66} \\ \hline B_{11} & B_{12} & B_{16} & D_{11} & D_{12} & D_{16} \\ B_{12} & B_{22} & B_{26} & D_{12} & D_{22} & D_{26} \\ B_{16} & B_{26} & B_{66} & D_{16} & D_{26} & D_{66} \end{array} \right] \begin{Bmatrix} \epsilon^o_{xx} \\ \epsilon^o_{yy} \\ \gamma^o_{xy} \\ \kappa_{xx} \\ \kappa_{yy} \\ \kappa_{xy} \end{Bmatrix} - \begin{Bmatrix} N_x \\ N_y \\ N_{xy} \\ M_x \\ M_y \\ M_{xy} \end{Bmatrix}^{HT}
\tag{7}
$$

$$
\begin{Bmatrix} Q_y \\ Q_x \end{Bmatrix} = \begin{bmatrix} A_{44} & A_{45} \\ A_{45} & A_{55} \end{bmatrix} \begin{Bmatrix} \gamma^o_{yz} \\ \gamma^o_{xz} \end{Bmatrix}
\tag{8}
$$

where

$$
(\{N\}, \{M\}) = \int_{-h/2}^{h/2} \{\sigma\} \cdot (1, z) \cdot dz
\tag{9}
$$

$$
\begin{Bmatrix} Q_y \\ Q_x \end{Bmatrix} = \int_{-h/2}^{h/2} \{\tau\} \cdot dz
\tag{10}
$$

$$
\{\sigma\} = (\sigma_{xx}, \sigma_{yy}, \tau_{xy})^T = [\overline{Q}]\langle\{\epsilon\} - \{\alpha\}\Delta T - \{\beta\}\Delta H\rangle
\tag{11}
$$

$$
\{\tau\} = \begin{Bmatrix} \tau_{yz} \\ \tau_{xz} \end{Bmatrix} = \begin{bmatrix} \overline{Q}_{44} & \overline{Q}_{45} \\ \overline{Q}_{45} & \overline{Q}_{55} \end{bmatrix} \{\gamma\}
\tag{12}
$$

$$
(A_{ij}, B_{ij}, D_{ij}) = \int_{-h/2}^{h/2} \overline{Q}_{ij} \cdot (1, z, z^2) \cdot dz
\tag{13}
$$

$$
(\{N\}^{HT}, \{M\}^{HT}) = \int_{-h/2}^{h/2} [\overline{Q}]\langle\{\overline{\alpha}\}\Delta T + \{\overline{\beta}\}\Delta H\rangle \cdot (1, z) \cdot dz
\tag{14}
$$

where

$\overline{Q}_{ij}$ = the transformed reduced stiffness [13];
$\{\overline{\alpha}\}$ = $(\overline{\alpha}_{xx}, \overline{\alpha}_{yy}, \overline{\alpha}_{xy})^T$ lamina coefficient of thermal expansion in x-y coordinates;
$\{\overline{\beta}\}$ = $(\overline{\beta}_{xx}, \overline{\beta}_{yy}, \overline{\beta}_{xy})^T$ lamina coefficient of hygroscopic expansion in x-y coordinates;
ΔT = temperature differential from cure temperature to test temperature; and
ΔH = percentage moisture weight gain.

For Sublaminate 3, the integrations in Eqs 9, 10, 13, and 14 should be carried out from $-h$ to h.

Governing Equations

Because all the strains and stresses are independent about the x-axis, a unit length of the laminate along the x-axis is considered. For all three sublaminates, the upper and lower surfaces are stress free. The governing equations and boundary conditions for the three sublaminates can be obtained from the principle of virtual work

$$\iint_{\Omega^1} \sigma_{ij}^1 \delta\epsilon_{ij}^1 \, dy \, dz + \iint_{\Omega^2} \sigma_{ij}^2 \delta\epsilon_{ij}^2 \, dy \, dz + \iint_{\Omega^3} \sigma_{ij}^3 \delta\epsilon_{ij}^3 \, dy \, dz = \int_{\Gamma^1+\Gamma^2+\Gamma^3} F_i \delta u_i \, ds \quad (15)$$

where Ω^1, Ω^2, Ω^3 represent the volumes of Sublaminates 1, 2, and 3, respectively, while Γ^1, Γ^2, and Γ^3 represent their respective boundaries. In order to distinguish similar quantities of the sublaminates, Superscripts 1, 2, and 3 are used to refer to the respective sublaminate. In the event of a quantity raised to a power, the quantity will be bracketed.

By following the procedures presented in Ref 9, the governing equations obtained from Eq 15 for each of the sublaminates may be simplified as

$$N_y^j = N_{xy}^j = M_y^j = Q_y^j = 0, \quad j = 1, 2, 3 \quad (16)$$

$$M_{xy,y}^j - Q_x^j = 0, \quad j = 1, 2, 3 \quad (17)$$

The boundary conditions and continuity conditions are

$$M_{xy}^3\big|_{y=-(b-a)} = 0 \quad (18)$$

$$M_{xy}^1\big|_{y=a} = M_{xy}^2\big|_{y=a} = 0 \quad (19)$$

$$M_{xy}^3\big|_{y=0} = M_{xy}^1\big|_{y=0} + M_{xy}^2\big|_{y=0} \quad (20)$$

$$\psi_x^1\big|_{y=0} = \psi_x^2\big|_{y=0} = \psi_x^3\big|_{y=0} \quad (21)$$

Solutions for the Displacement Functions

Substituting the stress resultants and moment couples given in Eqs 7 and 8 into Eq 16, and using Eq 6 yields

$$\left\{ \begin{array}{c} V^j_{,y} \\ U^j_{,y} + \Theta \cdot \bar{z}^j \\ \psi^j_{y,y} \end{array} \right\} = -[R^j] \left\{ \begin{array}{c} \epsilon_0 \\ \kappa \end{array} \right\} - \{\bar{R}^j\} \cdot (\psi^j_{x,y} + \Theta) + \{\hat{R}^j\}, \qquad j = 1, 2, 3 \qquad (22)$$

$$\psi^j_y + W^j_{,y} = -\frac{A^j_{45}}{A^j_{44}} [\psi^j_x - \Theta \cdot (y + \bar{y})], \qquad j = 1, 2, 3 \qquad (23)$$

where $[R^j]$, $\{\bar{R}^j\}$, and $\{\hat{R}^j\}$ are given in Eqs 55 through 57 of the Appendix. The remaining nonzero stress resultants and moment couples can be simplified as

$$\left\{ \begin{array}{c} N^j_x \\ M^j_x \\ M^j_{xy} \end{array} \right\} = [\xi^j] \left\{ \begin{array}{c} \epsilon_0 \\ \kappa \end{array} \right\} + \{\bar{\xi}^j\}(\psi^j_{x,y} + \Theta) + \{\hat{\xi}^j\}, \qquad j = 1, 2, 3 \qquad (24)$$

$$Q^j_x = \left[A^j_{55} - \frac{(A^j_{45})^2}{A^j_{44}} \right][\psi^j_x - \Theta \cdot (y + \bar{y})], \qquad j = 1, 2, 3 \qquad (25)$$

The matrix $[\xi^j]$ and vectors $\{\bar{\xi}^j\}$ and $\{\hat{\xi}^j\}$ are given in Eqs 58 through 60 of the Appendix.

The solution of ψ_x for each sublaminate can be obtained by applying Eq 17 to the three respective regions and making use of the boundary conditions given in Eq 18 through 21. The general form of ψ_x can be written as

$$\psi^j_x = I^j \cdot (e^{s^j \cdot y} + e^{2s^j \cdot \rho^j} \cdot e^{-s^j \cdot y}) + \frac{\eta^j}{s^j} e^{s^j \cdot \rho^j} \cdot e^{-s^j y} + \Theta \cdot (y + \bar{y}), \qquad j = 1, 2, 3 \quad (26)$$

where I^j represents the integration constant, ρ^j is replaced by $-(b - a)$ for Sublaminate 3, and by a for Sublaminate 1 and 2, respectively. The integration constant is different from each sublaminate. The three constants for the three sublaminates are given in Eq 61 of the Appendix. The characteristic root, s^j, and parameter, η^j, in Eq 26 are given by

$$s^j = \sqrt{\left[A^j_{55} - \frac{(A^j_{45})^2}{A^j_{44}} \right] \Big/ \bar{\xi}^j_3}, \qquad j = 1, 2, 3 \qquad (27)$$

$$\eta^j = \frac{\xi^j_{31}}{\bar{\xi}^j_3} \epsilon_0 + \frac{\xi^j_{32}}{\bar{\xi}^j_3} \kappa + 2\Theta + \frac{\hat{\xi}^j_3}{\bar{\xi}^j_3}, \qquad j = 1, 2, 3 \qquad (28)$$

Loading Conditions

The extension force, bending, and twisting moments for the entire laminate cross section are denoted by F_X, M_B, and M_T, respectively. These are expressed as

$$F_X = \int_{-(b-a)}^{0} N_x^3 \cdot dy + \int_{0}^{a} (N_x^1 + N_x^2) \cdot dy \tag{29}$$

$$M_B = \int_{-(b-a)}^{0} M_x^3 \cdot dy + \int_{0}^{a} [M_x^1 + M_x^2 + \bar{z}^1 \cdot N_x^1 + \bar{z}^2 \cdot N_x^2] \, dy \tag{30}$$

$$M_T = 2 \int_{-(b-a)}^{0} M_{xy}^3 \cdot dy + 2 \int_{0}^{a} [M_{xy}^1 + M_{xy}^2 + \bar{z}^1 \cdot N_{xy}^1 + \bar{z}^2 \cdot N_{xy}^2] \cdot dy \tag{31}$$

Substituting the stress resultants and moment couples from Eq 24 for the three sublaminates into Eqs 29 through 31 yields

$$\begin{Bmatrix} F_X \\ M_B \\ M_T \end{Bmatrix} = \{\Psi_{ij}\} \begin{Bmatrix} \epsilon_0 \\ \kappa \\ \theta \end{Bmatrix} + \{\Psi_j\}^{HT} \qquad [i, j = 1, 2, 3] \tag{32}$$

The parameters, Ψ_{ij}, in Eq 32 are functions of sublaminate stiffness coefficients (A_{ij}, B_{ij}, D_{ij}) and specimen geometry (a, b, h) and are given in Eq 72 of the Appendix. The parameters, Ψ_j^{HT}, represent the thermal and moisture contributions, and their expressions are given in Eq 73 of the Appendix.

The strain, ϵ_0, bending curvature, κ, and twist, Θ, can be written as a combination of the mechanical and hygrothermal contributions as

$$\begin{Bmatrix} \epsilon_0 \\ \kappa \\ \Theta \end{Bmatrix} = \begin{Bmatrix} \epsilon_0 \\ \kappa \\ \Theta \end{Bmatrix}^{M} + \begin{Bmatrix} \epsilon_0 \\ \kappa \\ \Theta \end{Bmatrix}^{HT} \tag{33}$$

where

$$\begin{Bmatrix} \epsilon_0 \\ \kappa \\ \Theta \end{Bmatrix}^{HT} = -[\Psi_{ij}]^{-1}\{\Psi_j\}^{HT} \tag{34}$$

For a laminate under torsional loading, $F_X = M_B = 0$, the induced extension strain and bending curvature can be written as

$$\epsilon_0 = \phi_1 \cdot \Theta + \phi_1^{HT} \tag{35}$$

$$\kappa = \phi_2 \cdot \Theta + \phi_2^{HT} \tag{36}$$

where

$$\left\{ \begin{matrix} \phi_1 \\ \phi_2 \end{matrix} \right\} = - \begin{bmatrix} \Psi_{11} & \Psi_{12} \\ \Psi_{21} & \Psi_{22} \end{bmatrix}^{-1} \left\{ \begin{matrix} \Psi_{13} \\ \Psi_{23} \end{matrix} \right\} \tag{37}$$

$$\left\{ \begin{matrix} \phi_1 \\ \phi_2 \end{matrix} \right\}^{HT} = - \begin{bmatrix} \Psi_{11} & \Psi_{12} \\ \Psi_{21} & \Psi_{22} \end{bmatrix}^{-1} \left\{ \begin{matrix} \Psi_1 \\ \Psi_2 \end{matrix} \right\}^{HT} \tag{38}$$

Parameters, ϕ_1 and ϕ_2, represent extension-twist and bending-twist couplings, respectively. The twisting moment can be obtained from Eq 32 as

$$M_T = \lambda \cdot (\Theta - \Theta^{HT}) \tag{39}$$

where

$$\lambda = \phi_1 \cdot \psi_{31} + \phi_2 \cdot \psi_{32} + \psi_{33} \tag{40}$$

The coefficient, λ, represents the laminate torsional stiffness.

Strain Energy and Strain Energy Release Rate

The strain energy per unit length of the laminate can be written as

$$\Pi = \frac{1}{2} \sum_{j=1}^{3} \left(\int_{\Omega^j} \langle \{\epsilon\} - \{\overline{\alpha}\} \cdot \Delta T - \{\overline{\beta}\} \cdot \Delta H \rangle^T \{\sigma\} \cdot dy \cdot dz \right.$$
$$\left. + \int_{\Omega^j} \{\gamma\}^T \{\tau\} \cdot dy \cdot dz \right) \tag{41}$$

Substituting Eq 11 into Eq 41, rearranging yields

$$\Pi = \frac{1}{2} \sum_{j=1}^{3} \int_{\Omega^j} (\{\epsilon\}^T \{\sigma\} + \{\gamma\}^T \{\tau\}) \cdot dy \cdot dz$$
$$- \frac{1}{2} \sum_{j=1}^{3} \int_{\Omega^j} \langle \{\overline{\alpha}\} \cdot \Delta T + \{\overline{\beta}\} \cdot \Delta H \rangle^T [\overline{Q}] \{\epsilon\} \cdot dy \cdot dz \tag{42}$$
$$+ \frac{1}{2} \sum_{j=1}^{3} \int_{\Omega^j} \langle \{\overline{\alpha}\} \cdot \Delta T + \{\overline{\beta}\} \cdot \Delta H \rangle^T [\overline{Q}] \langle \{\overline{\alpha}\} \cdot \Delta T + \{\overline{\beta}\} \cdot \Delta H \rangle \cdot dy \cdot dz$$

The last term in Eq 42 does not depend on delamination length, a, hence will not be considered further. The first term in Eq 42, Π_1, could be proved easily to be

$$\Pi_1 = \frac{1}{2} \cdot (\epsilon_0 \cdot F_X + \kappa \cdot M_B + \Theta \cdot M_T) \tag{43}$$

For a laminate under torsional loading, Eq 43 reduces to

$$\Pi_1 = \frac{1}{2} \cdot \Theta \cdot M_T = \frac{1}{2} \cdot \lambda \cdot \Theta \cdot (\Theta - \Theta^{HT}) \tag{44}$$

The second term in Eq 42, Π_2, may be written as

$$\Pi_2 = -\frac{1}{2} \Theta \cdot \Lambda_1 - \frac{1}{2} \Lambda_0 \tag{45}$$

Parameters Λ_1 and Λ_0 are functions of delamination length, a, and given in Eqs 80 and 81 of the Appendix.

For the laminate (as shown in Fig. 1) containing a planar delamination of length, a, that extends under a constant twist, Θ, no external work is performed as the delamination extends, and the total strain energy release rate may be calculated from

$$G_T = -\frac{\partial \Pi}{\partial a} \tag{46}$$

Substituting Eqs 44 and 45 into Eq 46 and differentiating yields

$$G_T = -\frac{1}{2} \lambda_{,a} \cdot \Theta^2 + \frac{1}{2} (\lambda_{,a} \cdot \Theta^{HT} + \lambda \cdot \Theta^{HT}_{,a} + \Lambda_{1,a}) \cdot \Theta + \frac{1}{2} \Lambda_{0,a} \tag{47}$$

In order to isolate the hygrothermal effects, the total strain energy release rate can be written in terms of mechanical twist by substituting Eq 33 into Eq 47 to yield

$$G_T = \frac{1}{2} g_2 \cdot (\Theta^M)^2 + \frac{1}{2} g_1 \cdot \Theta^M + \frac{1}{2} g_0 \tag{48}$$

where

$$g_2 = -\lambda_{,a} \tag{49}$$

$$g_1 = \lambda \cdot \Theta^{HT}_{,a} - \lambda_{,a} \cdot \Theta^{HT} + \Lambda_{1,a} \tag{50}$$

$$g_0 = \lambda \cdot \Theta^{HT}_{,a} \cdot \Theta^{HT} + \Lambda_{0,a} \tag{51}$$

Parameter g_2 represents the mechanical contribution to the strain energy release rate, while Parameter g_1 is the coefficient of the cross term representing the coupling of hygrothermal and mechanical contributions, and Parameter g_0 represents the hygrothermal contribution. Equation 47 indicates that there could be a nonzero strain energy release rate due to hygrothermal effects alone.

Mode III Layup Identification

The presence of extension-twist and bending-twist couplings in a laminate with arbitrary layup may induce Modes I and II contributions to the total strain energy release rate in addition to the Mode III contribution. The objective here is to identify the Modes I and II contributions to the total strain energy release rate in order to minimize them. The actual evaluation of Modes I and II components is not necessary. Instead, the closed-form solution given in Eq 49 may provide a convenient alternative to identify the presence of Modes I and II contributions. Substituting Eq 40 into Eq 49 yields

$$g_2 = -\Psi_{33,a} - (\phi_{1,a}\Psi_{31} + \phi_1\Psi_{31,a}) - (\phi_{2,a}\Psi_{32} + \phi_2\Psi_{32,a}) \tag{52}$$

The second and third terms in Eq 52 indicate the presence of Modes I and II contributions due to extension-twist and bending-twist couplings, respectively. Both these terms should be zero for pure Mode III layups. If for a particular layup these two terms are very small compared to the first term, this layup may be considered as a desirable Mode III layup. The $[90/(\pm45)_n/(\mp45)_n/90]_s$ ECT layups belong to this category. Two indicators are defined to help the identification of desirable Mode III layups as follows

$$\Xi_1 = \frac{(\phi_{1,a}\Psi_{31} + \phi_1\Psi_{31,a})}{\Psi_{33,a} + (\phi_{1,a}\Psi_{31} + \phi_1\Psi_{31,a}) + (\phi_{2,a}\Psi_{32} + \phi_2\Psi_{32,a})} \tag{53}$$

$$\Xi_2 = \frac{(\phi_{2,a}\Psi_{32} + \phi_2\Psi_{32,a})}{\Psi_{33,a} + (\phi_{1,a}\Psi_{31} + \phi_1\Psi_{31,a}) + (\phi_{2,a}\Psi_{32} + \phi_2\Psi_{32,a})} \tag{54}$$

where Ξ_1 and Ξ_2 are extension-twist and bending-twist coupling indicators, respectively. Note that these indicators are percentages of the second and third terms to the total term in Eq 52.

Results

First, the present analytical solution was compared with Refs *9* and *10* for the ECT layups $[90/(\pm45)_n/(\mp45)_n/90]_s$. The lamina properties are given in Table 1. The hygrothermal bending curvature and twist defined in Eq 33 vanish for these ECT layups because of the laminate and sublaminate symmetry. For these ECT layups, the present solution should be identical to

TABLE 1—*Elastic properties and configurational parameters of carbon/epoxy composites [8].*

Properties[a]		Configurational Parameters
$E_{11} = 165$ GPa	$\alpha_1 = -0.3 \times 10^{-6}/°C$	$b = 38.1$ mm (width)
$E_{22} = E_{33} = 10.3$ GPa	$\alpha_2 = 30 \times 10^{-6}/°C$	$t = 0.13$ mm (ply thickness)
$\mu_{12} = \mu_{13} = 5.5$ GPa	$\beta_1 = 0$	$h = $ half laminate thickness
$\mu_{23} = 5.5$ GPa	$\beta_2 = 0.2$	
$\nu_{12} = \nu_{13} = 0.28$	$\Delta T = -156°C$	
$\nu_{23} = 0.28$		

[a]These thermal and moisture expansion coefficients are adopted from a similar carbon/epoxy (IM6/SC1081) material [*14*] because such information was not available in Ref *8*.

that of Ref 9. Tables 2 and 3 show the comparisons of torsional stiffness and strain energy release rate parameter (g_2) predicted by the present solution and Refs 9 and 10. The present solution is identical to Ref 9 as seen in Tables 2 and 3. Also observed is the accuracy of the simplified solution given in Ref 10 for the ECT layups.

Altered ECT Layups

Second, the present analytical solution was used to investigate the residual thermal effect for several layups that were selected by altering the $[90/(\pm45)_3/(\mp45)_3/90]_s$ ECT layup and are presented in Table 4. The lamina properties from Table 1 were used to generate the analytical solutions and are presented in Table 5 for a delamination length $a/b = 0.3$. The temperature change, $\Delta T = -156°C$, was used. The moisture weight gain, ΔH, was assumed to be zero to isolate the residual thermal effects.

TABLE 2—*Comparison of torsional stiffness and strain energy release rate parameter, g_2, for the ECT layup $[90/(\pm45)_n/(\mp45)_n/90]_s$ with n = 3.*

n = 3, a/b	λ, N·m²			g_2, N·m		
	Present	Ref 9	Ref 10	Present	Ref 9	Ref 10
0.1	17.49	17.49	17.53	426.4	426.4	424.4
0.3	14.06	14.06	14.11	451.0	451.1	449.7
0.5	10.60	10.60	10.66	450.1	450.2	448.8
0.7	7.244	7.244	7.311	434.6	434.7	433.3
0.9	4.434	4.434	4.510	223.7	223.9	223.1

TABLE 3—*Comparison of torsional stiffness strain energy release rate parameter, g_2, for the ECT layup $[90/(\pm45)_n/(\mp45)_n/90]_s$ with n = 4.*

n = 4, a/b	λ, N·m²			g_2, N·m		
	Present	Ref 9	Ref 10	Present	Ref 9	Ref 10
0.1	37.19	37.19	37.24	859.7	859.8	855.9
0.3	29.82	29.82	29.88	985.9	986.0	985.7
0.5	22.27	22.27	22.36	976.5	976.6	976.3
0.7	15.13	15.13	15.23	890.7	890.8	890.5
0.9	9.852	9.852	9.964	360.0	360.2	359.6

TABLE 4—*Altered ECT layups and their representative codes.*

Code	Layup
L1	$[(\pm45)_3/(\mp45)_3/0/90]_s$
L2	$[(\pm45)_6/0/90]_s$
L3	$[45/(\pm45)_3/(\mp45)_3/90]_s$
L4	$[-45/(\pm45)_3/(\mp45)_3/90]_s$
L5	$[(\pm45_2)_3/0/90]_s$
L6	$[(\pm45_3)_2/0/90]_s$

TABLE 5—*Torsional stiffness, total energy release rate parameters, and bending-twist coupling indicator for the altered ECT layups at a/b = 0.3.*

Code	λ, N·m²	g_2, N·m	g_1, N	g_0, N·m⁻¹	Ξ_2, %
L1	16.39	567.4	0	0.040	−0.002
L2	16.31	561.7	0	0.040	−0.057
L3	16.06	526.5	0	0.278	−0.323
L4	16.54	531.0	0	0.277	−0.185
L5	15.91	549.5	0	0.040	−0.234
L6	15.24	529.2	0	0.041	−0.544

All of these altered layups have higher torsional stiffness (λ) than the original ECT layup when one compares the stiffness values in Table 5 to the corresponding value in Table 2. The thermal-mechanical coupling strain energy release rate parameter (g_1) is zero, and the thermal parameter (g_0) is negligible compared to the magnitude (~ 1000 N/m) of the Mode III toughness obtained from experiments [8]. The small amount of g_0 is due to the asymmetry in the sublaminates. The symmetry of the entire laminate, however, results in the coupling parameter, g_1, to diminish. The extension-twist coupling indicator (Ξ_1) vanishes for all the layups because there is no extension-twist coupling in these layups.

The bending-twist coupling indicator (Ξ_2) is very small for these layups as shown in Table 5 ($<0.6\%$). Both L1 and L2 layups have six pairs of $\pm 45°$ layers in each sublaminate, the bending-twist coupling indicator of L1 is one order of magnitude smaller than that of L2 due to the self-symmetry arrangement of the $\pm 45°$ layers in L1. The bending-twist coupling indicator is higher in the L3 layup than that in L4 layup where the outer-most 90° layer from the ECT layup has been changed to 45° and $-45°$, respectively. There are three and two pairs of $\pm 45°$ layers in each sublaminate of the L5 and L6 layups, respectively. The bending-twist coupling indicator is higher in L6 than that in L5, but still less than 0.6%. All these layups may be considered as desirable Mode III layups from the dominance of the Mode III strain energy release rate point of view. However, these layups may suffer from distortion after the curing process due to asymmetry in the sublaminates.

Antisymmetric Layups

The layups studied earlier are all symmetric laminates and show little thermal effect on the strain energy release rate. The analytical model developed in the second section is not limited to symmetric laminates. Next, a class of antisymmetric laminates are investigated for the residual thermal and moisture effects. These layups are antisymmetric laminates with $[\pm(\theta/(\theta - 90)_2/\theta)]_n$ layups. The sublaminates may be either symmetric (n = odd number) or antisymmetric (n = even number). This class of laminates exhibits extension-twist coupling and does not distort after curing.

Results for the $\theta = 30°$ and $n = 1$ to 4 layups are presented in Table 6 for a normalized delamination length, $a/b = 0.3$. The moisture weight gain, ΔH, was assumed to be zero when the results in Table 6 were generated. The torsional stiffness, strain energy release rate parameters, and extension-twist coupling indicator are presented in the table. The bending-twist coupling indicator vanishes because there is no bending-twist coupling. The thermal parameter, g_0, is zero for all the layups due to the symmetry or antisymmetry of the sublaminates. The presence of the coupling parameter, g_1, and the extension-twist coupling indicator are strong when the sublaminates are symmetric (n = 1 and 3). However, the significance of the parameter, g_1, and the Mode I indicator reduces as n increases. The moisture effect may also

TABLE 6—*Torsional stiffness, total energy release rate parameters, and extension-twist coupling indicator for $[\pm(30/-60_2/30)]_n$ layups at a/b = 0.3.*

n	λ, N·m²	g_2, N·m	g_1, N	g_0, N·m⁻¹	Ξ_1, %
1^a	0.247	5.244	8.078	0	−64.5
3	8.49	247.7	33.05	0	4.1
2^b	2.39	74.98	-0.5×10^{-5}	0	0.3×10^{-5}
4	19.5	599.4	−0.042	0	0.003

[a]$[(30/-60_2/30)\uparrow(-30/60_2/-30)]$. Note that $\uparrow$ indicates the interface with free edge delamination.
[b]$[(30/-60_2/30/-30/60_2/-30)\uparrow(30/60_2/30/-30/60_2/-30)]$.

be strong in these layups. Figure 4 shows the moisture effects on the coupling parameter, g_1, as a function of delamination length, a/b, for $n = 3$ with a fixed temperature change of $\Delta T = -156°C$. As the percentage of the moisture weight gain, ΔH, increases from zero, g_1 reduces and becomes zero (canceling the thermal effect) when $\Delta H = 0.0208$, irrespective of the delamination length. A further increase of ΔH changes the sign of g_1.

When the sublaminates are antisymmetric ($n = 2$ and 4), both the thermal-mechanical coupling parameter, g_1, and the extension-twist coupling indicator are very small. From this

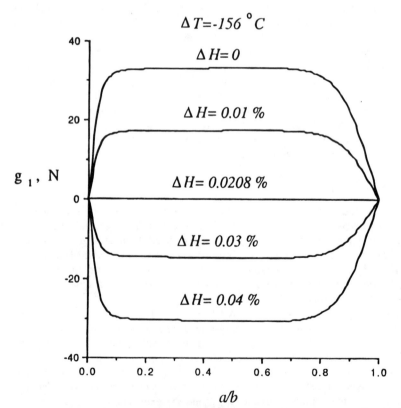

FIG. 4—*Coupling strain energy release rate parameter, g, as a function of normalized delamination length, a/b, for $[\pm(30/-60_2/30)]_3$ layup, with $\Delta T = 156°C$.*

point of view, these layups could be candidate Mode III layups to characterize the Mode III fracture toughness between two layers with different fiber orientations. Additional results for $\theta = 20°$, $40°$, $45°$, $50°$, $60°$, and $70°$ (when $n = 4$) are presented in Table 7 along with $\theta = 30°$. The torsional stiffness, (λ), strain energy release rate parameter (g_2), and the extension-twist coupling indicator are symmetric with respect to $\theta = 45°$, as seen from Table 7. Both g_1 and the extension-twist coupling indicator are very small and become identical to zero when $\theta = 45°$.

The torsional stiffness (λ) and strain energy release rate parameter (g_2) as a function of normalized delamination length (a/b) are plotted in Figs. 5 and 6 for $\theta = 30°$, $45°$. The torsional stiffness decreases as the delamination length increases, and reaches the torsional stiffness of the two sublaminates as the delamination completely separates the laminate into two pieces. The middle portion of the curve is almost linear. The curve becomes flat near both ends

TABLE 7—*Torsional stiffness, total energy release rate parameters, and extension-twist coupling indicator for* $[\pm(\theta/(90-\theta)_2/\theta)]_\phi$ *layups a a/b = 0.3.*

θ	λ, N·m²	g_2, N·m	g_1, N	Ξ_1, %
20	12.6	347.9	−0.004	0.0003
30	19.5	599.4	−0.042	0.003
40	24.3	745.1	−0.046	0.002
45	25.0	765.0	0	0
50	24.3	745.1	0.046	0.002
60	19.5	599.4	0.042	0.003
70	12.6	374.9	0.004	0.0003

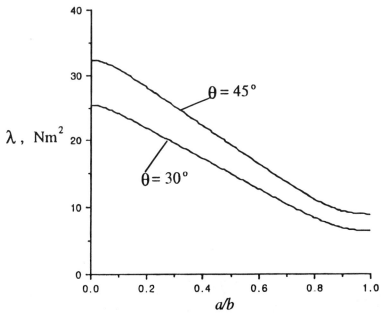

FIG. 5—*Torsional stiffness, λ, as a function of normalized delamination length, a/b, for* $[\pm(\theta/(\theta - 90)_2/\theta)]_4$ *layups ($\theta = 30°$, $45°$).*

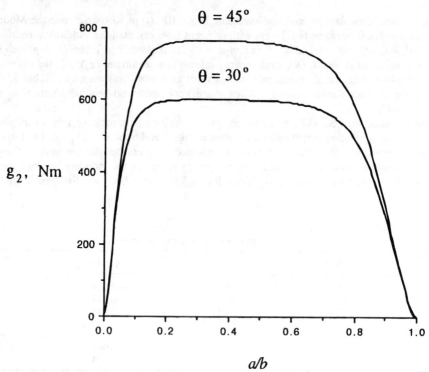

FIG. 6—*Strain energy release rate parameter, g_2, as a function of normalized delamination length, a/b, for $[\pm(\theta/(\theta - 90)_2/\theta)]_4$ layups ($\theta = 30°, 45°$).*

(a/b = 0 and 1). This indicates that both very small and very large delamination lengths are not desirable for Mode III delamination growth, because the available strain energy release rate to drive the delamination is proportional to the derivative of the torsional stiffness. Figure 6 is essentially the negative value of the derivative of the torsional stiffness. It is seen from Figs. 4 through 6 that the analytical solution gives meaningful results for the entire range of normalized delamination length (a/b = 0 to 1).

Conclusions

A shear deformation theory including hygrothermal effects is developed for the analysis of either symmetric or asymmetric laminates with midplane edge delamination under torsional loading. The torsional stiffness and total strain energy release rate are obtained in closed form. The hygrothermal effects on the strain energy release rate are identified by hygrothermal-mechanical coupling and pure hygrothermal contributions. Two indicators are defined to identify the presence of Modes I and II contributions in the total strain energy release rate.

The analysis of the $[90/(\pm45)_n/(\mp45)_n/90]_s$ ECT Mode III layups indicates that there are no hygrothermal effects on the strain energy release rate because of the laminate and sublaminate symmetry. Identical results to those obtained in Ref 9 are observed for these ECT layups. A further study of some altered ECT layups reveals that these layups have negligible residual thermal and bending-twist coupling effects even when the sublaminates above and below the delamination are not symmetric about their own midplanes. However, these layups may suffer from distortion after the curing process.

Another set of layups investigated is a class of antisymmetric laminates with $[\pm(\theta/(\theta - 90)_2/\theta)]_n$ layups. There is no bending-twist coupling effect in these layups. When n takes on odd numbers (1 and 3), both the thermal and extension-twist coupling effects on the total strain energy release rate may be strong. The relative significance of the thermal and extension-twist coupling effects reduces as n increases. The moisture effect may also be strong in these layups. However, the moisture effect tends to reduce the thermal effect and may cancel the thermal effect at some value of the percentage moisture weight gain irrespective of the delamination length.

When n takes on even numbers (2 and 4), both the hygrothermal and extension-twist coupling effects may be neglected. From this point of view, these layups are desirable Mode III layups to characterize the Mode III fracture toughness between two $\pm\theta$ layers. In particular, when $\theta = 45°$, both the hygrothermal and extension-twist coupling effects vanish irrespective of n.

Acknowledgment

This work was performed while Dr. Li was a National Research Council Resident research associate at NASA Langley Research Center.

APPENDIX

Definition of Parameters

In this Appendix, all the intermediate parameters needed to evaluate the torsional stiffness and strain energy release rate are summarized according to their sequence of appearance in the text.

The matrix $[R^j]$ and vectors $\{\overline{R}^j\}$ and $\{\hat{R}^j\}$ in Eq 22 are given by

$$[R^j] = \begin{bmatrix} A^j_{22} & A^j_{26} & B^j_{22} \\ A^j_{26} & A^j_{66} & B^j_{26} \\ B^j_{22} & B^j_{26} & D^j_{22} \end{bmatrix}^{-1} \begin{bmatrix} A^j_{12} & A^j_{12} \cdot \overline{z}^j + B^j_{12} \\ A^j_{16} & A^j_{16} \cdot \overline{z}^j + B^j_{16} \\ B^j_{12} & B^j_{12} \cdot \overline{z}^j + D^j_{12} \end{bmatrix}, \qquad j = 1, 2, 3 \qquad (55)$$

$$[\overline{R}^j] = \begin{bmatrix} A^j_{22} & A^j_{26} & B^j_{22} \\ A^j_{26} & A^j_{66} & B^j_{26} \\ B^j_{22} & B^j_{26} & D^j_{22} \end{bmatrix}^{-1} \begin{Bmatrix} B^j_{26} \\ B^j_{66} \\ D^j_{26} \end{Bmatrix}, \qquad j = 1, 2, 3 \qquad (56)$$

$$[\hat{R}^j] = \begin{bmatrix} A^j_{22} & A^j_{26} & B^j_{22} \\ A^j_{26} & A^j_{66} & B^j_{26} \\ B^j_{22} & B^j_{26} & D^j_{22} \end{bmatrix}^{-1} \begin{Bmatrix} N^j_y \\ N^j_{xy} \\ M^j_y \end{Bmatrix}^{HT}, \qquad j = 1, 2, 3 \qquad (57)$$

where $\overline{z}^1 = -\overline{z}^2 = h/2$ and $\overline{z}^3 = 0$.

The matrix, $[\xi^j]$, and vectors, $\{\overline{\xi}^j\}$ and $\{\hat{\xi}^j\}$, in Eq 24 are given by

$$[\xi^j] = \begin{bmatrix} A^j_{11} & A^j_{11} \cdot \bar{z}^j + B^j_{11} \\ B^j_{11} & B^j_{11} \cdot \bar{z}^j + D^j_{11} \\ B^j_{16} & B^j_{16} \cdot \bar{z}^j + D^j_{16} \end{bmatrix} - \begin{bmatrix} A^j_{12} & A^j_{16} & B^j_{12} \\ B^j_{12} & B^j_{16} & D^j_{12} \\ B^j_{26} & B^j_{66} & D^j_{26} \end{bmatrix} [R^j], \quad j = 1, 2, 3 \tag{58}$$

$$[\bar{\xi}^j] = \begin{Bmatrix} B^j_{16} \\ D^j_{16} \\ D^j_{66} \end{Bmatrix} - \begin{bmatrix} A^j_{12} & A^j_{16} & B^j_{12} \\ B^j_{12} & B^j_{16} & D^j_{12} \\ B^j_{26} & B^j_{66} & D^j_{26} \end{bmatrix} [\bar{R}^j], \quad j = 1, 2, 3 \tag{59}$$

$$[\hat{\xi}^j] = \begin{bmatrix} A^j_{12} & A^j_{16} & B^j_{12} \\ B^j_{12} & B^j_{16} & D^j_{12} \\ B^j_{26} & B^j_{66} & D^j_{26} \end{bmatrix} [\hat{R}^j] - \begin{Bmatrix} N^j_x \\ M^j_x \\ M^j_{xy} \end{Bmatrix}^{HT}, \quad j = 1, 2, 3 \tag{60}$$

The integration constant, I^j, in Eq 26 is given by

$$I^j = I_{jk}\eta^k, \quad j, k = 1, 2, 3 \tag{61}$$

The explicit expressions of I_{jk} are given in the following

$$\Delta \cdot I_{11} = \bar{\xi}^1_3 \left\{ 1 - \left[1 + \frac{\bar{\xi}^3_3}{\bar{\xi}^1_3} \frac{s^3}{s^1} th(s^3 \cdot (b - a)) + \frac{\bar{\xi}^2_3}{\bar{\xi}^1_3} \frac{s^2}{s^1} th(s^2 \cdot a) \right] e^{s^1 \cdot a} \right\} \tag{62}$$

$$\Delta \cdot I_{12} = \bar{\xi}^2_3 \{ 1 - [1 - th(s^2 \cdot a)] e^{s^2 \cdot a} \} \tag{63}$$

$$\Delta \cdot I_{13} = -\bar{\xi}^3_3 \{ 1 - [1 + th(s^3 \cdot (b - a))] e^{-s^3 \cdot (b-a)} \} \tag{64}$$

$$[1 + e^{2s^2 \cdot a}] \cdot I_{21} = I_{11} \cdot (1 + e^{2s^1 \cdot a}) + \frac{1}{s^1} e^{s^1 \cdot a} \tag{65}$$

$$[1 + e^{2s^2 \cdot a}] \cdot I_{22} = I_{12} \cdot (1 + e^{2s^1 \cdot a}) - \frac{1}{s^2} e^{s^2 \cdot a} \tag{66}$$

$$[1 + e^{2s^2 \cdot a}] \cdot I_{23} = I_{13} \cdot (1 + e^{2s^1 \cdot a}) \tag{67}$$

$$[1 + e^{-2s^3 \cdot (b-a)}] \cdot I_{31} = I_{11} \cdot (1 + e^{2s^1 \cdot a}) + \frac{1}{s^1} e^{s^1 \cdot a} \tag{68}$$

$$[1 + e^{-2s^3 \cdot (b-a)}] \cdot I_{32} = -I_{12} \cdot (1 + e^{2s^1 \cdot a}) \tag{69}$$

$$[1 + e^{-2s^3 \cdot (b-a)}] \cdot I_{33} = I_{13} \cdot (1 + e^{2s^1 \cdot a}) - \frac{1}{s^3} e^{-s^3 \cdot (b-a)} \tag{70}$$

where

$$\Delta = (1 + e^{2s^1 \cdot a})[s^3 \cdot \bar{\xi}_3^3 \cdot th(s^3 \cdot (b - a))$$
$$+ s^1 \cdot \bar{\xi}_3^1 \cdot th(s^1 \cdot a) + s^2 \cdot \bar{\xi}_3^2 \cdot th(s^2 \cdot a)] \tag{71}$$

The matrix, $[\Psi_{ij}]$, and vector, $\{\Psi_j^{HT}\}$, in Eq 32 are given by

$$[\Psi_{ij}] = \begin{bmatrix} \xi_{11}^1 + \xi_{11}^2 & \xi_{12}^1 + \xi_{12}^2 & 2\bar{\xi}_1^1 + 2\bar{\xi}_1^2 \\ \xi_{21}^1 + \xi_{21}^2 + \frac{h}{2}(\xi_{11}^1 - \xi_{11}^2) & \xi_{22}^1 + \xi_{22}^2 + \frac{h}{2}(\xi_{12}^1 - \xi_{12}^2) & 2\left[\bar{\xi}_2^1 + \bar{\xi}_2^2 + \frac{h}{2}(\bar{\xi}_1^1 - \bar{\xi}_1^2)\right] \\ 2\xi_{31}^1 + 2\xi_{31}^2 & 2\xi_{32}^1 + 2\xi_{32}^2 & 4\bar{\xi}_3^1 + 4\bar{\xi}_3^2 \end{bmatrix} \cdot a \tag{72}$$

$$+ \begin{bmatrix} \xi_{11}^3 & \xi_{12}^3 & 2\bar{\xi}_1^3 \\ \xi_{21}^3 & \xi_{22}^3 & 2\bar{\xi}_2^3 \\ 2\xi_{31}^3 & 2\xi_{32}^3 & 4\bar{\xi}_3^3 \end{bmatrix} \cdot (b - a) + \begin{bmatrix} \bar{\xi}_1^1 & \bar{\xi}_1^2 & \bar{\xi}_1^3 \\ \bar{\xi}_2^1 + \frac{h}{2}\bar{\xi}_1^1 & \bar{\xi}_2^2 - \frac{h}{2}\bar{\xi}_1^2 & \bar{\xi}_2^3 \\ 2\bar{\xi}_3^1 & 2\bar{\xi}_3^2 & 2\bar{\xi}_3^3 \end{bmatrix} \begin{bmatrix} f_1^1 & f_2^1 & 2f_3^1 \\ f_1^2 & f_2^2 & 2f_3^2 \\ f_1^3 & f_2^3 & 2f_3^3 \end{bmatrix}$$

$$\{\Psi_j^{HT}\} = \left\{ \begin{matrix} \hat{\xi}_1^1 + \hat{\xi}_1^2 \\ \hat{\xi}_2^1 + \hat{\xi}_2^2 + \frac{h}{2}(\hat{\xi}_1^1 - \hat{\xi}_1^2) \\ \xi_3^1 + \xi_3^2 \end{matrix} \right\} a + \left\{ \begin{matrix} \hat{\xi}_1^3 \\ \hat{\xi}_2^3 \\ \xi_3^3 \end{matrix} \right\} (b - a) \tag{73}$$

$$+ \begin{bmatrix} \bar{\xi}_1^1 & \bar{\xi}_1^2 & \bar{\xi}_1^3 \\ \bar{\xi}_2^1 + \frac{h}{2}\bar{\xi}_1^1 & \bar{\xi}_2^2 - \frac{h}{2}\bar{\xi}_1^2 & \bar{\xi}_2^3 \\ \bar{\xi}_3^1 & \bar{\xi}_3^2 & \bar{\xi}_3^3 \end{bmatrix} \begin{Bmatrix} f_1^{HT} \\ f_2^{HT} \\ f_3^{HT} \end{Bmatrix}$$

where

$$f_i^j = -\left(\frac{\xi_{3i}^1}{\xi_3^1} I_{j1} + \frac{\xi_{3i}^2}{\bar{\xi}_3^2} I_{j2} + \frac{\xi_{3i}^3}{\bar{\xi}_3^3} I_{j3}\right)(e^{s^j \cdot a} - 1)^2 - \frac{1}{s^j} \frac{\xi_{3i}^1}{\bar{\xi}_3^1} (e^{s^j \cdot a} - 1), \qquad i = 1, 2 \quad j = 1, 2 \tag{74}$$

$$f_i^j = \left(\frac{\xi_{3i}^1}{\bar{\xi}_3^1} I_{j1} + \frac{\xi_{3i}^2}{\bar{\xi}_3^2} I_{j2} + \frac{\xi_{3i}^3}{\bar{\xi}_3^3} I_{j3}\right)(e^{-s^j \cdot (b-a)} - 1)^2 + \frac{1}{s^j} \frac{\xi_{3i}^1}{\bar{\xi}_3^1} (e^{-s^j \cdot (b-a)} - 1), \qquad i = 1, 2 \quad j = 3 \tag{75}$$

$$f_3^j = -(I_{j1} + I_{j2} + I_{j3})(e^{s^j \cdot a} - 1)^2 - \frac{1}{s^j} (e^{s^j \cdot a} - 1), \qquad j = 1, 2 \tag{76}$$

$$f_3^j = (I_{j1} + I_{j2} + I_{j3})(e^{-s^j \cdot (b-a)} - 1)^2 + \frac{1}{s^j}(e^{-s^j \cdot (b-a)} - 1), \qquad j = 3 \tag{77}$$

$$f_j^{HT} = -\left(\frac{\hat{\xi}_3^1}{\overline{\xi}_3^1} I_{j1} + \frac{\hat{\xi}_3^2}{\overline{\xi}_3^2} I_{j2} + \frac{\hat{\xi}_3^3}{\overline{\xi}_3^3} I_{j3}\right)(e^{s^j \cdot a} - 1)^2 - \frac{1}{s^j}\frac{\hat{\xi}_3^j}{\overline{\xi}_3^j}(e^{s^j \cdot a} - 1), \qquad j = 1, 2 \tag{78}$$

$$f_j^{HT} = \left(\frac{\hat{\xi}_3^1}{\overline{\xi}_3^1} I_{j1} + \frac{\hat{\xi}_3^2}{\overline{\xi}_3^2} I_{j2} + \frac{\hat{\xi}_3^3}{\overline{\xi}_3^3} I_{j3}\right)(e^{-s^j \cdot (b-a)} - 1)^2 + \frac{1}{s^j}\frac{\hat{\xi}_3^j}{\overline{\xi}_3^j}(e^{-s^j \cdot (b-a)} - 1), \qquad j = 3 \tag{79}$$

The parameters, Λ_1, and vector, Λ_0, in Eq 45 are given by

$$
\begin{aligned}
\Lambda_1 = {} & \phi_1\left\{b(N_x^3)^{HT} - \sum_{j=1}^{3}[(N_y^j)^{HT}R_{11}^j + (N_{xy}^j)^{HT}R_{21}^j + (M_y^j)^{HT}R_{31}^j]\phi^j\right\} \\
& + \phi_2\left\{b(M_x^3)^{HT} - \sum_{j=1}^{3}[(N_y^j)^{HT}R_{12}^j + (N_{xy}^j)^{HT}R_{22}^j + (M_y^j)^{HT}R_{32}^j]\phi^j\right\} \\
& + \sum_{j=1}^{3}[(M_{xy}^j)^{HT} - (N_y^j)^{HT}\overline{R}_1^j - (N_{xy}^j)^{HT}\overline{R}_2^j - (M_y^j)^{HT}\overline{R}_3^j]\overline{\Psi}^j
\end{aligned}
\tag{80}
$$

$$
\begin{aligned}
\Lambda_0 = {} & \phi_1^{HT}\left\{b(N_x^3)^{HT} - \sum_{j=1}^{3}[(N_y^j)^{HT}R_{11}^j + (N_{xy}^j)^{HT}R_{21}^j + (M_y^j)^{HT}R_{31}^j]\phi^j\right\} \\
& + \phi_2^{HT}\left\{b(M_x^3)^{HT} - \sum_{j=1}^{3}[(N_y^j)^{HT}R_{12}^j + (N_{xy}^j)^{HT}R_{22}^j + (M_y^j)^{HT}R_{32}^j]\phi^j\right\} \\
& + \sum_{j=1}^{3}[(M_{xy}^j)^{HT} - (N_y^j)^{HT}\overline{R}_1^j - (N_{xy}^j)^{HT}\overline{R}_2^j - (M_y^j)^{HT}\overline{R}_3^j]\hat{\Psi}^j \\
& + \sum_{j=1}^{3}[(N_y^j)^{HT}\hat{R}_1^j + (N_{xy}^j)^{HT}\hat{R}_2^j + (M_y^j)^{HT}\hat{R}_3^j]\phi^j
\end{aligned}
\tag{81}
$$

where

$$\phi^1 = \phi^2 = a \qquad \phi^3 = b - a \tag{82}$$

$$\overline{\Psi}^j = -I_{jk} \cdot \overline{\eta}^k \cdot (e^{s^j \cdot a} - 1)^2 - \frac{\overline{\eta}^j}{s^j}(e^{s^j \cdot a} - 1) + 2a, \, j = 1, 2 \qquad k = 1, 2, 3 \tag{83}$$

$$\overline{\Psi}^j = I_{jk} \cdot \overline{\eta}^k \cdot (e^{-s^j(b-a)} - 1)^2 + \frac{\overline{\eta}^j}{s^j}(e^{-s^j \cdot (b-a)} - 1) + 2(b - a), \, j = 3 \qquad k = 1, 2, 3$$

$$\tag{84}$$

$$\hat{\Psi}^j = -I_{jk}\hat{\eta}^k (e^{s^j \cdot a} - 1)^2 - \frac{\hat{\eta}^j}{s^j} (e^{s^j \cdot a} - 1), j = 1, 2 \qquad k = 1, 2, 3 \qquad (85)$$

$$\hat{\Psi}^j = I_{jk} \cdot \hat{\eta}^k \cdot (e^{-s^j \cdot (b-a)} - 1)^2 + \frac{\hat{\eta}^j}{s^j} (e^{-s^j \cdot (b-a)} - 1), j = 3 \qquad k = 1, 2, 3 \qquad (86)$$

$$\overline{\eta}^k = \frac{\xi^k_{31}}{\overline{\xi}^k_3} \phi_1 + \frac{\xi^k_{32}}{\overline{\xi}^k_3} \phi_2 + 2, \qquad k = 1, 2, 3 \qquad (87)$$

$$\hat{\eta}^k = \frac{\xi^k_{31}}{\overline{\xi}^k_3} \phi^{HT}_1 + \frac{\xi^k_{32}}{\overline{\xi}^k_3} \phi^{HT}_2 + \frac{\hat{\xi}^k_3}{\overline{\xi}^k_3}, \qquad k = 1, 2, 3 \qquad (88)$$

References

[1] O'Brien, T. K. and Martin, R. H., "Round Robin Testing for Mode I Interlaminar Fracture Toughness of Composite Materials," *Journal of Composites Technology and Research*, Vol. 15, No. 4, Winter 1993, pp. 269–281.

[2] Russell, A. J., "On the Measurement of Mode II Interlaminar Energies," DREP Materials Report 82-0, Defense Research Establishment Pacific, Victoria, BC, Canada, 1982.

[3] Salpekar, S. A., Raju, I. S., and O'Brien, T. K., "Strain Energy Release Rate Analysis of the End-Notched Flexure Specimen Using the Finite Element Method," *Journal of Composites Technology and Research*, Vol. 10, No. 4, Winter 1988, pp. 133–139.

[4] O'Brien, T. K., Murri, G. B., and Salpekar, S. A., "Interlaminar Shear Fracture Toughness and Fatigue Thresholds for Composite Materials," *Composite Materials: Fatigue and Fracture, Second Volume, ASTM STP 1012*, P. Lagace, Ed., American Society for Testing and Materials, Philadelphia, 1989, pp. 222–250.

[5] Carlsson, L. A. and Gillespie, J. W., "Mode-II Interlaminar Fracture of Composites," *Application of Fracture Mechanics to Composite Materials*, Composite Materials Series, 6, Klaus Friedrich, Ed., Elsevier, Amsterdam, 1989, pp. 113–157.

[6] Tanaka, K., Kageyama, K., and Hojo, M., "Prestandardization Study on Mode II Interlaminar Fracture Toughness Test for CFRP in Japan," *Composites*, Vol. 26, No. 4, 1995, pp. 257–267.

[7] Martin, R. H., "Evaluation of the Split Cantilever Beam for Mode III Delamination Testing," *Composite Materials: Fatigue and Fracture, Third Volume, ASTM STP 1110*, T. K. O'Brien, Ed., American Society for Testing and Materials, Philadelphia, 1992, pp. 243–266.

[8] Lee, S. M., "An Edge Crack Torsion Method for Mode III Delamination Fracture Testing," *Composite Technology & Research*, Vol. 15, No. 3, Fall 1993, pp. 193–201.

[9] Li, J. and Wang, Y., "Analysis of a Symmetric Laminate with Mid-Plane Free Edge Delamination Under Torsion: Theory and Application to the Edge Crack Torsion (ECT) Specimen for Mode III Toughness Characterization," *Engineering Fracture Mechanics*, Vol. 49, No. 2, 1994, pp. 179–194.

[10] Li, J. and O'Brien, T. K., "Simplified Data Reduction Methods for the ECT Test for Mode III Interlaminar Fracture Toughness," *Journal of Composites Technology and Research*, Vol. 18, No. 2, April 1996, pp. 96–101.

[11] Love, A. E. H., *A Treatise of the Mathematical Theory of Elasticity*, 4th Ed., Dover Publication, New York, 1944, p. 173 and pp. 19–21.

[12] Lekhnitskii, S. G., *Theory of Elasticity of an Anisotropic Body*, Holden-Day, Inc., San Francisco, 1963, pp. 103–108.

[13] Vinson, J. R. and Sierakowski, R. L., *The Behavior of Structures Composed of Composite Materials*, Martinus Nijhoff Publishers, Dordecht, The Netherlands, 1986, pp. 46–47.

[14] Daniel, I. M. and Ishai, O., *Engineering Mechanics of Composite Materials*, Oxford University Press, Inc., New York, 1994, p. 35.

Basant K. Parida,[1] *Raghu V. Prakash,*[1] *Prakash D. Mangalgiri,*[2] *and K. Vijayaraju*[2]

Influence of Environmental and Geometric Parameters on the Behavior of Fastener Joints in Advanced Composites

REFERENCE: Parida, B. K., Prakash, R. V., Mangalgiri, P. D., and Vijayaraju, K., **"Influence of Environmental and Geometric Parameters on the Behavior of Fastener Joints in Advanced Composites,"** *Composite Materials: Fatigue and Fracture (Sixth Volume), ASTM STP 1285,* E. A. Armanios, Ed., American Society for Testing and Materials, 1997, pp. 432–451.

ABSTRACT: Single- and double-shear bearing strength of mechanically fastened joints in carbon fiber composite (CFC) systems have been studied and the analysis of results presented. CFC laminates made from unidirectional prepreg tapes as well as bidirectional fabric prepregs were tested in as-received condition at room temperature and under hot/wet environmental conditions after hygrothermal aging. The influence of geometric parameters like the specimen width-to-hole diameter ratio (w/d) and the specimen thickness-to-hole diameter ratio (t/d), on the bearing strength has been investigated. Bearing strength tests were performed with three different fastener bolt materials and, in all, over 300 specimens were tested.

Bearing stresses were evaluated at ultimate failure, at 2% hole deformation, at onset of nonlinearity, and at first load drop; and 2% offset bearing strength has been selected for the purpose of comparison of data. The influence of lamina configuration, mode of loading, hole-tolerance, and fastener bolt material on bearing strength has been investigated. The degradation of bearing strength in hygrothermally aged CFC specimens under hot/wet environmental conditions has been found to be around 25 to 30%, compared to the room temperature values.

KEYWORDS: fatigue (materials), fracture (materials), composite materials, carbon-fiber epoxy composite, single-shear bearing strength, double-shear bearing strength, hole deformation, hot/wet environment, environmental degradation

In view of the superior specific strength, stiffness, and durability over the metallic materials, carbon-fiber epoxy composites have found wider applications in aerospace structures over the last few decades. The weight fraction of composites to the total structural weight has increased considerably both in civil and military airframes in recent years [1]. However, though a few military aircrafts have been flying with primary structures made from composites, most civilian transport aircraft designers are still hesitant to use all-composite primary structural components, like the wings, fuselage, and empennage.

It is well recognized that in applications where the potential for damage is high, some fiber-reinforced composites may lose their weight advantage because of substantial strength loss with impact damage and environmental degradation [2–5]. Most analytical and empirical methods employed for strength and life prediction of metallic materials can not be applied to composites, primarily because of their nonhomogeneity, anisotropy, and relative brittleness.

[1]Scientist, Structural Integrity Division, National Aerospace Laboratories, Bangalore 560 017, India.
[2]Scientist, Aeronautical Development Agency, Vimanapura Post, Bangalore 560 017, India.

Composite laminates connected to other structural components through fastener joints exhibit wide variations in bearing strength and failure mode depending on the lamina configuration (layup/stacking sequence), geometric parameters, and environmental conditions [6,7]. Here, in accordance with the terminology of MIL-HDBK-17, Polymer Matrix Composites, Volume 1, Section 7 and the currently available ASTM Committee D30 Standard, D 5961M-96, "Standard Test Method for Bearing Response of Polymer Matrix Composite Laminates," bearing strength is defined as the value of the bearing stress occuring at a significant event on the bearing stress/bearing strain curve, and the various failure modes are as defined in the latter.

Since structural integrity of fastener joints significantly affects the overall performance of any composite structure, the influence of parameters-geometric, material, and environmental on the behavior of fastener/bolted joints has been addressed both analytically and experimentally by several researchers [7–13]. Based on their observations, it can be said that the strength/damage tolerance characteristics and failure mode of any mechanically fastened composite material would strongly depend upon the fiber/matrix system and layup configuration, besides a host of other parameters indicated earlier. In this paper, analysis of the results of an experimental investigation is presented with the objective of providing a better understanding of the influence of environmental and geometric parameters on the bearing strength of fastener joints in a carbon fiber/epoxy matrix composite system. Specific parameters investigated were: different layups, specimen width-to-hole diameter ratio (w/d), thickness-to-hole diameter ratio (t/d), fastener bolt material, environment (room temperature/as-received (RT/AR) and hot/wet (H/W) conditions, and mode of loading (single shear and double shear).

Material and Experimental Method

For the purpose of this investigation, specimens were fabricated out of T300/914C (Ciba, United Kingdom) carbon-epoxy unidirectional tapes with different layups and stacking sequence as well as from bi-directional 5 Harness Satin weave, 50% warp-50% weft carbon-fiber composite (CFC) fabric prepregs of same epoxy. The ply thickness of the unidirectional prepreg was nominally 0.15 mm and that of the fabric prepreg was 0.3 mm, with a fiber volume fraction of 0.6. Figure 1 shows the configuration of both single-shear and double-shear bearing specimens with dimensions. In all, three nominal thicknesses were used, namely, 2.4, 3.6, and 6.0 mm. The hole diameter used for fastener bolts was 6 mm with H8 tolerance (0.000 to 0.027 mm tolerance) throughout, excepting one series of specimens in which H11 tolerance (0.000 to 0.075 mm tolerance) was used, while 100° counter-sunk holes were used only for single-shear specimens. Three different fastener bolt materials were used in this investigation, namely, EN-24 (AISI 4340 equivalent), titanium alloy (Ti-6Al-4V), and maraging steel (M steel). For the specimens tested under hot/wet environment, moisture absorption levels, typically experienced by composite parts of aircraft in long-term operational environments, were simulated in the laboratory environmental chamber in an accelerated manner. As received specimens after fabrication were put inside an environmental chamber, together with small traveler coupons of identical thickness, which was maintained at 70°C and 85% relative humidity (RH) until saturation. This process took on an average 16 to 25 weeks and moisture absorption level attained in the range of 1.25 to 1.6% based on as received weight of specimens of different sizes. Figure 2 shows typical moisture absorption characteristics of a batch of CFC specimens at 70°C and 85% relative humidity.

The single-shear and double-shear bearing strength tests were performed in an automated servohydraulic test machine with hydraulic clamping of the specimen at the lower end and specially designed fixtures were used to attach the specimen at the upper end with the help of fastener bolts. An extensometer with a gage length of 12.5 mm and a travel of ±5.0 mm

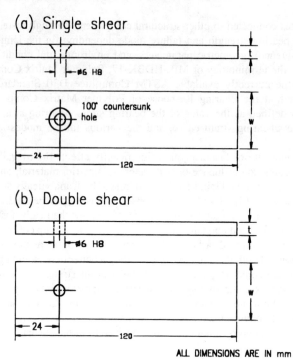

FIG. 1—*Details of CFC bearing test specimens with dimensions.*

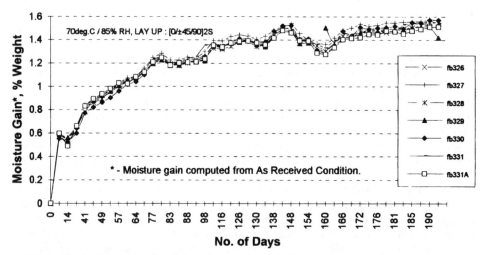

FIG. 2—*Typical moisture absorbtion characteristic of CFC specimen at 70°C and 85% RH.*

was employed for the purpose of measuring hole deformation in the specimen, by attaching one arm (without knife edge) to the supporting steel loading plate and the other arm with the knife edge strapped down to the CFC specimen. Figure 3 shows the schematic of the extensometer attachment and the test fixture configuration for single- and double-shear bearing test. All

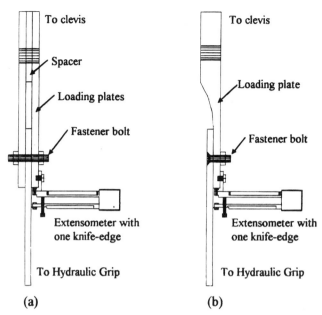

FIG. 3—*Schematic of the test fixtures showing extensometer attachment for bearing test: (a) double shear and (b) single shear.*

tests were performed under stroke (displacement) control condition with an average time to failure of around 90 s. The bearing load and hole deformation were continuously monitored as well as plotted in an X-Y plotter. The curve obtained in the X-Y plotter during the test represents the bearing stress/bearing strain response, where bearing stress is defined as the bearing load divided by the bearing area (diameter of the loaded hole multiplied by the thickness of the test coupon) and the bearing strain is defined as the normalized hole deformation, equal to the deformation of the bearing hole in the direction of the bearing load, divided by the nominal diameter of the hole. A typical X-Y recorder-plot of bearing load-versus-hole deformation is shown in Fig. 4. The hot/wet environmental tests were performed in the same test machine with the test fixture and moisture-saturated specimen enclosed within a split-type small environment chamber, into which saturated steam and hot-air sources were connected through insulated tubes. The mass flow of steam and temperature of hot air were regulated to achieve a temperature of 100°C and an 85% RH or above within the test chamber. Prior to the test, specimens were removed from the conditioning chamber and mounted to the test fixture within 3 to 5 min of their removal from the conditioning environmental chamber. On an average, it took about 10 to 15 min for the test chamber to attain a steady temperature (100°C ± 2°C) and humidity (RH ≥ 85%) condition, which was monitored with the help of a pair of hygrothermal probes. Figure 5 shows the photograph of a typical hot/wet environmental test setup. It may be noted that during both single-shear and double-shear test the bolts were only finger-tightened to avoid imposition of lateral constraint that might aid in the enhancement of bearing strength. Table 1 depicts the complete test matrix that was used during the present investigation, covering over 300 carbon-fiber composite specimens. The mean values and standard deviation of the ultimate bearing strength of each series are also indicated therein.

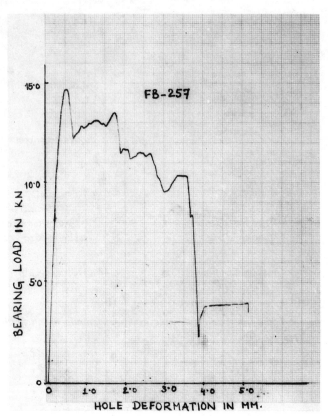

FIG. 4—*Typical X-Y plotter record of bearing load-versus-hole deformation for double-shear bearing, RT/AR test condition.*

Results and Discussions

For the purpose of computing bearing stress, nominal dimensions of specimens were used in the following equation

$$\sigma_{bi} = P_i/(d \cdot t) \tag{1}$$

where σ_{bi} and P_i correspond to the following definitions for the bearing stress, σ_b (MPa) and bearing load, $P(N)$, respectively, pertaining to significant events, i in the bearing response curve (Fig. 6). The hole nominal diameter is d (mm) and specimen nominal thickness is t (mm). σ_{bu}, P_u correspond to the ultimate bearing stress and ultimate bearing load, while σ_{b2}, P_2 correspond to those at 2% offset bearing strain (hole elongation along the bearing load direction, divided by hole nominal diameter). Similarly, σ_{bnl}, P_{nl} correspond to those at the onset of nonlinearity and σ_{ld}, P_{ld} correspond to first load drop that is very close to the load at first audible pop. Figure 6 shows a schematic of bearing load-versus-hole deformation curve including characteristic parameters (corresponding to significant events in the bearing response curve), defined earlier.

FIG. 5—*Photograph of H/W bearing test setup (test environment: 100°C and ≥ 85% RH).*

It may be noted that in order to obtain the load at 2% hole deformation, the slope of the linear portion of the load-versus-hole deformation curve was computed, considering a least square straight line fit of the data in the range of around 10 to 40% of the ultimate load and an offset line parallel to the above line was drawn at 2% bearing strain, that is at $0.02d =$ 0.12 mm hole deformation. The intersection of this offset line with the bearing load-versus-hole deformation curve provided the 2% offset bearing load (P_2), and the corresponding bearing stress thus computed, using Eq 1, represented the bearing stress at 2% hole deformation (σ_{b2}), which is also called the 2% offset bearing strength. It may however be noted that the hole deformation measured with the help of an extensometer did include the elastic deformation of a part of the loading plate, which incidentally was in the order of about 8 to 10 μm corresponding to the ultimate load or about 6 to 8 μm at 2% offset load that amounted to a maximum of around 5 to 6% of the 2% hole elongation. To this extent, the hole deformation is in error, but this is considered as tolerable. However, if the $0.02d$ offset line intersection indicated a load that was lower than a peak load to the left of this line (as shown in Fig. 6), then for the computation of σ_{b2}, that is, 2% offset bearing strength, the higher load corresponding to first load drop was used.

Figure 7 shows the bearing stresses, as defined earlier, of a typical CFC laminate, obtained using steel bolts in double shear under RT/AR and H/W condition. Note that the CFC laminate attribute 58/17/17/8 indicates the volume fractions of 0°, +45°, −45°, and 90° plies, respectively, and is expressed in percent of the total volume. From Fig. 7 it can be observed that the bearing stress at the onset of nonlinearity is rather low in all cases. The bearing stresses corresponding to first load drop and at 2% offset are almost identical under both RT/AR and H/W test conditions, but are slightly less than the corresponding ultimate bearing stress. The bearing stress data presented in this figure represent the average of the test data obtained from about three to five or more identical specimens in each case (see Table 1).

TABLE 1—Summary of single-shear and double-shear bearing tests performed on CFC specimens.

Serial No.	Test Type	Ply Volume Fraction	w/d	t/d	Ply Layup	Fastener Material	Text Environment	Mode of Failure	No. of Specimens	Ultimate Bearing Strength, MPa	
										Mean	SD
1	DS	50/25/25/0	3	0.6	$[\pm45/0_2/\pm45/0_3/\pm45/0]_s$	steel	RT/AR	bearing	5	743	28.57
2	DS	50/25/25/0	4	0.6	$[\pm45/0_2/\pm45/0_3/\pm45/0]_s$	steel	RT/AR	bearing	5	772	32.60
3	DS	50/25/25/0	5	0.6	$[\pm45/0_2/\pm45/0_3/\pm45/0]_s$	steel	RT/AR	bearing	5	918	26.23
4	DS	50/25/25/0	3	0.6	$[\pm45/0_2/\pm45/0_3/\pm45/0]_s$	titanium	RT/AR	bearing	5	743	23.57
5	DS	50/25/25/0	4	0.6	$[\pm45/0_2/\pm45/0_3/\pm45/0]_s$	titanium	RT/AR	shear	3	732	26.50
6	DS	50/25/25/0	5	0.6	$[\pm45/0_2/\pm45/0_3/\pm45/0]_s$	titanium	RT/AR	shear	5	786	17.15
7	DS	50/25/25/0	5	0.6	$[\pm45/0_2/\pm45/0_3/\pm45/0]_s$	steel	H/W	shear	5	506	14.61
8	DS	50/25/25/0	4	0.6	$[\pm45/0_2/\pm45/0_3/\pm45/0]_s$	steel	H/W	shear	5	631	30.70
9	DS	50/25/25/0	5	0.6	$[\pm45/0_2/\pm45/0_3/\pm45/0]_s$	steel	H/W	shear	5	703	36.14
10	DS	50/25/25/0	3	0.6	$[\pm45/0_2/\pm45/0_3/\pm45/0]_s$	titanium	H/W	shear	5	515	20.70
11	DS	50/25/25/0	4	0.6	$[\pm45/0_2/\pm45/0_3/\pm45/0]_s$	titanium	H/W	shear	5	597	22.90
12	DS	50/25/25/0	5	0.6	$[\pm45/0_2/\pm45/0_3/\pm45/0]_s$	titanium	H/W	shear	5	677	27.14
13	DS	58/17/17/8	3	0.6	$[\pm45/0_3/\pm45/0_3/90/0]_s$	steel	RT/AR	shear	4	682	24.28
14	DS	58/17/17/8	4	0.6	$[\pm45/0_3/\pm45/0_3/90/0]_s$	steel	RT/AR	shear	4	694	22.86
15	DS	58/17/17/8	5	0.6	$[\pm45/0_3/\pm45/0_3/90/0]_s$	steel	RT/AR	shear	5	741	28.10
16	DS	58/17/17/8	3	0.6	$[\pm45/0_3/\pm45/0_3/90/0]_s$	titanium	RT/AR	shear	5	719	12.77
17	DS	58/17/17/8	4	0.6	$[\pm45/0_3/\pm45/0_3/90/0]_s$	titanium	RT/AR	bearing	4	705	15.80
18	DS	58/17/17/8	5	0.6	$[\pm45/0_3/\pm45/0_3/90/0]_s$	titanium	RT/AR	shear	5	723	33.40
19	DS	58/17/17/8	3	0.6	$[\pm45/0_3/\pm45/0_3/90/0]_s$	steel	H/W	N/A	4	574	11.06
20	DS	58/17/17/8	5	0.6	$[\pm45/0_3/\pm45/0_3/90/0]_s$	steel	H/W	shear	5	556	16.60
21	DS	58/17/17/8	3	0.6	$[\pm45/0_3/\pm45/0_3/90/0]_s$	titanium	H/W	combined	5	584	19.01
22	DS	58/17/17/8	5	0.6	$[\pm45/0_3/\pm45/0_3/90/0]_s$	titanium	H/W	shear	3	538	14.86
23	DS	25/25/25/25	4	0.4	$[0/\pm45/90]_{3s}$	steel	RT/AR	tension	3	806	28.85
24	DS	25/25/25/25	5	0.4	$[0/\pm45/90]_{3s}$	steel	RT/AR	shear	5	750	29.7
25	DS	25/25/25/25	3	0.4	$[0/\pm45/90]_{3s}$	titanium	RT/AR	shear	5	674	21.37
26	DS	25/25/25/25	4	0.4	$[0/\pm45/90]_{3s}$	titanium	RT/AR	shear	4	591	58.04

TABLE 1—Summary of single-shear and double-shear bearing tests performed on CFC specimens (Continued).

Serial No.	Test Type	Ply Volume Fraction	w/d	t/d	Ply Layup	Fastener Material	Text Environment	Mode of Failure	No. of Specimens	Ultimate Bearing Strength, MPa Mean	SD
27	DS	25/25/25/25	5	0.4	$[0/\pm45/90]_{3s}$	titanium	RT/AR	shear	5	720	22.08
28	DS	25/25/25/25	3	0.4	$[0/\pm45/90]_{3s}$	steel	H/W	N/A	5	518	10.67
29	DS	25/25/25/25	4	0.4	$[0/\pm45/90]_{3s}$	steel	H/W	combined	4	539	40.05
30	DS	25/25/25/25	5	0.4	$[0/\pm45/90]_{3s}$	steel	H/W	shear	4	630	28.70
31	DS	25/25/25/25	3	0.4	$[0/\pm45/90]_{3s}$	titanium	H/W	tension	5	565	37.90
32	DS	25/25/25/25	4	0.4	$[0/\pm45/90]_{3s}$	titanium	H/W	combined	4	555	79.94
33	DS	25/25/25/25	5	0.4	$[0/\pm45/90]_{3s}$	titanium	H/W	shear	4	567	79.94
34	DS	25/25/25/25	5	0.6	$[0/\pm45/90]_{3s}$	titanium	RT/AR	N/A	5	858	43.05
35	DS	25/25/25/25	5	0.6	$[0/\pm45/90]_{3s}$	titanium	RT/AR	tension	12	897	58.65
36	DS	25/25/25/25	5	0.6	$[0/\pm45/90]_{3s}$	titanium	RT/AR	tension	8	961	40.80
37	DS	25/25/25/25	5	0.6	$[0/\pm45/90]_{3s}$	titanium	H/W	shear	6	775	51.34
38	DS	25/25/25/25	5	0.6	$[0/\pm45/90]_{3s}$	titanium	H/W	shear	12	795	89.12
39	DS	25/25/25/25	5	0.6	$[0/\pm45/90]_{3s}$	titanium	H/W	shear	8	722	82.45
40	DS	20/30/30/20	5	1.0	$[(\pm45)_3/(0/90)_2]_{2s}$	M.steel	RT/AR	combined	6	894	78.4
41	DS	30/20/20/30	5	1.0	$[(\pm45)_2/(0/90)_3]_{2s}$	M.steel	RT/AR	combined	5	818	47.1
42	DS	20/30/30/20	5	1.0	$[(\pm45)_3/(0/90)_2]_{2s}$	M.steel	H/W	combined	6	671	36.7
43	DS	30/20/20/30	5	1.0	$[(\pm45)_2/(0/90)_3]_{2s}$	M.steel	H/W	combined	6	669	37.5
44	DS	50/25/25/0	5	1.0	$[(\pm45)_3/(0/90)]_{2s}$	M.steel	RT/AR	combined	4	N/A	N/A
45	SS	25/25/25/25	5	0.6	$[0/\pm45/90]_{2s}$	titanium	RT/AR	shear	17	610	39.18
46	SS	25/25/25/25	5	0.6	$[0/\pm45/90]_{2s}$	titanium	H/W	bearing	16	479	24.9
47	DS	♣	5	0.8	$[0/90/0/90/\pm45/\pm45/\pm45]_s$	titanium	RT/AR	shear	5	665	70.9
48	DS	♣	5	0.8	$[0/90/0/90/\pm45/\pm45/\pm45]_s$	titanium	H/W	shear	5	526	70.6
49	DS	♣	5	0.8	$[0/90/0/90/\pm45/\pm45/\pm45]_s$	titanium	RT/AR	shear	5	766	17.6
50	SS	♣	5	0.8	$[0/90/0/90/\pm45/\pm45/\pm45]_s$	titanium	H/W	shear	4	589	51.8
51	SS	♣	5	0.8	$[0/90/0/90/\pm45/\pm45/\pm45]_s$	titanium	RT/AR	shear	5	655	39.5
52	DS	♣	5	0.8	$[0/90/0/90/\pm45/\pm45/\pm45]_s$	titanium	H/W	combined	5	620	23.9
53	DS		5	0.8	$[0/90]_{7s}$	titanium	RT/AR	bearing	5	662	13.3
54	DS		5	0.8	$[0/90]_{7s}$	titanium	H/W	tension	5	542	18.7
55	DS		5	0.8	$[\pm45]_{7s}$	titanium	RT/AR	combined	5	671	38.4
56	DS		5	0.8	$[\pm45]_{7s}$	titanium	H/W	tension	3	551	67.8

NOTE—DS → double shear; SS → single shear; ♣ → woven fabric specimens; RT/AR → room temperature / as received; H/W → hot wet; M.steel → maraging steel; and N/A → not available.

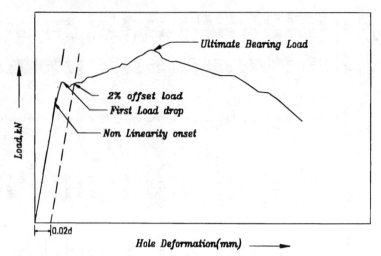

FIG. 6—*Schematic of bearing load-versus-hole deformation response and representation of various significant events.*

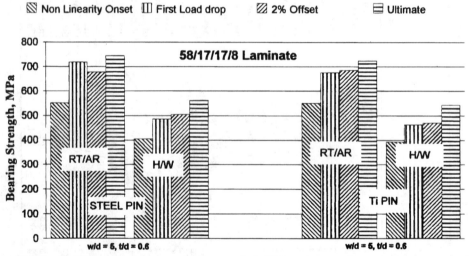

FIG. 7—*Variation of bearing strength for a typical CFC laminate under room temperature and H/W environments.*

Similar results pertaining to 2% offset strength showing the mean and standard deviation (SD) for a number of CFC laminates with different ply-volume-fractions are shown in Fig. 8. It can be seen that for the two laminates with ply-volume fractions of 50/25/25/0 and 58/17/17/8 with $w/d = 5.0$ and $t/d = 0.6$, the 2% offset bearing strength values exhibit considerable decrease under H/W environment as compared to the corresponding RT/AR values. In the matter of defining an allowable bearing strength for the design of fastener joints in composites, opinions differ among researchers, as evident from the literature [6,14–19]. Following the concepts of metallic materials, one might be tempted to use a certain factor of safety over the

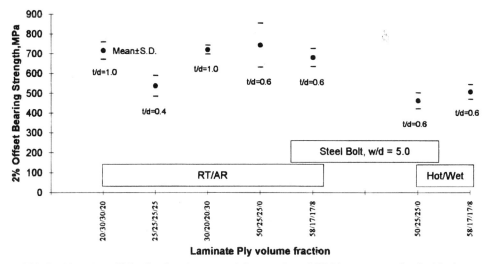

FIG. 8—*Variation of 2% offset bearing strength for a variety of CFC laminates under double shear.*

ultimate bearing strength. For instance, Johnson and Matthews [15] had suggested use of a factor of 2 based on the damage level corresponding to visible cracking. However, in view of the wide variations in ultimate bearing strength expected in composites, this would not be a suitable parameter for allowable design bearing strength. In some instances, damage have been observed at relatively lower loads [16,20,21]. For such ply layups, where damage can be clearly monitored, the concept of allowable bearing strength corresponding to an acceptable level of damage would appear to be akin to the proof stress or yield strength of metallic materials. Based on this limited damage level concept in composites, the various criteria such as the bearing stress corresponding to the onset of nonlinearity, first load-drop, first audible pop, 2% offset, etc. were considered. However, in the course of this investigation, practical difficulties of the following type were encountered in determining some of the preceding parameters:

1. For some specimens, the load-versus-hole deformation curve exhibited nonlinearity from very small loads and, in some instances, showed frequent changes in slope. This appears to be quite common to single-shear loading as experienced here and as reported by Bauer [21].
2. The load corresponding to the first audible pop and at first load drop was detectable in RT/AR tests. However, under H/W environmental test conditions, with the use of steam and hot air to maintain 100°C and RH ≥ 85% inside the test chamber, it was not possible to detect the audible pop. Hence, the bearing load corresponding to first load drop was considered as a reporting parameter.

Bauer [18,21] and Akay [16] have also argued in favor of selecting hold deformation-based offset load (2 to 4%) for computation of allowable bearing strength in composite fastener joints. However, 4% hole elongation may be considered to be on the higher side for generation of design data for applications in fatigue and fracture of composites. In fastener joints of metallic materials, 4% hole deformation may be tolerable in view of their better ductility and recovery characteristics upon unloading under dynamic/fatigue loading environment. However, because of the relative brittleness of composites, 4% hole elongation is expected to cause damage/delamination growth at the edge of the hole due to hammering, especially if the fastener joint is subjected to fatigue loading. Considering all of these factors, it was felt appropriate to identify the load corresponding to 2% offset bearing strain (normalized hole deformation) as the allowable bearing load, up to which the fastener joint can safely perform the intended task. This view is also supported by ASTM Committee D30 that has recently proposed 2% offset bearing stress in polymeric composites as the offset bearing strength in the new standard, ASTM D 5961M-96, "Test Method for Bearing Response of Polymer Matrix Composite Laminates."

Influence of Geometric Parameters

It is well known that the specimen geometry can seriously affect the magnitude of the bearing load at failure, as well as mode of failure in composites [*19,22–24*]. Significant nondimensionalized geometric parameters are, pitch distance ratio (specimen width-to-hole diameter ratio, w/d), thickness ratio (thickness-to-hole diameter ratio, t/d), edge distance ratio (ratio of edge distance from the center of bearing hole to the hole diameter, e/d). Based on the findings of several researchers indicated earlier, it was clear that for the layups considered in this investigation (Table 1), the effect of edge distance would be negligible for $e/d \geq 4$. Therefore, for all the specimens tested, $e/d = 4.0$ was maintained constant. The variation of other parameters considered were, $w/d = 3, 4, 5$ and $t/d = 0.4, 0.6, 0.8$, and 1.0. Figure 9 shows the variation of 2% offset bearing strength under double shear in a typical CFC laminate with w/d ratio, from which it is seen that the bearing strength is influenced both by width as well as environment. The mode of failure associated with a smaller w/d ratio is generally considered to be lateral (net tension) mode (shown in Fig. 10*a*), or shearout (Fig. 10*b*); where as the failure modes for larger w/d ratios appear to be the bearing mode (as shown in Fig. 10*c*) or the cleavage (combined) mode (Fig. 10*d*). However, depending on the combination of other parameters, the modes of failure in this investigation varied as shown in Table 1.

The influence of the laminate thickness on the strength behavior of fastener joints has been investigated here by varying the t/d ratio from 0.4 to 1.0 for a given hole diameter of 6.0 mm.

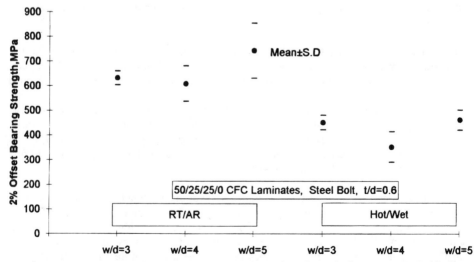

FIG. 9—*Comparison of double-shear bearing strength with pitch distance ratio (w/d) for a 50/25/25/0 CFC laminate under RT/AR and H/W conditions.*

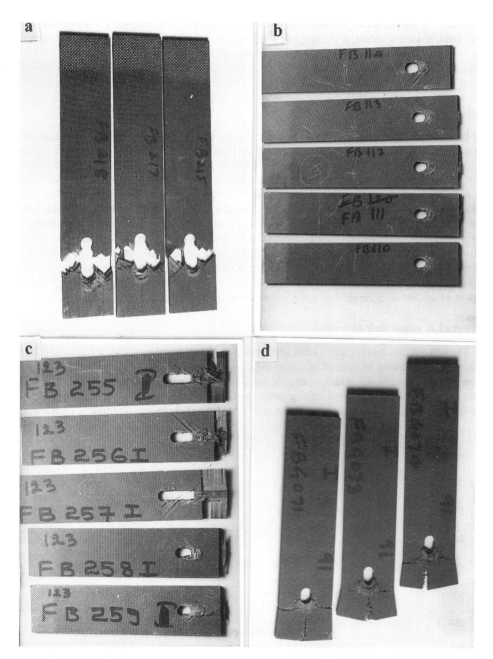

FIG. 10—*Characteristic modes of failure in double-shear bearing specimens, (a) lateral (net-tension) mode, (b) bearing mode, (c) shearout mode, (d) cleavage (mixed) mode.*

It is known that thinner laminates would show lower bearing strength compared to thicker laminates in double shear, primarily because, early failure in thin laminates is accentuated by the local brooming effect at the edge of the hole [*16,23*]. However, increasing the thickness to a large extent would ultimately cause failure of the fastener bolt. As a matter of fact, maraging steel bolts had to be used for 6-mm-thick specimens in order to avoid titanium and steel bolt failure.

Effect of Mode of Loading

It is to be expected that the bearing strength of a single-shear lap joint would be comparatively lower than that of a double-shear joint, mainly because of the loading eccentricity in the former case. For a typical quasi-isotropic laminate with 25/25/25/25 layup, with $w/d = 5.0$, $t/d = 0.6$, tested under both RT/AR and H/W environmental conditions with titanium fastener bolts, a comparison of single-shear and double-shear bearing strengths (2% offset strength) has been made in Fig. 11. From this figure, it is seen that the single-shear strengths are considerably lower than the corresponding double-shear strengths for the laminate tested. The strength degradation due to environment was almost similar for both single- and double-shear lap joints. However, it is observed that the scatter in single shear data, as depicted by SD in Fig. 11, appears to be smaller as compared to that of the double-shear joint, which is possibly due to the larger number of specimens (17 under RT/AR and 16 under H/W conditions) tested in single shear, as against an average of five specimens tested in the case of double-shear lap joints.

Effect of Fastener Bolt Material

Throughout this investigation, steel and titanium bolts of 6.0 mm diameter (including 100° countersunk head for bolts used in single shear) were used, except for only five series of 6.0 mm thick specimens, for which maraging steel bolts were used. From Fig. 7, it can be seen that for identical conditions of testing other than the fastener bolts, both under RT/AR and H/W conditions, the bearing strength for steel (AISI 4340) bolts are marginally higher than

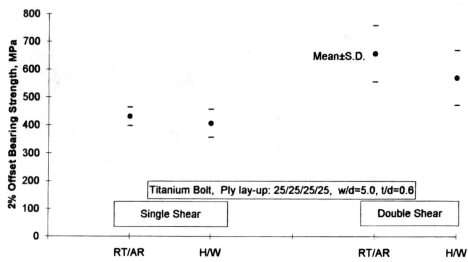

FIG. 11—*Comparison of single-shear and double-shear bearing stresses for a quasi-isotropic laminate under RT/AR and H/W condition.*

those for titanium bolts. Since the ply layups of the CFC laminates, for which maraging steel bolts were employed, were quite different (see Table 1), it was not possible to compare them.

Influence of Lamina Configuration

Six different types of laminates made from unidirectional prepreg tapes and five different types of laminates made from bidirectional fabric prepregs were used for specimen fabrication, as listed in Table 1. Figure 12 shows the 2% offset bearing strength for three extensively used laminates (that is, 50/25/25/0, 58/17/17/8, and 25/25/25/25 ply volume fractions) with different geometric parameters, bolt materials, and test environments. It was observed that all the ply layups exhibited a similar nature of variation in bearing strength for RT/AR and H/W environmental conditions, that is, H/W environment caused appreciable reductions in bearing strength irrespective of the geometric parameters/fastener bolt materials employed. However, in the case of laminates with different fabric layups, as shown in Fig. 13, the one with $[(\pm 45)_2/(0/90)_2]_S$ exhibited the highest room-temperature offset bearing strength ranging from around 500 to 575 MPa. The mean values including SD have been shown for all laminates both under room temperature and H/W conditions. From this figure, it can be seen that although RT/AR bearing strength values exhibited considerable variations between the laminates, those under H/W conditions showed rather smaller variations.

Effect of Hole Tolerance

It is generally believed that in fastener joints the magnitude of clearance between the bolt and the hole, which is dependent on the hole tolerance, may significantly influence the bearing strength. In the present study, this effect was examined with one batch of specimens made

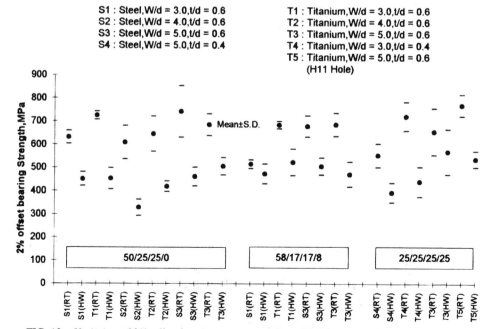

FIG. 12—*Variation of 2% offset bearing strength of three CFC laminates for different geometric parameters, fastener bolt materials, and environmental test conditions.*

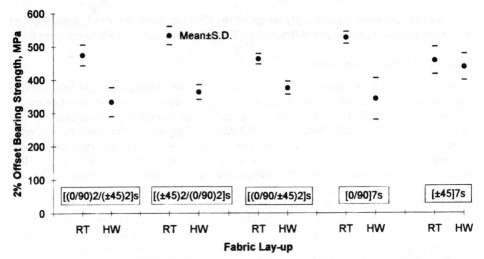

FIG. 13—*Comparison of 2% offset bearing strength in woven-fabric specimens of different layups under RT/AR and H/W condition.*

from $[0/\pm45/90]_{2S}$ ply layups, which had H11 hole tolerance as compared to H8 tolerance for all other batches of specimens (see Fig. 12). Considering a 2% offset bearing strength for H11 hole tolerance (ply volume fraction = 25/25/25/25, w/d = 5.0, t/d = 0.6, and titanium bolt case T5) with H8 hole tolerance and identical other parameters, case T3, it is seen that despite H11 tolerance being larger compared to H8, the bearing strength for H11 tolerance under RT/AR condition was about 10% higher. However, H/W bearing strength for H11 hole tolerance was almost the same as that for H8 tolerance. It is therefore felt that providing a marginally higher bearing hole tolerance (H11) does not significantly affect the double-shear bearing strength of fastener joints.

Effect of Environment

A hot and humid environment is known to degrade strength properties of carbon-fiber composites [6,16,25–27]. During this study, all hygrothermally aged specimens were tested for their H/W bearing strength at 100°C ± 2°C and RH ≥ 85%. The variation in 2% offset bearing strength due to H/W environment is shown in Figs. 12 and 13 for different angle-ply laminates and woven fabric laminates, respectively. From these figures, it is observed that the degradation in bearing strength is about 20 to 25% compared to RT/AR strength values. However, in the case of woven-fabric composite laminates, this drop in strength is somewhat lower for certain layups, like $[\pm45]_{7S}$ and $[(0/90/\pm45)_2]_S$, as seen from Fig. 13. The ratio of H/W-to-RT/AR 2% offset mean bearing strength values, considering a variety of laminates, including woven fabrics and bolt material (shown as S—for steel and T—for titanium) is shown in Fig. 14, from which it can be seen that considering all laminates/single-shear and double-shear data, the mean value of the ratio is around 0.70 to 0.75, with the lowest ratio reaching up to about 0.63. Figure 15 shows an identical comparison for ultimate bearing strength (in place of 2% offset strength) and it is seen that the influence of a H/W environment is almost similar, with a relatively lower scatter of data and the lowest ratio reaching to about 0.69. This implies that there is a 25 to 30% drop in bearing strength as compared to room temperature values due to environmental/hygrothermal effects. The primary cause for such

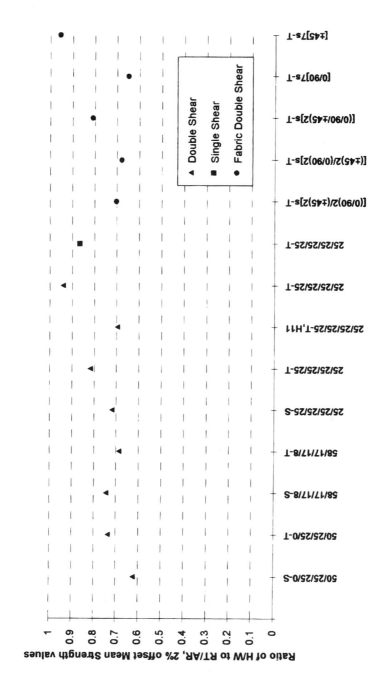

FIG. 14—*Effect of environment: ratio of H/W to RT/AR 2% offset bearing strength of CFC specimens.*

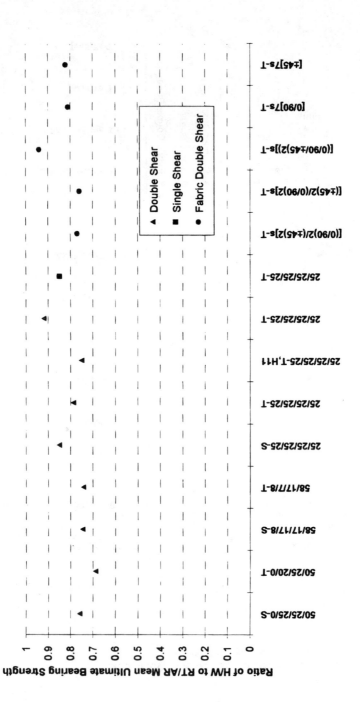

FIG. 15—*Effect of environment: ratio of H/W to RT/AR ultimate bearing strength of CFC specimens.*

strength degradation may be attributed to plasticizing of the epoxy matrix, which could possibly be the reason for the change in failure modes observed in some of the laminates. For example, Fig. 16 shows typical photographs of two sets of woven fabric failed specimens. It may be observed that identical woven fabric $[0/90]_{7S}$ specimens exhibited bearing mode failure (Fig. 16a) under RT/AR conditions, whereas those under H/W test conditions had failed in lateral (net tension) mode (Fig. 16b). This suggests a possible net-section matrix cracking/degradation of the interface bond strength between the hygrothermally-aged matrix and carbon fiber under a H/W test environment. The failure modes for the entire series of test specimens have been indicated in Table 1.

Conclusions

An investigation has been made to evaluate the influence of environmental and geometric parameters on the behavior of fastener joints in carbon-fiber composites, with an emphasis on the generation of relevant data for the design of fastener joints. Based on the analysis of test results, the following conclusions may be made:

1. In order to realize the maximum bearing strength in composite fastener joints, suitable values of geometric parameters need to be selected, which would avoid premature failures in the lateral or net-tension mode. From the test data analysis, within the constraints of the prescribed material system and test matrix, it has been observed that both under room temperature and hot/wet environmental conditions ($e/d = 4.0$, $w/d = 5.0$, and $t/d \geq 0.6$) adequate bearing strength is provided from the viewpoint of the fastener joint design.
2. For the typical quasi-isotropic laminate tested in this investigation, both under room temperature/as-received and hot/wet environments, it has been observed that the degrada-

FIG. 16—*Typical failure modes in woven-fabric CFC test specimens:* (a) *bearing under RT/AR test condition, and* (b) *lateral (net-tension) under H/W environmental condition.*

tion in offset bearing strength of the single-shear lap joint due to the environment was similar in magnitude to that of double-shear lap joint.

3. Considering a variety of CFC laminates, geometric parameters, and fastener bolt materials employed in this investigation, the degradation in the bearing strength of hygrothermally aged laminates under hot/wet test conditions is about 25 to 30% as compared to the corresponding room temperature values. It may therefore be recommended to use a factor of 0.7 to 0.75 on room temperature strength data to arrive at hot/wet strength allowables for the purpose of fastener joint design in this class of composite.

Acknowledgment

This study was supported by the National Team, C-Wing program. The authors thank Dr. K. N. Raju, Program Director, for his encouraging support and the authorities of HAL, Bangalore Complex, for fabrication of the specimens. Execution of this work was ably supported by engineers and technicians of M/s Comat Systems Solutions (P) Ltd., Bangalore. Specifically, the assistance of Kiran Bhojaraj, S. Pradeep, and H. V. Sumana is gratefully acknowledged.

References

[1] Sanger, K. B., Dill, H. D., and Kautz, E. F., "Certification of Testing Methodology for Fighter Hybrid Structure," *Composite Materials: Testing and Design (Ninth Volume), ASTM STP 1059*, S. P. Garbo, Ed., American Society for Testing and Materials, Philadelphia, 1990, pp. 34–37.
[2] Cantwell, W. J. and Morton, J., "An Assessment of the Residual Strength of an Impact Damaged Carbon Fiber Reinforced Epoxy," *Composite Structures*, Vol. 14, 1990, pp. 303–307.
[3] Berg, M., Gerharz, J. J., and Gokgol, O., "Consideration of Environmental Condition for the Fatigue Evaluation of Composite Airframe Structures," *Composite Materials: Fatigue and Fracture, Second Volume, ASTM STP 1012*, P. A. Lagace, Ed., American Society for Testing and Materials, Philadelphia, 1989, pp. 29–44.
[4] Anderson, B. W., "Factors for Designing of Military Aircraft's Structures in Carbon Fiber Reinforced Composites," *Advances in Fracture Research, Proceedings*, ICF-6, S. R. Valluri, et al., Eds., Pergamon Press, New York, 1984, pp. 607–622.
[5] Verette, R. M. and Labon, J. D., "Structural Criteria for Advanced Composites," AFFDL-TR-76-142, Air Force Flight Dynamics Laboratory, Dayton, OH, 1976.
[6] Ramkumar, R. L., "Bolted Joint Design," *Test Methods and Design Allowables for Fibrous Composites, ASTM STP 734*, C. C. Chamis, Ed., American Society for Testing and Materials, Philadelphia, 1981, pp. 376–395.
[7] Hart-Smith, L. J., "Bolted Joints in Graphite/Epoxy Composites," NASA CR-144899, National Aeronautics and Space Administration, NASA Langley Research Center, Hampton, VA, June 1976.
[8] Ramkumar, R. L., "Bolt-Bearing/By-Pass Study on Composite Laminates," Northrop corporation Report, NOR 78-154, Hawthrone, CA, Dec. 1978.
[9] Eisenmann, J. R., "Bolted Joint Static Strength Model for Composite Materials," NASA-TM-X-3377, National Aeronautics and Space Administration, NASA Langley Research Center, Hampton, VA, 1976.
[10] Harris, H. G., Ojalvo, I. U., and Hooson, R., "Stress and Deflection Analysis of Mechanically Fastened Joints," Technical report AFFDL-TR-70-49, Air Force Flight Dynamics Laboratory, Dayton, OH, May 1970.
[11] Agarwal, B. L., "A Computer Program for the Strength Prediction of Double Shear Composites Bolted Joints and Inates with Holes," Northrop Corporation Report, NOR 78-49, Hawthrone, CA, May 1978.
[12] Nuismer, R. J. and Labor, J. D., "Applications of the Average Stress Failure Criterion: Part1-Tension," *International Journal of Composite Materials*, Vol. 12, 1978, p. 250.
[13] Ramkumar, R. L. and Tossavienen, E. W., "Strength and Lifetime of Bolted Laminates," *Fatigue in Mechanically Fastened Composite and Metallic Joints, ASTM STP 927*, J. M. Potter, Ed., American Society for Testing and Materials, Philadelphia, 1986, pp. 251–273.

[14] Ramkumar, R. L. and Tossavainen, E. W., "Strength and Lifetime of Bolted Laminates," *Fatigue in Mechanically Fastened Composite and Metallic Joints, ASTM STP 927*, J. M. Potter, Ed., American Society for Testing and Materials, Philadelphia, 1986, pp. 251–273.

[15] Johnson, M. and Matthews, F. L., "Determination of Safety Factors for Use When Designing Bolted Joints in GRP," *Composites*, April 1979, pp. 73–76.

[16] Akay, M., "Bearing Strength of As-Cured and Hygrothermally Conditioned Carbon Fibre/Epoxy Composites Under Static and Dynamic Loading," *Composites*, Vol. 23, No. 2, March 1992, pp. 101–108.

[17] Garbo, S. P. and Ogonowsk, J. M., "Effect of Variences and Manufacturing Tolerances on the Design Strength and Life of Mechanically Fastened Composite Joints," AFWAL-TR-81-3041, Vol. I, Air Force Wright Aeronautical Laboratories, Dayton, OH, 1981.

[18] Bauer, J. and Mennle, E., "Comparison of Experimental Results and Analytically Predicted Data for Double Shear Fastened Joints," Behaviour and Analysis of Mechanically Fastened Joints in Composite Materials, *Advisory Group for Aerospace Research and Development, Conference Proceedings 427*, 1988.

[19] Crews, J. H., Jr. and Naik, R. V. A., "Failure Analysis of a Graphite/Epoxy Laminate subjected to Bolt Bearing Loads," NASA-TM-86297, NASA Langley Research Center, Hampton, VA, 1984.

[20] MIL-HDBK-17-1C, *Polymer Matrix Composites-Guidelines*, Vol. 1, Department of Defense, Washington, DC, 1988.

[21] Bauer, J., "Mechanism of Single Shear Joints," Behaviour and Analysis of Mechanically Fastened Joints in Composite Materials, *Advisory Group for Aerospace Research and Development, Conference Proceedings 427*, 1988.

[22] Collings, T. A., "Experimentally Determined Strength of Mechanically Fastened Joints," *Joining of Fibre Reinforced Plastics*, F. L. Matthews, Ed., Elsevier, London, 1987, p. 63.

[23] Matthews, F. L., "Joining of Composites," *Design with Advanced Composites*, L. N. Philips, Ed., Springer-verlag, Berlin, 1989, pp. 119–144.

[24] Collings, T. A., "The Strength of Bolted Joints in Multidirectional CFRP Laminates," *Composites*, Vol. 8, 1977, pp. 43–55.

[25] Collings, T. A., and Stone, D. E. W., "Hygrothermal Effects in CFRP Laminates: Strains Induced by Temperature and Moisture," *Composites*, Vol. 16, No. 4, 1985, pp. 307–316.

[26] Clark, G., Saunders, D. S., van Blaricum, T. J., and Richmond, M., "Moisture Absorption in Graphite/Epoxy Laminates," *Composites Science and Technology*, Vol. 39, 1990, pp. 355–375.

[27] Collings, T. A., "The Effect of Observed Climatic Conditions on the Moisture Equilibrium Level of Fibre Reinforced Plastics," *Composites*, Vol. 17, No. 1, Jan. 1986, pp. 33–41.

Alan R. Kallmeyer[1] and Ralph I. Stephens[2]

Creep Elongation of Bolt Holes Subjected to Bearing Loads in a Polymer Matrix Composite Laminate

REFERENCE: Kallmeyer, A. R. and Stephens, R. I., "Creep Elongation of Bolt Holes Subjected to Bearing Loads in a Polymer Matrix Composite Laminate," *Composite Materials: Fatigue and Fracture (Sixth Volume), ASTM STP 1285*, E. A. Armanios, Ed., American Society for Testing and Materials, 1997, pp. 452–467.

ABSTRACT: To assess the long-term durability of advanced composite joints, the localized creep response of a quasi-isotropic ([45/0/−45/90]$_{2s}$), graphite fiber/polymer matrix (G40-800/ 5260) composite laminate subjected to bolt-bearing loads at ambient and elevated temperatures was investigated. Monotonic tension tests and static creep tests were performed on single-hole bolted joints at temperatures of 23, 100, and 150°C. The influence of lateral constraint on the creep response was studied by performing creep tests with bolt clampup torques of 5.65 N·m and a "finger-tight" level. While the ultimate monotonic bearing strength was affected only marginally by temperature, substantial changes in the initial joint stiffness and the hole elongation at fracture were observed in the monotonic tests as the temperature was increased. In the creep tests, significant time-dependent hole elongations were observed at high bearing stress levels and temperatures, with the creep component of the hole elongation found to exceed 5% of the original hole diameter in one case. The lateral constraint had a large influence on the creep behavior of the joints. As the clampup torque was reduced from 5.65 N·m to a finger-tight level, the rate of hole elongation increased by a factor of 2 to 4, indicating the importance of maintaining high clampup conditions in bolted laminate joints. A simple empirical expression was proposed to model the creep behavior of the joints. This expression correlated the experimental data reasonably well and is useful for indicating trends in the creep response of bolted laminate joints.

KEYWORDS: composite materials, fracture (materials), laminates, bolted joints, creep (materials), elevated temperature, hole elongation, bearing strength, bolt torque

The high strength-to-weight and stiffness-to-weight ratios of fiber-reinforced, polymer-matrix composite (PMC) laminates have made them desirable for use in structural applications, such as in aircraft and spacecraft, where weight savings are crucial. These materials offer other benefits, such as corrosion resistance, high fatigue strength, and unique design flexibility. Despite these advantages, the viscous nature of most polymer resins has limited the use of PMCs in critical structural components that must operate in high temperature environments. However, with the increasing quest for performance and efficiency, the weight savings offered by PMCs are resulting in a greater desire to use these materials at temperatures that were previously avoided. Consequently, the need exists to better understand the behavior of these materials at elevated temperatures.

[1]Assistant professor, Department of Mechanical Engineering, North Dakota State University, Fargo, ND 58105.
[2]Professor, Department of Mechanical Engineering, University of Iowa, Iowa City, IA 52242.

In the past few years, considerable effort has been focused towards meeting this need. Much of this effort can be attributed to the High Speed Civil Transport (HSCT) program, with the goal of developing a commercial supersonic aircraft capable of attaining speeds in excess of Mach 2 while carrying 200 to 300 passengers [1]. Due to the high structural loads, elevated temperatures, and weight minimization requirements of such an aircraft, PMCs are being strongly considered for use in both primary and secondary components. Thus, it has become necessary to assess the long-term durability of these materials at elevated temperatures. These analyses must include the consideration of time dependence in the stress-strain response of the composite, such as the effects of creep and stress relaxation. Although past research has indicated that time-dependent effects in PMC laminates can be minimized by designing the laminate for fiber-dominated load paths [1–6], the presence of notches may dictate that time-dependent phenomena be considered due to the high stresses, free edge effects, and discontinuity of the fibers in the vicinity of the notches.

One such area of concern when using composite materials in structural applications is that of joining components together. Due to several inherent characteristics of laminated composites, adhesive bonds are often the preferred method of joining these materials. Some of the problems that arise in bolted connections include high stress concentrations around holes, low interlaminar and in-plane shear strengths, and low bearing strength due to localized fiber buckling [7]. However, because of the difficulties involved with inspection of the integrity of adhesive bonds and the complications involved in disassembly, it often becomes necessary to rely on mechanically fastened joints. To date, most research in the area of mechanically fastened joints has focused on the static and cyclic strength and failure prediction of these joints at ambient temperatures [8–15]. Although some investigations have been made concerning environmental effects on joint strength [15–19], relatively few have considered the long-term effects caused by the viscoelastic and viscoplastic behavior of the polymer resin at ambient or elevated temperatures. An analysis of the bolt clampup relaxation in graphite/epoxy bolted joints conducted by Shivakumar and Crews [19] demonstrated that the viscoelastic nature of the resin caused significant reductions in the bolt clampup force. However, a review of the literature found no research into the effects this viscoelastic behavior may have on the long-term strength, stiffness, or integrity of the bolted joints. The recent increase in emphasis placed on high temperature applications has made it necessary to thoroughly evaluate the behavior of laminate joints over a broad temperature range, including the effects of time-dependent deformations in the vicinity of fastener holes. Although catastrophic failure of the joint may not occur, excessive hole elongation may result in the inability of the joint to perform the required function.

The purpose of this paper is to present the results of an investigation into the localized creep response of a quasi-isotropic, graphite fiber-reinforced PMC laminate subjected to bolt-bearing loads at ambient and elevated temperatures (23, 100, and 150°C). The time-dependent elongations of fastener holes in the laminate were measured at various stress levels and bolt clampup torques at each temperature. The effects of these parameters on the creep behavior of the joints are discussed. A simple empirical expression is proposed that is useful for indicating trends in the creep response of bolted laminate joints.

Material and Test Specimen

The material used in this study was a graphite fiber (G40-800) reinforced, bismaleimide matrix (5260) composite. This material is presently under consideration for use in the HSCT program, where it is anticipated that it will experience operating temperatures up to 150°C. The glass transition temperature of the matrix listed by the manufacturer is approximately 220°C. Previous research [1–3] on a nearly identical composite (IM7 graphite fibers, 5260 matrix) has indicated that this material exhibits significant time-dependent stress-strain behavior when loaded in directions other than the fiber direction.

All experiments were conducted using the single-hole, bolted joint specimen shown in Fig. 1. The specimens were cut from a 16-ply, quasi-isotropic laminate with the layup [45/0/−45/90]$_{2s}$, and had a nominal thickness, t, of 2.16 mm. The bolt holes were drilled with a water-cooled, diamond grit tubular drill, with plexiglass placed beneath the laminate to avoid tear-out or delamination. The specimen width-to-hole diameter ratio ($w/d = 8$) and edge distance-to-hole diameter ratio ($e/d = 4$) were chosen to ensure that the initial failure mode would be local bearing failure of the material near the bolt, eliminating the net-tension and shearout modes of failures that have been found to occur for smaller values of w/d and e/d [8]. Prior to testing, all specimens were thoroughly dried by heating them to 120°C for 142 h, and were subsequently stored in a desiccator.

Experimental Procedures

The bolt-loading apparatus is shown in Fig. 2. The bolt was loaded in double shear by a pair of 6.35-mm-thick steel plates, with a friction type of grip reacting the load on the other end of the specimen. Each steel bolt had a nominal diameter of 6.35 mm, resulting in a snug fit between bolt and hole. Washers (6.35 mm inner diameter and 12.4 mm outer diameter) between the steel plates and specimen distributed the clampup torque load evenly around the bolt hole. Two levels of bolt clampup torque were used in this test program to study the effect of lateral constraint around the bolt hole: 5.65 N·m, a moderate clampup torque for a 6.35-mm bolt, and a "finger-tight" level, representing a worst-case scenario in which the lateral constraint is reduced to a very low level. To eliminate differences in the lateral constraint force at the elevated temperatures caused by differences in thermal expansion between the steel and composite, the clampup torque was applied after the specimen and bolt had reached the test temperature.

To measure the bolt-hole elongation, a strain gage-based extensometer was used to measure the relative displacement between a steel plate and a small, stiff wire (1.14 mm diameter) that was placed under the bolt along the lower edge of the hole, as shown in Fig. 2b. To provide clearance for the wire, a 3.2-mm-wide and 1.6-mm-deep slot was machined along the length of the bolt, and a 3.2-mm-wide slot was machined into the washers. The wire was held in place by a combination of frictional forces and high temperature epoxy. The extensometer,

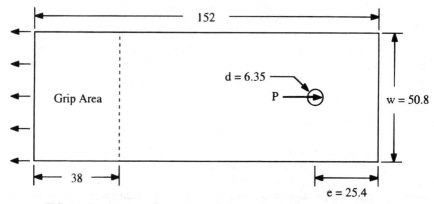

FIG. 1—*Specimen configuration and dimensions (all dimensions in mm).*

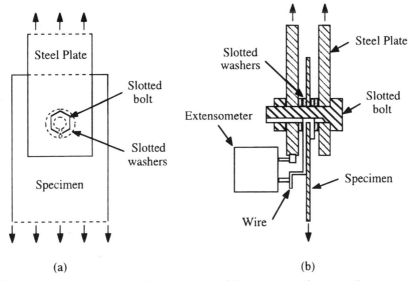

(a)   (b)

FIG. 2—*Bolt-loading apparatus: (a) front view and (b) cut-away side view with instrumentation.*

which had a gage length of 12.7 mm, was positioned with one knife edge on this wire and one edge on a circular stub protruding from the bottom of the steel plate. Assuming that the plate and bolt deflections were small, the measured displacement was approximately equal to the hole elongation. A similar experimental setup was used successfully by Crews and Naik [8,9] to measure bolt-hole elongations in graphite/epoxy laminates.

Prior to performing the creep tests, monotonic tension tests were performed at each test temperature (23, 100, and 150°C) to determine the ultimate bearing strength and hole elongation at fracture. All monotonic tests were conducted with a bolt clampup torque of 5.65 N·m. Constant-load creep tests were then performed at applied load levels ranging from 60 to 90% of the average ultimate monotonic bearing strength at each temperature. The creep tests were conducted with bolt clampup torques of 5.65 N·m and a finger-tight level. All experiments were performed using an 89-kN closed-loop electrohydraulic test system. A circulating air, convective heating environmental chamber was mounted on the load frame, in which the entire specimen and bolt-loading apparatus were placed. A thermocouple was inserted near the bolt between the specimen and steel plate to monitor the temperature, and a controller capable of maintaining the temperature to within ±1.5°C regulated the heating inside the chamber. The electrohydraulic test system was interfaced with a digital controller and personal computer, and could be operated in stroke (ram displacement), load, or extensometer (hole elongation) control. The monotonic tension tests were conducted in stroke control at a rate of 0.02 mm/s, and the hole elongation was measured as a function of load. The creep tests were conducted in load control. An initial loading rate of 2000 N/s was applied until the test load was reached, after which the load was held constant. Although electrohydraulic test systems are generally not intended for use in long-term creep loading situations, it was found that the digital controller maintained the load within ±1% of the intended level throughout the duration of the tests. During these tests, the hole elongation was monitored as a function of time for periods of up to 200 h.

Results and Discussion

Monotonic Tension Tests

A total of eight monotonic tension tests were performed: three each at 23 and 150°C and two at 100°C. Four of these tests were performed using a slightly different experimental setup, in which a linear variable differential transducer (LVDT) was used in place of the extensometer to measure the hole elongation. It was later determined that, due to the nature of the setup, the LVDT did not provide the accuracy needed for this study. As a result, the hole elongation data collected from these tests were considered unreliable, so valid hole elongation measurements were only obtained from the other four tests. However, the failure loads recorded from all tests were valid. Typical valid curves of nominal bearing stress, σ_b ($\sigma_b = P/td$, where P is the applied bolt load, d is the bolt diameter, and t is the laminate thickness), versus percent hole elongation, δ ($\delta = \Delta d/d \times 100$, where Δd is the hole elongation) at each temperature are shown in Fig. 3. During the recording of these curves, some initial nonlinearity was observed at low load levels (below approximately 15% of P_{max}) that was believed to be due to slight initial misalignment and flexing of the specimens in the early stages of the tests. Since this nonlinearity was not an accurate measure of the σ_b-versus-δ behavior, it was eliminated by linearizing the initial portion of the curves. This was accomplished by continuing the linear region of the σ_b-versus-δ curves through the small nonlinear region to the zero-

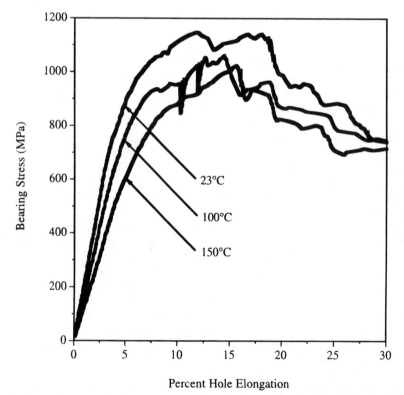

FIG. 3—*Typical curves of bearing stress versus percent hole elongation from monotonic tensile tests at each temperature.*

stress axis, and then shifting the data horizontally until the intercept was located on the origin. By performing the tests in stroke control, the resulting σ_b-versus-δ curves were very sensitive to localized damage and deformation near the bolt hole, and it was also possible to monitor the hole elongation and eventually remove the load from the specimens after the ultimate bearing strength had been reached. The load removal prevented the localized damage that occurred during loading from being concealed by the gross damage caused by complete specimen failure.

The influence of temperature on the monotonic tensile behavior of the joint can be seen clearly in Fig. 3. In previous studies, Gates [1–3] reported a negligible temperature effect in the longitudinal elastic modulus of the IM7/5260 composite in monotonic tension tests, and a small degradation in the transverse modulus below 150°C. As is evident in Fig. 3, a substantial decrease in the slope of the linear portion of the curves was found in this study as the temperature was increased from 23 to 150°C. This decrease in the initial slope of the monotonic curves would appear to indicate that the elastic response of this composite was sensitive to temperature. However, the degradation in the stiffness may be due in large part to the development of localized damage in the material in the bearing region near the bolt hole. Matrix cracking in the off-axis plies can initiate at low stress levels and lead to a reduction in the elastic modulus of the composite. An increase in the damage development with temperature would thus result in the stiffness degradation observed here. However, lacking additional data, a definitive assessment is difficult to make. The temperature sensitivity in the σ_b-versus-δ curves became much more apparent at higher stress levels, where the curves became nonlinear. This nonlinearity was likely due to the increased development of localized damage and plastic flow around the bolt hole. As the temperature was increased, the nonlinearity initiated at a lower stress level, and δ increased at a higher rate with increasing σ_b. These observations agree with the results reported by Gates [1–3], who found a significant increase in plastic flow with increasing temperature in the IM7/5260 material.

The range of values of the ultimate bearing strength, σ_{bu} (defined as the maximum bearing stress obtained in the monotonic tension tests), along with the average values of σ_{bu} and typical valid values of the percent hole elongation at fracture, δ_f (defined here as the percent hole elongation at the point of maximum bearing stress), are reported in Table 1. Also shown are typical values of the initial joint stiffness, κ, defined as σ_b/δ from the linear portion of the σ_b-versus-δ curves. The ultimate bearing strength was found to decrease by only about 10% as the temperature was increased from 23 to 150°C. However, significant variations in δ_f and κ were observed over the same temperature range (over 30% increase in δ_f and nearly 40% decrease in κ). Thus, in this composite, there appears to be a greater sensitivity to temperature in the deformation of the joint than in the strength of the joint.

Macroscopic examination of the specimens following removal of the load revealed that the damage initiated in the bearing region of the specimens, that is, in the region directly under the loaded portion of the bolt. The lateral constraint supplied by the washers prevented gross

TABLE 1—*Monotonic tension test results.*

Temperature, °C	Ultimate Bearing Strength, σ_{bu} (MPa)		Percent Hole Elongation at Fracture, δ_f	Typical Initial Joint Stiffness, $\kappa = \sigma_b/\delta$, MPa
	Range	Average		
23	1127 to 1194	1160	12	210
100	1058 to 1087	1070	14, 15	170
150	1007 to 1043	1020	16	130

delamination from occurring; however, crushing of the plies was observed in the bearing region between the washers. Once the ultimate strength had been reached, the failure mode switched to shearout, in which cracks developed in the specimen along either side of the bolt hole and propagated towards the end of the specimen, causing the bolt to be ripped through the material. This process also caused delamination of the outer plies in the region beyond the clampup washers.

Creep Tests

Creep tests were performed at each temperature (23, 100, and 150°C) and bolt clampup torque (5.65 N·m and finger-tight). At a torque level of 5.65 N·m, tests were performed at applied bearing stress levels ranging from 60 to 90% of the ultimate bearing strength, σ_{bu}. At the finger-tight torque level, tests were conducted at bearing stress levels of 60 and 70% of σ_{bu} (obtained from the monotonic tests at the 5.65 N·m torque level). The percent hole elongation, δ, measured during these tests, was decomposed into two components

$$\delta = \delta_\ell + \delta_c \tag{1}$$

where δ_ℓ is the component associated with loading up to the constant creep load, and δ_c is the time-dependent or creep component of the percent hole elongation, measured after the creep load had been reached. δ_c is shown as a function of time for typical creep tests in Figs. 4 through 8 at each temperature, bearing stress, and clampup torque. The values of δ_c at

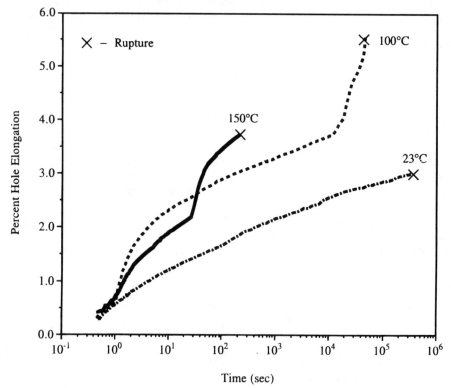

FIG. 4—*Creep component of percent hole elongation for bearing stress level of 90% σ_{bu} (5.65 N·m bolt clampup torque).*

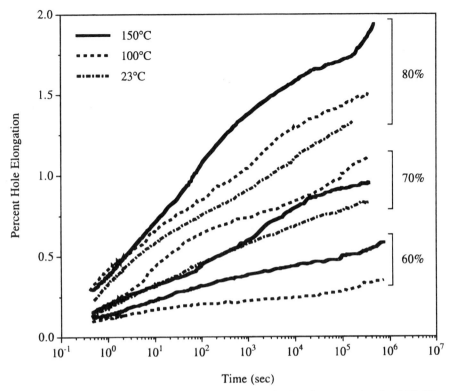

FIG. 5—*Creep component of percent hole elongation for bearing stress levels of 60, 70, and 80% σ_{bu} (5.65 N·m bolt clampup torque).*

rupture or after 100 h are also listed in Table 2. In all tests, the first value of δ_c was recorded approximately 0.5 s after the creep load was reached; thus, the creep response in the very early stages of the tests (0 to 0.5 s) was not obtained.

The creep curves at a stress level of 90% σ_{bu} and torque level of 5.65 N·m are shown in Fig. 4. At this stress, the specimens ruptured prior to 100 h at all three temperatures. The effect of temperature on the creep response of the joints is clearly evident. A substantial increase in the creep rate and a decrease in the rupture time were observed as the temperature was increased. At 150°C, rupture occurred in approximately 3 min, while at 100 and 23°C the rupture times were roughly 13 and 97 h, respectively. Significant time-dependent deformations occurred near the bolt hole, with values of δ_c ranging from 3.0 to 5.6% at the time of rupture. The smaller value of δ_c at rupture at 150°C, as compared to 100°C, was likely due to the rapid accumulation of damage near the bolt hole, which caused the specimen to rupture so quickly that large creep strains did not have sufficient time to develop.

The remaining creep curves at the 5.65 N·m torque level are shown in Fig. 5. No specimens ruptured prior to 100 h at the stress levels ranging from 60 to 80% σ_{bu}; however, a few tests ended prematurely due to equipment malfunction. A similar temperature effect was found at these stress levels, where, in general, the creep rate increased as the temperature was increased. The higher creep rate at 100°C, as compared to 150°C at the 70% σ_{bu} level, could be attributed to typical variations in material response or unknown experimental error. The influence of increasing the applied bearing stress at each temperature is also clearly shown in Fig. 5. Similar

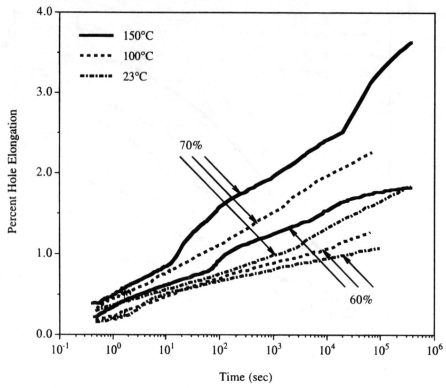

FIG. 6—*Creep component of percent hole elongation for bearing stress levels of 60 and 70%* σ_{bu} *(finger-tight bolt clampup torque).*

trends were observed in the creep response of the joints that were torqued to a finger-tight level, shown in Fig. 6.

The influence of bolt clampup torque on the creep response of the joints can be seen in Figs. 7 and 8. A substantial increase in the time-dependent hole elongation resulted as the torque was reduced from 5.65 N·m to a finger-tight level. It has been reported previously that reducing the bolt clampup torque results in a significant decrease in the ultimate bearing strength of laminates [9,10,15,17]. Furthermore, Crews [9] found that reducing the clampup torque strongly influenced the failure modes and caused an increase in hole elongation at fracture in monotonic tension tests. It is evident that the reduction in the lateral constraint force resulting from decreased bolt torque has a similar effect on the creep behavior of these joints. However, a reliable estimate of the influence of bolt clampup torque is difficult to obtain due to the relaxation of the clampup force that can occur in laminate bolted joints. Shivakumar and Crews [19] reported that the clampup force can relax by as much as 10% at room temperature and 25% at 121°C over 100 h in graphite/epoxy laminates due to viscoelastic deformations in the thickness direction. Nevertheless, despite the likely relaxation of the clampup force, it was found that the reduction in clampup torque investigated in this study led to an increase in the creep rate by a factor of 2 to 4, as shown in Figs. 7 and 8. This would indicate that the bolt clampup torque has a significant influence on the time-dependent response of laminate bolted joints.

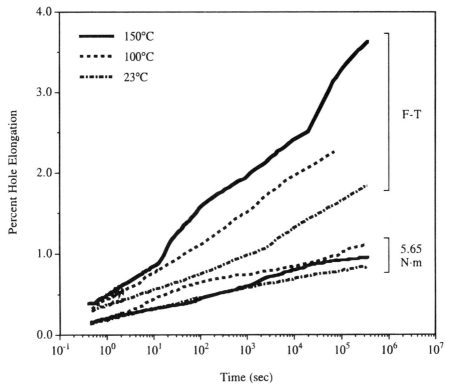

FIG. 7—*Influence of bolt clampup torque (finger-tight and 5.65 N·m) on time-dependent hole elongation at bearing stress level of 70% σ_{bu}.*

A comment should be made regarding duplication of the creep tests. Due to the number of specimens available, only seven of the creep tests were duplicated, as shown in Table 2. It was found that repeatability of the results ranged from very good to quite poor. The variation in the results at the 5.65 N·m torque level was fairly low, where scatter in the creep deformations were within 10% for three duplicate tests and approximately 30% for one duplicate test. However, at the finger-tight torque level, the scatter in the results was much larger, with variations in the creep rates ranging from approximately 18 to 100%. The poor repeatability is believed to be largely due to the difficulty of the experimental setup, particularly in applying the bolt clampup torque at the elevated temperatures. This was especially noticeable in the tests conducted at the lower clampup level, in which it was difficult to consistently apply the same finger-tight level of torque. As previously mentioned, the bolt clampup torque was found to strongly influence the creep response of the joints. Thus, significant scatter in the results at the finger-tight torque level does not seem unreasonable.

Another potential source of scatter in the creep results is the clearance between the bolt and hole. Although a snug fit was assumed in this research, the tolerances on the bolt diameter and hole diameter can result in small clearance levels between the bolt and hole. Using an inverse finite element analysis, Naik and Crews [14] found that even small clearances can have a substantial effect on the stresses and deformations near the hole under monotonic loading. For example, for a bolt-hole clearance of 1.6%, which is a typical clearance in a mechanically fastened joint in a composite material [14], the peak radial stresses increased

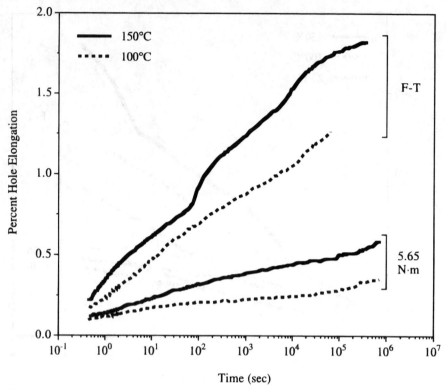

FIG. 8—*Influence of bolt clampup torque (finger-tight and 5.65 N·m) on time-dependent hole elongation at bearing stress level of 60% σ_{bu}.*

TABLE 2—*Experimentally obtained values of the creep component of percent hole elongation (δ_c) at rupture or after 100 h.*

Clampup Torque	σ_b/σ_{bu}	δ_c, 23°C	δ_c, 100°C	δ_c, 150°C
5.65 N·m	0.9	3.0[a] (97 h)	5.6[a] (13 h)	3.7[a] (3 min)
	0.8	1.4[b] (1.3 at 46 h)	1.5	1.9
		1.3	1.4	...
	0.7	0.8	1.1	0.9
		...	...	0.9
	0.6	...	0.3	0.5
		...	0.4	...
Finger-tight	0.7	1.8	2.5[b] (2.3 at 19 h)	3.6
		...	4.9	...
	0.6	1.1[b] (1.0 at 27 h)	1.4[b] (1.3 at 18 h)	1.8
		...	2.4	1.5

[a]Specimen ruptured in time indicated.
[b]Estimated from extrapolation. Test ended prematurely due to equipment malfunction, with value of δ_c shown at time of malfunction.

by 36% relative to the snug-fit case. Furthermore, at a bearing stress of 1000 MPa, the hole elongation increased by 16% relative to the snug-fit case. These results indicate that some scatter in hole elongation measurements in monotonic tests would be nearly unavoidable due to the slight variations in hole and bolt diameters that will exist in bolted joints, and these variations may also lead to scatter in creep measurements of hole elongation.

Through macroscopic examination of the specimens following creep loading, many of the same damage mechanisms were found as in the monotonic tests. Although the lateral constraint supplied by the washers prevented gross delamination from occurring near the bolt hole, localized crushing of the plies was evident in the bearing region between the washers directly beneath the bolt. This damage was more severe in the specimens tested at the lower clampup torque, where "brooming" of the plies (localized delamination and out-of-plane movement of the plies) near the hole and thickening of the laminate under the washers were observed. The increase in damage under the washers at the lower torque level may be largely responsible for the increased creep rate at the lower clampup torque. Significant damage accumulation in the 0° plies (longitudinal plies) would shift a greater percentage of the load to the off-axis plies that are more susceptible to time-dependent deformations [1–6].

Analysis

In the analysis of creep data, it has been found that the creep strain rate of many monolithic materials can be represented by an equation of the form

$$\dot{\epsilon}_c = A\sigma^m t^n \tag{2}$$

where A, m, and n are constants that are a function of temperature [20]. A number of researchers have also found that a power-law equation adequately represents the creep behavior of many fiber-reinforced, polymer-matrix composites [1–6]. Therefore, in this study, a similar relationship was assumed between the rate of percent hole elongation, $\dot{\delta}_c$, the normalized bearing stress, and time; that is

$$\dot{\delta}_c = A\left(\frac{\sigma_b}{\sigma_{bu}}\right)^m t^n \tag{3}$$

where σ_b is the applied bearing stress, and A, m, and n are constants that are a function of temperature and bolt clampup torque. The relaxation of the bolt clampup force was also found to be represented accurately by a power-law equation by Shivakumar and Crews [19], indicating another term could be included in Eq 3 to model this effect. However, since this was not a focus of the present study, no attempt was made to include the effect of the relaxation.

By observing Figs. 4 through 6, it can be seen that the majority of the creep curves for the bolted laminate joint were approximately linear when plotted on a semi-log scale, indicating a fairly constant rate of hole elongation with respect to log time. Using the relationship

$$\frac{d}{dt}\log_{10}t = \frac{1}{t\ln 10} = 0.4343t^{-1} \tag{4}$$

this would suggest a value of $n = -1$ in Eq 3, independent of temperature and bolt clampup torque. The values of A and m in Eq 3 were calculated by first applying a logarithmic curve fit to the creep curves in Figs. 4 through 6 to determine the logarithmic rate of the percent hole elongation at each temperature, bearing stress, and clampup torque; that is

$$\frac{d\delta_c}{d\,(\log t)} \tag{5}$$

These values were then plotted versus the normalized bearing stress (shown in Fig. 9) and fit with a power law expression of the form

$$\frac{d\delta_c}{d\,(\log t)} = A'\left(\frac{\sigma_b}{\sigma_{bu}}\right)^{m'} \tag{6}$$

The results of this procedure were then used, along with Eq 4 and the chain rule of differentiation, to determine the values of A and m in Eq 3 at each temperature and clampup torque. The values of A and m resulting from this procedure are presented in Table 3. It should be noted that the values obtained from the 90% σ_{bu} creep curves at 100 and 150°C (5.65 N·m torque level) were not included in Fig. 9 due to the nonlinear nature of these creep curves when plotted on a semi-log scale. It is believed this nonlinearity was due to significant localized damage in individual plies that accumulated during the tests, resulting in very high creep rates. Hence, the results shown in Table 3 for the elevated temperature tests are not valid for $\sigma_b/\sigma_{bu} > 0.8$.

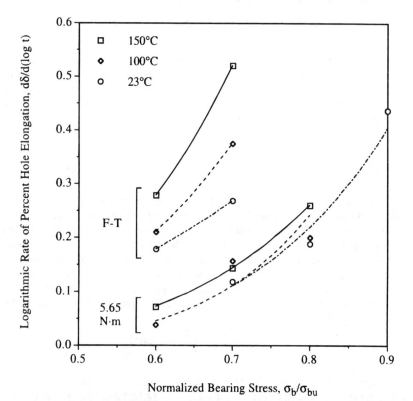

FIG. 9—*Logarithmic rate of percent hole elongation as a function of bearing stress at each temperature and bolt clampup torque.*

TABLE 3—*Joint creep rate parameters for Eq 3.*

Clampup Torque	Temperature, °C	A	m	n
5.65 N·m	23	0.30	5.20	−1
	100	0.39	5.94	−1
	150	0.31	4.51	−1
Finger-tight	23	0.30	2.67	−1
	100	0.62	3.74	−1
	150	0.97	4.09	−1

It is evident from Fig. 9 and Table 3 that a significant dependence on the temperature and clampup torque exists in the values of A and m. However, because of the small number of temperatures and clampup torques investigated in this study, no attempt was made to determine a functional relationship between the values of A or m and the temperature or torque level. Furthermore, due to the limited data available and the relatively high degree of variability between duplicate tests, the results shown in Fig. 9 and Table 3 should be considered useful primarily for indicating trends in the data. Several additional tests would be required before a high degree of confidence could be placed in the actual values generated in this study. Despite these limitations, the effectiveness of Eq 3 at modeling the creep response of the laminate joints was evaluated by calculating the percent hole elongation after 100 h for the 60, 70, and 80% σ_{bu} creep tests. Equation 3 was integrated and δ_c was calculated using a lower time limit of 0.1 s and the values of A, m, and n shown in Table 3. The values of δ_c calculated in this manner are shown in Table 4. In all but one case, the values obtained using Eq 3 were within 12% of the experimental values shown in Figs. 5 and 6. These results indicate that an expression of the form shown in Eq 3 is a reasonable model for predicting the creep response of the bolted laminate joints.

Summary and Conclusions

The results presented in this paper indicate that, even at ambient temperatures, the time-dependent behavior of PMC laminates should be considered in the design of mechanically fastened joints. Although time-dependent effects can be minimized by the reinforcing fibers, the discontinuity of the fibers and high stresses near a loaded fastener hole can cause significant localized time-dependent deformations and may eventually lead to rupture of the joint.

In the composite laminate studied here, the test temperature and bolt clampup torque were both found to have a substantial influence on the elongation of bolt holes subjected to bearing loads. In monotonic tension tests, the ultimate bearing strength decreased by approximately

TABLE 4—*Creep component of percent hole elongation (δ_c) after 100 h, calculated from Eq 3.*

Clampup Torque	σ_b/σ_{bu}	δ_c, 23°C	δ_c, 100°C	δ_c, 150°C
5.65 N·m	0.8	1.4	1.6	1.7
	0.7	0.7	0.7	0.9
	0.6	0.3	0.3	0.5
Finger-tight	0.7	1.7	2.5	3.4
	0.6	1.2	1.4	1.8

10% as the temperature was increased from 23 to 150°C, while the initial joint stiffness decreased by nearly 40% and the hole elongation at fracture increased by over 30%. A similar temperature effect was observed in the creep response of the bolted joints, where the time-dependent rate of hole elongation was found to increase by as much as a factor of 2 in some cases over the same temperature range. Decreasing the bolt clampup torque from 5.65 N·m to a finger-tight level resulted in an increase in the creep rate of hole elongation by a factor of 2 to 4. If a high bolt clampup torque is maintained in the joint, time-dependent deformations can likely be neglected at moderate bearing stress levels. However, if the joint experiences a reduction in the lateral constraint around the fastener hole through a loss of bolt clampup torque, significant creep deformations can occur near the hole, emphasizing the importance of maintaining high clampup conditions in bolted laminate joints.

A simple empirical expression was proposed to model the joint creep rate as a function of time and applied bearing stress. In most cases, the logarithmic creep rate of the joint was observed to be essentially constant in time, while a power-law relationship was used to model the dependence of the creep rate on the normalized bearing stress. The constants in this expression were found to be a function of temperature and bolt clampup torque. The expression proposed here models the experimental data reasonably well, and is useful for indicating trends in the creep response of bolted laminate joints.

Acknowledgments

The authors would like to express their appreciation to NASA Langley Research Center for donating the material used in this investigation and to the Iowa Space Grant Consortium for providing partial support for this project. Special thanks are also extended to Prof. W. Steven Johnson of the Georgia Institute of Technology and Dr. John H. Crews, Jr., and Dr. Thomas S. Gates of NASA Langley Research Center for their input to the various aspects of this research.

References

[1] Gates, T. S., "Effects of Elevated Temperature on the Viscoplastic Modeling of Graphite/Polymeric Composites," *High Temperature and Environmental Effects on Polymeric Composites, ASTM STP 1174*, C. E. Harris and T. S. Gates, Eds., American Society for Testing and Materials, Philadelphia, 1993, pp. 201–221.

[2] Gates, T. S., "Matrix-Dominated Stress/Strain Behavior in Polymeric Composites: Effects of Hold Time, Nonlinearity, and Rate Dependency," *Composite Materials: Testing and Design (Eleventh Volume), ASTM STP 1206*, E. T. Camponeschi, Jr., Ed., American Society for Testing and Materials, Philadelphia, 1993, pp. 177–189.

[3] Gates, T. S., "Experimental Characterization of Nonlinear, Rate-Dependent Behavior in Advanced Polymer Matrix Composites," *Experimental Mechanics*, Vol. 32, 1992, pp. 68–73.

[4] Mohan, R. and Adams, D. F., "Nonlinear Creep-Recovery Response of a Polymer Matrix and its Composites," *Experimental Mechanics*, Vol. 25, 1985, pp. 262–271.

[5] Yancey, R. N. and Pindera, M. J., "Micromechanical Analysis of the Creep Response of Unidirectional Composites," *Journal of Engineering Materials and Technology*, Vol. 112, 1990, pp. 157–163.

[6] Tuttle, M. E. and Brinson, H. F., "Prediction of the Long-Term Creep Compliance of General Composite Laminates," *Experimental Mechanics*, Vol. 26, 1986, pp. 89–102.

[7] Agarwal, B. D. and Broutman, L. J., *Analysis and Performance of Fiber Composites*, 2nd ed., Wiley, New York, 1990.

[8] Crews, J. H., Jr., and Naik, R. V. A., "Failure Analysis of a Graphite/Epoxy Laminate Subjected to Bolt-Bearing Loads," *Composite Materials: Fatigue and Fracture, ASTM STP 907*, H. T. Hahn, Ed., American Society for Testing and Materials, Philadelphia, 1986, pp. 115–133.

[9] Crews, J. H., Jr., "Bolt-Bearing Fatigue of a Graphite/Epoxy Laminate," *Joining of Composite Materials, ASTM STP 749*, K. T. Kedward, Ed., American Society for Testing and Materials, Philadelphia, 1981, pp. 131–144.

[*10*]   Eriksson, I., "On the Bearing Strength of Bolted Graphite/Epoxy Laminates," *Journal of Composite Materials*, Vol. 24, 1990, pp. 1246–1269.

[*11*]   Agarwal, B. L., "Static Strength Prediction of Bolted Joint in Composite Material," *AIAA Journal*, Vol. 18, 1980, pp. 1371–1375.

[*12*]   Chang, F. K., Scott, R. A., and Springer, G. S., "Failure Strength of Nonlinearly Elastic Composite Laminates Containing a Pin Loaded Hole," *Journal of Composite Materials*, Vol. 18, 1984, pp. 464–477.

[*13*]   Soni, S. R., "Failure Analysis of Composite Laminates with a Fastener Hole," *Joining of Composite Materials, ASTM STP 749*, K. T. Kedward, Ed., American Society for Testing and Materials, Philadelphia, 1981, pp. 145–164.

[*14*]   Naik, R. A. and Crews, J. H., Jr., "Stress Analysis Method for a Clearance-Fit Bolt Under Bearing Loads," *AIAA Journal*, Vol. 24, No. 8, 1986, pp. 1348–1353.

[*15*]   Ramkumar, R. L. and Tossavainen, E. W., "Strength and Lifetime of Bolted Laminates," *Fatigue in Mechanically Fastened Composite and Metallic Joints, ASTM STP 927*, J. M. Potter, Ed., American Society for Testing and Materials, Philadelphia, 1986, pp. 251–273.

[*16*]   Wilkins, D. J., "Environmental Sensitivity Tests of Graphite-Epoxy Bolt Bearing Properties," *Composite Materials: Testing and Design (Fourth Conference), ASTM STP 617*, American Society for Testing and Materials, Philadelphia, 1977, pp. 497–513.

[*17*]   Bailie, J. A., Duggan, M. F., Bradshaw, N. C., and McKenzie, T. G., "Design Data for Graphite Cloth Epoxy Bolted Joints at Temperatures up to 450 K," *Joining of Composite Materials, ASTM STP 749*, K. T. Kedward, Ed., American Society for Testing and Materials, Philadelphia, 1981, pp. 165–180.

[*18*]   Bailie, J. A., Fisher, L. M., Howard, S. A., and Perry, K. G., "Some Environmental and Geometric Effects on the Static Strength of Graphite Cloth Epoxy Bolted Joints," *Composite Structures*, I. H. Marshall, Ed., Applied Science Publishers, Englewood, NJ, 1981, pp. 63–78.

[*19*]   Shivakumar, K. N. and Crews, J. H., Jr., "Bolt Clampup Relaxation in a Graphite/Epoxy Laminate," *Long-Term Behavior of Composites, ASTM STP 813*, T. K. O'Brien, Ed., American Society for Testing and Materials, Philadelphia, 1983, pp. 5–22.

[*20*]   Kraus, H., *Creep Analysis*, Wiley, New York, 1980.

Testing and Failure Mechanisms

Nidal Alif[1] and Leif A. Carlsson[1]

Failure Mechanisms of Woven Carbon and Glass Composites

REFERENCE: Alif, N. and Carlsson, L. A., **"Failure Mechanisms of Woven Carbon and Glass Composites,"** *Composite Materials: Fatigue and Fracture (Sixth Volume), ASTM STP 1285*, E. A. Armanios, Ed., American Society for Testing and Materials, 1997, pp. 471–493.

ABSTRACT: Stress-strain responses in tension, compression, and shear of a five-harness satin-weave carbon/epoxy composite and a four-harness satin-weave glass/epoxy composite have been examined. Damage progression under tension was examined by optical microscopic inspection of the polished edges of the specimens. Models for elastic property and failure predictions of woven-fabric composites were examined and correlated with the experimental data. Damage inspection of the carbon/epoxy composite under tension revealed that the initial failure was cracking of pure matrix regions followed by transverse bundle cracking. Fill/weft debonding and longitudinal splits of the fill bundles occurred close to ultimate failure of the composite. The glass/epoxy composite displayed damage in the form of fill/weft debonding and longitudinal splits, but no transverse yarn cracking. The damage observed in both composites was confined to the region where ultimate failure occurred. Elastic properties of the composites were overall in good agreement with micromechanical predictions based on uniform strain, but failure stress predictions were less accurate.

KEYWORDS: woven-fabric composites, tension testing, compression testing, shear testing, damage progression, fatigue (materials), fracture (materials), composite materials

Woven-fabric composites (WFCs) have attracted considerable attention as alternatives to conventional laminates consisting of stacked unidirectional plies because WFCs are easier to handle and thus may reduce fabrication costs. Furthermore, the interlacing of the yarns from neighboring plies improves out-of-plane properties. Analysis of such composites, however, is still evolving. The basis for analysis of WFCs is the assumption of periodicity of the microstructure that enables identification of a unit cell (UC). Figure 1 shows an example of the UC for a five-harness satin-weave composite. As shown in Fig. 1, the microstructure of WFCs is expressed by several geometry parameters such as weave pattern, undulation length, yarn width, gap between adjacent yarns, number of yarns per unit length, etc. [1–3]. Figure 2 shows top views of UCs for plain weave, twill weave, and four-harness satin weave. The weave pattern is described by the minimum loom harness number, n_g [4].

Due to the complex microstructure of WFCs and the large variety of failure mechanisms, modeling of damage and failure requires further attention. Experimental studies of WFCs have revealed several microscopic failure mechanisms [2,4,5]. Initial failure of an orthogonal WFC subject to uniaxial tension is commonly cracking of a transversely oriented yarn [2]. Angle-interlock woven composites loaded in compression were examined by Cox et al. [4]. Compression failure occurred by formation of kink bands in the longitudinal yarns. Karayaka and Kurath [5] recently presented a microscopic investigation of failure mechanisms of WFCs

---

[1]Department of Mechanical Engineering, Florida Atlantic University, Boca Raton, FL 33431.

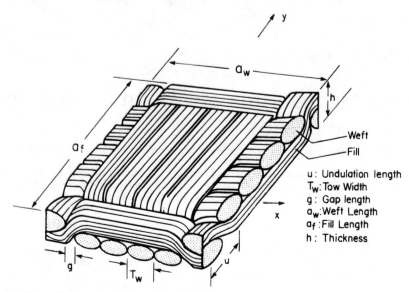

FIG. 1—*Unit cell (shown without pure matrix) for a five-harness satin-woven fabric composite.*

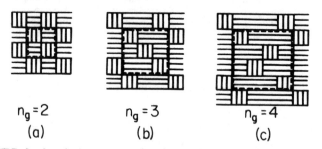

FIG. 2—*Interlacing patterns for plain and satin-weave composites.*

under tension, compression, and bending loads. They identified several localized failure mechanisms such as longitudinal yarn failure, fiber pull-out, kink bands, etc.

Modeling of fill yarn failure and post-failure response (knee behavior) has been discussed by Ishikawa and Chou [2]. Naik [6] developed an elastic and strength analysis applicable to both woven and braided composites. His analysis is based on a discretized curved-beam theory model to represent the undulations present in the fabric, and three-dimensional stress-averaging procedures. In addition to predicting elastic properties and thermal expansion coefficients, this model can predict stresses at failure initiation in pure matrix regions and transverse and longitudinal yarns based on maximum shear, maximum stress, and maximum strain theories, respectively. By this procedure, the sequence of failures can be predicted. The analysis is implemented in a computer code called the Textile Composite Analysis for Design (TEXCAD).

More progress has been made to model the elastic response of WFCs. Micromechanics analyses of the UC for such composites have led to several models, for example, the mosaic model of Ishikawa [1,7], the one-dimensional crimp model of Ishikawa and Chou [2], and the fiber inclination model of Naik and Shembekar [3]. The one-dimensional crimp model [2] considers the undulations approximately by incorporating geometrical shape functions of the yarns in a cross section of the UC. This model yields improved elastic property predictions compared to the mosaic model. The two-dimensional crimp model [3] developed for plain-weave composites incorporates the undulation in two dimensions. The UC is divided into sections cut parallel and perpendicular to the plane of the weave, and the elastic properties are calculated based on combinations of constant stress and strain models. Predictions of mechanical properties have been compared with experiments by Ishikawa et al. [8] and Naik and Shembekar [3].

This paper presents an experimental study on tension, compression, and shear mechanical properties and on damage development in tension of two WFCs with the same matrix, but different fibers (carbon and glass) and different yarn sizes and weave patterns (five- and four-harness satin weave). The experimentally measured elastic properties are compared with predicted values obtained using various mechanical models. Damage initiation and ultimate stresses measured for carbon/epoxy were compared with predictions obtained using the micro-mechanics strength analysis code (TEXCAD) developed by Naik [6].

Materials and Experimental Procedures

Two composites, namely, a Hercules five-harness satin-weave carbon/epoxy (A*370-5H) and a four-harness satin-weave glass/epoxy (120GL) were considered. Both composites employ Hercules's 3501-6 epoxy and have 0°/90° weave configurations. The fiber and resin properties are summarized in Table 1 [9,10]. The filament counts are 3K and 12K for the carbon and glass yarns, respectively.

Carbon/epoxy and glass/epoxy panels with the weft and fill yarns aligned with the x and y directions, Fig. 1, were laid up. The fill and weft yarns are identical, but this nomenclature is maintained throughout this paper to distinguish transversely and longitudinally oriented yarns. After layup, the panels were processed in an autoclave at a maximum cure temperature of 177°C at the Center for Composite Materials at the University of Delaware. The number of plies were 6 and 24 for the carbon/epoxy and glass/epoxy composites, respectively. The as-processed panel thicknesses were 2.08 and 1.95 mm, corresponding to ply thicknesses of 0.347 and 0.081 mm for carbon/epoxy and glass/epoxy, respectively.

The fiber volume fraction was determined by digestion of the matrix in hot concentrated nitric acid according to ASTM Test Method for Fiber Content of Resin-Matrix Composites by Matrix Digestion (D 3171-76). This method gives a measure of the overall (averaged) fiber volume fraction, V_f. Because woven-fabric composites contain regions of pure matrix, the fiber volume fraction of the reinforcing yarns, V_f^y, will exceed V_f. V_f^y can not be determined from the matrix digestion method, but it was estimated using the photo-micrographical method

TABLE 1—*Properties of fibers and matrix.*

Material	E_L, GPa	E_T, GPa	G_{LT}, GPa	G_{TT}, GPa	ν_{LT}	σ_t, MPa	ϵ_t, %	ρ, g/cm^3	α_L, 10^{-6}/°C	α_T, 10^{-6}/°C
AS4 Carbon	234	42	25	15.1	0.27	3930	1.60	1.79	−1.1	3.9
E-glass	72	72	27.7	27.7	0.3	2800	2.40	2.54	5.6	5.6
3501-6 Epoxy	4.2	4.2	1.56	1.56	0.35	69.0	1.70	1.265	30.0	30.0

NOTE—E_L and E_T are axial and transverse elastic moduli, G_{LT} and G_{TT} are axial and transverse shear moduli, ν_{LT} is major Poisson's ratio, σ_t and ϵ_t are ultimate stress and strain in tension, and ρ is density. α_L and α_T are the axial and transverse coefficients of thermal expansion.

[*11*]. With the densities of the constituents listed in Table 1, the matrix digestion method yields V_f = 0.69 and 0.56 for the carbon/epoxy and glass/epoxy composites, respectively. The corresponding yarn volume fractions were 0.71 and 0.59. The fiber volume fractions for the carbon/epoxy composite are relatively large (0.69 and 0.71), which may lead to brittle response of the composite.

The photomicrographic method was used also to determine weave geometry parameters such as the tow width, the gap between two adjacent yarns, the UC dimensions, and the undulation as schematically illustrated in Fig. 1. Figure 3 shows photographs of single-fabric plies isolated by acid digestion of the matrices. The geometrical parameters determined from micrographs and photographs such as Figs. 3 and 4 are listed in Table 2. The carbon/epoxy composite has much larger tow width and UC dimensions than the glass/epoxy composite. The undulation region for carbon/epoxy occupies 20% of the UC length, and 55% for glass/epoxy. Consequently, the glass/epoxy weave contains much more curved yarn elements than the carbon/epoxy weave.

Preparation of Test Specimens and Test Procedure

Tension, compression, and shear test specimens were machined from the panels using a diamond coated saw. The tension test specimens were straight-sided tabbed coupons, according to ASTM Test Method for Tensile Properties of Fiber-Resin Composites (D 3039-76), with a gage length of 153 mm. The tension tests employed wedge-action friction grips. Compression tests were performed using the IITRI (Illinois Institute of Technology Research) fixture, according to ASTM Test Method for Compressive Properties of Unidirectional or Crossply Fiber-Resin Composites (D 3410-87), and short-gage-length test specimens. Both faces of the compression specimens were strain gaged to verify that bending and buckling of the specimen did not occur. Shear testing employed the Iosepescu fixture modified by the composites group at the University of Wyoming [*12*] according to ASTM Practice for In-plane Shear Stress-Strain Response of Unidirectional Reinforced Plastics (D 3518-76). In addition to the standard tension test specimens, shorter end-tabbed test specimens (gage length = 76 mm) were prepared for examining damage progression during tensile loading. Both longitudinal edges of the specimen were carefully sanded and polished for subsequent microscopic observation. Figure 5 details the test specimen geometries, dimensions, and strain-gage configurations used. A minimum of three replicate specimens for each mechanical property test was used. Mechanical testing was performed using a 133-kN capacity Tinius-Olsen DS-50 test frame and a Megadac load and strain gage data acquisition system. Least-squares regression analysis was applied to determine the Young's and shear moduli. This method was applied to the data in the initial linear region of the stress versus strain curve.

Damage progression in the tensile specimens was monitored by examining edge-polished, strain-gaged tension specimens of 76 mm gage length (two replicates) in Unitron and Nikon metallographic microscopes. The tension specimens were subjected to incrementally increased loads. The examination procedure was as follows: the specimen was loaded to a predetermined load level, then unloaded, removed from the grips, and the edges were microscopically examined. All specimens were inspected at the edges only. It is assumed that damage initiates at the free edges because edge stresses are expected to be larger than in the interior. The loading was applied to examine the specimens over the entire stress-strain region. Figure 6 shows the loading/unloading history for the two composites. For each loading-unloading cycle, the longitudinal strain was monitored to detect any residual strains or anomalous stress-strain behavior of the specimen. As shown in Fig. 6*a*, the carbon/epoxy specimens were loaded to the following stress levels: 120, 200, 280, 330, 370, 420, 480, and 540 MPa. The glass/epoxy specimens were loaded to 105, 210, 280, and 320 MPa, see Fig. 6*b*.

FIG. 3—*Photographs of single fabric plies isolated by digestion of the matrix:* (a) *carbon/epoxy and* (b) *glass/epoxy.*

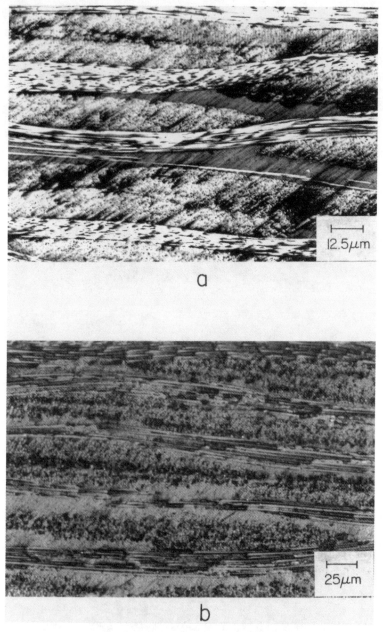

FIG. 4—*Micrographs of polished edge sections of the composites examined. (a) carbon/epoxy and (b) glass/epoxy.*

TABLE 2—*Weave geometry parameters for the composites.*

Material	Repeating Unit, n_g	Tow Width, mm	Gap, mm	Unit Cell Dimensions, mm	Undulation Length, mm	Yarn Spacing, mm
Carbon/epoxy	5	2.27	0.20	11.5 × 11.2	2.38	2.47
Glass/epoxy	4	1.35	0.15	2.80 × 2.90	1.52	1.50

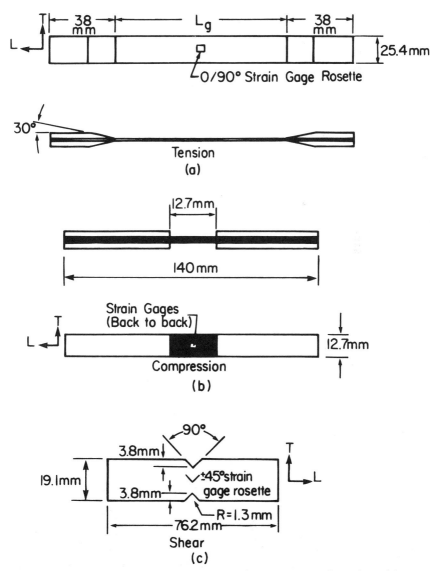

FIG. 5—*Test specimen geometries, dimensions, and strain gage configurations: (a) tension test specimen, (b) IITRI compression test specimen, and (c) Iosepescu shear specimen.*

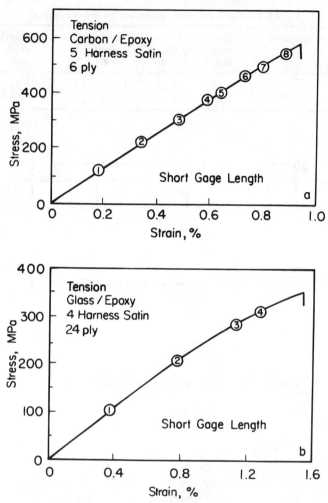

FIG. 6—*Loading/unloading program for damage progression studies in tension specimens: (a) carbon/epoxy and (b) glass/epoxy.*

Results and Discussion

Representative stress-strain curves for the composites are shown in Figs. 7 and 8. Strain values recorded on both faces of the compression specimens were similar, which indicates that eccentric loading or buckling were no issues. Stress-strain curves displayed are, therefore, averages calculated from strain readings on both sides. Carbon/epoxy specimens loaded in tension and compression exhibit an almost linear stress-strain relationship until failure, Figs. 7a and b, while the shear response is highly nonlinear, Fig. 7c. The glass/epoxy specimens displayed nonlinear responses in tension, compression, and shear, Fig. 8. The nonlinear response in tension is probably a result of the larger strains encountered for the glass/epoxy system. Compression and shear nonlinear responses may be due to the large curvature of the glass yarns, Table 2.

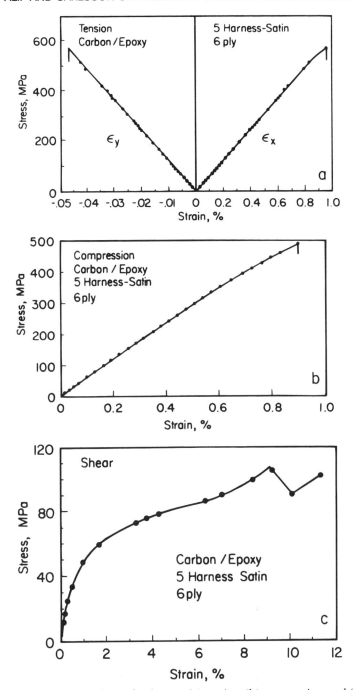

FIG. 7—*Stress-strain curves for carbon/epoxy:* (a) *tension,* (b) *compression, and* (c) *shear.*

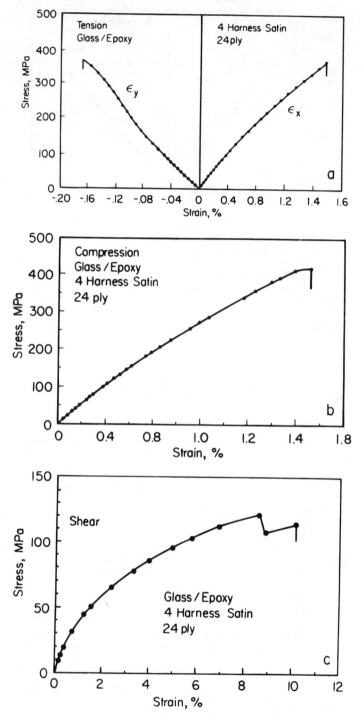

FIG. 8—*Stress-strain curves for glass/epoxy:* (a) *tension,* (b) *compression, and* (c) *shear.*

Figure 9 shows representative photographs of broken tension and compression specimens. The tension specimens failed in a narrow region by weft yarn failure. The compression specimens failed at inclined planes oriented approximately 45° to the loading direction that indicates a resolved shear-type of failure of the yarns and matrix.

Stress-strain curves in shear for both composites, Figs. 7c and 8c, show drops in the stress close to failure. Such drops have been observed in testing of unidirectional (0°) specimens [12] and are attributed to longitudinal splits at the notch root. Splitting was observed also in the WFCs examined here. Figure 10 shows photographs of failed carbon/epoxy and glass/epoxy specimens indicating splits and a vertical zone of shear yielding.

Table 3 summarizes the results from the mechanical property tests. The modulus, E_x, listed in Table 3 is the average of tension and compression moduli that were very close. Poisson's ratio for glass/epoxy is much larger than that of carbon/epoxy, quantifying that glass/epoxy contracts much more in the transverse direction than the carbon/epoxy composite. It is noted that the carbon fiber composite is stiffer and stronger in tension and compression than the glass-fiber composite. The strains to failure in tension and compression for the glass/epoxy composite are much larger than the corresponding values for the carbon/epoxy composite. The compression strength of the carbon/epoxy composite is about 15% less than the tension strength. For glass/epoxy, the tension and compression strengths are almost the same. The shear moduli of the two composites are similar. The shear strength was evaluated from the stress at the first load drop. Table 3 shows that the glass/epoxy composite is stronger in shear than the carbon/epoxy composite. Overall, the properties show very little scatter except ultimate tension strengths, indicating good quality laminates.

Damage Development Under Tension

Figure 11 schematically illustrates the various types of edge damage observed in the tension-loaded specimens, namely, matrix cracks, transverse cracks in the fill yarns, fill/weft debonding, and longitudinal splits in the fill yarns.

Damage progression observed in the carbon/epoxy composite is shown in Figs. 12 through 14. For the first two loading increments, 200 and 280 MPa, no damage was observed. At 330 MPa, minute cracks were observed in a pure matrix region between the fill yarns, close to a planar surface of the specimen. The first indication of yarn damage was observed at 370 MPa, Fig. 12a, when a matrix crack extended to become a transverse crack in the fill yarns. This crack was arrested at the fill/weft interface. After unloading the specimen from 370 MPa, a very small amount of residual strain (35 $\mu\epsilon$) was observed. No other cracks were observed along the edges of this particular specimen or in the replicate specimen loaded to the same stress level. At 420 MPa (not shown), several matrix cracks and fill yarn cracks were observed in the same region as before. Fill yarn cracks were arrested by the weft yarns. Fill/weft debonding and cracking between a pure matrix region and a fill yarn appeared at a stress of about 480 MPa, Fig. 12b. After unloading the specimen from 480 MPa, a residual strain of 85 $\mu\epsilon$ was recorded. Longitudinal splits (see Fig. 11) also occurred at 480 MPa in the fill yarns between the weft yarns as shown in Fig. 13. At 540 MPa, longitudinal splits and fill/weft debonds occurred in the region where initial damage was detected, Fig. 14. Notice that 540 MPa represents 92% of the failure stress, Table 3.

The longitudinal (horizontal) cracks are expected to initiate in a tension field, that is, by stress, σ_z. Such stress may develop in the narrow region between two curved weft yarns when they tend to straighten out under applied uniaxial tension as schematically illustrated in Fig. 15. Although the micrographs of horizontal cracks shown in Figs. 12b through 14 seem to support such a mechanism, further experimental and analytical work is needed to verify this hypothesis.

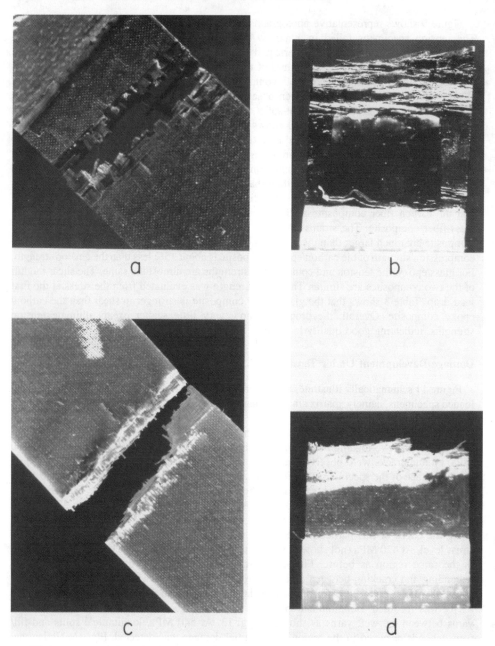

FIG. 9—*Photographs of broken tension and compression test specimens: (a) carbon/epoxy (tension), (b) carbon/epoxy (compression), (c) glass/epoxy (tension), and (d) glass/epoxy (compression).*

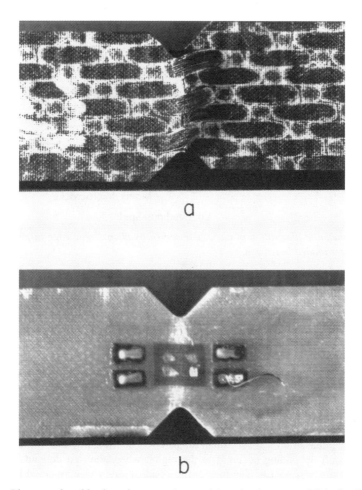

FIG. 10—*Photographs of broken shear specimens: (a) carbon/epoxy and (b) glass/epoxy.*

TABLE 3—*Mechanical properties for carbon/epoxy and glass/epoxy woven fabric composites.*

Property	Carbon/Epoxy	Glass/Epoxy
E_x, GPa	59.5 ± 0.5	26.6 ± 0.9
ν_{xy}	0.047 ± 0.001	0.144 ± 0.003
σ_t, MPa	584 ± 25	422 ± 13
ϵ_c, %	0.95 ± 0.05	1.85 ± 0.06
σ_c, MPa	491 ± 8	410 ± 5
ϵ_c, %	0.96 ± 0.06	2.0 ± 0.04
G_{xy}, GPa	4.96 ± 0.44	4.63 ± 0.09
τ_{ult}, MPa	99.0 ± 6	121 ± 0.2
γ_{ult}, %	9.05 ± 0.13	8.48 ± 0.06

NOTE—E_x represents Young's modulus, ν_{xy}, Poisson's ratio, σ_t and σ_c are tension and compression strengths, ϵ_t and ϵ_c are ultimate strains in tension and compression, G_{xy} is shear modulus, τ_{ult} is shear strength, and γ_{ult} is the ultimate shear strain. The $\pm$ range represents the standard deviation.

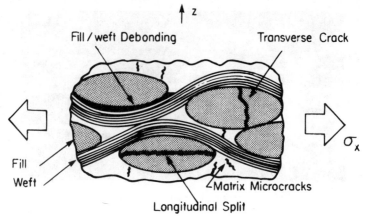

FIG. 11—*Schematic illustration of damage observed by microscopic edge inspection of woven fabric composites loaded in tension.*

No failures of the fibers in the weft yarns were observed prior to ultimate failure. Close to ultimate failure, damage accumulated in the damaged region of the gage section and the specimen broke by weft yarn failure, Fig. 9a.

Damage in the glass/epoxy composite was difficult to observe because glass and epoxy have similar colors. Figures 16 through 18 show micrographs of damage development observed. The first indication of failure, at 105 MPa, was pure matrix microcracks that occurred close to a planar surface of the specimen, Fig. 16. No residual strain was observed. At 210 MPa, longitudinal splits and fill/weft debonds were observed in the region of matrix microcracks, Fig. 17a. The residual strain was very small (65 $\mu\epsilon$). At 280 MPa, Fig. 17b, further longitudinal splits and fill/weft debonding occurred in the region where the first debond was observed. The residual strain was 95 $\mu\epsilon$. The increase in residual strain was expected because of the nonlinearity of the stress-strain curve past 210 MPa, Fig. 6b. At 320 MPa, Fig. 18 displays substantial longitudinal splits and longitudinally extended fill/weft debonds. Damage is also indicated by the nonlinear stress-strain behavior, Fig. 6b. The residual strain (after 320 MPa) was 145 $\mu\epsilon$.

Transverse cracks, Fig. 11, were not observed in the fill yarns prior to ultimate failure of the glass/epoxy composite. Formation of such cracks requires tension in a sufficient volume of the transverse yarns in planes parallel to the applied stress. It is recognized that the UC of the glass/epoxy weave is small, Table 2. There is thus not a large volume available for crack formation in the fill yarns of the glass/epoxy weave. Furthermore, the undulation length of the glass/epoxy composite occupies a much larger fraction of the UC length (55%) than for the carbon/epoxy composite (20%), Table 2. This feature is expected to lead to a local state of stress in the fill yarns different from uniaxial tension and, as such, does not favor transverse cracking. The low fiber volume fraction of the glass/epoxy composite may also contribute to increase its resistance to cracking.

The damage was confined to a small region as opposed to the distributed damage in the form of matrix cracking observed in conventional laminates [13]. When the stress approached its ultimate value, failure occurred in this narrow region of substantial longitudinal splits and fill/weft debonds.

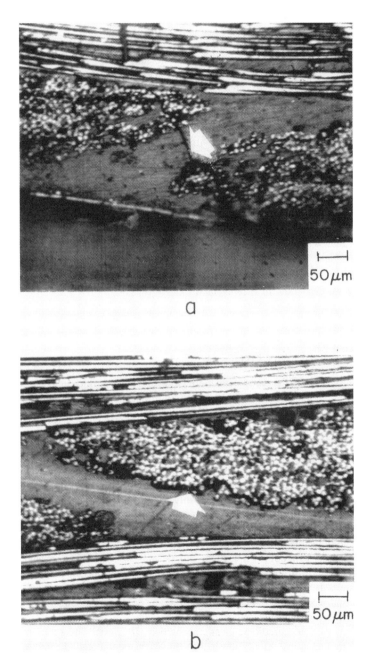

FIG. 12—*Photomicrographs of carbon/epoxy loaded to (a) σ_x = 370 MPa showing crack extended through matrix region and fill yarn, and (b) σ_x = 480 MPa showing a crack between a pure matrix region and a fill yarn.*

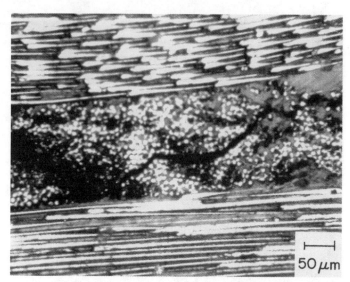

FIG. 13—*Photomicrograph of carbon/epoxy loaded to* σ_x = *480 MPa showing a longitudinal split in the fill yarn.*

Elastic Property and Failure Predictions

The elastic properties of the composites were predicted using the mosaic [1], one-dimensional crimp model [2], and a modification [14] of the two-dimensional laminate model for the plain-weave composites of Naik and Shembekar [3] and Naik's TEXCAD code [6]. The Naik and Shembekar analysis [3] was extended to predict the elastic constants for satin-weave composites [14]. The calculations of the elastic properties of any satin-weave composite were performed based on an approximate representation of the UC, Ref 14. The UC is in this treatment approximated by a combination of plain-weave cells and cross-ply laminates. Shape functions specified in Ref 3 were used to model and crimped regions of the UC, while the straight region of the UC was considered as a cross-ply laminate. The integrations to establish the average in-plane compliance and stiffness constants [3] of the UC, hence, were modified to include weaves with an arbitrary harness number, n_g (Fig. 2). Elements of the UC were assembled in parallel/series and series/parallel configurations as suggested by Naik and Shembekar in their plain-weave model [3]. The mosaic model was assembled in a parallel configuration corresponding to uniform strain. The one-dimensional crimp model was assembled both in parallel and series (uniform stress). The TEXCAD code of Naik [6] is based on a parallel configuration of the constituents. All predictions are based on the material and geometry properties of the composites listed in Tables 1 and 2.

Table 4 summarizes the predicted elastic properties. The series and series-parallel models predict moduli that are substantially lower than the experimental results listed in Table 3. The parallel model predictions are more reasonable. The moduli predicted from the modified two-dimensional parallel-series model correlate well with the measured values. Naik's code TEXCAD [6] similarly provides predictions in good agreement with measured data for the carbon/epoxy composite.

The damage initiation during tension loading of the five-harness carbon/epoxy composite was analyzed using the computer code, TEXCAD [6]. TEXCAD, however, could not be

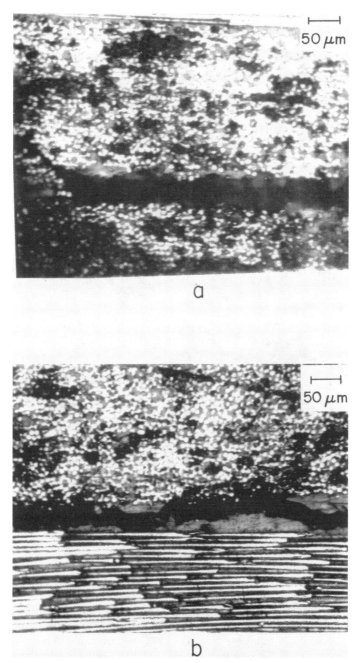

FIG. 14—*Photomicrographs of carbon/epoxy loaded to 540 MPa showing* (a) *a longitudinal split in the fill yarn and* (b) *a fill/weft debond.*

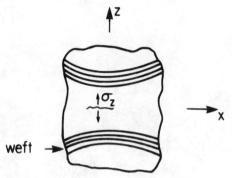

FIG. 15—*Schematic of mechanism for horizontal crack initiation by stress, σ_z.*

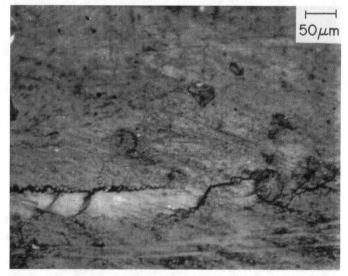

FIG. 16—*Photomicrograph of glass/epoxy loaded to $\sigma_x = 105$ MPa showing cracks in a pure matrix region.*

applied to the four-harness satin glass/epoxy composite, since this code is limited to plain, five-harness, and eight-harness weaves. In order to perform strength calculations, TEXCAD requires material properties listed in Table 1 and geometry parameters listed in Table 2 for the carbon/epoxy weave. The analysis, furthermore, requires the actual distance between the centers of two neighboring yarns (2.32 mm), the fiber count of the yarns (3×10^3), the overall fiber volume fraction, v_f (0.69), and the crimp angle (26°). Twelve segments were used to discretize the undulation region to achieve a converged solution. To predict the thermal residual stresses due to cool-down following cure, a $-150°C$ temperature difference was applied to bring the composite from the cure temperature (177°C) to ambient laboratory temperature. The thermal shrinkage strain of the composite was 0.027% in the x and y directions.

Table 5 summarizes the predicted and measured stress values for damage initiation and ultimate strength for the carbon/epoxy composite. The computer code predicts that the initial failure occurs in the transverse yarn, followed by matrix failure and ultimate failure when the

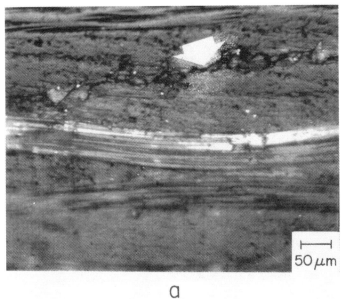

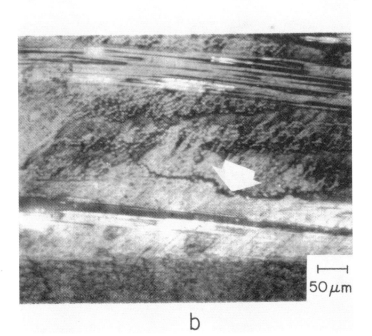

FIG. 17—*Photomicrographs of glass/epoxy showing longitudinal fill splits.* (a) $\sigma_x = 210$ *MPa and* (b) $\sigma_x = 280$ *MPa.*

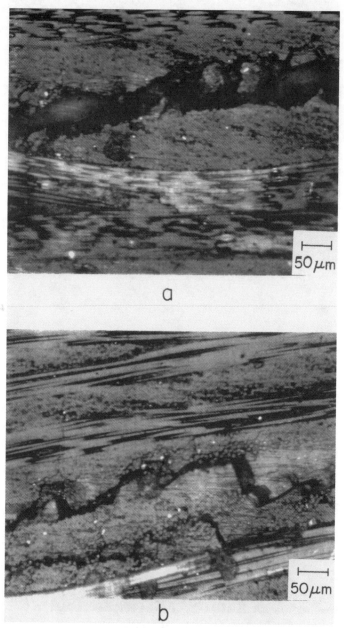

FIG. 18—*Photomicrograph of glass/epoxy loaded to* $\sigma_x = 320$ *MPa showing* (a) *a longitudinal split in fill yarn and* (b) *a longitudinal fill split and fill/weft debonding.*

TABLE 4—*Predicated elastic constants for carbon/epoxy and glass/epoxy.*

Model	Carbon/Epoxy			Glass/Epoxy		
	E_x, GPa	G_{xy}, GPa	v_{xy}	E_x, GPa	G_{xy}, GPa	v_{xy}
Mosaic [1]	70.6	3.51	0.038	33.1	4.93	0.162
1D-S [2]	41.6	3.46	0.038	19.5	3.67	0.119
Mod. 2D-PS	61.1	4.75	0.046	27.0	4.26	0.147
Mod. 2D-SP	53.5	3.2	0.039	19.7	3.79	0.133
Naik [6]	63.4	4.66	0.039	...	...	...

NOTE—P = parallel and S = series configuration. Mod. 2D model is an extension of the model of Naik and Shembekar [3] to satin weaves [14].

TABLE 5—*Stresses at initiation of damage and ultimate strength for carbon/epoxy obtained from TEXCAD analysis [6] and experiments.*

Property	Prediction	Experiment
Transverse failure stress, MPa	288	370 ± 25
Matrix failure stress, MPa	336	330 ± 25
Ultimate strength, MPa	720	584 ± 25
Strain to failure, %	1.2	0.95 ± 0.05

weft yarns fail. The actual failure sequence, however, was matrix failure followed by transverse yarn and weft yarn failures, see Table 5. The predicted stress and strain at ultimate failure are about 25% larger than the experimental values, while the pure matrix failure prediction agrees with the measured value. The stress predicted at transverse yarn failure is about 22% below the measured value. The predicted compression and shear strengths were 630 and 78 MPa, respectively, that are 28 and 22% above and below the measured values (Table 3), respectively.

The tension strength of the composites may also be compared with a simple rule of mixtures estimate of the strength based on the fact that 50% of the yarns are aligned in the loading direction and neglecting the crimp present in the yarns

$$\sigma_t = v_f \sigma_t^y / 2 \qquad (1)$$

where v_f is the overall fiber volume fraction and σ_t^y is the tension strength of the fibers (Table 1).

The ultimate strengths estimated for carbon/epoxy and glass/epoxy are 1275 and 784 MPa, respectively, that exceed the experimental results by about 85%. The substantial over-predictions of strength are expected to be related to the stress concentrations in the fibers due to the undulations present.

Conclusions

Stress-strain responses in tension, compression, and shear of five-harness satin-weave carbon/epoxy and four-harness satin-weave glass/epoxy composites have been examined. Of particular interest was the damage evolution during tensile loading of the two composites and modeling of elastic properties and failure stresses.

The two composites employed the same epoxy matrix, but different fibers and fabric geometries and unit cell sizes. The tension strengths of both composites were substantially below a rule of mixtures estimate, which tends to give reasonable strength predictions for conventional composites. Over-predictions of strength indicate that the crimps present in the longitudinal fiber tows induce stress concentrations and the associated reduction of strength. The strength in compression was practically the same as that in tension for glass/epoxy, and 15% less than that in tension for carbon/epoxy.

Damage inspection of the carbon/epoxy composite under tension revealed that the initial failure was cracking of pure matrix regions followed by transverse yarn cracking. Fill/weft debonding and longitudinal splits of the fill bundles occurred close to ultimate failure of the composite that involved failure of the fibers. In the glass/epoxy composite, pure matrix cracking, fill/weft debonding, and longitudinal splits (but no transverse yarn cracking) were observed prior to ultimate (fiber) failure. For both composites, the damage was confined to the region where ultimate failure occurred. This is in contrast to the situation in conventional laminates where distributed damage in the form of matrix cracks is observed. The localized damage in the woven-fabric composites seems to indicate that structural inhomogeneities are present and that they govern the failure process.

Elastic properties and failure progression were analyzed using some contemporary models and modifications thereof. It was shown that models based on series configurations of the constituents produce very conservative estimates of the elastic moduli. The parallel models, especially the modified two-dimensional parallel-series and Naik's beam on the elastic foundation model predicted moduli in close agreement with experimental data. Damage progression and failure analysis of the carbon/epoxy composite were performed using Naik's computer code (TEXCAD). The predictions were reasonable considering the complexity of the microstructure, but much less accurate than the elastic property predictions. The same conclusion applied to the prediction of compression and shear strengths of the carbon/epoxy composite.

Acknowledgments

The authors are grateful for the support of this work by the Department of Mechanical Engineering, Florida Atlantic University. We are also grateful to Dr. John W. Gillespie, Jr., and Mr. Touy Thiravong at the Center for Composite Materials at the University of Delaware for a supply of materials and the processing of the panels examined in this work. Mr. Paul Sip and Mr. Lawrence Silverstein of FAU helped us machine the specimens, and Mrs. Rosemarie Chiucchi and Mrs. Teresa Parez typed all versions of this manuscript. Finally, we would like to thank Dr. Rajiv Naik of Analytical Services and Materials, Inc., Hampton, Virginia, for providing us with the computer code and assistance in running it. The reviewers (unknown) provided useful comments.

References

[1]   Ishikawa, T., "Anti-Symmetric Elastic Properties of Composite Plates of Satin Weave Cloth," *Fiber Science and Technology*, Vol. 15, 1981, pp. 127–145.

[2]   Ishikawa, T. and Chou, T.-W., "One Dimensional Micromechanical Analysis of Woven Fabric Composites," *AIAA Journal*, Vol. 21, 1983, pp. 1714–1721.

[3]   Naik, N. K. and Shembekar, P. S., "Elastic Behavior of Woven Fabric, Composites: I-Lamina Analysis," *Journal of Composite Materials*, Vol. 26, 1992, pp. 2196–2225.

[4]   Cox, B. N., Dadkhah, S., Inman, R., Morris, W., and Zupon, J., "Mechanisms of Compressive Failure in 3D Composites," *Acta Metallurgica*, Vol. 40, 1992 , pp. 3285–3298.

[5]   Karayaka, M. and Kurath, P., "Deformation and Failure Behavior of Woven Fabric Composite Laminates," *Journal of Engineering Materials and Technology*, Vol. 116, 1994, pp. 222–232.

[6]   Naik, R. A., "Analysis of Woven and Braided Fabric Reinforced Composites," NASA Contractor Report 194930, June 1994.

[7]   Ishikawa, T. and Chou, T.-W., "Stiffness and Strength Properties of Woven Fabric Composites," *Proceedings*, Fourth International Conference on Composite Materials (ICCM-4), Japan Society for Composite Materials, Tokyo, 1982, pp. 489–496.

[8]   Ishikawa, T., Matsushima, M., and Hayashi, Y., "Experimental Confirmation of the Theory of Elastic Moduli of Fabric Composites," *Journal of Composite Materials*, Vol. 19, 1985, pp. 443–458.

[9]   Hercules Advanced Materials and Systems Company, Product Data sheets H050–610/Misc/sjb, 1990.

[10] *Engineered Materials Handbook, Vol. 1, Composites*, ASM International, Metals Park, OH, 1989.

[11] Carlsson, L. A. and Pipes, R. B., *Experimental Characterization of Advanced Composite Materials*, Prentice-Hall, Englewood Cliffs, NJ, 1987.

[12] Walrath, D. E. and Adams, D. F., "Verification and Application of the Iosepescu Shear Test Method," Report, UWMB-DR-401-103-1, University of Wyoming, Laramie, 1984.

[13] Masters, J. E. and Reifsnider, K. L., "An Investigation of Cumulative Damage in Quasi-Isotropic Graphite/Epoxy Laminates," *Damage in Composite Materials, ASTM STP 775*, K. L. Reifsnider, Ed., American Society for Testing and Materials, Philadelphia, 1982, pp. 40–62.

[14] Alif, N. and Carlsson, L. A., "Two-Dimensional Elastic Model for Satin Weave Fabric Composites," in preparation.

Dianne Benson,[1] *Prasanna Karpur,*[1] *David A. Stubbs,*[1] *and Theodore E. Matikas*[1]

Evaluation of Damage Evolution and Material Behavior in a Sigma/Ti-6242 Composite Using Nondestructive Methods

REFERENCE: Benson, D., Karpur, P., Stubbs, D. A., and Matikas, T. E., "**Evaluation of Damage Evolution and Materials Behavior in a Sigma/Ti-6242 Composite Using Nondestructive Methods,**" *Composite Materials: Fatigue and Fracture (Sixth Volume), ASTM STP 1285*, E. A. Armanios, Ed., American Society for Testing and Materials, 1997, pp. 494–515.

ABSTRACT: Correlations between damage, as it evolves under simulated service conditions, and the results produced from nondestructive evaluation (NDE) techniques are useful in establishing successful life prediction methodologies in metal-matrix composites. Traditional characterization techniques provide limited information on the failure mechanisms in metal-matrix composites because of the complexities caused by the inhomogeneous, anisotropic nature of these materials. In addition, the currently used destructive techniques yield only qualitative information on the internal damage of composites. Very little quantitative information exists correlating the internal damage with property changes in the material such as stiffness, elongation, and residual strength. This research effort correlated NDE results with the residual tensile strength of a six-ply, unidirectional BP Sigma-1240 SiC/Ti-6Al-2Sn-4Zr-2Mo composite after being isothermally fatigued. Baseline tension and fatigue curves were initially generated since minimal information on this particular metal-matrix composite was available in the literature. Information obtained from these tests was used to pinpoint load levels and interruption points for subsequent interrupted fatigue tests. The following nondestructive evaluation techniques were used to evaluate the test specimens before and after fatigue testing: (1) scanning acoustic microscopy, (2) oblique incidence shear wave scanning, (3) reflector plate ultrasonic scanning, (4) immersion surface wave scanning, (5) in situ surface and longitudinal waves and, (6) X-ray radiography. Following the interrupted fatigue tests, the composite specimens were nondestructively evaluated again prior to the residual tension tests to determine the residual strength. Scanning electron microscopy and metallography were used in the correlation and verification of fatigue damage. From these results, the immersion surface wave technique proved to be the most promising method for correlating damage with the residual tensile strength for this particular composite. This paper presents the results from each of the NDE techniques and examines the correlation among the techniques, other destructive methods, and the residual tensile strength.

KEYWORDS: nondestructive evaluation, metal-matrix composites, residual tensile strength, isothermal fatigue, scanning acoustic microscopy, oblique incidence shear waves, reflector plate inspection, in situ surface acoustic waves, in situ longitudinal acoustic waves, fatigue (materials), fracture (materials), composite materials

[1]Graduate student, associate professor and research engineer, associate research engineer and associate research engineer, respectively, University of Dayton Research Institute, 300 College Park Ave., Dayton, OH 45469–0126.

Continuous-fiber metal-matrix composites (MMCs) have a multitude of potential applications in situations requiring light-weight, high stiffness materials possessing high temperature capability [1]. Some of the potential applications for these materials are high-performance aerospace vehicles, advanced aircraft engines, missiles, advanced supersonic transports, and advanced fighter aircraft [2]. Since all of these applications involve cyclic loads that can lead to a decrease in load carrying capability, frequent inspection and monitoring of these materials for detection and sizing of flaws or other types of damage are necessary to ensure structural integrity [3,4].

In the past, information regarding the damage mechanisms occurring in a material was obtained by observing the macroscopic mechanical response of material specimens subjected to forces (static or cyclic), temperatures (static or cyclic), and environments (oxidizing gas, turbine engine exhaust, etc.) representative of the target application. Typical mechanical responses monitored include changes in stiffness, elongation, and residual tensile strength. In addition to the mechanical response, metallographic examination of the material as well as microscopic inspection or photography of the specimen surface were used to reveal oxidation, cracking, or other accumulated damage. These traditional methods proved useful for understanding the propagation of self-similar cracks in both aerospace and automotive structures. In addition, information gained from inspections can be used to determine how often a component needs to be inspected to detect growing cracks before they reach a critical size and cause failure of the structure as a whole [5].

Unfortunately, many of the traditional inspection techniques provide somewhat limited information when applied to metal-matrix composites because of the inhomogeneous, anisotropic nature of composites. Damage in the new advanced materials evolves in more subtle forms than a dominant crack that can be quantified primarily through measurements made on the surface of the material. In some tests, a dominant crack is observed on the surface of the composite, but distributed damage can also strongly influence the life of the composite [6–12]. A crack can be bridged either by fibers or ductile material that at elevated temperatures can be degraded by environmental attack [13,14]. In addition, fibers fail within the material, microcracks form in the matrix [15,16], and matrix/fiber debonding occurs. Since these forms of damage are not readily observable or measurable, obtaining information on these typical forms of damage from bulk averaged measurements and other commonly used techniques for established materials is extremely difficult.

Existing nondestructive evaluation (NDE) techniques need to be evaluated, and new experimental capabilities need to be developed to inspect metal-matrix composites and to provide quantitative data because quantitative data is essential for developing methodologies in life prediction studies [2,9,17]. A review of the literature revealed only a few studies that quantitatively assessed the residual strength of metal-matrix composites after expending a certain percentage of the proposed fatigue life [18]. Therefore, the main objectives of this research effort were to evaluate various NDE methods to study the evolution of isothermal fatigue damage and to correlate this information with the residual strength of the composite. Such correlations between damage, as it evolves under simulated service conditions, and the characterization results from NDE techniques are necessary to produce successful life prediction methodologies.

Background

Nondestructive evaluation methods can be used to evaluate the integrity of a material without compromising its mechanical properties. Each of the NDE techniques used in this study is described in the following paragraphs.

Nondestructive Evaluation Techniques

High-Frequency Scanning Acoustic Microscopy (SAM)—Ultrasonic scanning acoustic microscopy (SAM) is a nondestructive method used for quantifying material elastic properties, detecting surface and subsurface crack initiation and growth, and assessing fiber-matrix interfacial damage [19]. Acoustic microscopy uses an ultrasonic beam diameter that is smaller than the fiber diameter allowing for evaluation of microscopic and macroscopic variations. The scanning acoustic microscope was developed by Quate et al. [20] for the nondestructive evaluation of integrated circuits [21] and has been extensively studied by Briggs et al. [22,23]. The primary contrast mechanism in a SAM is the presence of leaky Rayleigh waves that are very sensitive to local mechanical properties of the materials under study. Since the generation and propagation of the leaky Rayleigh waves are affected by the material properties, imaging of even subtle changes of the mechanical properties is possible. The SAM can be used to understand the flaw initiation and growth mechanisms at the surface as well as the subsurface depths when the transducer is suitably defocused. In general, delaminations, fiber/matrix debonds, matrix cracking, bunched fibers, broken fibers, voids, and fiber orientation have been detected and verified using this temperature [5]. Materials studies that have used successfully this technique are described in greater detail in literature [24–27].

Oblique Incidence Shear Waves—This technique can be used to characterize the fiber-matrix bond rigidity and load transfer efficiency in composites [21,28]. This method produces shear wave propagation in the composite through mode conversion of the incident longitudinal energy at the water/composite interface. The use of this particular method has some advantages compared to other NDE techniques. First of all, resolution capabilities are enhanced since the shear wave velocity is lower than the longitudinal wave velocity for a given frequency. Second, a shear wave incident on the interface between the matrix and the fiber applies stresses that are tangential to the fiber circumference. This method has been used in monitoring the deterioration of the fiber-matrix interface due to elevated temperature tests [5] and evaluating fiber alignment and porosity levels in a composite [29].

Reflector Plate Ultrasonic Scanning—This technique is similar to conventional through-transmission ultrasonic scanning but uses a reflector plate instead of a receiving transducer. During scanning of the test specimen, ultrasonic waves pass through a test specimen to a glass "reflector plate" beneath the specimen. The waves reflect off the plate and then travel through the specimen a second time before returning to the transducer. The transducer is scanned in a raster pattern acquiring data at regularly spaced X, Y locations. The amplitude of the gated, reflected signals are plotted as a function of X and Y locations to produce a C-scan. This technique has been used to screen out defective and improperly made test samples prior to material behavior studies, thus reducing data scatter due to manufacturing defects. Since this technique is sensitive to changes in material density and elastic modulus [30], reflection plate inspection has also been used in identifying damage produced during cyclic loading [31].

Ultrasonic Surface Waves—Surface wave (or Rayleigh wave) techniques provide a useful, nondestructive evaluation of near-surface material damage. Surface waves can only penetrate the surface of a material to a depth of approximately one wavelength and are extremely sensitive to the presence of small surface or subsurface cracks. Attenuation of the surface wave is dependent upon the amount of scattering caused by cracks, material grains, other surface anomalies, as well as absorption by the material. The change in attenuation and velocity of surface waves can be used as a good indication of possible changes in the surface and subsurface areas of the material due to cracking and property gradients [32–34]. Immersion pulse-echo ultrasonic inspection using surface waves was used during this research effort to produce C-scan-type images of the specimens. In addition, in situ contact surface waves were used, as proposed by MacLellan [31], to monitor progressive damage.

X-Ray Radiography—X-ray radiography is based on the differential absorption of penetrating electromagnetic radiation. Unabsorbed X-rays passing through the part produce an image correlating to variations in thickness or density and is recorded on photographic film. In general, radiography can detect only features that have an appreciable thickness in a direction parallel to the radiation beam. The ability to detect planar discontinuities such as cracks depends on proper orientation of the part to obtain the optimum X-ray absorption differences. An advantage of radiography is the ability to detect flaws located well below the surface of the part [35]. X-ray radiography was selected for its capability to image fiber alignment and material abnormalities oriented perpendicular to the material surface as well as its potential in detecting cracks oriented parallel to the X-ray beam. X-ray radiography has been used successfully to detect fiber swimming and misalignment in MMCs [30].

Materials and Equipment

Material

The material system evaluated during this study consists of unidirectional BP Sigma SM-1240 silicon carbide (SiC) fibers in a Ti-6Al-2Sn-4Zr-2Mo matrix. The six-ply composite was manufactured by Howmet[2] and was determined to have a fiber volume percentage of 24.5 ± 0.2%. Sigma SM1240 is a C/TiB$_2$ coated SiC fiber produced by BP Metal Composites Ltd.[3] The SiC is chemical vapor deposited onto a tungsten filament substrate. The fiber has a nominal diameter of 100 μm (0.004 in.), and the duplex protective coating is approximately 2 μm thick. Due to the poor thermal shock resistance of the outer TiB$_2$ coating, which causes fiber degradation during composite manufacture, Howmet developed a protective coating for the fiber to reduce this problem. The matrix material, Ti-6Al-2Sn-4Zr-2Mo, is described as a near-a $a + b$ alloy that has good mechanical heat resistance [36].

The composite was produced by plasma melting the titanium alloy powder to deposit the matrix material around a fiber array precision wrapped on a mandrel. Monotape layups were produced subsequently by cutting and arranging the fiber-reinforced "monotapes." Multilayered fiber-reinforced composite panels were produced by hot consolidation of monotape layups using hot isostatic pressing. This method reportedly offers the advantage of improved fiber spacing control over conventional methods of titanium-matrix composite (TMC) fabrication [37].

Specimens were cut from the consolidated, unidirectional plate by abrasive water jet into dog-bone-shaped test specimens (Fig. 1). All specimens were mechanically tested with the load applied in the longitudinal, or fiber, direction.

Ultrasonic Test Equipment

The ultrasonic data acquisition and imaging system used for reflection plate inspection and immersion surface wave scanning consisted of a five-axis mechanical scanning system with 0.025-mm minimum step size (the actual resolution of the system is dependent on the ultrasonic frequency used and is generally larger than the step size), broadband ultrasonic spike pulser/

[2]Howmet Corporation, Operhall Research Center, Whitehall, MI.
[3]BP Metals Composites, Ltd., Farnborough, UK.

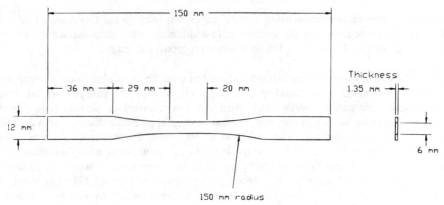

FIG. 1—*Schematic of dog-bone-shaped fatigue specimens used during this study.*

receiver, and a 100-MHz, 8-bit signal digitizer. Data acquisition and imaging were controlled by a computer with custom software. Information about the transducers used during the scans is listed in Table 1. The glass plate used during reflection plate inspection was 18 mm thick.

The ultrasonic data acquisition and imaging system used for oblique incidence shear wave scanning and acoustic microscopy also consisted of a five-axis mechanical scanning system with 0.025-mm resolution. However, the broadband pulser/receiver used had a wider bandwidth and a shorter pulse necessary for high frequency scanning. In addition, a 2-GHz 8-bit digitizer was used. As mentioned previously, data acquisition and imaging were controlled by a digital computer with custom software. Information about the high frequency transducers used during the scans is also listed in Table 1.

Equipment used to generate, receive, and digitize ultrasonic signals during in situ ultrasonic testing consisted of a broadband (35-MHz) ultrasonic spike pulser/receiver and a personal computer equipped with a 100-MHz 8-bit resolution data acquisition board for digitization of the ultrasonic signal [31]. Surface wave transducers and wedges were necessary for in situ surface wave monitoring (Fig. 2a), and in situ longitudinal wave testing required a fatigue test frame with grips machined specially for placement of the ultrasonic transducers at the ends of the specimens (Fig. 2b). Broadband contact transducers possessing a center frequency of 10 MHz were used for both in situ surface wave and longitudinal wave monitoring. Mode conversion wedges were specially manufactured by Panametrics[4] to produce surface waves in titanium matrix composites. The primary couplant used to provide good acoustic coupling

TABLE 1—*Transducer information.*

Scan Type	Transducer Frequency, MHz	Diameter, mm	Focal Length, mm	Theoretical −6dB Focal Spot Size, mm
Immersion surface wave	10	12.7	76.2	0.92
Reflection plate	25	6.35	50.8	0.49
Oblique incidence shear wave	50	6.35	12.7	0.062
Acoustic microscopy	100	6.35	5.08	0.012

[4]Panametrics, Waltham, MA.

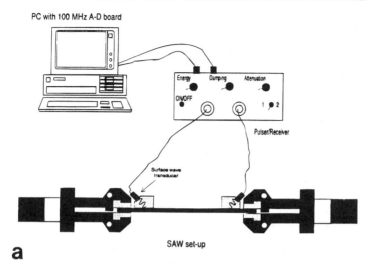

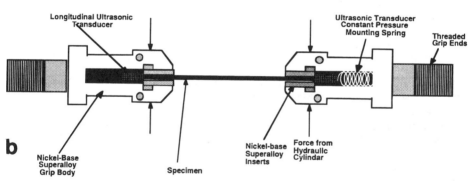

FIG. 2—(a) *Experimental setup for in situ surface wave testing.* (b) *Experimental setup for in situ longitudinal wave testing.*

during testing was Vaseline Petroleum Jelly. It was used because its viscous nature prevents significant evaporation of the couplant over time intervals of 100 h or more. An alternate couplant, THERMOSONIC, a high temperature (0 to 500°F) couplant manufactured by Echo Ultrasound,[5] was used during in situ longitudinal wave characterization. All in situ tests were conducted at room temperature.

A standard film-based X-ray system was used to take the X-ray radiographs. Typical energies were 60 to 80 keV with 5-mA current. Exposure times range from 30 to 60 s, and high-resolution film was used. The system was set up to give a 1:1 specimen-size-to-image—size exposure. Previous work showed that this system could image individual fibers in MMCs [30].

[5]Echo Ultrasound, Reedsville, PA.

Mechanical Test Equipment

Isothermal fatigue tests were conducted on a horizontal test frame incorporating a pneumatic ram for load control. The test system was positioned horizontally to improve temperature control and to allow for proper extensometry mounting. A 25-kN load cell was used, and loads were controlled to within 0.1 kN. Specimens were positioned horizontally in precisely aligned, hydraulically actuated, rigid grips [37–39]. Gripping pressure was approximately 60 MPa. A symmetric, triangular load cycle was generated by a personal computer using control software developed by the University of Dayton [40]. Axial strain was acquired throughout the tests with a 12.7-mm gage length, high-temperature, MTS extensometer containing quartz extension rods.

For the 500°C fatigue tests, the specimens were heated using radiant energy, quartz lamp, heaters. Two heating units were used, each containing four tungsten filament quartz lamps. One heater was positioned above the top surface of the specimen and the other placed below, and each lamp was paired with another to form four controllable heating zones. A uniform temperature profile ($\pm$3°C) was maintained throughout a 25-mm region centered along the length of the specimen. The quartz lamp outputs were controlled by commercial four-zone, digital, temperature controllers. Four Type K thermocouples welded to the top and bottom surfaces of the specimen were used for temperature sensing. A more detailed description is provided by Hartman et al. [37–39]. This heating system produced a temperature of 500 $\pm$ 3°C in the specimen gage section for the duration of the tests.

Procedures

Baseline Tension and Fatigue Tests

Since the literature contains minimal information on the Sigma/Ti-6242 composite system, baseline tension and fatigue curves were generated. Two tension tests were conducted at room temperature, and another two were tested at 500°C. This temperature was chosen since it represents the upper limit at which Ti-6242 is typically used [36]. The tests were run in load control at a rate of 10 MPa/s. Information obtained from these tests was used in the selection of load levels and interruption points for subsequent fatigue tests.

Baseline isothermal fatigue tests were conducted at room temperature and 500°C as depicted in Fig. 3. All tests were tension-tension fatigue, run in load-control with a triangular waveform, a stress ratio of 0.1, and a frequency of 0.01 Hz. Six baseline fatigue tests were conducted at each temperature. The maximum applied stress for each test was chosen as a percentage of the baseline ultimate tensile strength at that temperature: 60, 65, 72, 80, and 90%. The stress ratio was chosen to ensure consistency with previous work done on similar titanium matrix composites, and the frequency was selected to ensure a uniform loading profile since pneumatic-actuated fatigue systems are limited in this regard at higher frequencies.

Interrupted Isothermal Fatigue Tests

The maximum applied tensile stress for all interrupted fatigue tests was 65% of the ultimate tensile stress at the corresponding temperature. This stress level was chosen to yield a fatigue life that did not exceed 10 days due to time constraints. The temperatures, frequencies, and stress ratios were consistent with the baseline tests. Baseline curves, changes in modulus, and in situ surface wave data were all used in the selection of appropriate interruption points for each specimen. The interruption points relative to fatigue lives of baseline specimens tested at the same stress level are shown in Fig. 4. The in situ surface wave technique was used to monitor progressive damage throughout the room temperature tests [31]. Some of the room

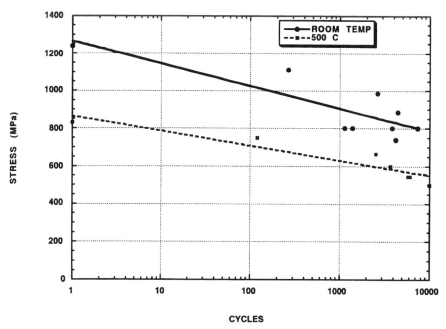

FIG. 3—*Baseline isothermal fatigue curves for Sigma/Ti-6242 composite at room temperature and 500°C (frequency = 0.01 and H_2/R = 0.1).*

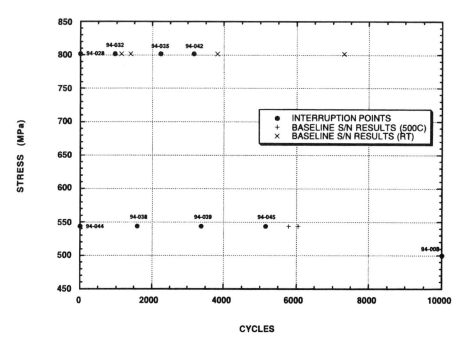

FIG. 4—*Cycles accumulated prior to interruption or failure for fatigue specimens used in this study.*

temperature specimens were interrupted during testing, ultrasonically C-scanned in immersion tanks, and then reinstalled in the fatigue fixture for additional cycling if minimal damage was evident.

One isothermal fatigue test was conducted at room temperature to monitor longitudinal waves traveling the length of the specimen. A horizontal, servohydraulic test frame with specially machined grips for placement of the contact transducers at the ends of the specimen (Fig. 2b) was used. This test was tension-tension fatigue, run in load-control with a triangular waveform at a stress ratio of 0.1 and a frequency of 1 Hz. As with the interrupted fatigue tests, the maximum applied tensile stress was 65% of the ultimate tensile strength.

Nondestructive evaluation of the interrupted specimens was performed to characterize damage such as matrix cracking, fiber bridging, or cracked fibers. The following methods were used to evaluate each specimen before and after fatigue testing: high-frequency scanning acoustic microscopy, oblique incidence shear waves, reflector plate ultrasonic scanning, immersion surface waves, and X-ray radiography.

Following the nondestructive evaluation of the test specimens, tension tests were conducted to determine residual strength. All tests were run in load control at a rate of 10 MPa/s at room temperature.

Failure Analysis

After testing, scanning electron microscopy, metallography, and other destructive methods were used to characterize fatigue damage. Qualitative and quantitative data obtained from fatigue tests, nondestructive evaluations, and residual tension tests were correlated with the observations made during destructive analyses.

Results and Discussion

Feasibility of Nondestructive Techniques for Evaluating Damage Evolution and Material Behavior

Reflector Plate Ultrasonic Scanning—Reflector plate C-scans of Specimen 94-047 at various points in its fatigue life are shown in Fig. 5. This specimen was fatigue tested at room temperature and the testing was interrupted three times during fatigue cycling (1000, 1965, and 3822 cycles at a maximum applied stress of 800 MPa) and ultrasonically scanned (in an immersion tank, off the load frame). The C-scans were calibrated such that the full-scale amplitudes (white in these C-scans) in the color-coded scales represented the level of ultrasonic transmission in a Ti-6-4 specimen of similar thickness. Slight differences in amplitude from one image to the next represent typical variances in the calibration process. Regions of attenuation of the ultrasound oriented perpendicular to the specimen axis are apparent at all stages of testing and do not appear to change significantly during testing. These regions are possibly caused by localized bunching of fibers along the width of the specimen as indicated in Fig. 6 (metallograph of the edge of the sample). Many of the specimens fatigue tested during this study failed adjacent to one of these attenuated regions. No other anomalies were evident in the reflection plate scans.

Immersion Surface Waves—Immersion surface wave scans of room temperature fatigue specimens interrupted prior to failure are shown in Fig. 7. All scans were calibrated such that the full-scale amplitudes (black in these C-scans) in the color-coded scales represent the level of reflection from the polished edge of a Sigma/Ti-6242 calibration block. Attenuation was reduced by 12 dB prior to scanning of the actual specimens to increase detection sensitivity. None of the specimens showed any evidence of surface or subsurface damage prior to testing.

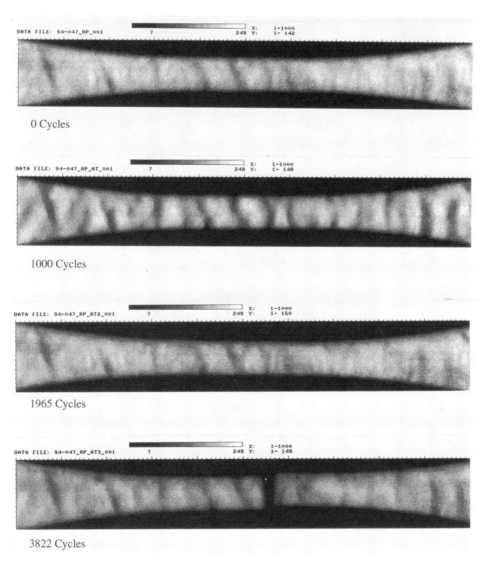

0 Cycles

1000 Cycles

1965 Cycles

3822 Cycles

FIG. 5—*Reflector plate C-scans of Specimen 94-047 at various points in its fatigue life.*

However, cracks formed (all the cracks nucleated at the edges) during room temperature fatigue cycling in Specimens 94-035 (2237 cycles) and 94-042 (3168 cycles) as evidenced by the immersion surface wave scans (black regions along the edges of the samples in Fig. 7). The 500°C interrupted fatigue samples, on the other hand, revealed no signs of significant damage after being interrupted. One exception is a 500°C baseline sample tested at a maximum applied stress of 500 MPa. This particular sample (94-008) was removed from the fatigue fixture after exceeding 10 000 cycles due to time constraints. An immersion surface wave scan of this sample revealed several surface and subsurface cracks.

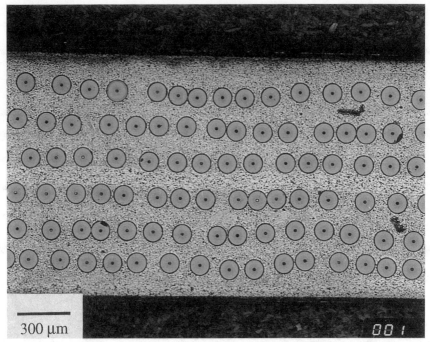

FIG. 6—*Microphotograph of transverse cross section of a Sigma/Ti-6242 composite specimen in an attenuated region detected during reflection plate inspection.*

High-Frequency Scanning Acoustic Microscopy—The 100-MHz transducer used during this study produced Rayleigh waves that penetrated the composite to a depth of approximately 0.03 mm; however, the outer layer of matrix material for this composite measured 0.160 mm. Since the imaging of fibers was desirable, longitudinal waves produced by the acoustic microscope were monitored instead. Since the full-scale amplitude (black in these C-scans) represents high levels of reflection, the fibers in the first ply of the gage section appear as dark lines in the C-scan of a 500°C fatigue specimen shown in Fig. 8. Fiber alignment in the first ply was successfully evaluated using this method. In addition, some possible fiber breaks and surface cracks were detected.

Oblique Incidence Shear Waves—Oblique incidence shear wave C-scans of two failed specimens, Specimen 94-046 (8563 cycles) and Specimen 94-047 (3822 cycles), are shown in Fig. 9. Both specimens were tested at room temperature with a maximum applied stress of 800 MPa, at frequencies of 1 Hz and 0.01 Hz, respectively. This technique was "somewhat successful" in evaluating fiber alignment and detecting some surface cracks; however, details were far less apparent than those attained with acoustic microscopy. This method failed to identify any changes in the fiber-matrix interface during fatigue cycling. The shortcomings of this method may be attributed to the small diameter of the Sigma fiber (104 mm) used in the composite. Other studies that have been successful in using this technique typically involved fibers with larger diameters such as SCS-6 (142 mm). Lower frequency transducers (25 MHz) were originally used in this study; however, since the wavelength of the resulting shear wave was larger than the Sigma fiber diameter, good resolution was difficult to attain. As the

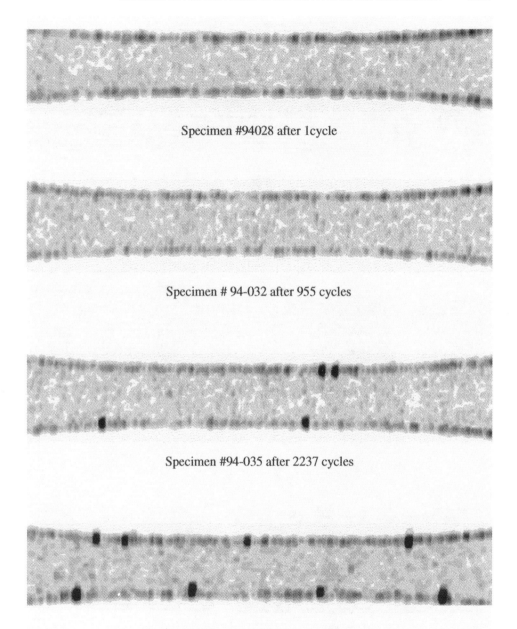

Specimen #94028 after 1cycle

Specimen # 94-032 after 955 cycles

Specimen #94-035 after 2237 cycles

Specimen #94-042 after 3168 cycles

FIG. 7—*Immersion surface wave C-scans of room temperature fatigue specimens interrupted after the designated number of cycles.*

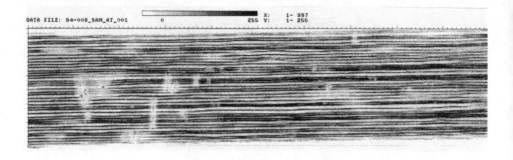

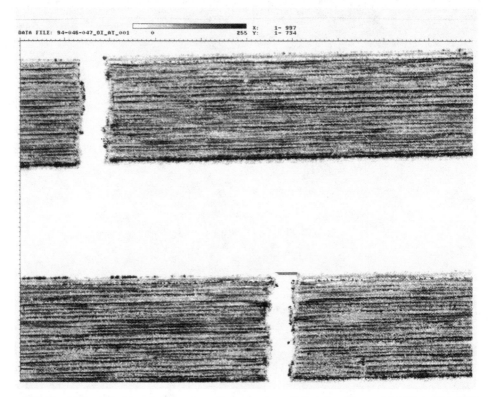

FIG. 9—*Oblique incidence shear wave C-scans of two failed Specimens, 94-046 and 94-047, tested at room temperature.*

transducer frequency was increased to 50 MHz, the wavelength decreased to 92 mm, but attenuation of the shear wave signal increased, which hindered data acquisition. In addition to these difficulties, the undulating nature of the fibers made detection and proper gating of the ultrasonic signal extremely difficult.

In Situ Surface and Longitudinal Waves—In situ surface wave results were similar to those obtained by MacLellan [*31*]. A large, initial decrease of the pitch-catch ultrasonic amplitude was typically seen during the first few cycles and may be an indication of fiber/matrix debonding. Some specimens subsequently displayed an increase in amplitude. This observation was also made by MacLellan [*32*], although the actual cause of the observation is still being investigated at this time. Following this slight increase, the transmitted amplitude gradually decreased until failure occurred. The gradual decrease in surface wave amplitude is believed to be due to reflection and scattering of the ultrasound from damage developing in the material as cycles are applied. A surface wave amplitude plot for Specimen 94-027 is shown in Fig. 10. This specimen, which was cycled at a maximum applied stress of 740 MPa at room temperature, failed after 4191 cycles.

Some difficulties encountered when using this technique may have affected the results. First of all, the transmitted surface wave amplitude was extremely sensitive to slight movements of the wedges. In addition, the potential for error exists during the manual alignment of wedges to maximize the transmitted signal. These practices may have contributed to the variability present in the surface wave amplitude plots of specimens tested under identical conditions. An alignment fixture is recommended for future testing to ensure standardization.

In situ longitudinal wave results for Specimen 94-046 tested at room temperature at a frequency of 1 Hz are shown in Fig. 11 (note the *y*-axis scales). A comparison between changes in longitudinal wave amplitude and modulus (measured using extensometer displacement measurements) yielded similar results; however, the normalized results show that the modulus decreased by about 3% prior to failure, whereas the longitudinal wave amplitude decreased by 17% prior to failure. The longitudinal wave amplitude method is clearly more sensitive to property changes or damage occurring in the material under study, or both.

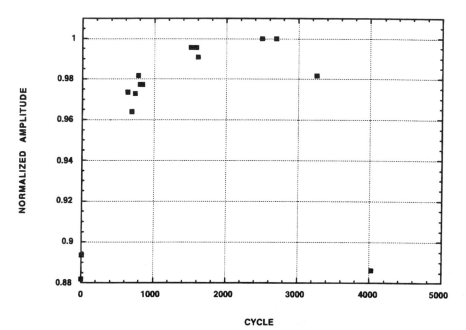

FIG. 10—*In situ surface wave amplitude plot for Specimen 94-027.*

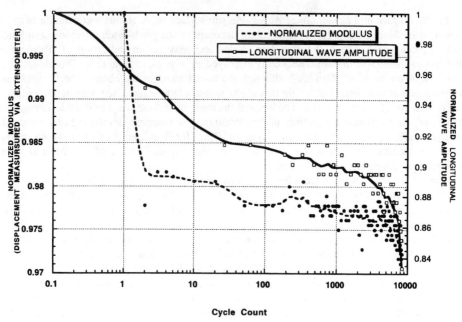

FIG. 11—*In situ longitudinal wave amplitude versus mechanically measured modulus of Specimen 94-046 (65% ultimate tensile strength/room temperature/1 Hz).*

X-Ray Radiography—Regions of low fiber density and fiber displacement were easily detected in the X-ray radiographs taken during this study. Additionally, cracks were apparent in the Sigma/Ti-6242 unidirectional composites using magnification. The cracks were detectable because the fiber breaks appeared as gaps in the tungsten core, and all specimens were unidirectional, which facilitated detection.

Summary—In general, reflector plate method was successful in identifying high fiber density regions caused by bunched fibers. The immersion surface wave technique, on the other hand, succeeded in detecting surface and subsurface cracks. Scanning acoustic microscopy, oblique incidence shear waves, and X-ray radiography were effective in evaluating fiber alignment and some favorably oriented fatigue cracks.

Correlating Observed Damage with Residual Tensile Strength

Table 2 lists the residual tensile strengths and moduli of the interrupted specimens. This information is also shown graphically in Fig. 12. The only specimens showing significant reductions in tensile strength were Specimen 94-035 (2237 cycles), Specimen 94-042 (3168 cycles), and Specimen 94-008 (10 000 cycles). These results correspond well with the findings of the immersion surface wave scans that revealed the presence of surface and sub-surface cracks in these samples. The reduction in tensile strength does not appear to be related to the number of cracks detected; rather, crack size seems to be more indicative of residual tensile strength in this particular composite. Specimens showing no evidence of surface or subsurface cracking possessed residual tensile strengths comparable with the baseline values.

TABLE 2—*Room temperature residual tensile strengths and moduli of interrupted fatigue specimens.*

Specimen	Fatigue Stress, MPa	Temperature, °C	Cycles Accumulated	Residual Tensile Strength, MPa	Mechanically Measured Modulus, GPa
94-030	⋯	23	0	1243	173
94-048	⋯	23	0	1226	180
94-028	800	23	1	1317	188
94-032	800	23	955	1213	180
94-035	800	23	2 237	929	178
94-042	800	23	3 168	960	177
94-044	540	500	1	1283	190
94-038	540	500	1 583	1263	182
94-039	540	500	3 369	1273	176
94-045	540	500	5 144	1193	186
94-008	540	500	10 000	759	183

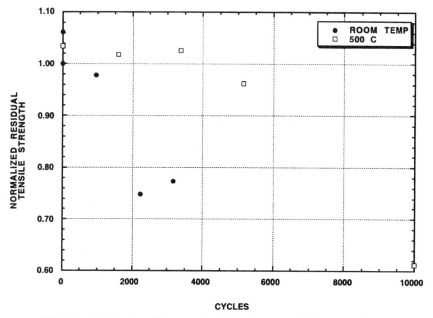

FIG. 12—*Residual tensile strengths of the interrupted fatigue specimens.*

Damage Mechanisms Involved in Producing Indications During Nondestructive Evaluation

Scanning electron microscopy (SEM) of the fracture surfaces of baseline specimens revealed fatigue cracks that initiated at the fiber/matrix interface and propagated radially outward as depicted in Fig. 13. This failure mechanism was detected in both the room temperature and 500°C specimens. Failure of the matrix surrounding some fibers apparently preceded fiber failure and subsequent overload. Fibers near the edges are more susceptible because constraints to failure are reduced once the matrix crack reaches an edge. Metallographic analyses of

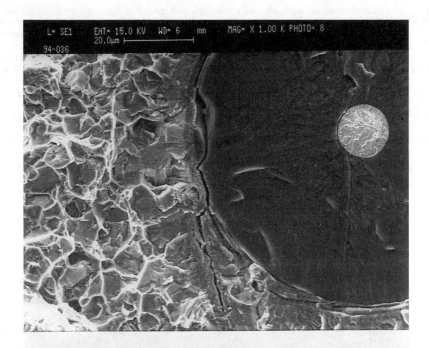

FIG. 13—*Fractographs of Specimen 94-036 showing fatigue emanating from the fiber/matrix interface.*

damaged regions were consistent with SEM findings as shown in Fig. 14. Cracks propagating away from the fiber/matrix interface are evident.

In order to verify damage detected during nondestructive evaluations, scanning electron microscopy and metallography were used to evaluate all interrupted specimens after the residual tension tests. Little, if any, fatigue damage was detected on the fracture surfaces of specimens that possessed a residual tensile strength near 100%. Minimal fatigue damage was observed near the fiber/matrix interface of Specimen 94-045 (5114 cycles, 500°C) that displayed a slight decrease in tensile strength. On the other hand, significant fatigue damage was detected on the fracture surfaces of specimens that displayed a reduction in tensile strength after fatigue: Specimen 94-035 (2237 cycles, RT), Specimen 94-042 (3168 cycles, RT), and Specimen 94-008 (10 000 cycles, 500°C, 500 MPa). Most of the visible fatigue damage was located near the outer surface of the specimens; however, fatigue damage at the fiber/matrix interface was also present. These findings correspond well with the NDE results as well as the residual tension tests.

Conclusions

The usefulness of ultrasonic nondestructive evaluation to assess fatigue damage in a $[0]_6$ Sigma-1240/Ti-6242 composite has been demonstrated through correlation of immersion and in situ ultrasonic data with residual tensile strength for the test conditions used in this study. Immersion surface wave scanning proved to be one of the most promising methods for correlating fatigue damage with the residual tensile strength for the composite used in this

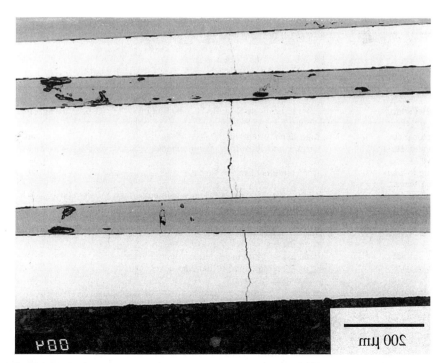

FIG. 14—*Microphotograph showing cracks present in Specimen 94-029 following fatigue failure at 500°C.*

study as summarized in Tables 3 and 4. The only interrupted specimens showing significant reductions in tensile strength were those found to contain surface or subsurface cracks during scanning. Acoustic microscopy, oblique incidence shear wave, and X-ray radiography techniques proved to be useful in evaluating fiber displacement and locating favorably oriented cracks. Although reflection plate inspection was unsuccessful in identifying damage produced during cyclic loading, slight variations in fiber density due to fiber bunching were detected prior to mechanical testing. In situ surface wave and longitudinal wave methods appeared to be more sensitive to property changes or damage or both occurring in the material than the mechanically measured modulus. Scanning electron microscopy and metallography were used

TABLE 3—*Correlation of immersion surface wave results with residual tensile strength.*

Specimen	Fatigue Stress, MPa	Temperature °C	Total of Number of Fatigue Cycles	Residual Tensile Strength, MPa	Number of Cracks Detected in Gage Section	Maximum Crack Length, mm (Side A + Side B)	Σ Crack Lengths, mm (Side A + Side B)
94-028	800	23	1	1317	0	0	0
94-032	800	23	955	1213	0	0	0
94-035	800	23	2 237	929	4	1.9	4.6
94-042	800	23	3 168	960	8	1.3	3.4
94-044	540	500	1	1283	0	0	0
94-038	540	500	1 583	1263	0	0	0
94-039	540	500	3 369	1273	0	0	0
94-045	540	500	5 144	1193	0	0	0
94-008	540	500	10 000	759	10	2.4	11.9

TABLE 4—*Summary of defects and damage revealed by nondestructive evaluation.*

NDE Technique	Indications Revealed by Technique	Relevant Specimens
Reflector plate	high density regions	all
Immersion surface wave	surface and subsurface fatigue cracks	94-035, 94-042 94-008
Oblique incidence shear wave	fiber alignment in the first ply	all
	surface fatigue cracks >1 mm in length	94-008, 94-047
Scanning acoustic microscopy	fiber alignment in the first ply (touching or bunched fibers, missing fibers)	all
	surface fatigue cracks	94-042, 94-008
	fiber breaks	94-042
In situ surface waves	decrease in pitch catch ultrasonic amplitude may be indicative of damage developing in material	94-027, 94-032, 94-035, 94-042, 94-047
In situ longitudinal waves	sensitive to property changes and/or damage occurring in the material under study	94-046
X-ray radiography	regions of low and high fiber density	all
	displaced fibers	94-030
	fatigue cracks containing cracked fibers	94-035, 94-042, 94-008
Fluorescent penetrant inspection	surface breaking cracks	94-035 94-042 94-008

to verify fatigue damage detected using these methods. Information obtained from nondestructive evaluations has been used to facilitate early detection of damage during fatigue testing of MMCs. However, it should be noted that the damage types and mechanisms in MMCs will vary for different types of cyclical loading, and, therefore, different types of NDE methods will be suitable to detect these damage types. Further research is essential to bring out the usefulness of each NDE method to detect various types of damages caused by different types of fatigue testing methods.

Acknowledgments

Funding for this project was provided through the Air Force Office of Scientific Research Grant F49620-93-1-0461DEF, program manager, Dr. Walter F. Jones; partial support from Air Force Contracts F33615-91-C-5606 and F33615-94-C-5213. All work was performed in the Materials Directorate, Wright Laboratory, at Wright-Patterson Air Force Base, Ohio.

References

[1] Gabb, T. P., Gayda, J., and MacKay, R. A., "Isothermal and Nonisothermal Fatigue Behavior of a Metal Matrix Composite," *Journal of Composite Materials*, Vol. 24, 1990, pp. 667–686.

[2] Johnson, W. S., "Fatigue of Continuous Fiber Reinforced Titanium Matrix Composites," *Proceedings*, Engineering Foundation Conference, MCE Publications, Santa Barbara, CA, 1991, pp. 357–377.

[3] Nayfeh, A. H., Crane, R. L., and Hoppe, W. C., "Reflection of Acoustic Waves from Water/Composite Interfaces," *Journal of Applied Physics*, Vol. 55, 1984, pp. 685–689.

[4] Larson, J. M., Russ, S. M., and Jones, J. W., "Possibilities and Pitfalls in Aerospace Applications of Titanium Matrix Composites," *Proceedings*, NATO AGARD Conference on Characterization of Fiber Reinforced Titanium Metal Matrix Composites, Bordeaux, France, 1993, pp. 1–21.

[5] Karpur, P., Stubbs, D. A., Matikas, T. E., Blodgett, M. P., and Krishnamurthy, S., "Ultrasonic Nondestructive Characterization Methods for the Development and Life Prediction of Titanium Matrix Composites," *Proceedings*, NATO AGARD Conference on Characterization of Fiber Reinforced Titanium Matrix Composites, Bordeaux, France, 1993, pp. 13–1 to 13–12.

[6] Jira, J. R. and Larsen, J. M., "Crack Bridging Behavior in Unidirectional SCS-6/Ti-24Al-11Nb Composite," *Fatigue '93*, J. P. Bailon, I. J. Dicksons, Eds., Engineering Material Advancement Services Ltd., Ecole Polytechnique, Montreal, Canada, Vol. 2, 1993, pp. 1085–1090.

[7] Castelli, M .G. and Bartolotta, P. E., Jr., "Thermomechanical Testing of High Temperature Composites: Thermomechanical Fatigue (TMF) Behavior of SiC(SCS-6)/Ti 15-3," *Composite Materials: Testing and Design, (Tenth Volume) ASTM STP 1120, G. C. Grimess, Ed.*, American Society for Testing and Materials, Philadelphia, 1992, pp. 70–86.

[8] Neu, R. W. and Roman, I., "Acoustic Emission Monitoring of Damage in Metal Matrix Composites Subjected to Thermomechanical Fatigue," *Composites Science and Technology*, Vol. 52, 1994, pp. 1–8.

[9] Neu, R. W., "A Mechanistic Thermomechanical Fatigue Life Prediction Model for Metal Matrix Composites," *Fatigue and Fracture of Engineering Materials and Structures*, Vol. 16, 1993, pp. 811–828.

[10] Russ, S. M., Rosenberger, A. H., and Stubbs, D. A., "Isothermal Fatigue of a SCS-6/Ti- 22Al-23Nb Composite in Air and Vacuum," *Proceedings*, ASME Summer Annual Meeting, American Society of Mechanical Engineers, New York, 1995.

[11] Johnson, W. S., "Mechanisms Controlling Fatigue Damage Development in Continuous Fiber Reinforced Metal Matrix Composites," *Advances in Fracture Research-ICF7*, K. Salama, K. Ravi-Chandar, D. M. R. Taplin, P. R. Raos, Eds., Pergamon Press, New York, 1989, pp. 897–905.

[12] Chan, K. S. and Davidson, D. L., "Fatigue Crack Growth in Fiber-Reinforced Metal-Matrix Composites," *Fatigue of Advanced Materials*, 1990.

[13] Nicholas, T. and Russ, S. M., "Elevated Temperature Fatigue Behavior of SCS-6/Ti-24Al-11Nb," *Material Science Engineering*, Vol. A153, 1992, pp. 514–519.

[14] Kortyna, B. R. and Ashbaugh, N. E., "Fatigue Characteristics of a Titanium Aluminide Composite at Elevated Temperature," *Titanium Aluminide Composites—Proceedings from Titanium Aluminide Composite Workshop*, P. R. Smith, S. J. Balsone, and T. Nicholass, Eds., Report No. WL-TR-91-4020, Wright Laboratory/Wright Patterson AFB, OH, 1991.

[15]   John, R. and Ashbaugh, N. E., "Fatigue Crack Growth in Ceramics and Ceramic Matrix Composites," *Cyclic Deformation, Fracture, and Nondestructive Evaluation of Advanced Materials, ASTM STP 1157*, M. R. Mitchell and O. Bucks, Eds., American Society for Testing and Materials, Philadelphia, 1992, pp. 28–51.

[16]   Butkus, L. M., Holmes, J. W., and Nicholas, T., "Thermomechanical Fatigue Behavior of a Silicon Carbide Fiber-Reinforced Calcium Alluminosilicate Composite," *Journal*, American Ceramic Society, Vol. 76, 1993, pp. 2817–2825.

[17]   John, R., Jira, J. R., and Ashbaugh, N. E., "Analysis of Bridged Fatigue Cracks in Unidirectional SCS-6/Ti-24Al-11Nb Composite," *Fatigue '93*, J. P. Bailon and I. J. Dicksons, Eds., Engineering Material Advancement Services Ltd., Ecole Polytechnique, Montreal, Canada, Vol. 2, 1993, pp. 1091–1096.

[18]   Castelli, M. G., *Life Prediction Methodology for Titanium Matrix Composites, ASTM STP 1253*, W. S. Johnson, J. M. Larsen, and B. N. Cox, Eds. American Society for Testing and Materials, Philadelphia, 1995, pp. 412–431.

[19]   Karpur, P., Matikas, T. E., and Blodgett, M. P., "Acoustic Microscopy as a Tool for Fiber-Matrix Interface Evaluation," *Proceedings*, First International Conference on Composites Engineering (ICCE/1), New Orleans, D. Huis, Ed., 1994, pp. 253–254.

[20]   Quate, C. F., Atalar, A., and Wickramasinghe, H. K., "Acoustic Microscopy with Mechanical Scanning—A Review," *Proceedings*, Institute of Electrical and Electronic Engineers, Inc., New York, Vol. 67, 1979, pp. 1092–1114.

[21]   Karpur, P., Matikas, T., and Krishnamurthy, S., "Matrix-Fiber Interface Characterization in Metal Matrix Composites Using Ultrasonic Imaging of Fiber Fragmentation," *Proceedings*, American Society for Composites, Seventh Technical Conference, Technomic, Lancaster, PA, 1992, pp. 420–427.

[22]   Briggs, G. A. D., *Acoustic Microscopy*, Oxford University Press, Oxford, 1992.

[23]   Lawrence, C. W., Briggs, G. A. D., Scruby, C. B., and Davies, J. R. R., "Acoustic Microscopy of Ceramic-Fibre Composites; Part I: Glass-matrix Composites," *Journal of Materials Science*, Vol. 28, 1993, pp. 3635–3644.

[24]   Bertoni, H. L., "Raleigh Waves in Scanning Acoustic Microscopy," *Raleigh Wave Theory and Application*, The Royal Institution, London, Vol. 2, 1985, pp. 274–290.

[25]   Blatt, D., Karpur, P., Matikas, T. E., Blodgett, M. P., and Stubbs, D. A., "Elevated Temperature Degradation and Damage Mechanisms of Titanium Based Metal Matrix Composites," *Proceedings*, American Society for Composites, Eight Technical Conference on Composite Materials, Cleveland, 1993.

[26]   Blodgett, M. P., Matikas, T. E., Karpur, P., Jira, J. R., and Blatt, D., "Ultrasonic Evaluation of Fiber-Matrix Interfacial Degradation of Titanium Matrix Composites Due to Temperature and Mechanical Loading," 20th Annual Review of Progress in Quantitative Nondestructive Evaluation, Vol. 13B, D. O. Thompson and D. E. Chimentis, Eds., Plenum, Bowdoin College, Brunswick, ME, 1993, pp. 1213–1219.

[27]   Karpur, P., Matikas, T. E., Blodgett, M. P., Jira, J. R., and Blatt, D., in *Special Applications and Advanced Techniques for Crack Size Determination, ASTM STP 1251*, J. J. Ruschau and J. K. Donald, Eds., American Society for Testing and Materials, Philadelphia, 1995, pp. 130–146.

[28]   Matikas, T. E. and Karpur, P., "Ultrasonic Reflectivity Technique for the Characterization of Fiber-Matrix Interface in Metal Matrix Composites," *Journal of Applied Physics*, Vol. 74, 1993, pp. 228–236.

[29]   Bashyam, M., "Ultrasonic NDE for Ceramic- and Metal-Matrix Composite Material Characterization," *Review of Progress in Quantitative Nondestructive Evaluation*, Vol. 10B, D. O. Thompson and D. E. Chimentis, Eds., Plenum Press, New York, 1991, pp. 1423–1430.

[30]   Stubbs, D. A. and Russ, S. M., "Examination of the Correlation Between NDE-Detected Manufacturing Abnormalities and Thermomechanical Fatigue Life," *Proceedings*, Structural Testing Technology at High Temperature—II, Society for Experimental Mechanics, Inc., Ojai, CA, 1993, pp. 165–173.

[31]   MacLeallan, P. T., Master's thesis, University of Dayton, 1993.

[32]   Testa, A. J. and Burger, C. P., "A Measurement of Crack Depth by Changes in the Frequency Spectrum of a Rayleigh Wave," *Proceedings*, Novel NDE Methods for Materials, B. B. Raths, Ed., Metallurgical Society of AIME, 1982, pp. 91–108.

[33]   Karpur, P. and Resch, M. T., in *Review of Progress in Quantitative Nondestructive Evaluation*, Vol. 10A, D. O. Thompson and D. E. Chimenti, Eds., Plenum Press, New York, 1991, pp. 757–764.

[34]   Achenbach, J. D., Fine, M. E., Komsky, I., and McGuire, S., "Ultrasonic Wave Technique to Assess Cyclic-Load Fatigue Damage in Silicon-Carbide Whisker Reinforced 2124 Aluminum Alloy Composites," *Cyclic Deformation, Fracture and Nondestructive Evaluation of Advanced Materials,*

ASTM STP 1157, M. R. Mitchell and O. Bucks, Eds., American Society for Testing and Materials, Philadelphia, 1992, pp. 241–250.

[35] Deiter, G., *Engineering Design: A Materials and Processing Approach*, McGraw-Hill Book Co., New York, 1983.

[36] *The Physical Metallurgy of Titanium Alloys*, E. W. Collins, Ed., American Society for Metals, Metals Park, OH, 1984.

[37] Hartman, G. A., Zawada, L., and Russ, S., "Techniques for Elevated Temperature Tensile Testing of Advanced Ceramic Composite Materials," *Proceedings*, Fifth Annual Hostile Environment and High Temperature Measurements Conference, Society for Experimental Mechanics, Costa Mesa, CA, 1988, pp. 31–38.

[38] Hartman, G. A. and Russ, S., "Techniques for Mechanical and Thermal Testing of Ti$_3$Al/SCS-6 Metal Matrix Composites," *Metal Matrix Composites: Testing, Analysis and Failure Modes, ASTM STP 1032*, W. S. Johnson, Ed., American Society for Testing and Materials, Philadelphia, 1989, pp. 43–53.

[39] Hartman, G. A. and Buchanan, D. J., "Methodologies for Thermal and Mechanical Testing of TMC Materials," AGARD Report 796, NATO AGARD Characterization of Fiber Reinforced Titanium Matrix Composites Bordeaux, France, 1993, pp. 12–1 to 12–9.

[40] Hartman, G. A. and Ashbaugh, N. E., "A Fracture Mechanics Test Automation System for a Basic Research Laboratory," *Applications of Automation Technology to Fatigue and Fracture Testing, ASTM STP 1092*, A. A. Braun, N. E. Ashbaugh, and F. M. Smith, Eds., American Society for Testing and Materials, Philadelphia, 1990, pp. 95–112.

Kevin L. Koudela,[1] *Larry H. Strait,*[1] *Anthony A. Caiazzo,*[1] *and*
Karin L. Gipple[2]

Static and Fatigue Interlaminar Tensile Characterization of Laminated Composites

REFERENCE: Koudela, K. L., Strait, L. H., Caiazzo, A. A., and Gipple, K. L., **"Static and Fatigue Interlaminar Tensile Characterization of Laminated Composites,"** *Composite Materials: Fatigue and Fracture (Sixth Volume), ASTM STP 1285,* E. A. Armanios, Ed., American Society for Testing and Materials, 1997, pp. 516–530.

ABSTRACT: Spool and curved-beam specimens were evaluated to determine the viability of using either one or both of these configurations to characterize the static and fatigue interlaminar tensile behavior of carbon/epoxy laminates. Unidirectional curved-beam and quasi-isotropic spool specimens were fabricated, nondestructively inspected, and statically tested to failure. Tension-tension fatigue tests were conducted at 10 Hz and an *R*-ratio ($\sigma_{min}/\sigma_{max}$) equal to 0.1 for each specimen configuration. The interlaminar tensile strength of the spool specimen was 12% larger than the strength obtained using curved-beam specimens. In addition, data scatter associated with spool specimens was significantly less than the scatter associated with curved-beam specimens. The difference in data scatter was attributed to the influence of the fabrication process on the quality of the laminates tested. The fatigue limit at 10^7 cycles for both specimen types was shown to be at least 40% of the average interlaminar tensile strength. Based on the results of this study, it was concluded that either the spool or the curved-beam specimens can be used to characterize the interlaminar tensile static and fatigue behavior of carbon/epoxy laminates. However, to obtain the most representative results, the test specimen configuration should be selected so that the specimen fabrication process closely simulates the actual component fabrication process.

KEYWORDS: composite materials, curved-beam specimens, spool specimens, interlaminar tensile strength, delamination, fatigue (materials), fracture (materials)

In recent years, composite materials have been gaining acceptance as load-bearing members for marine structures because of their high specific strengths and moduli, inherent corrosion resistance, and potential for vibration and acoustic performance improvements. Marine structures are often thick and may exhibit complex curvatures. Laminates used in such components may experience an excessive buildup of interlaminar tensile and shear stresses. These stresses coupled with singular stress states developed at or near ply drop-offs, free edges, areas of local impact damage, manufacturing defects and transverse matrix cracks may cause component delamination.

Both fracture mechanics- and strength-based approaches have been proposed to predict delamination. It has been shown that when singular stress states are present, a fracture mechanics-based approach is most appropriate. Fracture mechanics-based methodologies [1–4] assume that delaminations will occur when the available strain energy release rate (G) is equal to or

[1] Research associate, research associate, and research assistant, respectively, Materials Science Department, Applied Research Laboratory, State College, PA 16804.
[2] Materials engineer, Naval Surface Warfare Center, Carderock Division, Bethesda, MD. 20084-5000.

greater than the critical strain energy release rate (G_c), where G_c is determined from coupon fracture tests. If the interlaminar stresses are finite (that is, nonsingular), then a strength-based approach can be used [5]. Strength-based methodologies assume that delaminations will occur when the interlaminar stress (or strain) is greater than or equal to the interlaminar strength (or failure strain) determined from an appropriate coupon strength test. In the current work, discussion is limited to strength-based characterization of interlaminar tensile behavior.

Currently, the only standardized strength-based interlaminar tension test method is ASTM Test Method for Tensile Strength of Flat Sandwich Constructions in Flatwise Plane (C 297-61)(1988); however, several other strength-based interlaminar tension test procedures have been proposed and evaluated. A spool specimen was first proposed by Mandell et al. [6] to generate the interlaminar tensile strength (ILTS) for glass/epoxy laminates. Mandell's specimen design was modified by Lagace and Weems [7] to characterize the ILTS of carbon/epoxy materials. The spool specimens discussed were machined from thick composite panels with either rectangular or circular cross-sectional geometries. Regardless of cross-sectional shape, the center of the specimen was necked down to force failure to occur in the gage-section. The authors evaluated several laminate stacking sequences and found that the ILTS was independent of specimen layup. They obtained interlaminar tensile strengths that were 75 to 85% of the laminate's in-plane transverse tensile strength.

Hiel et al. [8] introduced curved-beam specimens to characterize the ILTS of composite laminates. The authors evaluated two candidate specimen geometries; the semicircular and elliptical curved-beam. For each configuration, a maximum ILTS was generated in the specimen apex by applying load to the free ends of the specimen. Strengths developed from modified (necked down at the apex) semicircular beam specimens separated into two groups. The "strong" group produced ILTS up to 99% of the reported in-plane transverse tensile strength for the same material. The "weak" group produced ILTS of only 50% of the in-plane transverse strength. The decrease in ILTS was attributed to slight manufacturing irregularities in the specimen apex. The elliptical specimens were designed to eliminate the need for gage-section modifications and improve the consistency of the results. Interlaminar strengths equivalent to 194% of the material's in-plane transverse tensile strength were obtained using the elliptical specimen geometry. Hiel et al. speculated that the large ILTS occurred because only a small volume of material located at the specimen apex was subjected to the peak interlaminar tensile stress. They concluded that this may have represented a size effect but that further investigation was needed to provide an appropriate explanation. In addition, they performed limited fatigue testing using the elliptical beam specimen and demonstrated a fatigue strength at 10^6 cycles equal to 50% of the ILTS.

Unidirectional right angle coupons were proposed by other researchers [9–11] as an alternative to semicircular and elliptical specimen configurations. Martin [9] used a strength-based criterion to predict delamination onset and fracture mechanics techniques to predict the failure sequence of AS4/3501-6 unidirectional right-angle coupons. The failure load corresponding to the onset of delamination was over-predicted by 15% using an ILTS equivalent to the measured in-plane transverse tensile strength. Jackson and Martin [10] extended the right-angle coupon studies to assess the effects of loading arm length, inner radius, thickness, and width on the ILTS. Specimen width and loading arm length had little effect on the measured ILTS. Differences in laminate quality made it difficult to determine the effects of inner radius and thickness. The highest quality 16-ply-thick specimens had an ILTS equivalent to 120 to 140% of the in-plane transverse tensile strength. Limited fatigue data for this specimen was reported by Martin and Jackson [11]. Unidirectional AS4/3501-6 specimens were fatigued at 5 Hz and an R-ratio $(\sigma_{min}/\sigma_{max})$ equal to 0.1. Runout occurred at approximately 43 to 86% of the ILTS between 10^6 and 10^7 cycles.

In the current work, spool and curved-beam specimens were evaluated to determine the viability of using either one or both of these configurations to characterize the static and fatigue interlaminar tensile behavior of carbon/epoxy laminates. Closed-form solutions and finite-element analyses were used to design unidirectional quasi-isotropic spool and curved-beam specimens. IM7/977-2 specimens were fabricated, nondestructively inspected, and statically tested to failure. Upon completion of the static tests, tension-tension fatigue tests were conducted at 10 Hz and an R-ratio equal to 0.1. The specimen designs, fabrication methods, and test results are presented in subsequent sections.

Specimen Design

The quasi-isotropic spool specimen design in shown in (Fig. 1) where the plies are oriented perpendicular to the length direction of the specimen. The interlaminar normal and shear stresses are identified in the figure as σ_{nn} and τ_{nt}, respectively. Numerical shape optimization coupled with two-dimensional finite element analysis (FEA) was used to develop an optimum spool specimen design. The optimization and FEA were conducted using the I-DEAS Master Series 2.1 software package. The objective of the optimization was to determine the dimensions of the specimen that minimized the interlaminar tensile stress concentration so that failure was confined to the gage section. The finite element model used for the numerical shape optimization consisted of 821 elements with a total of 1806 degrees of freedom. The quasi-isotropic material properties used for the optimization are listed in (Table 1). The in-plane properties listed in the table were determined using standard ASTM test methods, the interlaminar shear properties were obtained from Iosipescu (V-notched beam) tests, and the interlaminar tensile modulus was assumed to be equivalent to the material's in-plane transverse tensile modulus.

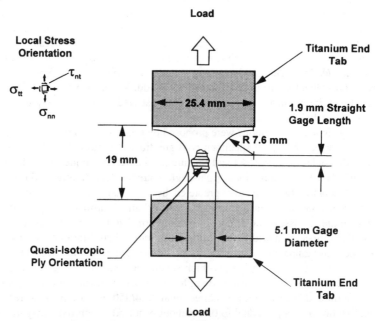

FIG. 1—*Spool specimen design, final dimensions.*

TABLE 1—*Unidirectional and quasi-isotropic IM7/977-2T material properties.*

Material Property	Unidirectional, GPa	Quasi-isotropic, GPa
E_{11}	162	59.7
E_{22}	9.31	59.7
E_{33}	9.31	9.31
G_{12}	5.45	22.7
G_{13}	5.45	3.86
G_{23}	3.31	3.86
v_{12}[a]	0.27	0.31
v_{13}[a]	0.27	0.28
v_{23}[a]	0.42	0.28

[a]Unitless values.

The gage-section diameter and blend radius were used as design variables during the shape optimization, and a geometric constraint was applied so that the minimum gage-section diameter was forced to be greater than or equal to 5.1 mm. The specified minimum gage diameter was the smallest diameter that could be machined without distorting the specimen cross section. The optimal specimen dimensions are shown in (Fig. 1). As shown in the figure, the gage-section diameter was equal to the minimum allowable (5.1 mm) and the blend radius converged to 7.6 mm. The maximum diameter of the specimen was 25.4 mm with a gage length equal to 1.9 mm. The interlaminar tensile stress concentration defined as the peak stress divided by the nominal stress (failure load divided by the gage-section cross-sectional area) for the optimized design was 1.11. The stress concentration occurred just above the point of tangency between the blend radius and the straight gage-section.

The unidirectional curved-beam specimen design is shown in (Fig. 2) where the plies are aligned parallel to the circumferential direction. The material properties for the unidirectional

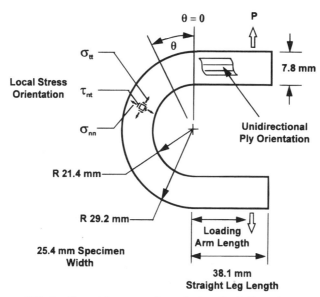

FIG. 2—*Curved-beam specimen design, final dimensions.*

IM7/977-2 curved-beam specimen are listed in (Table 1). The ILTS of the curved beam cannot be measured directly from coupon strength tests. Therefore, closed-form elasticity solutions or finite element analyses or both are required to relate the external load to the maximum interlaminar tensile stress. A substantial amount of analytical work has been devoted to the study of interlaminar stresses developed in curved-beams subject to arbitrary end loads and moments [12–15]. In the current study, a closed-form elasticity solution based on the solution developed by Ko et al. [15] was used to determine the in-plane and interlaminar stress distributions for a given external load.

The elasticity solution was used to perform parametric studies to determine the optimal dimensions of the curved-beam specimen. The results of the parametric study demonstrated that increasing the radius to thickness (R/t) ratio increased the ratio of peak interlaminar shear (τ_{nt}) to peak interlaminar tensile (σ_{nn}) stress at θ equal to $0°$, focused the peak σ_{nn} near the specimen apex (θ equal to $90°$), and increased the ratio of the peak in-plane circumferential stress (σ_{tt}) to σ_{nn}. It was also demonstrated that increasing the straight leg (loading arm) length increased the specimen opening displacement, decreased the peak τ_{nt} to peak σ_{nn} ratio at θ equal to $0°$, produced a more uniform σ_{nn} distribution along the specimen circumference, and decreased the σ_{tt} to σ_{nn} ratio. An optimal curved-beam design should (1) have a small opening displacement to facilitate fatigue testing, (2) possess a small τ_{nt} to σ_{nn} ratio at $\theta = 0°$ to preclude interlaminar shear failure, (3) have a large σ_{nn} gradient along the circumference of the specimen to force failure in the specimen apex, and (4) have a small maximum σ_{tt} to σ_{nn} ratio in the apex to preclude premature in-plane failure. A specimen configuration that satisfied this design is shown in Fig. 2. The curved-beam had an outer radius equal to 29.2 mm, an inner radius equal to 21.4 mm, a straight leg length of 38.1 mm, and a width of 25.4 mm.

Specimen Fabrication

A 19.0-mm-thick (104 plies) quasi-isotropic panel with a $[(0/\pm45/90)_s]_{13}$ layup was manufactured from IM7/977-2 prepreg for the spool specimen. Room temperature vacuum and autoclave debulks were performed alternately after each 16-ply quasi-isotropic sublaminate was layed up. The autoclave debulks were performed under 0.69 MPa of pressure at 66°C. Each debulk cycle lasted 1 h. The debulk cycles were included to ensure that the panel was properly consolidated and pre-bled prior to the final autoclave cure. The panel was cured for 3 h at 177°C under 1.0 MPa of pressure. After cure, the panel was nondestructively inspected and machined to the dimensions shown in Fig. 1.

Unidirectional curved-panels 457 mm long by 7.8 mm thick (48 plies) were fabricated from IM7/977-2 prepreg material. The prepreg material was placed on a male layup tool with a radius of 19.0 mm. The unidirectional material was aligned with the circumferential direction of the layup tool. The curved-panel was room temperature vacuum debulked for 1 h after the 12th and 36th plies were applied to the male tool. The curved-panel was autoclave debulked for 1 h under 0.69 MPa of pressure at 66°C after the 24th and 48th plies were applied. The pre-consolidated laminate was removed from the male tool and placed in a female tool. The laminate was formed to the female tool and pre-mixed silicone RTV 630 material was poured into the end-dammed female tool/curved-laminate assembly and allowed to cure for 24 h at room temperature. The assembly was vacuum bagged and autoclave cured for 3 h at 177°C under a pressure of 0.69 MPa. After cure, the curved-panel was nondestructively inspected and machined into 25.4-mm-wide curved-beam specimens.

Spool Specimen Experimental Setup

Prior to testing, pre-machined titanium end tabs were bonded to the spool specimen. For the static tests, the threaded-ends of the end tabs were interfaced with threaded adapters that attached to self-aligning grips mounted in a screw-driven test frame as shown in Fig. 3. The specimens were loaded under displacement control at a cross-head displacement rate of 1.27 mm/min. Prior to conducting the static tests, a spool specimen was instrumented with four strain gages placed 90° apart around the circumference of the gage-section. The specimen was loaded to 50% of the expected failure load, and strain readings from each gage were recorded. Virtually no differences were detected between the four strain values indicating that the load train was properly aligned. Once proper loading alignment was demonstrated, each specimen was instrumented with a single CEA-06-062AQ-350 strain gage oriented parallel to the specimen axis. A computerized data acquisition system was used to record the load and strain data. The ILTS was determined by dividing the measured failure load by the gage cross-sectional area and then multiplying the resultant value by the stress concentration factor (1.11) determined from the finite element analysis.

To use this data reduction technique, the accuracy of the finite element model was first demonstrated. This was accomplished by comparing measured surface strains in the specimen gage-sections to the predicted gage-section strains for a given static load. The results of the comparison for the five static specimens are listed in Table 2. As shown in the table, the predicted strains were within ±4% of the measured values. In addition, the finite element model predicted that failure would occur just above the point of tangency between the blend radius and the gage section. In all instances, specimen failure occurred at this location.

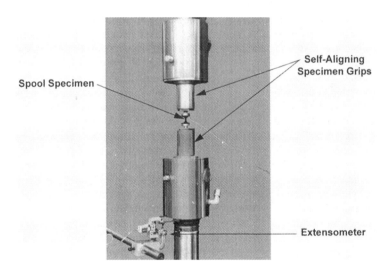

FIG. 3—*Spool specimen test setup.*

TABLE 2—*Experimental validation of the spool specimen finite-element analysis results.*

Specimen	P, N	ϵ^{meas}, $\mu\epsilon$	ϵ^{FEA}, $\mu\epsilon$	Difference, %
1	1575	8600	8765	1.9
2	1437	7700	7998	3.9
3	1717	9600	9558	−0.4
4	1632	9400	9087	−3.3
5	1637	lost gage	9112	⋯

The fatigue specimens were mounted in a closed-loop servohydraulic test frame using a fixturing setup identical to the static tests. The specimens were fatigue cycled under load control at 10 Hz. All specimens were tested at an R-ratio equal to 0.1 at maximum fatigue stress levels ranging from 40 to 60% of the ILTS. The specimens were air cooled during the fatigue tests. The fatigue specimens were originally instrumented with an axial strain gage; however, in every case the strain gage disbonded early in the test. As a result, extensometers were configured as deflectometers in lieu of the strain gages. A computerized data acquisition system was used to continuously monitor load and deflection versus cycle data. Upon completion of the fatigue tests, the specimens that did not fail were statically tested to failure to determine the residual strength of the specimen. The residual strength tests were conducted in the same manner as the static tests.

Curved-Beam Specimen Experimental Setup

A special steel-loading fixture was designed for the curved-beam specimens to preclude having to load the specimens with male rod ends bolted to the straight legs of the specimen [8]. The curved-beam specimens were inserted in the fixture to allow for 50.8- and 76.2-mm-long loading arms, respectively. The fixture was attached to a screw-driven static load frame and the specimens were loaded at a rate equal to 1.27 mm/min. An extensometer monitored specimen opening displacements. Hoop strains were recorded using CEA-06-125UW-350 strain gages mounted at the specimen apex (θ equal to 90°) on the outer radius top and inner radius bottom surfaces of the specimen. A CEA-06-062UW-350 strain gage was placed across the thickness at the apex on one side of the specimen. The other side of the specimen was coated with typewriter correction fluid to improve delamination visualization. Both the stresses (strains) and opening displacements at failure were calculated from the elasticity solution by scaling the stresses and displacements predicted for a unit point load by the experimentally determined failure loads. Prior to using the elasticity solution to calculate the ILTS, the accuracy of the solution was assessed by comparing the experimentally measured opening displacements and top and bottom circumferential strains with the corresponding predicted values for each of the measured failure loads. The results of the comparison are shown in Table 3. As shown in the table, the maximum absolute error in the deflection prediction was less than 11%. The maximum errors in the hoop tension and compression strains were less than 4 and 5%, respectively. Based on these results, the elasticity solution was considered acceptable for the ILTS calculations.

The servohydraulic system used for the curved-beam fatigue tests is shown in Fig. 4. Specimen insertion in the special loading fixture was identical to that for static testing except that the loading arm length was always equal to 50.8 mm. The use of the shorter loading arm was required to minimize the specimen opening displacement. The displacement was minimized so that a 10 Hz cycle rate could be achieved with the existing hydraulic system. Fatigue testing

TABLE 3—*Experimental validation of the curved-beam specimen elasticity results.*

Load, N	δ, mm			ϵ^{TEN}, $\mu\epsilon$			ϵ^{COM}, $\mu\epsilon$		
	Measured	Pre-dicted	Difference, %	Measured	Pre-dicted	Difference, %	Measured	Pre-dicted	Difference, %
2620	⋯	2.41	⋯	7763	7523	−3.1	−5995	−6012	0.3
2753	2.49	2.51	0.8	8195	7906	−3.5	−6609	−6319	−4.4
3091	2.69	2.84	5.6	9116	8877	−2.6	−7237	−7094	−2.0
1837	1.52	1.68	10.5	5288	5275	−0.2	−4102	−4216	2.8
2264	1.88	2.08	10.6	6461	6501	0.6	−5089	−5196	2.1

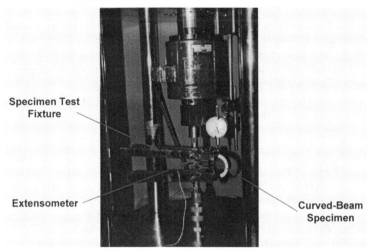

FIG. 4—*Curved-beam specimen test setup.*

was performed in load control at an R-ratio equal to 0.1 and a maximum load corresponding to 50% of the ILTS. The specimens were air cooled during the fatigue tests. Load deflection data were monitored every 100 000 cycles. Upon completion of the fatigue tests, the specimens that did not fail were statically tested to failure to determine the residual strength of the specimen. The residual strength tests were conducted in the same manner as the static tests.

Spool Specimen Test Results

The ILTS for five spool specimens are listed in Table 4. The average ILTS was 87.7 MPa with a standard deviation equal to 5.4 MPa. The average elongation at failure was 8825 $\mu\epsilon$. In all instances, failure occurred immediately above the intersection of the straight gage section and blend radius. The load-strain curves exhibited bilinear behavior where an increase in compliance (decrease in modulus) was observed at approximately 50 to 60% of the specimen failure strain. The increase in compliance was attributed to nonlinear deformation of the toughened resin system especially along resin-rich interlayers. The authors observed a similar nonlinear compliance increase (although much smaller) during in-plane transverse tensile testing of IM7/977-2 thin laminates.

The average interlaminar tensile strength was 14% larger than the reported in-plane transverse tensile strength for the same material. This trend is consistent with the behavior of notch sensitive materials where strength is assumed to be a function of the volume of material being

TABLE 4—*Spool specimen static test results.*

Specimen	P, N	σ^{PEAK}, MPa	ϵ^{meas}, $\mu\epsilon$
1	1574	86.0	8600
2	1437	79.2	7700
3	1717	93.6	9600
4	1632	90.3	9400
5	1637	89.4	lost gage
Average	1601	87.7	8825

stressed. For these materials, strength decreases as the volume of material stressed increases. O'Brien and Salpekar [16] were able to relate the strengths of various specimen configurations using a Weibull shape parameter and the specimen volumes using a Weibull scaling law. For this study, the Weibull scaling law, shown in Eq 1, was used to relate the in-plane transverse tensile strength (S_{tt}) to the ILTS (S_{nn}) of the spool specimen

$$\frac{S_{nn}}{S_{tt}} = \left(\frac{V_{tt}}{V_{nn}}\right)^{1/\beta}$$

(1)

where β is the Weibull shape parameter calculated from a two-parameter Weibull statistical distribution, V_{tt} is the volume of the in-plane transverse tensile specimen, and V_{nn} is the volume of the spool specimen under peak interlaminar tensile stress. For the spool specimen, V_{nn} was determined from FEA as the volume of material with interlaminar stresses that were within 6% (one standard deviation) of the peak interlaminar stress. The volume corresponding to this criterion was 56.6 mm³. The volume for the in-plane transverse tensile specimen with a uniform stress state was determined for a 229- by 25.4- by 2.0-mm-thick (12 plies) specimen. The resultant volume was 11 575 mm³. Substituting S_{tt} equal to 75.4 MPa and a Weibull shape parameter equal to 38.5 for the in-plane transverse tensile data (determined from a previous study) into Eq 1 and solving for S_{nn} resulted in an ILTS equal to 86.5 MPa. The prediction was within 2% of the measured ILTS. Although, the scaling law accurately predicted the average ILTS of the spool specimens, the scatter of the ILTS data as indicated by its associated Weibull shape parameter (β equal to 23.6) was significantly greater than the corresponding in-plane data scatter (β equal to 38.5). The difference in scatter may be attributed to (1) variations in material quality between the two specimen types, (2) the small sample sizes used to calculate the shape parameters, or (3) inherent differences between the materials transverse and interlaminar microstructures. It should be noted that the Weibull scaling law predictions are valid only if the spool and in-plane Weibull parameters are identical. The existence (or not) of a single-shape parameter may be demonstrated with additional testing of in-plane and spool specimens. However, this additional testing was beyond the scope of the current study.

A plan view photomicrograph of a typical spool specimen failure surface is shown in Fig. 5. Failure occurred across two plies producing a rough and jagged failure surface. Failure initiated along an interface (resin interlayer) between two plies, sheared through an adjacent ply until it hit the next resin interlayer and propagated along the interface until complete separation of the specimen occurred. The crack was arrested at the adjacent ply interface because at that point, to continue through the thickness, the crack would have had to shear through fibers in the adjacent ply. The adjacent ply was always oriented at 45° to the sheared ply. The through-the-thickness shear plane is shown in Fig. 5 as a horizontal striation across the specimen that separates the two adjacent plies. All but one static spool specimen had a failure surface similar to one shown in Fig. 5. However, in the specimen with the lowest interlaminar tensile strength, it appeared that failure occurred at an interface between two plies with no evidence of shearing through the thickness of an adjacent ply. In that instance, there was evidence of a local delamination covering approximately 30% of the specimen diameter.

The results of the spool fatigue tests are listed in Table 5. Specimens were tested at 40, 45, 50, and 60% of the ILTS. Both the number of cycles and residual strengths (for specimens that did not fail during the fatigue tests) are presented in the table. Runout was achieved after 10^7 cycles for two specimens tested at a maximum stress of 35.1 MPa (40% of the ILTS). One specimen failed prematurely at this stress level due to the presence of a macroscopically observable void in the gage-section. Two additional specimens tested at the 40% load level experienced no failure up to 10^6 cycles. The tests were stopped after 10^6 cycles, and static

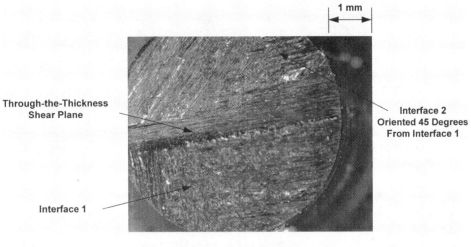

1 mm

Through-the-Thickness
Shear Plane

Interface 2
Oriented 45 Degrees
From Interface 1

Interface 1

Failure Surface From the
Gage-Section Diameter

FIG. 5—*Photomicrograph of the gage-section failure surface of a typical spool specimen.*

TABLE 5—*Spool specimen fatigue test results.*

	Maximum Fatigue Stress, percent of static ultimate strength							
	40%		45%		50%		60%	
Specimen	Cycles	Residual Strength, MPa	Cycles	Residual Strength, MPa	Cycles	Residual Strength, MPa	Cycles	Residual Strength, MPa
1	1.0×10^7	74.1	2.601×10^6	79.2	37 830	⋯	127 130	⋯
2	103 990	⋯	2.745×10^6	82.4	8 120	⋯	1 700	⋯
3	1.0×10^7	80.0	1.0×10^6	83.4	23 310	⋯	40 240	⋯
4	1.0×10^6	66.4	1.0×10^6	67.8	1.0×10^6	84.3	320	⋯
5	1.0×10^6	81.4	⋯	⋯	1.0×10^6	89.4	1.0×10^6	90.2
Average		75.5		78.2		86.8		90.2

[a]Macroscopically observable flaw in the gage-section.

residual strength tests were performed. The average residual strength for the specimens tested at 40% of the average static strength was 75.5 MPa, which was 14% less than the ILTS.

None of the four specimens tested at a maximum stress of 39.5 MPa (45% of the ILTS) experienced fatigue failure up to 10^6 cycles. Once again, the specimen fatigue tests were truncated after 10^6 cycles. The residual strength corresponding to each specimen is shown in Table 5. The average residual strength was 78.2 MPa, which was 11% less than the ILTS. Three of the five specimens tested at a maximum stress of 43.9 MPa (50% of the ILTS) failed in less than 100 000 cycles. The two remaining specimens, tested to 10^6 cycles, did not fail. The average residual strength for the two specimens was 86.8 MPa, which was 99% of the ILTS. All but one of the five specimens tested at a maximum fatigue load of 52.6 MPa (60% of the static ILTS) failed before 130 000 cycles. The remaining specimen, tested to 10^6 cycles, did not fail. The residual interlaminar tensile strength of the specimen was 90.2 MPa, which was 3% greater than the ILTS. To obtain a better estimation of residual strength change as a function of maximum fatigue stress, additional tests must be performed. However, based on the limited fatigue and residual strength testing, it appears that there is very little specimen wear-out prior to specimen failure.

Photomicrographs of the failure surfaces for the fatigued specimens were similar to the failure surfaces of the static specimens. The failure surfaces were rough and jagged and failure generally occurred across two or three plies. Like the static specimens, some of the fatigue specimens had relatively large intralaminar resin-rich inclusions or local delaminations located near the center of the specimen. For those cases, the failure surface appearance changed and lower fatigue strengths were manifest.

Curved-Beam Specimen Test Results

The interlaminar tensile strengths for ten curved-beam specimens are presented in Table 6. The interlaminar tensile strengths generated for the curved-beam specimens with loading arm lengths of 50.8 and 76.2 mm were pooled in order to determine the average ILTS and the Weibull-shaped parameter. A k-sample Anderson Darling test [17] that compares the k-sample Anderson-Darling statistic to a critical value was used to justify pooling the data. The average ILTS of the curved-beam specimens was 77.3 MPa with a standard deviation equal to 12.1 MPa. The ILTS of the curved-beam specimen was 12% less than the ILTS of the spool specimen and 3% greater than the in-plane transverse tensile strength. In addition, the scatter in the data was significantly greater than the scatter observed for the spool specimens as indicated by the smaller Weibull-shaped parameter. The shape parameter for the curved-beam static data was equal to 7.4.

Specimen failures were always catastrophic and occurred symmetrically about the specimen apex (θ equal to 90°). A post-test visual inspection indicated the presence of a single major delamination occurring between 20 to 50% of the specimen thickness that was consistent with the elasticity solution that predicted failure to occur at 45% of the specimen thickness. The major delamination was often accompanied by multiple smaller delaminations that covered the majority of the specimen circumference. In most cases, the load-strain curves for the in-plane tensile (inner radius surface), in-plane compression (outer radius surface), and interlaminar tensile strains were linear to failure. However, for two of the specimens there was a slight load drop prior to specimen failure. The load drop was attributed to the formation of minor delaminations in the specimen apex.

TABLE 6—*Curved-beam static test results.*

Specimen	P, N	ILTS, MPa
1	2669	81.8
2	2811	86.2
3	3158	96.8
4	1913[a]	58.6
5	2357	72.2
6	2193	67.2
7	3136	72.0
8	3385	77.7
9	4048	92.9
10	2962	68.0
Average	2863	77.3

[a]Potential fixturing problem.

The ILTS of the curved-beam was also predicted using the Weibull scaling law presented in Eq 1. The volume of the curved-beam used in the equation included interlaminar stresses that were within 16% (one standard deviation) of the peak stress. The resultant volume was 1284 mm^3 where the volume was calculated assuming an included angle of 40°, an inner radius of 23.4 mm, a thickness of 3.1 mm, and a 25.4 mm width. As noted previously, the volume of the in-plane transverse tensile specimen was 11 575 mm^3. Substituting the average transverse tensile strength of 75.4 MPa and the measured shape parameter for the in-plane transverse tensile data (β equal to 38.5) into Eq 1 and solving for S_{nn} yielded a predicted ILTS equal to 79.8 MPa. The prediction was within 4% of the measured ILTS. Although, the scaling law accurately predicted the average ILTS of the curved-beam specimen, once again the scatter of the ILTS data as indicated by the Weibull shape parameter (β equal to 7.4) was significantly greater than the corresponding in-plane data scatter.

A dark-field photomicrograph of a typical curved-beam apex cross section is shown in Fig. 6. As shown here, multiple delaminations exist across the thickness of the specimen. The major delaminations tend to congregate between 25 and 75% of the specimen thickness. The cracks tend to branch out at the locations of large intralaminar resin-rich areas. Although the resin interlayers were less distinct for the curved beams as compared to the spool specimen, there was a significant increase in the size and number of intralaminar resin-rich areas. The increase in resin-rich areas manifested itself as local reductions in fiber volume fraction and may have increased the data scatter of the curved-beam test results. Image analyses were performed using a LECO image analyzer to confirm this observation. The results of the analyses indicated that average fiber volume across the thickness for a given specimen varied from 57 to 72%.

Limited fatigue testing was conducted for the unidirectional curved-beam specimen. The specimens were fatigue tested using a loading arm length equal to 50.8 mm. The shorter loading arm was used to minimize the specimen opening displacements so that a 10-Hz cycle rate could be achieved. Four specimens were tested at approximately 50% of the ILTS. The first two specimens failed after 800 000 and 43 000 cycles, respectively. During the first specimen fatigue test, the steel loading fixture broke. The specimen was remounted in a new clamping fixture and failed after an additional 100 000 cycles. The specimen failure may have

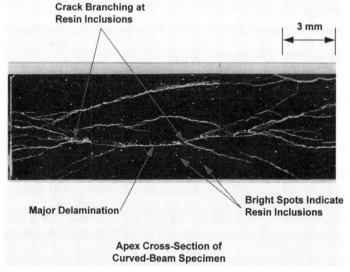

FIG. 6—*Photomicrograph of a typical failed curved-beam apex cross section.*

been affected by the fixture break, although the compliance of the specimen did not change following failure of the fixture. Image analyses of the second specimen indicated a much lower average fiber volume fraction relative to the other specimen (54% as compared to 62% by volume).

Specimens 3 and 4 were tested to 10^6 cycles. After 10^6 cycles, the fatigue tests were stopped and residual strength tests were conducted. The resulting residual strengths were 72.7 MPa (6% less than the ILTS) and 54.9 MPa (29% less than the ILTS), respectively. Specimen 3 failed in the apex with multiple delaminations distributed throughout the specimen thickness. Specimen 4 failed outside the apex in what appeared to be a compression shear-type failure mode. Photomicrographs of the fatigue specimens were similar to the static specimen photomicrographs. Like the static specimens, there were a significant number of small intralaminar inclusions. The variation in resin volume fractions due to the presence of the inclusions may have had a significant effect on the specimen residual strength.

Summary and Conclusions

In the current work, spool and curved-beam specimens were evaluated to determine the viability of using either one or both of these configurations to characterize the interlaminar tensile static and fatigue behavior of carbon/epoxy laminates. Closed-form solutions and finite element analyses were used to design unidirectional curved-beam and quasi-isotropic spool specimens. Specimens were fabricated, nondestructively inspected, and statically tested to failure. The correlated analytical and finite element models were then used in conjunction with the experimental results to calculate the static and residual interlaminar tensile strengths. A Weibull scaling law was used to predict the ILTS of the spool and curved-beam tensile specimens. Upon completion of the static tests, tension-tension fatigue tests were conducted at 10 Hz and an R-ratio equal to 0.1. Based on the results of this investigation the following conclusions are offered.

1. The major advantage of using spool specimens is that the peak interlaminar tensile stresses are very localized thereby confining specimen failure to a small region. The major disadvantage of using spool specimens is the cost and time associated with specimen fabrication.
2. The major advantage of the curved-beam specimen is that they can be fabricated using processes similar to those used to fabricate full-scale components. The major disadvantage of the curved-beam specimen is that the volume of the maximum stressed material is large. As a result, failure may occur over a large included angle about the specimen apex and may occur over a large portion of the specimen thickness.
3. The scatter in the ILTS data is significantly greater than the corresponding in-plane data scatter. This increase in data scatter may be attributed to variations in material quality between the specimen types, the small sample sizes used to calculate the shape parameters, or differences between the material's transverse and interlaminar microstructures. As a result, a single Weibull-shaped parameter (as required by the Weibull scaling law) may not be appropriate for these specimen types. More curved-beam, spool, and in-plane specimens need to be tested before the existence (or not) of a single Weibull-shaped parameter can be demonstrated for IM7/977-2 laminates.
4. The fatigue results demonstrated that the interlaminar tensile fatigue endurance limit for the IM7/977-2 material was at least 40% of the ILTS. Based on the limited fatigue testing, it appears that there is little specimen wear-out prior to fatigue failure.
5. Either curved-beam or optimized spool specimens can be used to characterize the interlaminar tensile behavior of carbon/epoxy laminates. However, to obtain the most

representative results, the test specimen should be selected so that the specimen fabrication process closely simulates the actual component fabrication process.

Acknowledgments

The authors wish to thank Art Spero and Ed Robinson for supporting this work under Prime Contract No. N00039-92-C-0100. Special thanks are also due to Pat Connelly from Cincinnati Testing Labs for spool specimen fixturing and test support, Dan Mouer and Dave Dreese for specimen fabrication, and Clark Moose for NDI. Ms. Gipple is grateful for the support of Ivan Caplan, the Ship and Submarine Materials Technology Plan Manager for the Office of Naval Research (ONR), in the area of ILTS testing that provided the technical background for the curved-beam testing. Additional thanks go to Bonnie Wells, Joe Waskey, and Rob Tregoning for curved-beam testing, and to Al Brandemarte for the photomicrographs and image analysis.

References

[1]   Johnson, W. S. and Mall, S., "A Fracture Mechanics Approach for Designing Adhesively Bonded Joints," *Delamination and Debonding of Materials, ASTM STP 876*, W. S. Johnson, Ed., American Society for Testing and Materials, Philadelphia, 1985, pp. 189–199.

[2]   O'Brien, T. K., "Generic Aspects of Delamination in Fatigue of Composite Materials," *Journal*, American Helicopter Society, Vol. 32, No. 1, Jan. 1987, pp. 13–18.

[3]   O'Brien, T. K., "Towards a Damage Tolerance Philosophy for Composite Materials and Structures," *Composites Materials: Testing and Design (Ninth Volume), ASTM STP 1059*, S. P. Garbo, Ed., American Society for Testing and Materials, Philadelphia, 1990, pp. 7–33.

[4]   Sun, C. T. and Kelly, S. R., "Failure in Composite Angle Structures—Part II: Onset of Delamination," *Journal of Reinforced Plastics and Composites*, Vol. 7, May 1988, pp. 233–244.

[5]   Martin, R. H. and Jackson, W. C., "Damage Prediction in Cross-Plied Curved Composite Laminates," NASA Technical Memorandum 104089, Hampton, VA, July 1989, p. 22.

[6]   Mandell, J. F., McKenna, G. B., and McGarry, F. J., "Interlaminar Strength and Toughness of Fiberglass Laminates," *Proceedings*, Twenty-ninth Annual Technical Conference, Paper 13-C, The Society of the Plastics Industry, Washington, DC, 1974.

[7]   Lagace, P. A. and Weems, D. B., "A Through-Thickness Strength Specimen for Composites," *Test Methods and Design Allowables for Fibrous Composites: Second Volume, ASTM STP 1003*, C. C. Chamis, Ed., American Society for Testing and Materials, Philadelphia, 1989, pp. 197–207.

[8]   Hiel, C. C., Sumich, M., and Chappell, D. P., "A Curved Beam Test Specimen for Determining the Interlaminar Tensile Strength of a Laminated Composite," *Journal of Composite Materials*, Vol. 25, July 1991, pp. 854–868.

[9]   Martin, R. H., "Delamination Failure in a Unidirectional Curved Composite Laminate," *Composite Materials: Testing and Design (Tenth Volume), ASTM STP 1120*, G. C. Grimes, Ed., American Society for Testing and Materials, Philadelphia, 1992, pp. 365–383.

[10] Jackson, W. C. and Martin, R. H., "An Interlaminar Tensile Strength Specimen," *Composite Materials: Testing and Design (Eleventh Volume), ASTM STP 1206*, E. T. Camponeschi, Ed., American Society for Testing and Materials, Philadelphia, 1993, pp. 333–354.

[11] Martin, R. H. and Jackson, W. C., "Damage Prediction in Cross-Plied Curved Composite Laminates," *Composite Materials: Fatigue and Fracture (Fourth Volume), ASTM STP 1156*, W. W. Stinchcomb and N. E. Ashbaugh, Eds., American Society for Testing and Materials, Philadelphia, 1993, pp. 105–126.

[12] Khandoker, S. I. and Kozik, T. J., "Improvement in the Calculation of Through the Thickness Stress in Laminated Heterogeneous Beams," *Proceedings*, ASME Pressure Vessels and Piping Conference, Pittsburgh, PA, 1988, pp. 159–166.

[13] Ko, W. L., "Delamination Stresses in Semicircular Laminated Composite Bars," NASA Technical Memorandum 4026, Edwards, CA, 1988.

[14] Kedward, K. T., Wilson, R. S., and McLean, S. K., "Flexure of Simply Curved Composite Shapes," *Composites*, Vol. 20, No. 6, Nov. 1989, pp. 527–536.

[15] Ko, W. L. and Jackson, R. H., "Multilayer Theory for Delamination Analysis of a Composite Curved Bar Subjected to End Forces and End Moments," NASA Technical Memorandum 4139, Edwards, CA, 1989.

[16] O'Brien, T. K. and Salpekar, S., "Scale Effects on the Transverse Tensile Strength of Graphite Epoxy Composites," *Composite Materials: Testing and Design (Eleventh Volume), ASTM STP 1206*, E. T. Camponeschi, Jr., Ed., American Society for Testing and Materials, Philadelphia, 1993.

[17] *Engineered Materials Handbook: Composites*, Vol. 1, ASM International, Metals Park, OH, 1987, pp. 304–305.

Levon Minnetyan[1] and Christos C. Chamis[2]

The Compact Tension, C(T), Specimen in Laminated Composite Testing

REFERENCE: Minnetyan, L. and Chamis, C. C., **"The Compact Tension, C(T), Specimen in Laminated Composite Testing,"** *Composite Materials: Fatigue and Fracture (Sixth Volume), ASTM STP 1285*, E. A. Armanios, Ed., American Society for Testing and Materials, 1997, pp. 531–550.

ABSTRACT: Use of the compact tension, C(T), specimens in laminated composites testing is investigated by considering two examples. A new computational methodology that scales up constituent material properties, stress, and strain limits to the structure level is used to evaluate damage propagation stages as well as the structural fracture load. Damage initiation, growth, accumulation, progressive fracture, and ultimate fracture modes are identified. Specific dependences of C(T) specimen test characteristics on laminate configuration and composite constituent properties are quantified.

KEYWORDS: composite materials, computational simulation, damage, degradation, durability, fracture, laminates, structural degradation, fracture (materials)

Nomenclature

$E_{\ell ii}$ composite modulus in the i direction
$K_{\ell 12\alpha\beta}$ directional interaction factor
$S_{\ell ij\alpha}$ composite stress limit corresponding to the sign of stress, α
$\sigma_{\ell ij}$ ply stress in the material directions, ij
$\nu_{\ell ij}$ Poisson's ratio with respect to the ij directions

Design considerations with regard to the durability of fiber composite structures require an a priori evaluation of damage initiation and fracture propagation mechanisms under expected loading and service environments. Concerns for safety and survivability of critical components require a quantification of the structural fracture resistance under loading.

Inherent flexibilities in the selection of constituent materials and the laminate configuration make composites more capable of fulfilling structural design requirements. However, those same design flexibilities render the assessment of composite structural response and durability more elaborate, prolonging the design process. It is difficult to design and certify a composite structure because of the complexities in predicting the overall congruity and performance of fibrous composites under various loading and hygrothermal conditions.

[1]Associate professor, Department of Civil and Environmental Engineering, Clarkson University, Potsdam, NY 13699-5710.
[2]Senior aerospace scientist, Structures Division, National Aeronautics and Space Administration, Lewis Research Center, Cleveland, OH 44135.

Laminated composite design practice has been based on extensive testing with attempts to apply formal fracture mechanics concepts to interpret test results. In certain cases, interpretation of laminated composite test data via fracture mechanics has been satisfactory. However, in most cases, fracture mechanics methods have significantly underpredicted the strength of fiber composites. Reconciliation of test results with fracture mechanics has required significant modifications of effective fracture toughness and specific laminate-configuration-dependent effective stress concentration field parameters. Additionally, required adjustments of fracture mechanics parameters have had to be reassessed with every change in constituent and laminate characteristics.

The proposed standard, E24.07.02, on the translaminar fracture of composites includes C(T) specimen testing to generate experimental data with regard to crack propagation, similar to ASTM Test Method for Plane-Strain Fracture Toughness of Metallic Materials (E 399-90) standard for metals. However, even when C(T) test data were available for a given laminate, many questions on the damage progression characteristics would not be answered easily and sensitive material parameters would be difficult to identify. The complete evaluation of laminated composite fracture requires an assessment of ply and subply level damage/fracture processes.

The present approach by-passes traditional fracture mechanics to provide an alternative evaluation method, conveying to the design engineer a detailed description of damage initiation, growth, accumulation, and propagation that would take place in the process of ultimate fracture of a fiber composite structure. Results show in detail the damage progression sequence and structural fracture resistance during different degradation stages. This paper demonstrates that computational simulation, with the use of established material modeling and finite element modules, adequately tracks the damage growth and subsequent propagation to fracture for fiber composite specimens.

For the purpose of the present study, the following terminology is used to describe the various stages of degradation in the composite structure: (1) "damage initiation" refers to the start of damage induced by loading that the composite structure is designed to carry; (2) "damage growth" is the progression of damage from the location of damage initiation to other regions; (3) "damage accumulation" is the increase in the amount of damage in the damaged regions with additional damage modes becoming active; (4) "damage propagation" is the rapid progression of damage to other regions of the structure; and (5) "structural fracture" is the ultimate disintegration of the specimen.

Methodology

The behavior of fiber composite structures under loading is rather complex, especially when possible degradation with preexisting damage and damage propagation to structural fracture is to be considered. Because of the numerous possibilities with material combinations, laminate configuration, and loading conditions, it is essential to have an integrated and effective computational capability to predict the behavior of composite structures for any loading, geometry, composite material combinations, and boundary conditions. The predictions of damage initiation, growth, accumulation, and propagation to fracture are important in evaluating the load-carrying capacity and reliability of composite structures. The CODSTRAN (COmposite Durability STRuctural ANalysis) computer code [1] has been developed for this purpose. CODSTRAN is able to simulate damage initiation, damage growth, and fracture in composites under various loading and environmental conditions. The simulation of progressive fracture by CODSTRAN has been verified to be in reasonable agreement with experimental data from tensile coupon tests on graphite/epoxy laminates [2]. Recent additions to CODSTRAN have enabled the investigation of the effects of composite degradation on structural response [3], composite damage induced by dynamic loading [4], composite structures global fracture toughness [5], effect of hygrothermal environment on durability [6], damage progression in composite shells subjected to internal pressure [7], an overall evaluation of progressive fracture in polymer-matrix composite structures [8], the durability of stiffened composite shell panels under combined loading [9], and damage progression in composite shell structures for expeditious and efficient structural design [10]. The purpose of this paper is to describe the application of CODSTRAN to simulate progressive fracture in fiber composite compact tension, C(T), specimens, taking into account damage initiation/propagation mechanisms.

CODSTRAN is an integrated, open-ended, stand-alone computer code consisting of three modules: composite mechanics, finite element analysis, and damage progression modeling. The overall evaluation of composite structural durability is carried out in the damage progression module [1] that keeps track of composite degradation for the entire structure. The damage progression module relies on ICAN [11] for composite micromechanics, macromechanics and laminate analysis, and calls a finite element analysis module that uses anisotropic thick shell elements to model laminated composites [12].

Figure 1 shows a schematic of the computational simulation cycle in CODSTRAN. The ICAN composite mechanics module is called before and after each finite element analysis. Prior to each finite element analysis, the ICAN module computes the composite properties from the fiber and matrix constituent characteristics and the composite layup. The finite element analysis module accepts the composite properties that are computed by the ICAN module at each node and performs the analysis at each load increment. After an incremental finite element analysis, the computed generalized nodal force resultants and deformations are supplied to the ICAN module that evaluates the nature and amount of local damage, if any, in the plies of the composite laminate. Individual ply failure modes are assessed by ICAN using failure criteria associated with the negative and positive limits of the six ply-stress components in the material directions as follows

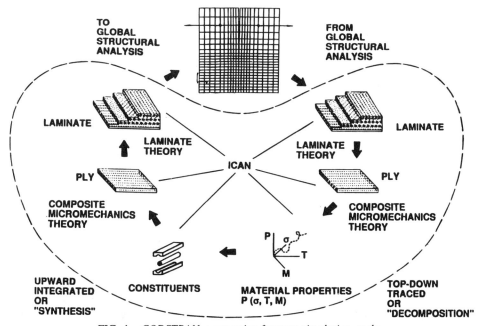

FIG. 1—*CODSTRAN progressive fracture simulation cycle.*

$$S_{\ell11C} < \sigma_{\ell11} < S_{\ell11T} \tag{1}$$

$$S_{\ell22C} < \sigma_{\ell22} < S_{\ell22T} \tag{2}$$

$$S_{\ell33C} < \sigma_{\ell33} < S_{\ell33T} \tag{3}$$

$$S_{\ell12(-)} < \sigma_{\ell12} < S_{\ell12(+)} \tag{4}$$

$$S_{\ell23(-)} < \sigma_{\ell23} < S_{\ell23(+)} \tag{5}$$

$$S_{\ell13(-)} < \sigma_{\ell13} < S_{\ell13(+)} \tag{6}$$

The stress limits in Eqs 1 through 6 are computed by the micromechanics equations in ICAN, based on constituent stiffness, strength, and fabrication process parameters. The equations used for ply stress limits are given in Ref 11. If ply damage is predicted by Eq 1, ply stiffness is reduced to zero at the damaged node. On the other hand, if ply damage is predicted by Eqs 2 through 6, only the matrix stiffness is degraded and the longitudinal tensile stiffness of fibers is retained. In addition to the failure criteria based on the stress limits, interply delamination due to relative rotation of the plies, and a modified distortion energy (MDE) failure criterion that takes into account combined stresses is considered. The MDE failure criterion is expressed as

$$F = 1 - \left[\left(\frac{\sigma_{\ell11\alpha}}{S_{\ell11\alpha}} \right)^2 + \left(\frac{\sigma_{\ell22\beta}}{S_{\ell22\beta}} \right)^2 - K_{\ell12\alpha\beta} \frac{\sigma_{\ell11\alpha}}{S_{\ell11\alpha}} \frac{\sigma_{\ell22}}{S_{\ell22}} + \left(\frac{\sigma_{\ell12S}}{S_{\ell12S}} \right)^2 \right] \tag{7}$$

where α and β indicate tensile or compressive stress, $S_{\ell11\alpha}$ is the local longitudinal strength in tension or compression, $S_{\ell22\alpha}$ is the transverse strength in tension or compression, and

$$K_{\ell12\alpha\beta} = \frac{(1 + 4\nu_{\ell12} - \nu_{\ell13})E_{\ell22} + (1 - \nu_{\ell23})E_{\ell11}}{[E_{\ell11}E_{\ell22}(2 + \nu_{\ell12} + \nu_{\ell13})(2 + \nu_{\ell21} + \nu_{\ell23})]^{1/2}} \tag{8}$$

The MDE failure criterion is obtained by modifying the usual distortion energy failure criterion that predicts combined stress failure in isotropic materials. The modification takes into account the significant differences in the stress limits of the longitudinal and transverse directions of an orthotropic composite ply. Each component of ply stress is normalized with respect to its limiting strength. No relationship is assumed between normal and shear strengths. The directional interaction factor, $K_{\ell12\alpha\beta}$, defined by Eq 8 reduces to unity for homogeneous isotropic materials. The MDE criterion has been demonstrated to be a good predictor of combined stress failure in composites. Details of the MDE criterion, as well as other options for the assessment of local failure in composites are given in Ref 13.

The MDE failure criterion becomes active in the majority of cases during computational simulation of progressive damage. If the failure predicted by the MDE criterion is not accompanied by a specific failure mode given by Eqs 1 to 6, then the type of failure is assessed by comparison of the magnitudes of the squared terms in Eq 7. Depending on the dominant term in the MDE failure criterion, fiber failure or matrix failure is assigned. The generalized stress-strain relationships are revised locally according to the composite damage evaluated after each finite element analysis. The model is automatically updated with a new finite element mesh having reconstituted properties, and the structure is reanalyzed for further deformation and damage. If there is no damage after a load increment, the structure is considered to be in equilibrium and an additional load increment is applied leading to possible damage growth, accumulation, or propagation. Simulation is continued until global fracture, when the specimen is broken into two pieces.

Graphite/Epoxy Specimen

The structural example for this case consists of a C(T) specimen made of AS-4 graphite fibers in a low-modulus high-strength (LMHS) toughened epoxy matrix. The fiber and matrix properties are obtained from a databank [11] of composite constituent material properties resident in CODSTRAN. The corresponding fiber and matrix properties are given in Tables 1 and 2.

TABLE 1—*AS-4 graphite fiber properties.*

Number of fibers per end = 10 000
Fiber diameter = 0.00762 mm (0.300E-3 in.)
Fiber density = 4.04E-7 kg/m^3 (0.063 lb/in.^{3})
Longitudinal normal modulus = 227 GPa (32.90E + 6 psi)
Transverse normal modulus = 13.7 GPa (1.99E + 6 psi)
Poisson's ratio (ν_{12}) = 0.20
Poisson's ratio (ν_{23}) = 0.25
Shear modulus (G_{12}) = 13.8 GPa (2.00E + 6 psi)
Shear modulus (G_{23}) = 6.90 GPa (1.00E + 6 psi)
Longitudinal thermal expansion coefficient = -1.0E-6/°C (-0.55E-6/°F)
Transverse thermal expansion coefficient = 1.0E-5/°C (0.56E-5/°F)
Longitudinal heat conductivity = 83.7 J-m/s/m^2/°C (4.03 Btu in./h/in.^{2}/°F)
Transverse heat conductivity = 8.37 J-m/s/m^2/°C (0.403 Btu-in./h/in.^{2}/°F)
Heat capacity = 712 J/kg/°C (0.17 Btu/lb/°F)
Tensile strength = 3723 MPa (540 ksi)
Compressive strength = 3351 MPa (486 ksi)

TABLE 2—*LMHS toughened epoxy matrix properties.*

Matrix density = 3.20E-7 kg/m^3 (0.0430 lb/in.^{3})
Normal modulus = 2.41 GPa (350 ksi)
Poisson's ratio = 0.43
Coefficient of thermal expansion = 1.03/°C (0.57E-4/°F)
Heat conductivity = 1.802 J-m/s/m^2/°C (8.68E-3 Btu-in./h/in.^{2}/°F)
Heat capacity = 1046 J/kg/°C (0.25 Btu/lb/°F)
Tensile strength = 121 MPa (17.5 ksi)
Compressive strength = 242 MPa (35.0 ksi)
Shear strength = 93.4 MPa (13.5 ksi)
Allowable tensile strain = 0.08
Allowable compressive strain = 0.15
Allowable shear strain = 0.1
Allowable torsional strain = 0.1
Void conductivity = 4.67 J-m/s/m^2/°C (0.225 Btu-in./h/in.^{2}°F)
Glass transition temperature = 180°C (350°F)

The LMHS matrix properties were representative of the 977-2 resin. The 977-2 toughened-epoxy resin matrix has been designed for space applications since it does not become brittle at low temperatures. At room temperatures, the stress limits are comparable to those of usual epoxy resin. However, the straining capability is considerably higher and the material is less stiff than standard epoxy resin.

The fiber volume ratio was 60%. The laminate structure consisted of thirty-six 0.133 mm (0.00525 in.) plies, resulting in a composite thickness of 4.80 mm (0.189 in.). The laminate configuration was $[0_2/90]_{6s}$. The 0° plies were in the direction of loading and the 90° plies were perpendicular to the load direction. The C(T) specimen, as shown in Fig. 2, had a height of $2H = 30$ mm (1.18 in.), an effective width (distance between load line and back face) of $W = 25.0$ mm (0.984 in.), a notch slot height of $2h = 0.3$ mm (0.012 in.), and a distance between load line and notch tip of $a = 12.5$ mm (0.492 in.). A computational model of the specimen was prepared using 420 rectangular thick-shell elements with 467 nodes as shown in Fig. 3. Pin holes were not modeled in the finite element representation of the specimen to enable nodal support and loading. The finite element model was configured to have a nodal point at the center of each pin hole. One of the load points was restrained in all degrees of freedom except for θ_z. The other load point was restrained only in the D_y, D_z, θ_x, θ_y directions but allowed freedom in D_x and θ_z directions. A concentrated tensile load was applied in the D_x direction. The load was increased gradually.

Figure 4 shows N_x generalized stress contours under a 1779-N (400-lb) loading, prior to damage initiation. The generalized stresses are defined as the through-the-thickness stress resultants per unit width of the laminate. The maximum tensile stresses were concentrated at the tip of the preexisting notch. These generalized N_x tensile stresses produce longitudinal tension in the 0° plies and transverse tension in the 90° plies. There was a distinct compression zone at the back of the specimen opposite the notch due to compressive N_x stresses.

Figure 5 shows N_y generalized stress contours under the 1779-N (400-lb) loading. The maximum tensile stresses were concentrated at the tip of the preexisting notch. The tensile N_y stresses produce longitudinal tension in the 90° plies and transverse tension in the 0° plies. There were also distinct compression zones at the top and bottom of the specimen above and below the notch tip due to compressive N_y stresses.

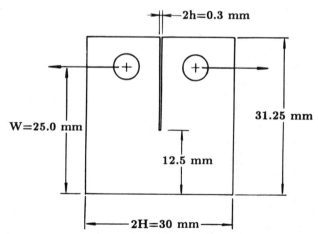

FIG. 2—*Graphite/epoxy $[0_2/90]_{6s}$ C(T) specimen—t = 4.80 mm (0.189 in.), 2H = 30 mm (1.18 in.), W = 25 mm (0.984 in.), 2h = 0.3 mm (0.012 in.), and a = 12.5 mm (0.492 in.).*

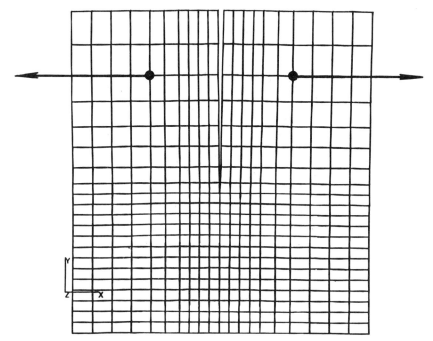

FIG. 3—*Graphite/epoxy $[0_2/90]_{6s}$ C(T) specimen finite element model—467 nodes, 420 quadrilateral elements.*

Figure 6 shows N_{xy} generalized shear stress contours under the 1779-N (400-lb) loading. The maximum shear stresses were concentrated half way between the notch tip and back corners of the specimen. The N_{xy} shear stresses produce in-plane shear stresses in both the 0° and the 90° plies.

Generalized stress contours with stress concentrations shown in Figs. 4, 5, and 6 indicate possible locations of damage initiation and progression. In a traditional fracture mechanics approach, the stress concentrations at the notch tip would be modeled as singularities to calibrate the fracture toughness. In the present computational simulation approach, effects of the through-the-thickness generalized stresses were assessed by evaluation of ply local stresses and stress limits in a composite mechanics module via laminate theory and ply micromechanics equations. Modeling the degradation of composite properties at the notch tip and elsewhere via computational simulation enabled the assessment of damage progression modes, without the assumption of a through-the-thickness stress singularity based on original material properties. Computational simulation showed damage initiation at 2091 N (470 lb) due to transverse tensile failures in the 90° plies at the notch tip. Damage growth continued gradually at the vicinity of the notch tip by transverse tensile failures of both the 90° as well as the 0° plies until a load of 3114 N (700 lb) was reached. It should be noted that transverse tensile failures in the 90° plies were caused by N_x generalized stresses whereas transverse tensile failures in the 0° plies were caused by the N_y generalized stresses. Above 3114 N (700 lb), new damage zones were formed due to the compressive failures of 0° plies at the back face and in-plane shear failures half way between the notch tip and back corners of the specimen.

Compressive failures of the 0° plies at the back face began at the surface of the specimen. The ply compressive failures were in the compressive shear mode, mainly controlled by matrix shear strength and shear modulus. This mode of failure is also called delamination failure [11]

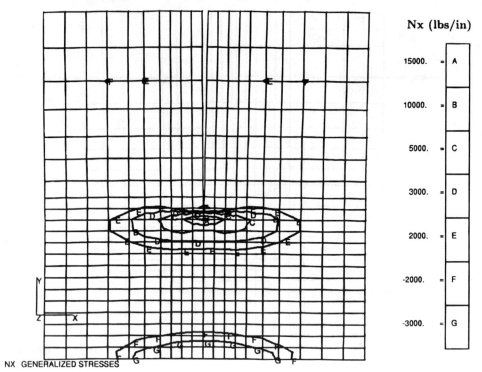

FIG. 4—N_x *generalized stresses under 1779-N (400-lb) load—graphite/epoxy $[0_2/90]_{6s}$ C(T)* *specimen.*

as it is typically followed by delamination of the failed plies. As plies were failed at critical nodes, the corresponding stiffness coefficients associated with the failure modes were reduced to zero in the computational model; thereby accounting for stress redistributions through the thickness of the laminate. At 3745 N (842 lb), additional damage zones were formed due to new compressive failures of 0° plies at the back face, compressive failures of the 90° plies at the top and bottom, and enlargement of matrix shear failure zones half way between the notch tip and back corners of the specimen. The notch tip remained at its original position as there were no fiber failures at the notch tip. When loading was increased to 3767 N (847 lb), damage increased significantly with through-the-thickness fracture of the compression and shear failure zones, but still without notch extension. The simulated ultimate load was reached at 3870 N (870 lb) due to the coalescence of shear and compressive damage zones into the notch tip and disintegration of the specimen.

Figure 7 shows the physical locations of (1) damage initiation at the notch tip by matrix cracking due to ply transverse tensile, $\sigma_{\ell22T}$, failures; (2) $\sigma_{\ell12S}$ in-plane shear failures; (3) $\sigma_{\ell11C}$ longitudinal compression failures in 0° plies at the back face; and (4) $\sigma_{\ell11C}$ longitudinal compression failures in 90° plies at the sides of the specimen. In-plane shear failures at Locations 2 in Fig. 7, controlled by the failure model defined in Eq 4, were the most significant factor affecting the overall damage progression characteristics and the ultimate load for this specimen. As the in-plane shear failures caused by the $\sigma_{\ell12}$ stresses occurred, the matrix stiffness of the failed plies was reduced to zero and the computational simulation cycle was repeated. The diminished shear capacity of the plies with matrix failures caused the stress

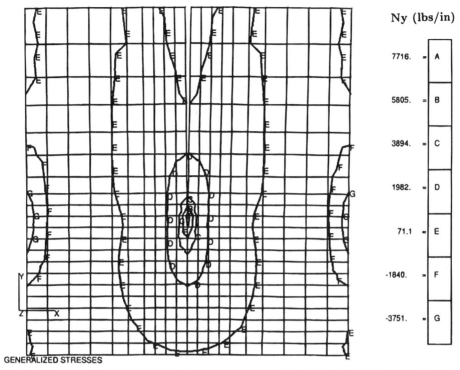

FIG. 5—N_y generalized stresses under 1779-N (400-lb) load—graphite/epoxy $[0_2/90]_{6s}$ C(T) specimen.

redistributions that resulted in the longitudinal compressive failures of the 0° plies at the back face of the specimen.

Figure 8 shows the damage progression with applied loading, indicating that the rate of damage progression increases considerably after the 3114-N (700-lb) loading is exceeded. The scalar damage variable shown in Fig. 8 is derived from the total volume of the composite material affected by the various damage mechanisms. This scalar damage variable is useful for assessing the overall degradation of a given structure under a prescribed loading condition. The rate of increase in the overall damage during composite degradation may be used as a measure of structural propensity for fracture. Computation of the overall damage variable has no interactive feedback on the detailed simulation of composite degradation. In this paper, the overall damage variable is defined simply as the ratio of the volume of damaged plies to the total volume of the composite specimen. The procedure by which the overall damage variable is computed is given in Ref 5.

The global damage energy release rate (DERR) is defined as the rate of work done by external forces during structural degradation, with respect to the produced damage [5]. DERR can be used to evaluate structural resistance against damage propagation at different stages of loading. Low DERR levels indicate that degradation takes place without a significant resistance by the structure. On the other hand, high DERR levels are due to well-defined stages of overall structural resistance to damage propagation. Figure 9 shows the DERR as a function of loading, indicating significant DERR levels at damage initiation, at the beginning of the damage propagation phase, and at the ultimate load. The first DERR peak occurs at

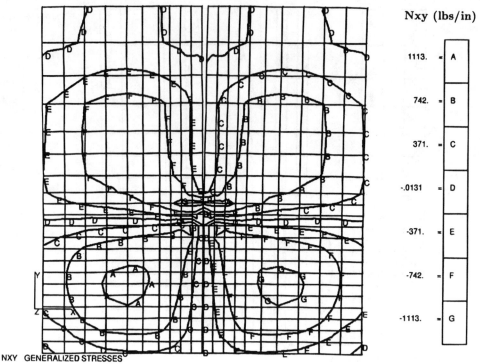

FIG. 6—N_{xy} *generalized stresses under 1779-N (400-lb) load—graphite/epoxy* $[0_2/90]_{6s}$ *C(T) specimen.*

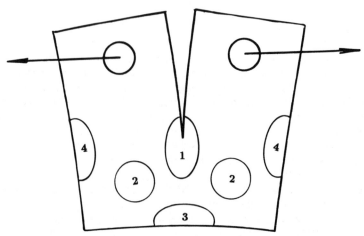

FIG. 7—*Damage modes and locations—graphite/epoxy* $[0_2/90]_{6s}$ *C(T) specimen: (1) matrix cracking at the notch tip, (2) shear failure zones, (3) compression failure at the back face, and (4) compression failures at the sides.*

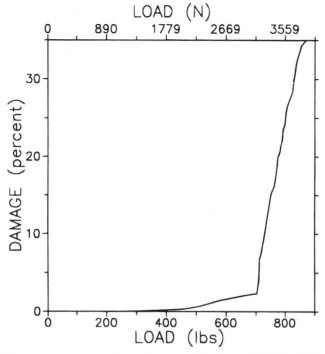

FIG. 8—*Damage progression with loading—graphite/epoxy [0₂/90]₆ₛ C(T) specimen.*

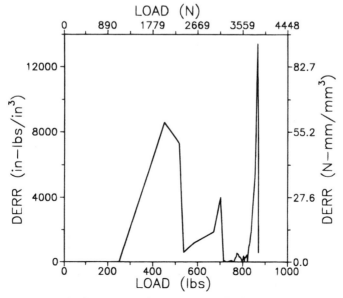

FIG. 9—*Damage energy release rate with loading—graphite/epoxy [0₂/90]₆ₛ C(T) specimen.*

damage initiation under a 2091-N (470-lb) loading, indicating a well-defined damage initiation load. The second peak occurs at a 3114-N (700-lb) load, immediately prior to ply longitudinal compressive failures at the back face of the specimen. After this load, DERR is reduced to very low levels indicating the specimen exerts no significant resistance against further degradation. The third and final peak occurs at the ultimate load when a significant DERR level is experienced indicating the sudden energy release at the ultimate load.

Figure 10 shows load versus the COD displacement, indicating that the damage initiation and growth stages prior to the 3114-N (700-lb) loading are not discernable from the load-COD relationship for this case. Above the 3114-N (700-lb) loading, softening of the load-displacement relationship is due to extensive damage accumulation, including longitudinal compressive failures of the 0° plies in the back face of the specimen.

A displacement-controlled laboratory test reached 3959 N (890 lb) at the peak load for this specimen and overall test response was consistent with computational simulations. However, the damage initiation and growth stages were not discernable during test observations. Computational simulation was able to fill in the degradation details that were needed to completely assess fracture characteristics. Computer simulation also indicated that the ultimate load was most sensitive to the matrix shear strength for this specimen.

Ceramic-Matrix Composite Specimen

Progressive fracture of a ceramic-matrix fiber composite C(T) specimen was computationally simulated. The composite system consisted of silicon-carbide, SiC (Nicalon), fibers in an aluminosilicate glass (1723) matrix. The fiber volume ratio was 45%. The laminate structure consisted of twelve 0.213-mm (0.00837-in.) thick plies, resulting in a composite thickness of 2.55 mm (0.100 in.). The laminate configuration was $[0/90]_{3s}$. The C(T) specimen, as shown in Fig. 11, had a half-height of $H = 23.95$ mm (0.943 in.), a width (distance between load line and back face) of $W = 40.13$ mm (1.58 in.), a half-height of notch slot of $h = 1.68$ mm (0.066 in.), and a distance between load line and notch tip of $a = 18.12$ mm (0.713 in.). The specimen complied with ASTM E 399 specifications.

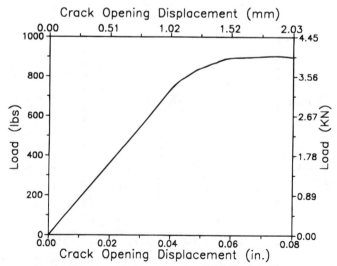

FIG. 10—*Load-displacement relationship—graphite/epoxy $[0_2/90]_{6s}$ C(T) specimen.*

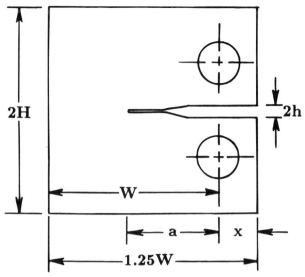

FIG. 11—*Ceramic-matrix composite C(T) specimen*—t = 2.55 mm (0.100 in.), H = 23.95 mm (0.943 in.), W = 40.13 mm (1.58 in.) h = 1.68 mm (0.066 in.), and a = 18.12 mm (0.713 in.).

The finite element model was made containing 161 nodes and 130 quadrilateral thick shell elements, as shown in Fig. 12, with 0° plies oriented in the loading direction. As in the previous example, pin holes were not modeled in the finite element representation of the specimen to enable nodal support and loading. The finite element model was configured to have a nodal

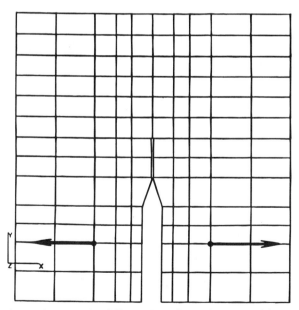

FIG. 12—*Ceramic-matrix composite C(T) specimen finite element model—SiC/glass[0/90]$_{3s}$ 161 nodes, 130 quadrilateral elements.*

point at the center of each pin hole. One of the load points was restrained in all degrees of freedom except for θ_z. The other load point was restrained only in the D_y, D_z, θ_x, θ_y directions but allowed free movement in D_x, and θ_z directions. A concentrated tensile load was applied in the D_x direction. The load was increased gradually. Figure 13 shows Ply 1 (0° ply) longitudinal stress contours under a 445-N (100-lb) loading, prior to damage initiation. The maximum tensile stresses were concentrated at the tip of the preexisting notch. There was a distinct compression zone at the back face of the specimen opposite the notch.

Test data on the SiC/glass C(T) specimen with the same geometry, composite ply constituents, and symmetric cross-ply laminate configuration were presented in Ref *14*. The simulated specimen, with an *a/W* ratio of 0.452, was modeled after Specimen 88C23-6 in Ref *14*. Reported observations during experiments indicated the specimen response to loading was in the brittle mode.

Fiber and matrix elastic properties used in the computational simulation were taken from Ref *14*. However, Ref *14* did not report the constituent material strengths. To enable assessment of damage progression characteristics for the specimen and to evaluate the sensitivity of specimen fracture to fiber strength, an effective in situ fiber strength of 276 MPa (40 ksi) was calibrated. Calibration of material strength via parametric mapping to enable computational simulation of brittle fracture with a reasonably sized finite element model is detailed in Ref *15*. As noted in Ref *15*, the effective strength, σ_e, depends on the ratio of the finite element size to the size of the inelastic process zone at the notch tip. The effective strength is obtained by calibrating the specific finite element model with the experimental data. The compressive strength of SiC fibers is expected to be significantly higher than the tensile strength. However, since the fracture of the SiC/glass C(T) specimen was controlled by fiber tensile strength at the notch tip, and no specific compressive strength data were available, fiber compressive strength was assumed to be the same as tensile strength in the computational simulation. Also, since the stress-free temperature was not reported in Ref *14*, residual stresses were not explicitly modeled in the computational simulation. Therefore, the calibrated strengths used for the

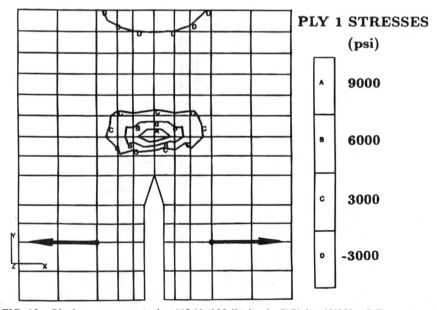

FIG. 13—*Ply 1* $\sigma_{\ell 11}$ *stresses under 445-N (100-lb) load—SiC/glass[0/90]$_{3s}$ C(T) specimen.*

simulation of this specimen included the effects of residual stresses due to fabrication temperatures.

The constituent properties used in the computational simulation of progressive fracture are given in Tables 3 and 4.

CODSTRAN simulation of the SiC/glass C(T) specimen indicated a damage initiation load of 667 N (150 lb). Initial damage was in the form of fiber fractures by longitudinal failure of the 0° plies at the notch tip. When the load was further increased, fiber fractures and matrix cracking due to excessive transverse ply stresses spread to all plies at the notch tip and immediately ahead of it. Computational simulation was carried out to show the details of progressive damage and fracture propagation in the composite structure up to the global fracture of the C(T) specimen. Global fracture was simulated at 1320 N (297 lb), breaking the C(T) specimen into two pieces.

Figure 14 shows the simulated relationship between structural damage and the applied loading. The damage initiation stage corresponds to the development of a damage zone at the notch tip by longitudinal tensile fractures of the 0° plies. Fiber fractures in the 0° plies were immediately followed by the fracture of the 90° plies. The damage initiation stage was concluded

TABLE 3—*SiC (Nicalon) fiber properties.*

Number of fibers per end = 1
Fiber diameter = 0.012 mm (0.472E-3 in.)
Fiber density = 2.07E-6 kg/m^3 (0.278 lb/in.^{3})
Longitudinal normal modulus = 200 GPa (29.0E + 6 psi)
Transverse normal modulus = 200 GPa (29.0E + 6 psi)
Poisson's ratio (ν_{12}) = 0.30
Poisson's ratio (ν_{23}) = 0.30
Shear modulus (G_{12}) = 77.2 GPa (11.2E + 6 psi)
Shear modulus (G_{23}) = 77.2 GPa (11.2E + 6 psi)
Longitudinal thermal expansion coefficient = 4.57E-6/°C (2.54E-6/°F)
Transverse thermal expansion coefficient = 4.57E-6/°C (2.54E-6/°F)
Longitudinal heat conductivity = 15.6 J-m/s/m^2/°C (0.75 Btu-in./h/in.^{2}/°F)
Transverse heat conductivity = 15.6 J-m/s/m^2/°C (0.25 Btu-in./h/in.^{2}/°F)
Heat capacity = 503 J/kg/°C (0.12 Btu/lb/°F)
Tensile strength = 276 MPa (40 ksi)
Compressive strength = 276 MPa (40 ksi)

TABLE 4—*Glass (1723) matrix properties.*

Matrix density = 6.70E-7 kg/m^3 (0.090 lb/in.^{3})
Normal modulus = 85.5 GPa (12 400 ksi)
Poisson's ratio = 0.22
Coefficient of thermal expansion = 6.70E-6/°C (1.21E-5/°F)
Heat conductivity = 1.081 J-m/s/m^2/°C (0.052 08 Btu-in./h/in.^{2}/°F)
Heat capacity = 7115/kg/°C (0.17 Btu/lb/°F)
Tensile strength = 207 MPa (30.0 ksi)
Compressive strength = 207 GPa (300 ksi)
Shear strength = 207 MPa (30.0 ksi)
Allowable tensile strain = 0.0073
Allowable compressive strain = 0.0073
Allowable shear strain = 0.0124
Allowable torsional strain = 0.0124
Void conductivity = 4.67 J-m/s/m^2/°C (0.225 Btu-in./h/in.^{2}/°F)
Glass transition temperature = 816°C (1500°F)

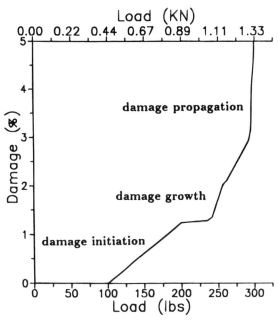

FIG. 14—*Damage propagation with loading—SiC/glass[0/90]₃ₛ C(T) specimen.*

at a 890-N (200-lb) load by the formation of a damage/fracture zone at the tip of the original notch. After the damage initiation stage, the loading was increased to 1.045 kN (235 lb) without additional damage. When the load was increased beyond 1.045 kN, damage growth occurred by compressive fractures of the 0° plies at the back face of the specimen opposite the notch. Compressive failure at the back face of the specimen caused, in turn, tensile failures at the damage zone at the notch tip. The process of alternating compressive and tensile failures was continued until the final damage propagation stage was reached when compressive and tensile fracture zones coalesced and the C(T) specimen was broken into two pieces.

Figure 15 shows the simulated progression of fracture alternately from the tensile zone at the notch tip and from the compressive zone at the back face of the SiC/glass specimen. The damage progression stages are labeled in Fig. 15 in numerical order. Stage 1 corresponds to damage initiation by fiber fractures at the notch tip at 667-N (150-lb) loading. Stage 2 corresponds to damage progression by fiber fractures and notch extension at 890 N (200 lb). Stage 3 corresponds to compressive failures ($\sigma_{\ell11C}$) at the back face of the specimen under 956 to 988 N (215 to 222 lb). Stage 4 marks additional fiber tensile fractures and notch extension. Stage 5 corresponds to growth of compressive failures from the back of the specimen. Stage 6 corresponds to additional compressive failure growth accompanied by the fiber fractures at the tensile regions under 1210 to 1317-N (272 to 296-lb) loading. Immediately following Stage 6, global fracture occurs at 1321 N (297 lb) as the tensile and compressive failure regions are fully coalesced.

Figure 16 shows the DERR as a function of the applied tensile loading on the ceramic-matrix fiber composite C(T) specimen. The DERR for damage initiation was relatively small, indicating low resistance to damage initiation under tensile loading. However, after the damage initiation stage, DERR reached considerably higher levels, indicating greater structural resistance against damage propagation prior to global fracture.

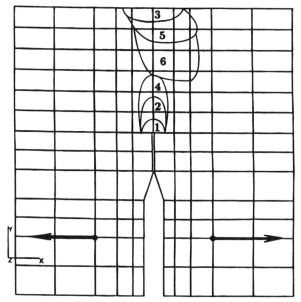

FIG. 15—*Damage progression sequence and locations—SiC/glass[0/90]₃ₛ C(T) specimen.*

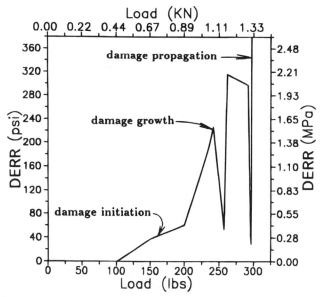

FIG. 16—*Damage energy release rates with loading—SiC/glass[0/90]₃ₛ C(T) specimen.*

Figure 17 shows the load versus displacement relationship for the SiC/glass C(T) specimen. The transition from damage initiation to damage growth is not easily discernible from the load-displacement relationship. However, as the damage propagation stage begins, the load-displacement behavior becomes highly nonlinear. At the structural fracture stage, the displacement increases without any increase in the loading. The global fracture load of 1320 N (297 lb) obtained by computational simulation was slightly under the 1344-N (302-lb) fracture load observed experimentally, as reported in Ref *14*.

Computational simulation also indicated that damage initiation and progression is very sensitive to the fiber tensile strength for this specimen. Figure 18 shows the variation in the structural fracture load and the damage initiation load as functions of the fiber tensile strength. In particular, if the fiber strength increases above the 276 MPa (40 ksi) effective strength of SiC fibers, the fracture load increases at a very steep rate. Computational simulation results indicate that if effective fiber strength increases above 276 MPa (40 ksi), the in-plane shear failure mode begins to participate in the progressive fracture process. As the fracture progresses, matrix stiffness degradation due to in-plane shear failures allows a redistribution of fiber stresses ahead of the crack tip, causing a more ductile response. On the other hand, if the effective fiber strength is at or below 276 MPa (40 ksi), fracture progression is controlled mainly by the ply longitudinal fracture mode, accompanied by ply transverse tensile failures. When effective fiber strength was taken as 276 MPa (40 ksi), damage initiation at 667 N (150 lb) loading was controlled by Eq 1, predicting longitudinal tensile failures in 0° plies at the notch tip. After damage initiation, damage growth was controlled by Eq 7 in the 90° plies. On the other hand, if effective fiber strength were to be increased to 42 ksi, damage initiation and growth would occur at 75.75 kg (167 lb). Both the damage initiation as well as the damage growth modes were controlled by Eq 7. Sensitivity of the fracture progression mode and the resulting ultimate load to fiber strength explains some of the difficulty in fitting experimental

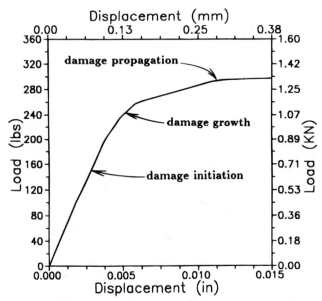

FIG. 17—*Load-displacement relationship—SiC/glass[0/90]₃ₛ C(T) specimen.*

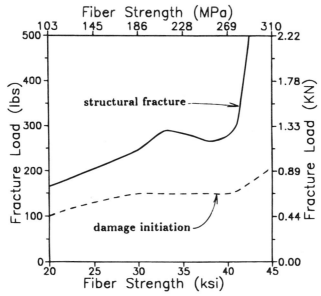

FIG. 18—*Effect of fiber strength on fracture—SiC/glass[0/90]$_{3s}$, C(T) specimen.*

data to a traditional fracture toughness parameter for the room temperature fracture of similar ceramic-matrix C(T) specimens [*14*].

Concluding Remarks

The significant results from this investigation in which computational simulation was used to evaluate damage growth and propagation to fracture for graphite/epoxy and SiC/glass fiber composite C(T) specimens are as follows:

1. Computational simulation, with the use of established composite mechanics and finite element modules, can be used to predict the influence of an existing notch, as well as loading, on the safety and durability of fiber composite structures.
2. Computational simulation adequately tracks the damage growth and subsequent propagation to fracture for fiber composite C(T) specimens.
3. Computational simulation can be used prior to testing to identify locations and modes of composite damage that need be monitored by proper instrumentation and inspection of the specimen during a laboratory experiment.
4. Interpretation of experimental data can be facilitated significantly by detailed results from a computational simulation.
5. Computational simulation provides detailed information on damage initiation and progression mechanisms, as well as identifying sensitive material parameters affecting structural fracture.
6. The demonstrated procedure is flexible and applicable to all types of constituent materials, structural geometry, and loading. Hybrid composite structures, composites containing homogeneous materials such as metallic layers, as well as binary composites can be simulated.
7. Fracture toughness parameters such as the structural fracture load are identifiable for any specimen or structure by the demonstrated method.

8. Computational simulation represents a new global approach that may be used for progressive damage and fracture assessment in design investigations.

References

[1] Chamis, C. C. and Smith, G. T. "Composite Durability Structural Analysis," NASA TM-79070, NASA-Lewis Research Center, Cleveland, OH, 1978.

[2] Irvine, T. B. and Ginty, C. A., "Progressive Fracture of Fiber Composites," *Journal of Composite Materials*, Vol. 20, March 1986, pp. 166–184.

[3] Minnetyan, L., Chamis, C. C., and Murthy, P. L. N., "Structural Behavior of Composites with Progressive Fracture," *Journal of Reinforced Plastics and Composites*, Vol. 11, No. 4, April 1992, pp. 413–442.

[4] Minnetyan, L., Murthy, P. L. N., and Chamis, C. C., "Progression of Damage and Fracture in Composites under Dynamic Loading," NASA TM-103118, NASA-Lewis Research Center, Cleveland, OH, April 1990.

[5] Minnetyan, L., Murthy, P. L. N., and Chamis, C. C., "Composite Structure Global Fracture Toughness via Computational Simulation," *Computers & Structures*, Vol. 37, No. 2, 1990, pp. 175–180.

[6] Minnetyan, L., Murthy, P. L. N., and Chamis, C. C., "Progressive Fracture in Composites Subjected to Hygrothermal Environment," *International Journal of Damage Mechanics*, Vol. 1, No. 1, Jan. 1992, pp. 60–79.

[7] Minnetyan, L., Chamis, C. C., and Murthy, P. L. N., "Damage and Fracture in Composite Thin Shells," NASA TM-105289, NASA-Lewis Research Center, Cleveland, OH, Nov. 1991.

[8] Chamis, C. C., Murthy, P. L. N., and Minnetyan, L., "Progressive Fracture of Polymer Matrix Composite Structures: A New Approach," NASA TM-105574, NASA-Lewis Research Center, Cleveland, OH, Jan. 1992.

[9] Minnetyan, L., Rivers, J. M., Murthy, P. L. N., and Chamis, C. C., "Structural Durability of Stiffened Composite Shells," *Proceedings*, Thirty-third SDM Conference, Dallas, 13–15 April 1992, Vol. 5, pp. 2879–2886.

[10] Minnetyan, L. and Murthy, P. L. N., "Design for Progressive Fracture in Composite Shell Structures," *Proceedings*, Twenty-fourth International SAMPE Technical Conference, Toronto, Canada, 20–22 Oct. 1992, Society for the Advancement of Materials and Process Engineering, pp. T227–T240.

[11] Murthy, P. L. N. and Chamis, C. C., *Integrated Composite Analyzer (ICAN): Users and Programmers Manual*, NASA Technical Paper 2515, NASA-Lewis Research Center, Cleveland, OH, March 1986.

[12] Nakazawa, S., Dias, J. B., and Spiegel, M. S., *MHOST Users' Manual*, NASA-Lewis Research Center, MARC Analysis Research Corp., April 1987.

[13] Chamis, C. C., "Failure Criteria for Filamentary Composites," *Composite Materials Testing and Design: ASTM STP 460*, American Society for Testing and Materials, Philadelphia, 1969, pp. 336–351.

[14] Coker, D. and Ashbaugh, N. E., "Characterization of Fracture in [0/90]$_{3s}$ SiC/1723 Composites," Report No. WL-TR-91-4119, Materials Directorate, Wright Laboratory, Air Force Systems Command, Wright-Patterson AFB, OH, March 1992.

[15] Minnetyan, L. and Chamis, C. C., "Pressure Vessel Fracture Simulation," *Fracture Mechanics: Twenty-fifth Volume, ASTM STP 1220*, F. Erdogan, Ed., American Society for Testing and Materials, Philadelphia, 1995, pp. 671–684.

Ming Wu,[1] *Sudhakar V. Reddy,*[2] *and Dale Wilson*[3]

Design and Testing of Z-Shaped Stringer-Stiffened Compression Panels—Evaluation of ARALL, GLARE, and 2090 Materials

REFERENCE: Wu, M., Reddy, S. V., and Wilson, D., **"Design and Testing of Z-Shaped Stringer-Stiffened Compression Panels—Evaluation of ARALL, GLARE, and 2090 Materials,"** *Composite Materials: Fatigue and Fracture (Sixth Volume), ASTM STP 1285,* E. A. Armanios, Ed., American Society for Testing and Materials, 1997, pp. 551–563.

ABSTRACT: A design study was conducted to determine the potential weight savings and performance increase from advanced metallic materials for wing skin panels. The materials included aluminum lithium 2090-T83, ARALL-3 (aramid-reinforced aluminum laminate) and GLARE-2 (glass-aluminum-reinforced epoxy). An executive business jet airplane was taken as a baseline in the design trade study. This wing has mechanically attached stringers to stiffen the panel against compressive and shear loading. The advanced skin materials were designed into an advanced wing box; advantage was taken of the increase in strength and stiffness. The design, fabrication, and testing of the test panels were done at Textron Aerostructures. Two 2090-T83 aluminum-lithium skins with 7075-T6511 extruded Z-shaped stringers bonded to them were used for the evaluation of the upper wing cover structure. One panel had five bays, the other four. The respective widths were 59.39 cm (23.38 in.) and 49.48 cm (19.48 in.). The overall length was 125.43 cm (49.38 in.) for all panels. Two ARALL-3 and two GLARE-2 compression panels with the same type of stringers were designed to evaluate the lower wing cover structure. The laminate panels were all five-stringer bays 78.74 cm (31 in.) in width.

The study confirmed that a weight savings in the order of 10 to 15% can be achieved with panels made with these advanced materials. The compression tests showed that all test panels failed in column bending and the predicted critical loads compared to those from the tests were conservative. The tests also validated the design methodology.

KEYWORDS: composite materials, fatigue (materials), fracture (materials), wing skin panels, aramid-reinforced aluminum laminate, glass-aluminum reinforced epoxy, aluminum-lithium materials, stringer-stiffened compression panels

Within the aerospace industry, there is a constant objective to develop more efficient and more economical aircraft. It is well known that the more prohibitive costs of air travel are associated with fuel consumption and the service losses incurred during downtime maintenance. Obviously, fuel consumption is greatly affected by weight. Thus, there is a search for new materials and construction techniques that offer substantial weight savings. Since downtime is largely associated with aircraft inspections, efforts are aimed at developing more reliable materials that would allow greater intervals between routine inspections, and ultimately, greater

[1]Engineer, Mechanics and Materials Department, Failure Analysis Associates, Inc., Menlo Park, CA 94025.
[2]Research and development engineer, Textron Aerostructures, Nashville, TN 37202.
[3]Professor, Mechanical Engineering Department, Tennessee Technological University, Cookeville, TN 38505.

service life. As a result, the increased research and use of fiber-reinforced composite materials and other nonconventional materials have been seen in the last three decades. Because of high strength-to-weight ratios, composite structures offer weight savings of up to 30% when compared to components based on more traditional engineering materials. Additionally, composites that offer good fatigue resistance are especially attractive for cyclically loaded fuselage skins and wing surfaces.

When utilizing fiber-reinforced composite materials and other new materials, it is necessary for the design engineer to make additional material considerations. Conventional theories sometimes can not be applied directly to these new materials. Therefore, one of the main steps in the development and application of new materials is testing. Test results are essential for the design engineer to decide upon the applicability of new materials to verify conventional theories. The purpose of this study was to investigate the compression characteristics of ARALL-3, GLARE-2,[4] and 2090 aluminum-lithium (Al-Li) material through large-scale structural compression tests on the stringer-stiffened panels.

Developed in the early 1980s, and currently under license for manufacture by Structural Laminates Company, ARALL, illustrated in Fig. 1, and GLARE represent a family of composite materials. They incorporate adhesively bonded laminated sheets with thin, high-strength aluminum alloy sheets and strong unidirectional or woven aramid or glass fibers impregnated with a thermoset or thermoplastic resin. Stretching and rolling of the material after curing can be performed to achieve desirable compressive residual stresses. The final properties of these laminates are highly dependent on the variables of the material. They can be tailored for many different applications by varying fiber-resin systems, aluminum alloys, sheet gages, stacking sequences, fiber orientations (such as uniaxial and cross ply), surface preparation techniques, and by the degree of post-cure stretching and rolling. Several ARALL and GLARE laminates are listed in Table 1 (only ARALL-3 and GLARE-2 were tested in this study). ARALL-3 and GLARE-2 laminates were developed primarily as fatigue-resistant materials, and there is

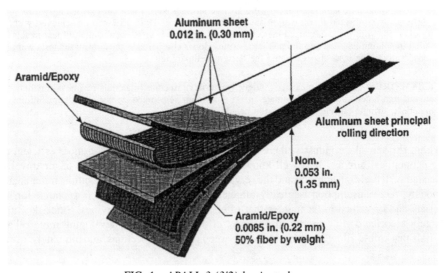

FIG. 1—*ARALL-3 (3/2) laminate layup.*

---

[4]ARALL-3 and GLARE-2 are registered trademarks of Structural Laminates Company, New Kensington, Pennsylvania.

TABLE 1—*ARALL and GLARE laminates.*

Laminates	Alloy	Composite	Orientation	Post Stretched
ARALL-1	7075-T6	aramid/epoxy	UD[a]	none
ARALL-2	2024-T3	aramid/epoxy	UD	none
ARALL-3	7475-T76	aramid/epoxy	UD	0.4%
ARALL-4	2025-T8	aramid/epoxy	UD	none
GLARE-1	7075-T7	glass/epoxy	UD	0.4%
GLARE-2	2024-T3	glass/epoxy	UD	none
GLARE-3	2024-T3	glass/epoxy	cross ply(50/50)	none
GLARE-4	2024-T3	glass/epoxy	cross ply(70/30)	none

[a]UD = unidirectional.

considerable research available documenting their fatigue and fracture properties [*1–4*]. However, since it is hoped that these materials will find extensive use in the future construction of aircraft, there is a need for characterization of the compression behavior of these laminates, especially through large-scale structural tests.

Experimental Program

Test Specimens

The compression panels were designed using a computer program called ZSPAN [*5*]. Written in FORTRAN language, ZSPAN is a computerized analysis of Z-stiffened panels that has been used successfully on various wing designs. This program uses theory and empirical data from various sources including NACA (prior to NASA) reports. The input required to perform analysis is the structural geometry, material allowables, and applied loading of a given panel. The output is section properties, allowable stresses, margins of safety, and the type of failure for each structural element. The analysis applies to most metals, and the skin and stringer need not be from the same family. The stringers are of extruded Z-sections and are attached to the padded skin by rivets or bolts (bonding can be simulated by closer spacing of fasteners that suppresses failure modes associated with fasteners such as wrinkling and inter-rivet buckling). Further information on various inputs, outputs, and method of solution can be obtained from Ref. *6*.

Aluminum-lithium alloy, 2090, is 11% stiffer and 8% lighter than 7075 aluminum alloy. Hence, the design of upper wing panels using this material can yield weight savings of at least 8%. ARALL-3 consists of thin 7475-T761 aluminum alloy sheets of 0.031 cm (0.012 in.) thickness alternating with aramid-fiber epoxy prepreg layers of 0.022 cm (0.0085 in.) thickness embedded in a special resin matrix. Similarly, GLARE-2 is made up of 2024-T3 aluminum alloy layers of 0.031 cm (0.012 in.) thickness alternating with glass-fiber epoxy prepreg layers of 0.025 cm (0.010 in.) thickness. Both ARALL and GLARE have different layups depending on the structural design need. The outer layers of these laminates are always aluminum sheets, and the layups are always symmetric. The ARALL-3 laminate, shown in Fig. 1 has five layers and is denoted as (3/2) consisting of three layers of aluminum and two layers of aramid. The primary advantage of these laminates, as stated previously, is the significant improvement in fatigue and crack growth properties compared to conventional aluminum alloys. ARALL-3 and GLARE-2 also offer higher strength-to-weight ratios over aluminum. The properties of these materials are tabulated in Table 2. The 7075-T7351 and 2024-T351 aluminum alloy plates are conventional upper and lower wing skin materials. These properties along with actual dimensions of the test specimens were used in failure load predictions.

In order to evaluate the new materials (that is, 2090 Al-Li for upper skin; ARALL-3 and GLARE-2 for lower skins), the wing of an executive business jet was taken as a baseline and then a trade study was conducted. The stringer areas were redesigned using these new materials from inboard to outboard end to show the weight savings over the baseline configuration. The number, height, and material of the stringers were kept constant. The skin was replaced by 2090 Al-Li, ARALL-3, and GLARE-2. The stringer and skin dimensions were then varied to match the strength and stiffness of the baseline design.

A total of two panels were tested for the upper cover configuration. One panel (Panel 1) was representative of the wing upper cover areas where thick gages for skin and stringer were used, while the other one (Panel 2) represented areas where thinner gages were used. These panels consisted of 7075-T6511 Z-extrusions bonded to 2090-T83 sheet by using AF163 structural adhesive film with 122°C (250°F) cure temperature. Panel 2 was five bays in width whereas Panel 1 consisted of only four bays to accommodate the test machine capability. To account for the difference between the end fixity coefficient, c, assumed to be 1.00 for the analytical wing stress analysis, and assumed to be 3.75 for the test [8,9], the test panel lengths were obtained as follows

$$\text{test specimen length} = \text{rib spacing} \times (c)^{0.5} = 64.77 \times (3.75)^{0.5}$$

$$= 125.43 \text{ cm}$$

$$= 49.38 \text{ in.} [10]$$

where rib spacing = 64.77 (25.5 in.) (typical on wing). The c-value of 3.75 is commonly accepted when the specimens are tested with the ends of the panels flat against the heads of the compression test machine.

Four lower cover specimens were tested in compression. Two panels (one representing a thinner gage region, Panel 3, and the other representing a thicker gage region, Panel 4) consisted of 7075-T6511 Z-extrusions bonded to ARALL-3 (5/4 layup) sheet laminates, and two more panels (one thinner, Panel 5, and one thicker, Panel 6) consisted of the same stringers bonded to GLARE-2 (4/3 layup) sheet laminates. The AF163 was also used on all panels to adhesively bond the stringers to the skin. The specimens were five stringer bays in width and their lengths were calculated using $c = 3.75$ as shown earlier.

Instrumentation

Strain measurements were obtained by various strain gages located on the test panels. Each compression specimen had a total of 12 uniaxial strain gages to record strains in the loading direction and two deflection gages. The deflection gages were attached to the skin side of the specimen and were located near the two edge stringers, midway between the loaded edges as illustrated in Fig. 2. The strain gage readings were recorded to make sure that the panels were loaded uniformly. Panel geometry with skin and stringer gages is documented in Table 3.

Test Procedure and Results

Six compression test specimens (Panel 1 through Panel 6) were manufactured and assembled at Textron Aerostructures. Assemblies were then inspected to record actual dimensions. These

TABLE 2—Material properties for test verification [7].

	Stringer, 7075 T6511 extr QQ-A-200/11 t = 1.90 to 3.81 cm (0.750 to 1.490 in.)	Upper Skin		Lower Skin		
		7075 T7351 plate QQ-A-250/12 t = 1.270 to 2.540 cm (0.500 to 1.000 in.)	2090 T83 sheet AMS 4251 t = 0.320 to 0.640 cm (0.126 to 0.249 in.)	2024 T351 plate QQ-A-259/4 t = 1.27 to 2.54 cm (0.500 to 1.000 in.)	ARALL-3 5/4 sheet AMS 4302A t = 0.239 cm (0.094 in.)	GLARE-2 4/3(5/4) sheet t = 0.198 to 0.254 cm (0.078 to 0.100 in.)
$F tu$, MPa (ksi) L	652.5 (94.7)	482.3 (70)	584.3 (84.8)	447.9 (65)	853.7 (123.9)	1194.0 (173.3)
$F ty$, MPa (ksi) L	583.6 (84.7)	406.5 (59)	545.7 (79.2)	344.5 (50)	618.0 (89.7)	386.5 (56.1)
$F cy$, MPa (ksi) L	583.7 (84.7)	399.6 (58)	491.3 (71.3)	282.5 (41)	323.8 (47)	407.2 (59.1)
$F su$, MPa (ksi)	337.6 (49)	268.7 (39)	288.7 (41.9)	261.8 (41)	235.6 (34.2)	...
$F bru$, MPa (ksi)						
θ/D = 1.5	840.6 (122)	730.3 (106)	689.0 (100)	675.2 (98)	565.0 (82)	...
θ/D = 2.0	1047.3 (152)	937.0 (136)	868.1 (126)	826.8 (120)	578.8 (84)	709.7 (103)
$E t$, GPa (msi) L	71.7 (10.4)	71.0 (10.3)	77.9 (11.3)	73.7 (10.7)	63.4 (9.2)	64.8 (9.4)
$E c$, GPa (msi) L	73.7 (10.7)	73.0 (10.6)	80.0 (11.6)	75.1 (10.9)	63.4 (9.2)	64.8 (9.4)
G, GPa (msi)	27.6 (4)	26.9 (3.9)	...	27.6 (4)	15.9 (2.3)	...
μ	0.33	0.33	...	0.33	0.35	...
n	26	20	20	9	5.8	...
w, kg/cm³ (lb/in.³)	0.0605 (0.10)	0.0605 (0.101)	0.0547 (0.093)	0.0589 (0.1)	0.0477 (0.081)	0.0530 (0.09)
cm t, (in.)	...	...	...	...	0.239 (0.094)	0.198 (0.204) 0.078 (0.100)
Cost, $/kg ($/lb)	5.5 (2.5)	8.8 (4)	33 (15)	6.6 (3)	127.4 (58)	123.2 (56)

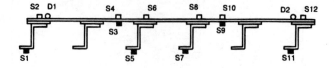

Section AA

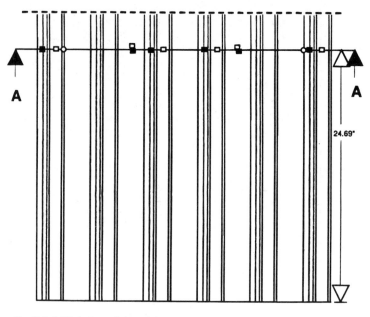

■   Uniaxial Strain Gage - Stringer Side   S : Strain Gage
□   Uniaxial Strain Gage - Skin Side   D : Deflection Gage
○   Deflection Gage - Skin Side

FIG. 2—*Locations of strain and deflection gages on the test panel.*

dimensions were documented for the predictions of failure loads using the ZSPAN program. Specimens were mounted in the compression test fixture as shown in Fig. 3. The fixture was designed and built by the Engineering Laboratory at Textron Aerostructures specifically to test the panels of this size. The gages were connected to the data acquisition system, and zero load readings were recorded. The test was conducted at room temperature in laboratory air environment. Gage readings were taken in steps of 89 kN (20 kips) until failure of the panel. The test results (failure loads) are documented in Table 4.

Test Verification

The loads and strain gage readings for each panel test were documented. To account for the fact that the specimens had unequal dimensions due to manufacturing tolerances, the failure load of each bay was computed, six times for six different bays, and then added to obtain the

TABLE 3—*Panel dimensions.*

Panel No.	Skin Material	b, cm (in.)	t_{sk}, cm (in.)	t_{pad}, cm (in.)	W_{pad}, cm (in.)	W_{af}, cm (in.)	W_{fl}, cm (in.)	h, cm (in.)	t_{st}, cm (in.)
1[a]	2090 Al-Li	49.48 (19.48)	0.305 (0.120)	0.424 (0.167)	3.76 (1.48)	2.79 (1.10)	1.52 (0.60)	4.85 (1.91)	0.254 (0.100)
2	2090 Al-Li	60.40 (23.78)	0.152 (0.060)	0.361 (0.142)	3.25 (1.28)	2.29 (0.90)	1.52 (0.60)	3.66 (1.44)	0.152 (0.060)
3	ARALL-3	78.74 (31.00)	0.239 (0.094)	0.478 (0.188)	2.54 (1.00)	2.29 (0.90)	1.91 (0.75)	3.66 (1.44)	0.152 (0.060)
4	ARALL-3	80.77 (31.80)	0.239 (0.094)	0.478 (0.188)	4.57 (1.80)	2.79 (1.10)	1.91 (0.75)	4.85 (1.91)	0.254 (0.100)
5	GLARE-2	78.74 (31.00)	0.198 (0.078)	0.452 (0.178)	2.54 (1.00)	2.29 (0.90)	1.91 (0.75)	3.66 (1.44)	0.152 (0.060)
6	GLARE-2	80.77 (31.80)	0.198 (0.078)	0.452 (0.178)	4.57 (1.80)	2.79 (1.10)	1.91 (0.75)	4.85 (1.91)	0.254 (0.100)

[a]Five stringers (four bays) only.

ultimate failure load of the total panel using the ZSPAN program. The bay dimension inputs for ZSPAN were from the actual test panel measurements. The end bays were designed to be shorter than the rest of the bays, so the edges do not fail prematurely. Since the ZSPAN predicted load was based on all equal bays, the total load was multiplied by 0.95 to account for the endbay reduction in area.

Predicted failure loads using ZSPAN

$$P = (P1 + P2 + P3 + P4 + P5 + P6)\ (0.95)$$

where P1, P2, P3, P4, P5, and P6 represent individual bay failure loads and are the output of the program (Table 4), and Spacing is the individual bay width. The comparisons between test failure loads and predicted failure loads for these compression panels are tabulated in Table 5. All compression panels failed in column bending as shown in Fig. 4. It is noted that the bond between skin and stringers did not have any effect on the failure mode.

Design and Weight Study

The trade study was performed to assess the weight savings that these new materials may provide when used in wing-type structures. The wing of an executive business jet currently in production was chosen as the baseline structure for the study due to its conventional aluminum construction and the availability of the design data at Textron Aerostructures. The primary focus of the trade study was to provide the estimate of the weight savings that is

FIG. 3—*Compression test setup.*

possible when substituting 2090 Al-Li, ARALL-3, and GLARE-2 in lieu of standard aluminum plate materials. Stringer material was kept the same (7075-T6511 extrusion) throughout the study.

Baseline Structure Description

The baseline structural component used in this study represents a typical wing of a business jet. The wing has a 185.42-cm (73-in.) span, a root chord of 513.08 cm (202 in.), and a tip

TABLE 4—*Predicted failure load for compression test panels.*

Panel	P1,[a] kN/m (lb/in.)	P2, kN/m (lb/in.)	P3, kN/m (lb/in.)	P4, kN/m (lb/in.)	P5, kN/m (lb/in.)	P6, kN/m (lb/in.)	Spacing, m (in.)	ARR[b]	Σ^P, kN (lb)
1[c]	1744 (9960)	1822 (10 400)	1634 (9325)	1618 (9235)	1665 (9500)	···	0.11 (4.5)	0.95	8483 (206 996)
2	683 (3900)	672 (3835)	769 (4390)	739 (4215)	726 (4147)	672 (3835)	0.11 (4.5)	0.95	4261 (103 977)
3	471 (2690)	562 (3210)	493 (2815)	552 (3150)	615 (3510)	517 (2950)	0.15 (6.0)	0.95	3210 (104 453)
4	908 (5180)	774 (4415)	916 (5230)	899 (5132)	891 (5083)	882 (5035)	0.15 (6.0)	0.95	5270 (192 785)
5[d]	···	···	···	···	···	···	···	···	···
6	1006 (5740)	938 (5355)	1045 (5965)	1056 (6025)	910 (5193)	958 (5468)	0.15 (6.0)	0.95	5913 (192 352)

[a]Failure load for Bay 1.
[b]ARR = area reduction ratio.
[c]Panel 1 has only four bays due to the load capability of the test machine.
[d]Panel 5 was tested with improper setup. Results are invalid.

TABLE 5—*Compression test panel results versus analysis.*

Panel No.	Actual Failure Load, kN (lb)	Actual Failure Mode	ZSPAN Predicted Load, kN (lb)	ZSPAN Predicted Failure Mode	Difference, %
1	1111.78 (249 839)	column bending	921.132 (206 996)	column bending	−17.1
2	592.896 (133 235)	column bending	462.698 (103 977)	column bending	−21.9
3	556.967 (125 161)	column bending	464.816 (104 453)	column bending	−16.5
4	993.018 (223 150)	column bending	857.893 (192 785)	column bending	−13.6
5[a]	···	···	···	···	···
6	1069.055 (240 237)	column bending	855.966 (192 352)	column bending	−19.9

[a]Panel 5 was tested with improper setup. Results are invalid.

chord of 200.66 cm (79 in.). It has a single cell torque box with integrally stiffened front and rear beams made from 7075-T7351. The upper cover (7075-T7351) consists of a single panel to which both Z- and Hat-section stringers are mechanically fastened. The lower cover consists of three panels (2024-T351) to which Z-section stringers are mechanically fastened. The 11.43-cm (4.50-in.) stiffener spacing is used for the upper cover and 15.24 cm (6.00 in.) for the lower cover. The lower panels are spliced by means of inverted "J" section stringers. Access to the wing interior is provided by access holes located at the lower cover's center panel. All the ribs are integrally stiffened and are spaced at approximately 66.04 cm (26 in.) apart.

Weight Savings Evaluation

One optimization tool, ZSPAN, was extensively used during the trade study. ZSPAN analyzes Z-stiffened panels given the material properties, geometry, and loads. Section properties,

FIG. 4—*Failed compression test specimen.*

allowable stresses, margins of safety, and the type of failure are the output. The program code predicts failure loads for a given input against rupture, yielding, column instability, local buckling, torsional bucking, and wrinkling. Panel configurations for the upper cover were re-sized using ZSPAN with 7075-T7351 skin replaced by 2090 Al-Li. Percentage weight savings are plotted in Figs. 5 and 6. A typical thicker region (Stringer 10) on the wing is presented in Fig. 5, and a typical thinner region (Stringer 18) is presented in Fig. 6 (STGR stands for stringer). Similarly, panel configurations for lower cover (Stringer 6 and Stringer 14) were re-sized with 2024-T351 skin replaced by ARALL-3 and GLARE-2. Percentage weight savings are presented in Figs. 7 and 8. These locations were chosen, such that they included thinner and thicker gage regions of the wing covers and the results could be projected to represent the whole wing. New skin thicknesses were varied from the baseline design to obtain optimum design for similar strength/failure loads.

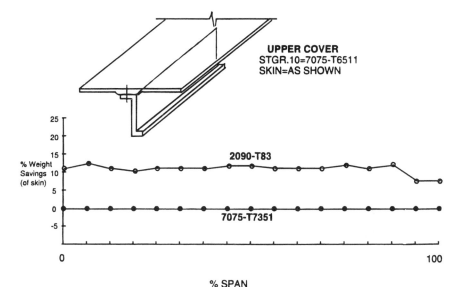

FIG. 5—*Weight study results for upper cover configuration (thick area).*

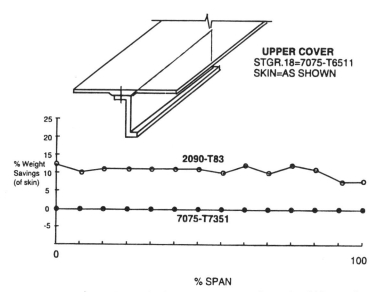

FIG. 6—*Weight study results for upper cover configuration (thin area).*

Based on the preceding results, the weight savings for the entire wing were projected. Replacement of the upper skin by 2090 Al-Li resulted in weight savings of 113.4 kg (250 lb) or 10% of the baseline structure. For the lower cover, ARALL-3 showed the highest weight savings of 170.1 kg (374 lb), which is about 15% of the skin weight, and GLARE-2 showed a reduction of 124.8 kg (275 lb) or 11%. The lower weight savings at the outboard end reflect the minimum gage requirements for machining. The summary of weight savings is presented in Table 6.

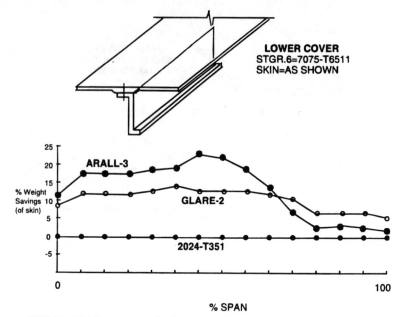

FIG. 7—*Weight study results for lower cover configuration (thick area).*

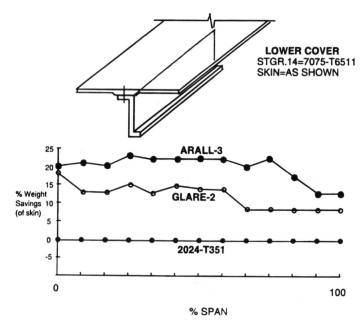

FIG. 8—*Weight study results for lower cover configuration (thin area).*

TABLE 6—*Weight savings summary.*

	Weight Saved	% of Baseline
Upper cover (2090 AL-Li)	113.4 kg (250 lb)	10% of skin
Lower cover (ARALL-3)	170.1 kg (375 lb)	15% of skin
Lower cover (GLARE-2)	124.8 kg (275 lb)	11% of skin

Conclusions

Six Z-stiffened compression panels using advanced materials such as aluminum lithium, ARALL, and GLARE representing upper and lower wing covers were fabricated and tested to failure. These panels were designed using a computer program called ZSPAN. The test results indicate that the ZSPAN-predicted loads are conservative in all cases, the largest discrepancy being 21.9%. The test validity of the design methodology confirms the weight savings in the order of 10 to 15% of the skin that can be achieved by replacing the conventional aluminum alloys by these new materials in a wing box.

Further work is required to carefully evaluate the manufacturing and material costs and trade studies performed using the data. The overall advantages of using these new materials have to be considered not only from the performance and weight saving aspects, but also from the material and manufacturing standpoint.

References

[1] Gunnink, J. W. and Schee, P. A. V. D., "Design of the ARALL F-27 Lower Wing Skin Fatigue Panel," *Proceedings*, Fourth International Conference on Composite Structures (ICCS4), Paisley, Scotland, 1987.

[2] Chen, D., "Bulging of Fatigue Crack in a Pressurized Aircraft Fuselage," Ph.D. thesis, Delft University of Technology, Delft, The Netherlands, Jan. 1991.

[3] Wu, M., "Residual Strength Prediction of Center Notched Unstiffened and Bonded Stringer Stiffened ARALL-3 and GLARE-2 Panels through The R-Curve Approach," Ph.D. thesis, Tennessee Technological University, Cookeville, TN, 1994.

[4] Wu, M., Reddy, S. V., and Wilson, D. A., "Fatigue and Residual Strength Investigation of ARALL-3 and GLARE-2 Panels with Bonded Stringers," *Proceedings*, International Symposium on Advanced Structural Integrity Methods for Airframe Durability and Damage Tolerance, NASA/FAA, Hampton, VA, May 1994.

[5] Reddy, S. V., "Z-Stiffened Panel Analysis (ZSPAN)," Report No. STED-TR-85-0124, Textron Aerostructures, Nashville, TN, 11 Aug. 1987.

[6] Johnston, G. S., "Stringer Panel Analysis Methods," Optimization Inc., Englewood, FL, 1985.

[7] "Metallic Materials and Elements for Aerospace Vehicle Structures," *Military Handbook, MIL-HDBK-5F*.

[8] Hickman, W. and Dow, N., "Data on the Compressive Strength of 75 S-T6 Aluminum Alloy Flat Panels with Longitudinal Extruded Z-Section Stiffeners," NACA Technical Note 1829, National Advisory Committee on Aeronautics, Washington, DC, March 1949.

[9] Hickman, W. and Dow, N., "Data on the Compressive Strength of 75 S-T6 Aluminum Alloy Flat Panels Having Small Thin Widely Spaced, Longitudinal Extruded Z-Section Stiffeners," NACA Technical Note 1978, Washington, DC, Nov. 1949.

[10] Bruhn, E. F., "Analysis and Design of Flight Vehicle Structures," Tri-State Offset Co., Indianapolis, IN, 1973.

Author Index

Subject Index

A

Adhesive nonlinearity, unidirectional fiber composites, 203
Aluminum-lithium materials, 551
Analytical global-local model, free-edge delamination, 381
Aramid-reinforced aluminum laminate, 249, 551
Assessment methods, composite materials, recreational structures, 45

B

Beams, composite materials, 70
BEARBY analysis code, 225
Bearing-bypass interaction, composite bolted joints, 225
Bearing strength, creep elongation of bolt holes, polymer matrix composites, 452
Biaxial loading
  composite joints, bypass strength, 225
  graphite/epoxy composites, damage tolerance, 9
Bicycle frames, assessment methods for composite damage, 45
Bolted composite joints, bypass strength, 225
Bolt holes, creep elongation, polymer matrix composites, 452
Bolt torque
  composite bolted joints, 225
  creep elongation of bolt holes, polymer matrix composites, 452
Brittle matrix composites, crack extension modeling, 364
Buckling, composite materials, 70
  damage, 45
  under cyclic loading, 162
Bypass strength, composite bolted joints, 225

C

Carbon and glass-fiber composites, unidirectional fiber composites, 203
Carbon/epoxy laminates, static and fatigue interlaminar tensile strength, 516
Carbon fiber-reinforced epoxy composites
  delamination growth under cyclic loading, 162

fastener joints, effects of environmental and geometric parameters, 432
  multidirectional, 283
Collapse, composite materials, 70
Compact tension specimen, laminated composite testing, 531
Compliance calibration, mixed-mode bending specimen, 305
Compression panels, Z-shaped stringer-stiffened, 551
Compressive loading
  composite damage, 45
  cross-ply composite plates, 143
  free-edge delamination, 381
  woven-fabric composites, 471
Compressive strength, unidirectional fiber composites, 203
Computational simulation, laminated composite testing, 531
Crack extension
  modeling, chopped-fiber composites, 364
  mode separation, composite delamination problems, 324
Crack growth law, cross-ply composite plates, 143
Crack propagation
  chopped-fiber composites, 364
  interlaminar fatigue cracks, 126
  titanium-matrix composite, 103
  unstable, composite materials, 70
Creep, bolt holes subject to bearing loads, 452
Cross-ply compression, unidirectional fiber composites, 203
Curved-beam specimens, static and fatigue interlaminar tensile characterization, 516
Cyclic loading, delamination growth prediction, 162

D

Damage initiation, composite materials, 70
Damage progression, woven-fabric composites, 471
Damage tolerance, graphite/epoxy composites, 9
Data analysis, tensile and compressive strengths, 203
Degradation, laminated composite testing, 531